Understanding Animal Genetics

Understanding Animal Genetics

Edited by **Dominic Fasso**

R CALLISTO
REFERENCE

New York

Published by Callisto Reference,
106 Park Avenue, Suite 200,
New York, NY 10016, USA
www.callistoreference.com

Understanding Animal Genetics
Edited by Dominic Fasso

International Standard Book Number: 978-1-63239-646-4 (Hardback)

Printed in the United States of America.

Contents

Preface

I am honored to present to you this unique book which encompasses the most up-to-date data in the field. I was extremely pleased to get this opportunity of editing the work of experts from across the globe. I have also written papers in this field and researched the various aspects revolving around the progress of the discipline. I have tried to unify my knowledge along with that of stalwarts from every corner of the world, to produce a text which not only benefits the readers but also facilitates the growth of the field.

This book aims to understand the field of animal genetics. It has gained prominence due to the recent advances in this area. Animal genetics is predominantly a branch of biology although it borrows its elements from several disciplines such as life sciences, etc. It primarily studies heredity and variation in animals. It also delves into the core areas of genetics such as resistance to disease, morphological defects, underdevelopment of individual organs, trait inheritance, etc. The extensive content of this book provides the readers with a thorough understanding of the subject. Those in search of information to further their knowledge of this field will be greatly assisted by this book.

Finally, I would like to thank all the contributing authors for their valuable time and contributions. This book would not have been possible without their efforts. I would also like to thank my friends and family for their constant support.

Editor

Coronary risk in relation to genetic variation in *MEOX2* and *TCF15* in a Flemish population

Wen-Yi Yang[1], Thibault Petit[1,2], Lutgarde Thijs[1], Zhen-Yu Zhang[1], Lotte Jacobs[1], Azusa Hara[1], Fang-Fei Wei[1], Erika Salvi[3], Lorena Citterio[4], Simona Delli Carpini[4], Yu-Mei Gu[1], Judita Knez[1], Nicholas Cauwenberghs[1], Matteo Barcella[3], Cristina Barlassina[3], Paolo Manunta[5], Giulia Coppiello[6], Xabier L. Aranguren[6], Tatiana Kuznetsova[1], Daniele Cusi[3], Peter Verhamme[6], Aernout Luttun[6] and Jan A. Staessen[1,7*]

Abstract

Background: In mice MEOX2/TCF15 heterodimers are highly expressed in heart endothelial cells and are involved in the transcriptional regulation of lipid transport. In a general population, we investigated whether genetic variation in these genes predicted coronary heart disease (CHD).

Results: In 2027 participants randomly recruited from a Flemish population (51.0 % women; mean age 43.6 years), we genotyped six SNPs in *MEOX2* and four in *TCF15*. Over 15.2 years (median), CHD, myocardial infarction, coronary revascularisation and ischaemic cardiomyopathy occurred in 106, 53, 78 and 22 participants. For SNPs, we contrasted CHD risk in minor-allele heterozygotes and homozygotes (variant) *vs.* major-allele homozygotes (reference) and for haplotypes carriers (variant) *vs.* non-carriers. In multivariable-adjusted analyses with correction for multiple testing, CHD risk was associated with *MEOX2* SNPs ($P \leq 0.049$), but not with *TCF15* SNPs ($P \geq 0.29$). The *MEOX2* GTCCGC haplotype (frequency 16.5 %) was associated with the sex- and age-standardised CHD incidence (5.26 *vs.* 3.03 events per 1000 person-years; $P = 0.036$); the multivariable-adjusted hazard ratio [HR] of CHD was 1.78 (95 % confidence interval, 1.25–2.56; $P = 0.0054$). For myocardial infarction, coronary revascularisation, and ischaemic cardiomyopathy, the corresponding HRs were 1.96 (1.16–3.31), 1.87 (1.20–2.91) and 3.16 (1.41–7.09), respectively. The *MEOX2* GTCCGC haplotype significantly improved the prediction of CHD over and beyond traditional risk factors and was associated with similar population-attributable risk as smoking (18.7 % *vs.* 16.2 %).

Conclusions: Genetic variation in *MEOX2*, but not *TCF15*, is a strong predictor of CHD. Further experimental studies should elucidate the underlying molecular mechanisms.

Keywords: Clinical genetics, Coronary heart disease, *MEOX2*, Population science, *TCF15*, Translational research

Background

Endothelial cells lining the microvasculature constitute the interface between the circulating blood and tissues [1]. They differentiate to acquire the molecular, morphological and functional characteristics required for proper organ function [1]. In the heart, endothelial cells play an active role in the transport of fatty acids, the principal energy source for the continuously beating muscle [1, 2]. Using microarray profiling on endothelial cells isolated from the heart, brain, and liver of mice, we recently identified a specific genetic signature for heart endothelial cells, including MEOX2/TCF15 heterodimers as novel transcriptional determinants [3]. This signature was largely shared with skeletal muscle and adipose tissue endothelium and was enriched in genes encoding fatty acid transport-related proteins [3]. Using gain- and loss-of-function approaches, we showed that MEOX2/TCF15 mediates fatty acid uptake in heart endothelial cells, in part, by driving endothelial CD36 and lipoprotein lipase (LPL) expression and thereby facilitating fatty acid transport across cardiac endothelial cells [3].

LPL is expressed at the luminal endothelial surface of arteries and capillaries and hydrolyses circulating lipoprotein

* Correspondence: jan.staessen@med.kuleuven.be
[1]Research Unit Hypertension and Cardiovascular Epidemiology, KU Leuven Department of Cardiovascular Sciences, University of Leuven, Kapucijnenvoer 35, Box 7001, BE-3000 Leuven, Belgium
[7]R & D VitaK Group, Maastricht University, Maastricht, The Netherlands
Full list of author information is available at the end of the article

triglycerides into free fatty acids and glycerol [4]. Local [4] and systemic [5] dysregulation of lipid metabolism and endothelial dysfunction [6, 7] are hallmarks of coronary atherosclerosis and long precede clinically overt disease. These observations suggest that genetic predisposition plays an important role in the pathogenesis of coronary heart disease (CHD) [6, 7]. In view of our recent observations of enriched expression of MEOX2 and TCF15 in heart endothelial cells [3], we hypothesised that genetic variation in the genes encoding these transcription factors might be associated with coronary risk. To test this hypothesis, we analysed data accumulated since 1985 in a Flemish population study [8, 9].

Methods

Study population

The Flemish Study on Environment, Genes and Health Outcomes (FLEMENGHO) complies with the Helsinki declaration for research in human subjects and the Belgian legislation for the protection of privacy (http://www.privacycommission.be). The Ethics Committee of the University of Leuven approved the study. Recruitment for the FLEMENGHO study started in 1985 [8, 9]. From August 1985 to November 1990, a random sample of the households living in a geographically defined area of Northern Belgium was investigated with the goal to recruit an equal number of participants in each of six strata by sex and age (20–39, 40–59, and ≥60 years). All household members aged 20 years or older were invited, if the quota of their sex-age group had not yet been met. From June 1996 until January 2004 recruitment of families continued using the former participants (1985–1990) as index persons and including teenagers. The participants were repeatedly followed up. In all study phases, we used the same standardised methods to measure blood pressure and to administer questionnaires. The participation rate at enrolment was 78.0 %. At each contact, participants gave or renewed informed written consent.

Of 3343 enrolled participants, we excluded 1316 from analysis, because blood stored in the biobank was exhausted with no material left for genotyping ($n = 521$), because of DNA degradation ($n = 314$), because at enrolment they were less than 20 years old ($n = 372$), or because one or more of the six MEOX2 or four TCF15 SNPs were unavailable ($n = 109$). Thus, the number of participants statistically analysed totalled 2027.

Measurements at baseline

Trained nurses measured the participants' anthropometric characteristics and blood pressure. Body mass index was weight in kilograms divided by the square of height in meters. Blood pressure was the average of five consecutive auscultatory readings obtained with a standard mercury sphygmomanometer after participants had rested in the

sitting position for at least 5 min. Hypertension was a blood pressure of at least 140 mm Hg systolic or 90 mm Hg diastolic, or use of antihypertensive drugs. The nurses also administered a standardised questionnaire inquiring about each participant's medical history, smoking and drinking habits, and intake of medications. Plasma glucose and serum total and high-density lipoprotein (HDL) cholesterol and serum creatinine were measured by automated methods in certified laboratories. Diabetes mellitus was a fasting or random plasma glucose level exceeding 7.0 or 11.1 mmol/L, or use of antidiabetic agents [10].

Ascertainment of coronary events

FLEMENGHO received ethical approval. The database was registered with the Privacy Commission. These legal requirements being fulfilled, we could ascertain the vital status of participants at annual intervals until 06 December 2012 via the Belgian Population Registry. In addition, we could obtain the International Classification of Disease codes for the immediate and underlying causes of death from the Flemish Registry of Death Certificates. For 1853 participants, we collected information on the incidence of non-fatal endpoints either via face-to-face follow-up visits with repeated administration of the same standardised questionnaire as used at baseline ($n = 1521$) or via a structured telephone interview ($n = 332$). Follow-up data were available from one visit in 360 participants, from two in 304, from three in 436, and from four or more in 421 participants.

Trained nurses used the International Classification of Diseases to code incident cases of CHD. Two investigators blinded with regard to the genotypic results adjudicated all coronary events against the medical records of general practitioners or hospitals. Coronary events included sudden death, fatal and non-fatal myocardial infarction, acute coronary syndrome requiring hospitalisation, ischaemic cardiomyopathy, and surgical or percutaneous coronary revascularisation. In the outcome analyses, we only considered the first event within each category.

Genotyping

Ethics approval and informed consent covered genotyping. After DNA extraction from peripheral blood [11], SNPs were genotyped using the TaqMan® OpenArray™ Genotyping System (Life Technologies, Foster City, CA). All DNA samples were loaded at 50 ng/mL and amplified according to the manufacturer's instructions. For analysis of the genotypes, we used autocalling methods, as implemented in the TaqMan Genotyper software version 1.3 (Life Technologies). Next, genotype clusters were evaluated manually with the sample call rate set above 0.90. Sixteen duplicate samples gave 100 % reproducibility for all 64 SNPs on the custom made array, including the genes of interest in the current article [12].

MEOX2 (75601 base pairs) maps to chromosome 7 (p22.1–p21.3). To select the *MEOX2* SNPs to genotype, we first reviewed all SNPs in this gene, including the flanking regions, as available in the Illumina 1 M Duo and OmniExpress arrays (San Diego, CA). We excluded SNPs with a minor allele frequency of less than 1 % and those that were in high linkage disequilibrium ($r^2 \geq 0.80$). Next, based on the availability of SNPs on the TaqMan OpenArray Genotyping System, we selected 12 tagging SNPs (rs6946099, rs10777, rs7800473, rs13438001, rs12056299, rs7787043, rs758297, rs4532497, rs10263561, rs6959056, rs740566, rs1050290) that are in high linkage disequilibrium ($r^2 \geq 0.80$) with 92 neighbouring SNPs (Additional file 1: Figure S1 and Table S1), but were not in high linkage disequilibrium ($r^2 < 0.80$) with one another. The 12 selected SNPs covered the entire gene with extension into the 3' and 5' flanking regions. We excluded six SNPs with a successful genotyping call rate of less than 0.98. Finally, we retained six *MEOX2* SNPs (rs10777, rs12056299, rs7787043, rs4532497, rs6959056, and rs1050290) in the analysis (Additional file 1: Table S2) that are in linkage disequilibrium ($r^2 > 0.80$) with 23 other SNPs (Additional file 1: Table S1). *TCF15* (6602 base pairs) maps to chromosome 20p13. We genotyped five SNPs covering the whole gene (rs282152, rs6116745, rs282162, rs3761308 and rs12624577), but excluded rs282152, because the SNP call rate was less than 0.98 (Additional file 1: Figure S2).

Statistical analysis

For database management and statistical analysis, we used SAS software, version 9.3 (SAS Institute, Cary, NC). For comparison of means and proportions, we applied the large sample z-test or ANOVA and Fisher's exact, respectively. We tested Hardy-Weinberg equilibrium in unrelated founders, using the exact statistics available in the PROC ALLELE procedure of the SAS package. For analysis of single SNPs, we combined the least frequent homozygous group with heterozygous subjects. We tested linkage disequilibrium and reconstructed haplotypes using the SAS procedures PROC ALLELE and PROC HAPLOTYPE. To check for consistency, we repeated haplotype construction accounting for pedigree information using SHAPEIT version 2 (http://mathgen.stats.ox.ac.uk/genetics_software/shapeit/shapeit.html [13]).

We compared the incidence of coronary endpoints in relation to genetic variants, using (i) rates standardised by the direct method for sex and age (<40, 40–59, ≥60 years) and (ii) the cumulative incidence derived from Cox models adjusted for sex and age. Next, we assessed the prognostic value of the genetic variants in multivariable-adjusted Cox regression. We checked the proportional hazard assumption by applying a Kolmogorov-type supremum test as implemented in the ASSESS statement of the PROC PHREG procedure. To account for family clusters, we used the

PROC SURVIVAL procedure of the SUDAAN 11.0.1 software (Research Triangle Institute, NC). In this procedure, non-independence among family members was taken into account by including family as a random effect along with other covariables as fixed effects. We analysed genotypes and haplotypes using major allele homozygotes and non-carriers as the reference group, respectively. We adjusted *P* values for the associations between outcomes and genetic variants, using the Benjamini and Hochberg false discovery rate [14] according to the number of SNPs retained in the analysis.

We computed the positive predictive value of the risk carrying *MEOX2* haplotype *GTCCGC* as $(R \times D)/([G/100] \times [R -1] + 1)$, where R is the multivariable-adjusted hazard ratio, D is the incidence of CHD in the whole population, and G is the prevalence of the *GTCCGC* haplotype [15]. The attributable risk is given by $([R -1] \times 100)/R$ and the population-attributable risk by $([G/100] \times [R -1] \times 100)/([G/100] \times [R -1] + 1)$ [15]. Finally, we assessed the power of the *MEOX2 GTCCGC* haplotype to predict CHD over and beyond classical risk factors, using the integrated discrimination improvement (IDI) and the net reclassification improvement (NRI), as described by Pencina and colleagues for survival data [16].

Results

Baseline characteristics

All 2027 participants were White Europeans, of whom 1034 (51.0 %) were women. The study population consisted of 332 singletons and 1695 related subjects, belonging to 49 single-generation families and 191 multi-generation pedigrees. Age averaged (±SD) 43.6 ± 14.3 years, blood pressure 125.0 ± 15.4 mm Hg systolic and 76.2 ± 9.5 mm Hg diastolic, body mass index 25.7 ± 4.3 kg/m^2, and total cholesterol 5.49 ± 1.15 mmol/L. Among all participants, 486 (24.0 %) had hypertension, of whom 214 (44.0 %) were on antihypertensive drug treatment, 33 (1.6 %) had diabetes mellitus, and 41 (2.0 %) reported a history of CHD. Previous coronary complications included angiographically proven coronary stenosis, myocardial infarction, and coronary revascularisation in 8 (0.4 %), 11 (0.5 %) and 22 (1.1 %) patients, respectively. Of 1034 women and 993 men, 277 (26.8 %) women and 328 (33.0 %) men were smokers, and 168 (16.3 %) women and 418 (42.1 %) men reported intake of alcohol. In smokers, median tobacco use was 15 cigarettes per day (interquartile range, 10 to 20 cigarettes per day). In drinkers, the median alcohol consumption was 14 g per day (8 to 26 g per day).

Table 1 lists the baseline characteristics of participants according to CHD incidence. Most risk factors differed in the expected direction between cases and non-cases. However, compared with non-cases, the prevalence of smoking was not different in patients with incident CHD (29.6 % *vs.* 36.8 %; *P* = 0.13), while the prevalence

Table 1 Baseline characteristics of participants by incident CHD

Characteristic	Non-cases	Cases	All
N°	1921	106	2027
N° with characteristics (%)			
Women	1006 (52.4)	28 (26.4)‡	1034 (51.0)
Current smoker	566 (29.6)	39 (36.8)	605 (29.9)
Drinking alcohol	565 (29.4)	21 (19.8)*	586 (28.9)
Diabetes mellitus	26 (1.4)	7 (6.6)‡	33 (1.6)
Hypertension	432 (22.5)	54 (50.9)‡	486 (24.0)
Treated hypertension	189 (9.8)	25 (23.6)‡	214 (10.6)
History of CHD	30 (1.6)	11 (10.4)‡	41 (2.0)
Mean of characteristic (±SD)			
Age, years	42.8 ± 14.1	57.4 ± 11.3‡	43.6 ± 14.3
Body mass index, kg/m^2	25.6 ± 4.3	27.1 ± 3.9‡	25.7 ± 4.3
Waist-to-hip ratio	0.84 ± 0.09	0.91 ± 0.08‡	0.85 ± 0.09
Systolic blood pressure, mm Hg	124.5 ± 15.2	135.3 ± 15.8‡	125.0 ± 15.4
Diastolic blood pressure, mm Hg	76.0 ± 9.4	78.6 ± 9.9†	76.2 ± 9.5
Heart rate, beats per minute	69.2 ± 9.5	69.9 ± 9.0	69.3 ± 9.5
Total cholesterol, mmol/L	5.46 ± 1.14	6.06 ± 1.15‡	5.49 ± 1.15
HDL cholesterol, mmol/L	1.40 ± 0.39	1.16 ± 0.33‡	1.37 ± 0.39
Total-to-HDL cholesterol ratio	4.21 ± 1.58	5.75 ± 2.30‡	4.29 ± 1.66
Serum creatinine, μmol/L	90.2 ± 16.8	104.3 ± 19.4‡	91.1 ± 16.8
Plasma glucose, mmol/L	5.03 ± 1.27	5.48 ± 2.29‡	5.05 ± 1.35

HDL cholesterol refers to the serum concentration of high-density lipoprotein cholesterol. Diabetes mellitus was a fasting or random plasma glucose level exceeding 7.0 or 11.1 mmol/L, or use of antidiabetic agents. Hypertension was a blood pressure of ≥140 mm Hg systolic or ≥90 mm Hg diastolic or use of antihypertensive drugs. Significance of the differences between non-cases and cases: * $p \leq 0.05$; † $p \leq 0.01$; ‡ $p \leq 0.001$

of drinking was lower among cases (29.4 % *vs.* 19.8 %; $P = 0.036$). Heart rate at baseline was similar in participants without and with incident CHD (69.2 *vs.* 69.9 beats per minute; $P = 0.43$).

Incidence of events
Over a median follow-up of 15.2 years (5th to 95th percentile interval, 5.7 to 27.1 years), 106 new coronary events occurred, 24 fatal and 82 non-fatal. Coronary events comprised 12 fatal and 34 non-fatal myocardial infarcts and 7 sudden deaths. There were 78 patients who underwent surgical ($n = 29$) or percutaneous ($n = 56$) coronary revascularisation. Coronary events also included 5 fatal and 17 non-fatal cases of ischaemic cardiomyopathy.

Analyses of single SNPs in *MEOX2* and *TCF15*
Additional file 1: Table S2 describes the position and the SNPs retained in the analysis and the allele and genotype frequencies in 825 unrelated founders. The six SNPs in *MEOX2* and the four SNPs in *TCF15* complied with Hardy-Weinberg equilibrium ($0.30 \leq P \leq 0.80$). In the whole study population (Additional file 1: Table S3), the frequencies of the minor alleles ranged from 21.1

to 43.0 % for *MEOX2* and from 9.6 to 38.5 % for *TCF15*. The prevalence of minor allele homozygotes ranged from 4.4 to 17.8 % for *MEOX2*, and from 0.7 to 15.9 % for *TCF15*.

The sex- and age-standardised incidence rates of coronary events associated with the *MEOX2* SNPs appear in Additional file 1: Table S4. Compared with major allele homozygotes, minor allele carriers experienced a higher CHD incidence except for rs6959056. The sex- and age-adjusted cumulative incidence of coronary events (Fig. 1) showed significant association ($P \leq 0.012$) with the *MEOX2* SNPs except for rs1050290 ($P = 0.058$). There were no differences in these estimates between homozygous and heterozygous minor allele carriers ($0.23 \leq P \leq 0.98$) except for rs12056299 ($P = 0.014$). For all coronary events combined, the sex- and age-standardised incidence rates ($0.11 \leq P \leq 0.39$) and the sex- and age-adjusted cumulative incidence ($0.11 \leq P \leq 0.71$) did not differ among minor allele carriers and major allele homozygotes of the four *TCF15* SNPs.

Next, we accounted for family clusters and adjusted the hazard ratios for baseline characteristics, including sex, age, body mass index, systolic pressure, the total-to-

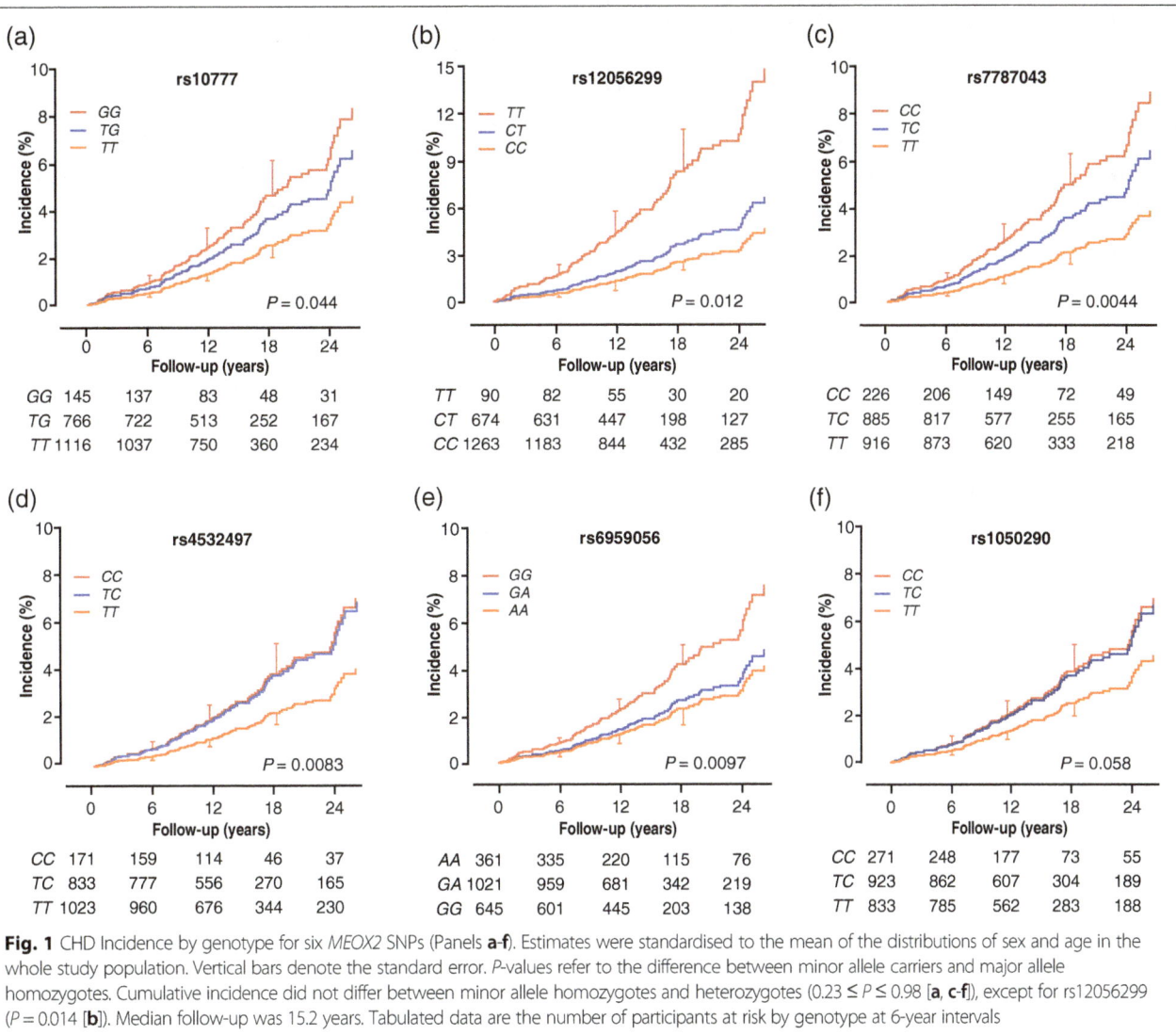

Fig. 1 CHD Incidence by genotype for six *MEOX2* SNPs (Panels **a-f**). Estimates were standardised to the mean of the distributions of sex and age in the whole study population. Vertical bars denote the standard error. *P*-values refer to the difference between minor allele carriers and major allele homozygotes. Cumulative incidence did not differ between minor allele homozygotes and heterozygotes ($0.23 \leq P \leq 0.98$ [**a**, **c-f**]), except for rs12056299 ($P = 0.014$ [**b**]). Median follow-up was 15.2 years. Tabulated data are the number of participants at risk by genotype at 6-year intervals

HDL cholesterol ratio, smoking and drinking, and anti-hypertensive drug treatment. Compared with homozygotes of the major allele, for rs10777, rs12056299, rs7787043, rs4532497, and rs1050290, CHD risk was higher in minor allele carriers, whereas the opposite was the case for rs6959056 (Table 2). These findings remained consistent with correction for multiple testing (Table 2) and after excluding patients who had a history of CHD at baseline (Additional file 1: Table S5).

For the four *TCF15* SNPs, the multivariable-adjusted hazard ratios modelling the CHD risk of minor allele carriers *vs.* major allele homozygotes did not reach significance ($0.75 \leq$ hazard ratio ≤ 1.45; $0.072 \leq P \leq 0.52$). However, Additional file 1: Figure S3 shows interaction ($P = 0.011$) between rs12624577 in *TCF15* and rs4532497 in *MEOX2*. Among *C* allele carriers of *MEOX2* rs4532497, the hazard ratio expressing the risk of the *C* allele relative to the

TT genotype in *TCF15* was 2.44 (95 % confidence interval, 1.38−4.29; $P = 0.0021$), whereas among MEOX2 *TT* homozygotes the corresponding hazard ratio was 0.75 (0.37−1.51; $P = 0.42$).

Analysis of *MEOX2* haplotypes

Using the expectation-maximisation algorithm as implemented in the PROC HAPLOTYPE procedure of the SAS software version 9.3, three haplotypes of *MEOX2* SNPs had a frequency of over 10 % and were carried through in the analysis. With letters referring to the alleles in rs10777, rs12056299, rs7787043, rs4532497, rs6959056, and rs1050290 in *MEOX2*, these haplotypes were *TCTTAT* (27.5 %), *TCTTGT* (26.4 %), and *GTCCGC* (16.5 %). For all coronary events, rates standardised for sex and age (5.26 *vs.* 3.03 events per 1000 person-years; Additional file 1: Table S6*)* and cumulative incidence

Table 2 Multivariable-adjusted hazard ratios for CHD by *MEOX2* SNPs

SNP Event	N° events/at risk		Hazard ratio	P	P_{BH}
	Minor allele carriers	Major allele homozygotes			
rs10777	**GG + TG**	**TT**			
All coronary events	59/911	47/1116	1.56 (1.06–2.29)	0.024	0.029
Myocardial infarction	31	22	1.75 (1.01–3.03)	0.045	0.054
Coronary revascularisation	42	36	1.42 (0.90–2.24)	0.13	0.14
Ischaemic cardiomyopathy	16	6	4.68 (1.86–11.77)	0.0011	0.0066
rs12056299	**TT + CT**	**CC**			
All coronary events	51/764	55/1263	1.68 (1.17–2.41)	0.0052	0.016
Myocardial infarction	28	25	2.05 (1.20–3.51)	0.0087	0.042
Coronary revascularisation	37	41	1.59 (1.02–2.49)	0.041	0.082
Ischaemic cardiomyopathy	14	8	3.37 (1.49–7.62)	0.0036	0.011
rs7787043	**CC + TC**	**TT**			
All coronary events	74/1111	32/916	1.72 (1.13–2.62)	0.011	0.023
Myocardial infarction	38	15	1.82 (1.02–3.27)	0.044	0.054
Coronary revascularisation	54	24	1.68 (1.02–2.77)	0.041	0.082
Ischaemic cardiomyopathy	16	6	2.22 (0.84–5.86)	0.11	0.20
rs4532497	**CC + TC**	**TT**			
All coronary events	66/1004	40/1023	1.80 (1.22–2.66)	0.0031	0.016
Myocardial infarction	33	20	1.88 (1.03–3.42)	0.040	0.054
Coronary revascularisation	50	28	1.88 (1.20–2.96)	0.0062	0.037
Ischaemic cardiomyopathy	14	8	1.97 (0.80–4.83)	0.14	0.20
rs6959056	**AA + GA**	**GG**			
All coronary events	59/1382	47/645	0.62 (0.42–0.92)	0.017	0.025
Myocardial infarction	27	26	0.52 (0.31–0.88)	0.014	0.042
Coronary revascularisation	46	32	0.71 (0.45–1.12)	0.14	0.14
Ischaemic cardiomyopathy	12	10	0.75 (0.33–1.70)	0.49	0.49
rs1050290	**CC + TC**	**TT**			
All coronary events	73/1194	33/833	1.50 (1.00–2.26)	0.049	0.049
Myocardial infarction	36	17	1.48 (0.81–2.70)	0.20	0.20
Coronary revascularisation	55	23	1.59 (0.99–2.56)	0.055	0.083
Ischaemic cardiomyopathy	16	6	1.97 (0.75–5.16)	0.17	0.20

Numbers of events do not add up, because only the first event in each category was analysed. Hazard ratios (95 % confidence interval) express the risk of minor allele carriers *vs.* major allele homozygotes, account for family clusters, and were adjusted for baseline characteristics including sex, age, body mass index, systolic pressure, total-to-HDL cholesterol ratio, smoking and drinking, and antihypertensive drug treatment. P and P_{BH} indicate the significance of the hazard ratios without and with Benjamini-Hochberg's correction for multiple testing

estimates adjusted for sex and age (Additional file 1: Figure S4), both before ($P \leq 0.012$) and after adjustment for multiple testing ($P \leq 0.036$), showed significant association with *GTCCGC*. In multivariable-adjusted analyses (Table 3), *GTCCGC* carriers, compared to non-carriers, had a 78 % higher CHD risk. The *p*-values before and after correction for multiple testing were 0.0018 and 0.0054, respectively. Myocardial infarction, coronary revascularisation and ischaemic cardiomyopathy all showed significant association with the *GTCCGC* haplotype. These findings were not materially different when we reconstructed haplotypes

accounting for pedigree information (Additional file 1: Table S7) or after excluding patients who had a history of CHD at baseline (Additional file 1: Table S8).

With adjustments applied as before, the positive predictive value and the attributable and population-attributable CHD risks associated with *GTCCGC* were 7.5, 43.1 and 18.7 %, respectively. For smoking, analysed as reference, the corresponding estimates (unadjusted for genetic risk) were 7.2, 39.3 and 16.2 %, respectively. Table 4 shows that in all participants and in those without CHD at entry, adding the *GTCCGC* haplotype to the basic model

Table 3 Multivariable-adjusted hazard ratios for CHD by *MEOX2* haplotypes

Haplotype Event	N° events/at risk		Hazard ratio	P	P_BH
	Carrier	Non-carrier			
TCTTAT					
All coronary events	40/951	66/1076	0.73 (0.49–1.11)	0.14	0.21
Myocardial infarction	20	33	0.77 (0.42–1.41)	0.40	0.56
Coronary revascularisation	30	48	0.75 (0.47–1.19)	0.22	0.22
Ischaemic cardiomyopathy	6	16	0.56 (0.21–1.51)	0.25	0.25
TCTTGT					
All coronary events	46/937	60/1090	0.90 (0.58–1.39)	0.63	0.63
Myocardial infarction	26	27	1.19 (0.67–2.11)	0.56	0.56
Coronary revascularisation	30	48	0.74 (0.46–1.20)	0.22	0.22
Ischaemic cardiomyopathy	9	13	0.59 (0.24–1.43)	0.24	0.25
GTCCGC					
All coronary events	43/614	63/1413	1.78 (1.24–2.56)	0.0018	0.0054
Myocardial infarction	23	30	1.96 (1.16–3.31)	0.012	0.036
Coronary revascularisation	33	45	1.87 (1.20–2.91)	0.0058	0.017
Ischaemic cardiomyopathy	11	11	3.16 (1.41–7.09)	0.0053	0.016

Numbers of events do not add up, because only the first event in each category was analysed. Letters coding the haplotypes refer to the rs10777, rs12056299, rs7787043, rs4532497, rs6959056 and rs1050290 alleles (see Additional file 1: Table S1 and S2). Haplotypes were reconstructed using the expectation-maximisation algorithm as implemented in the PROC HAPLOTYPE procedure of the SAS software version 9.3. Hazard ratios (95 % confidence interval) express the risk associated with carrying *vs.* not carrying a haplotype, account for family clusters, and were adjusted for baseline characteristics including sex, age, body mass index, systolic pressure, total-to-HDL cholesterol ratio, smoking and drinking, and antihypertensive drug treatment. P and P_BH indicate the significance of the hazard ratios without and with Benjamini-Hochberg's correction for multiple testing

including traditional risk factors improved ($0.016 \leq P \leq 0.056$) IDI and NRI.

Discussion

To our knowledge, our study is the first to relate in a general population CHD incidence to genetic variation in *MEOX2* and *TCF15*, two transcription factors that are highly expressed by cardiac endothelium and that in a heterodimeric fashion interfere with cardiac energy metabolism by driving endothelial CD36 and LPL expression, thereby facilitating fatty acid transport across the cardiac endothelium [3]. The key finding of our current study was that the risk of advanced CHD was associated with genetic variation in *MEOX2*, as captured by six tagging SNPs. On the other hand, genetic variation in *TCF15*, coding for the heterodimeric partner of MEOX2, was not associated with the incidence of coronary events. Nonetheless, the CHD risk associated with *MEOX2* rs4532497 was

confined to *TCF15* rs12624577 variant allele carriers, which might reflect the known heterodimeric action picked up in our experimental studies [3]. Although our current study firmly established an association between CHD risk and genetic variation in *MEOX2*, the molecular mechanisms underlying this relation need further clarification in experimental studies back translating our epidemiological findings. For now, working hypotheses might be developed along two lines respectively involving disturbed lipid handling [5, 17–19], a key mechanism in atherosclerosis, or the involvement of MEOX2 in the angiogenic responses to stressors [20–24] or in the migration or proliferation of endothelial and vascular smooth muscle cells [25, 26].

A large-scale genome-wide association study identified 46 significant lead SNPs associated with CHD. Twelve showed a significant association with a lipid trait. Variation in the *MEOX2* gene was not among these SNPs, but *LPL* (rs264) was [27]. LPL catalyses the hydrolysis of

Table 4 Improvement in predicting CHD events by adding haplotype *GTCCGC* to the basic model

Study group	Integrated discrimination improvement		Net reclassification improvement	
	%Δ (95 % CI)	P	%Δ (95 % CI)	P
All (n = 2027)	0.81 (−0.02 to 1.65)	0.056	21.7 (2.5 to 40.8)	0.026
Free of CHD at entry (n = 1986)	1.15 (0.17 to 2.12)	0.021	24.9 (4.7 to 45.3)	0.016

%Δ is the percentage change (95 % confidence interval). The basic model includes the baseline covariables sex, age, body mass index, systolic pressure, total-to-HDL cholesterol ratio, smoking and drinking, and antihypertensive drug treatment. The integrated discrimination improvement is the difference between the discrimination slopes of the basic model and the basic model extended with the *GTCCGC* haplotype. The discrimination slope is the difference in predicted probabilities between non-cases and cases. The net reclassification improvement is the sum of the percentages of participants correctly reclassified to non-cases and cases triglycerides in plasma triglyceride-rich lipoproteins, chylomicrons and very low density lipoproteins at the capillary endothelial cell surface, providing free fatty acids and glycerol as energy source for tissues [19]. Genetic variation in *LPL* is associated with the levels of circulating LPL activity [5], the plasma concentration of triglycerides [5, 18] and HDL cholesterol [5, 18] and in some [17], albeit not all [18], studies with the risk of CHD. Parenchymal cells in adipose, skeletal and cardiac muscle widely express LPL throughout the body [19]. Cardiac endothelial cells have a particular expression profile including *MEOX2*, a gene that facilitates fatty acid transport into and through heart endothelial cells. As for genetic variation in *LPL* [5, 18], we hypothesised that dysregulation of lipid transport in cardiac endothelial cells might increase CHD risk. Important in this regard is that in our current study, in contrast to the studies on genetic variants of *LPL* [5, 18], we did not find any association of the circulating lipid levels with *MEOX2* variants (data not shown), pointing to a local coronary rather than a systemic underlying mechanism.

Moving to the second hypothetical pathophysiological pathway, MEOX2 is also known as growth arrest-specific homeobox (GAX) [20], During embryonic development, the three muscle lineages express GAX [21]. In adult life, vascular smooth muscle cells also express GAX [21]. Mitogenic stimuli, such as platelet-derived growth factor and angiotensin II or injury of the endothelium [22], inhibit GAX expression, whereas growth arrest signals, such as serum deprivation of cultured cells, enhance its expression and negatively regulate the cell cycle [23]. Observations in transfected cells also point to *MEOX2* as a potentially important regulatory gene inhibiting not only the angiogenic response of endothelial cells to pro-angiogenic factors, but also their response to chronic inflammatory stimulation that normally activates NF-κB [24]. Inflammatory pathways identified in a network analysis of 233 candidate genes play key roles in development of coronary atherosclerosis [27]. These observations [22–24, 27] may offer an alternative explanation why in our current study coronary risk was associated with genetic variation in *MEOX2*. Chen and Gorski did an in silico search for micro-RNA binding sites in the *GAX* 5'UTR and identified consensus sites for multiple candidate micro-RNAs, of which only miR-130a was expressed in proliferating endothelial cells [26]. miR-130a was largely responsible for the down-regulation of GAX expression in response to mitogens and pro-angiogenic factors and antagonised the antiangiogenic activity of GAX [26].

To our knowledge, previously published GWAS results did not demonstrate association between coronary heart disease and *MEOX2*. However, these GWAS studies relied on comparing cases and controls drawn from heterogeneous sources [27–29] or on a retrospective cross-sectional analysis of patients referred for coronary angiography [30]. GWAS case–control studies offer the opportunity for searching for association between CHD and densely distributed SNPs across the whole genome in large numbers of patients and controls. Such studies require significance levels of 10^{-6} to 10^{-8}. In contrast, our study was prospective and population-based and tested a prior hypothesis involving only six SNPs in *MEOX2* and four in *TCF15*. We did therefore not rely on such extreme *P*-values, but applied the Benjamin-Hochberg approach for multiple testing. Admittedly, our sample size was smaller than in the GWAS studies. This is particularly relevant for the interaction between *MEOX2* rs4532497 and *TCF15* rs12624577 (Additional file 1: Figure S3), a finding, which although in line with our experimental findings [3] can only be considered as hypothesis generating. Future population-based research projects might address this issue.

We performed the annotation of the genomic context surrounding the SNPs retained for *MEOX2*, based on the data of the ENCODE project (http://www.genome.gov/encode). As shown in Additional file 1: Table S1, rs10777 and rs1050290 map into the 3'UTR and 5'UTR regions, respectively, whereas rs12056299, rs7787043, rs4532497 and rs6959056 are in the first intron of *MEOX2*. Moreover, according to Ensemble 75 annotation (http://www.ensembl.org/Homo_sapiens/Info/Annotation) and GENCODE 22 (http://www.gencodegenes.org/releases/22.html), rs4532497 and rs6959056 also map into the ENSG00000237070 (AC005550.3) antisense non-protein coding gene. In particular, rs4532497 maps into intron 1, whereas rs6959056 maps into exon 4 of ENSG00000237070. rs6959056 and rs1050290 fall into a promoter regulatory region (ENST00000622287) both in human umbilical vein endothelial cells and in human dermal fibroblasts, where MEOX2 is expressed. Further functional studies are needed to investigate the possible modulation of MEOX2 expression. Moreover, rs6959056 in exon 4 of the non-coding transcript ENST00000451240 could affect gene function, although no information on the ENSG00000237070 gene is currently available.

TCF15, also known as Paraxis, is a member of the Twist subfamily of basic helix-loop-helix transcription factors that regulate specification of mesodermal derivatives during vertebrate embryogenesis [31]. TCF15 primes pluripotent cells for differentiation [32]. During dermomyotome formation in *Xenopus laevis*, TCF15 directly activates the expression of MEOX2 [31]. Our experimental studies demonstrated that the MEOX2/TCF15 heterodimer facilitates the transport of fatty acids across cardiac endothelial cells and that in mice haplodeficiency in these genes results in impaired contractility of cardiomyocytes and heart failure [3]. In our population study, we therefore also searched for association between the incidence of well-documented heart failure and variation in the *MEOX2* and *TCF15* genes. Several reasons may explain why such associations were not detected. Indeed, heart failure is a heterogeneous disease caused by a multitude of instigators, including ischaemic or valvular heart disease, comorbidities, or risk factors such as hypertension. Moreover, the diagnosis of heart failure depends on the clinical interpretation of a combination of signs and symptoms, that are difficult to recognise [33, 34].

The present study must be interpreted within the context of some potential limitations. First, even though our analysis was hypothesis-driven based on published evidence from experimental studies [3], we adjusted significance levels for multiple testing according to the number of SNPs tested, using the Benjamini and Hochberg false discovery rate [14]. Even applying the most stringent approach described by Bonferroni did not remove the significance for rs12056299 ($P = 0.031$), rs45324977 ($P = 0.019$) and haplotype GTCCGC ($P = 0.0054$) in relation to CHD. On the other hand, we did not consider applying a correction for multiple testing based on the four coronary endpoints. Indeed, such events are highly correlated. Multiple testing is therefore not indicated, because each new test does not provide an independent opportunity for a type-I error [35]. Second, although an observational study cannot prove causation, the Bradford-Hill criteria [36] suggest that the association between coronary risk and genetic variation in MEOX2 might be causal, taking into account (i) the strength and consistency of the association across different SNPs; (ii) temporality, genetic variability preceding the event; (iii) plausibility based on the experimental studies [3]; and (iv) the analogy observed with genetic variability in LPL [5, 17–19]. Third, only few genes regulated by MEOX2 and TCF15 are currently known. In this regard, it is important to note that we noticed that of the genes overexpressed in cardiac endothelial cells, those most upregulated included genes involved in lipid homeostasis, including LPL [3]. Finally, as is common in many population studies, follow-up was inconsistent, with varying numbers of follow-up visits across participants. In addition, participants without blood sample, compared with those included in the analyses (Additional file 1: Table S9), were slightly older (49.4 vs. 43.6 years) and had a higher systolic blood pressure (129.5 vs. 125.0 mm Hg), resulting in a higher prevalence of hypertension (40.0 vs. 24.0 %). We cannot ascertain whether these factors might have biased our analyses.

The clinical implications of our current findings can be gauged by the observation that the attributable and population-attributable CHD risks were similar for the MEOX2 GTCCGC carrying state and smoking. Several investigators proposed the use of genetic risk scores based on genome-wide association studies to stratify for the probability of CHD [37, 38]. In the Framingham Heart Study [37], a score consisting of 13 SNPs did not refine the prediction of CHD or cardiovascular disease, but led to modest improvements in risk reclassification. In contrast, in the Rotterdam Study [38], a score based on 152 SNPs was associated with incident CHD, but did not enhance risk prediction. SNP discovery based on prevalent rather than incident CHD might explain these discrepancies [38]. In our current study, genetic variation in MEOX2 improved IDI and NRI over and beyond the basic model including traditional CHD risk factors.

Conclusion

Our current study based on a predefined hypothesis generated by data from our experimental studies [3], identified genetic variation in the transcription factor MEOX2 gene as a novel risk factor for CHD in a white population. However, further experimental studies are required to back-translate our epidemiological observations into underlying molecular mechanisms. Elucidation of these pathways might reveal new targets for the prevention and treatment of CHD.

Additional file

Additional file 1: Table S1. Common tagging SNPs in MEOX2. **Table S2.** MEOX2 and TCF15 SNPs and allele and genotype frequencies in unrelated founders. **Table S3.** MEOX2 and TCF15 allele and genotype frequencies in 2027 analysed participants. **Table S4.** Sex- and age-standardised CHD rates by MEOX2 SNPs. **Table S5.** Hazard ratios for CHD by MEOX2 SNPs in participants free of CHD at baseline. **Table S6.** Sex- and age-standardised CHD rates by MEOX2 haplotypes. **Table S7.** Hazard ratios for CHD by MEOX2 haplotypes reconstructed while accounting for pedigree information. **Table S8.** Hazard ratios for CHD by MEOX2 haplotypes in participants free of CHD at baseline. **Table S9.** Baseline characteristics of participants without blood left for genotyping compared with those included in the analyses. **Figure S1.** Plot of the MEOX2 gene and flanking regions on chromosome 7. **Figure S2.** Plot of the TCF15 gene and flanking regions on chromosome 20. **Figure S3.** Interaction between TCF15 rs12624577 and MEOX2 rs4532497. **Figure S4.** Incidence of coronary endpoints, myocardial infarction and coronary revascularisation in MEOX2 GTCCGC carriers and non-carriers.

Abbreviations

CHD: Coronary heart disease; CD36: Cluster of differentiation 36; DNA: Deoxyribonucleic acid; ENCODE: Encyclopaedia of DNA elements; GAX: Growth arrest-specific homeobox; GENCODE: Encyclopaedia of genes and genes variants; FLEMENGHO: Flemish Study on Environment, Genes and Health Outcomes; GWAS: Genome-wide association study; HDL: High density lipoprotein; IDI: Integrated discrimination improvement; LPL: Lipoprotein lipase; MEOX2: Mesenchyme homeobox 2; miRNA: Micro Ribonucleic acid; NF-κB: Nuclear factor of kappa light polypeptide gene enhancer in B cells; NRI: Net reclassification improvement; SNP: Single nucleotide polymorphism; TCF15: Transcription factor 15; UTR: Untranslated regions.

Competing interests

The authors declare that they have no competing interests.

Authors' contributions

JAS coordinated the Flemish Study on Environment, Genes and Health Outcomes and managed funding. LT and LJ coordinated the construction and updates of the master database and the management of the biobank. MB, LC, SDC, and ES did the genotyping. CB, DC and PM supervised DNA extraction and amplification, set up the genotyping procedures and managed quality control of genotyping. NC, Y-MG, TP, F-FW, Z-YZ collected phenotypic data and participated in the quality control of the database. AH and TK collected outcome data. W-YY and JAS did the statistical analysis with guidance provided by LT. W-YY and JAS wrote the first draft of the manuscript. XLA, GC, AL and PV assisted in translating the basic science data. All authors interpreted the results, commented on successive drafts of the manuscript and approved the final version.

Acknowledgements

The authors gratefully acknowledge the clerical assistance of Annick De Soete and Renilde Wolfs and the contribution of Linda Custers, Marie-Jeanne Jehoul, Daisy Thijs and Hanne Truyens in data collection at the field centre. The European Union (HEALTH-2011.2.4.2-2-EU-MASCARA, HEALTH-F7-305507 HOMAGE, and the European Research Council Advanced Researcher Grant-2011-294713-EPLORE) and the Fonds voor Wetenschappelijk Onderzoek Vlaanderen, Ministry of the Flemish Community, Brussels, Belgium (G.0881.13 and G.088013) currently support the Studies Coordinating Centre in Leuven. Hypothesis-generating studies at the Centre for Molecular and Vascular Biology were supported by the European Research Council Starting Gran-2007-203291-IMAGINED and the Fonds voor Wetenschappelijk Onderzoek Vlaanderen, Ministry of the Flemish Community, Brussels, Belgium (G.0393.12). The funding source had no role in study design, data extraction, data analysis, data interpretation, or writing of the report. The corresponding author had full access to all the data in the study and had responsibility for the decision to submit for publication.

Author details

[1]Research Unit Hypertension and Cardiovascular Epidemiology, KU Leuven Department of Cardiovascular Sciences, University of Leuven, Kapucijnenvoer 35, Box 7001, BE-3000 Leuven, Belgium. [2]Cardiology, Department of Cardiovascular Sciences, University of Leuven, Leuven, Belgium. [3]Genomics and Bioinformatics Platform at Filarete Foundation, Department of Health Sciences and Graduate School of Nephrology, Division of Nephrology, San Paolo Hospital, University of Milan, Milan, Italy. [4]Division of Nephrology and Dialysis, IRCCS San Raffaele Scientific Institute, University Vita-Salute San Raffaele, Milan, Italy. [5]School of Nephrology, University Vita-Salute San Raffaele, Milan, Italy. [6]Centre for Molecular and Vascular Biology, Department of Cardiovascular Sciences, University of Leuven, Leuven, Belgium. [7]R & D VitaK Group, Maastricht University, Maastricht, The Netherlands.

References

1. Aird WC. Phenotypic heterogeneity of the endothelium: I. Structure, function, and mechanisms. Circ Res. 2007;100:158–73.
2. Stanley WC, Recchia FA, Lopaschuk GD. Myocardial substrate metabolism in the normal and failing heart. Physiol Rev. 2005;85:1093–129.
3. Coppiello G, Collantes M, Sirerol-Piquer MS, Vandenwijngaert S, Schoors S, Swinnen M, et al. Meox2/Tcf15 heterodimers program the heart capillary endothelium for cardiac fatty acid uptake. Circulation. 2015;131:815–26.
4. Goldberg IJ. Lipoprotein lipase and lipolysis: central roles in lipoprotein metabolism and atherogenesis. J Lipid Res. 1996;37:693–707.
5. Tang W, Apostel G, Schreiner PJ, Jacobs Jr DR, Boerwinkle E, Fornage M. Associations of lipoprotein lipase gene polymorphisms with longitudinal plasma lipid trends in yound adults: The Coronary Artery Risk Development in Young Adults (CARDIA) Study. Circ Cardiovasc Genet. 2010;3:179–86.
6. Treasure CB, Manoukian SV, Klein JL, Vita JA, Nabel EG, Renwick GH, et al. Epicardial coronary artery responses to acetyl choline are impaired in hypertensive patients. Circ Res. 1992;71:776–81.
7. Celermajer DS, Sorensen KE, Bull C, Robinson J, Deanfield JE. Endothelium-dependent dilation in the systemic arteries of asymptomatic subjects relates to coronary risk factors and their interaction. J Am Coll Cardiol. 1994;24:1468–74.
8. Staessen JA, Wang JG, Brand E, Barlassina C, Birkenhäger WH, Herrmann SM, et al. Effects of three candidate genes on prevalence and incidence of hypertension in a Caucasian population. J Hypertens. 2001;19:1349–58.
9. Stolarz-Skrzypek K, Kuznetsova T, Thijs L, Tikhonoff V, Seidlerová J, Richart T, et al. Fatal and nonfatal outcomes, incidence of hypertension and blood pressure changes in relation to urinary sodium excretion in White Europeans. JAMA. 2011;305:1777–85.
10. Expert Committee on the Diagnosis and Classification of Diabetes Mellitus. Report of the expert committee on the diagnosis and classification of diabetes mellitus. Diabet Care. 2003;26 Suppl 1:5–20.
11. Citterio L, Simonini M, Zagato L, Salvi E, Delli Carpini S, Lanzani C, et al. Genes involved in vasoconstriction and vasodilation affect salt-sensitive hypertension. PLoS One. 2011;6:e19620.
12. Liu YP, Gu YM, Thijs L, Knapen MHJ, Salvi E, Citterio L, et al. Inactive matrix gla protein is causally related to adverse health outcomes: a Mendelian randomization study in a Flemish population. Hypertension. 2015;65:463–70.
13. O'Connell J, Gurdasani D, Delaneau O, Pirastu N, Ulivi S, Cocca M, et al. A general approach for haplotype phasing across the full spectrum of relatedness. PLoS Genet. 2014;10:e1004234.
14. Benjamini Y, Hochberg Y. Controlling the false discovery rate: a practical and powerful approach to multiple testing. J Royal Stat Soc B. 1995;57:289–300.
15. Holtzman NA, Marteau TM. Will genetics revolutionize medicine? N Engl J Med. 2000;343:141–4.
16. Pencina MJ, D'Agostino Sr RB, D'Agostino Jr RB, Vasan RS. Evaluating the added predictive ability of a new marker: from area under the ROC curve to reclassification and beyond. Stat Med. 2008;27:157–72.
17. Nordestgaard BG, Albigaard S, Wittrup HH, Steffensen R, Jensen G, Tybjærg-Hansen A. Heterozygous lipoproteinase lipase deficiency. Circulation. 1997;96:1737–44.
18. Wittrup HH, Tybjærg-Hansen A, Nordestgaard BG. Lipoprotein lipase mutations, plasma lipids and lipoproteins, and the risk of ischemic heart disease. A meta-analysis. Circulation. 1999;99:2901–7.
19. Li Y, He PP, Zhang DW, Zheng XL, Cayabyab FS, Yin WD, et al. Lipoprotein lipase: from gene to atherosclerosis. Atherosclerosis. 2014;237:597–608.
20. Cantile M, Schiavo G, Terracciano L, Cillo C. Homeobox genes in normal and abnormal vasculogenesis. Nutr Metab Cardiovasc Dis. 2008;18:651–8.
21. Skopicki HA, Lyons GE, Schatterman G, Smith RC, Andrés V, Schirm S, et al. Embryonic expression of the Gax homeodomain protein in cardiac, smooth, and skeletal muscle. Circ Res. 1997;80:452–62.
22. Weir L, Chen D, Pastore C, Isner JM, Walsh K. Expression of gax, a growth arrest homeobox gene, is rapidly down-regulated in the rat carotid artery during the proliferative response to balloon injury. J Biol Chem. 1995;270:5457–61.
23. Smith RC, Branellec D, Gorski DH, Guo K, Perlman H, Dedieu JF, et al. p21CIP1-mediated inhibition of cell proliferation by overexpression of the gax homeodomain gene. Genes Dev. 1997;11:1674–89.
24. Chen Y, Rabson AB, Gorski DH. MEOX2 regulates nuclear factor-κB activity in vascular endothelial cells through interactions with p65 and IκBβ. Cardiovasc Res. 2010;87:723–31.
25. Tabas I, García-Cardeña G, Owens GK. Recent insights into the cellular biology of atherosclerosis. J Cell Biol. 2015;209:13–22.
26. Chen Y, Gorski DH. Regulation of angiogenesis through a microRNA (miR-130a) that down-regulates antiangiogenic homeobox genes GAX and HOXA5. Blood. 2008;1217–26.
27. The CARDIoGRAMplusCD4 Consortium. Large-scale association analysis identifies new risk loci for coronary heart disease. Nat Genet. 2013;45:25–33.
28. Ripatti S, Tikkanen E, Orho-Melander M, Havulinna AS, Silander K, Sharma A, et al. A multilocus genetic risk score for coronary heart disease: case–control and prospective cohort analyses. Lancet. 2010;376:1393–400.
29. Mehta NN. Large-scale association analysis identifies 13 new susceptibility loci for coronary artery disease. Circ Cardiovasc Genet. 2011;4:327–9.
30. Saade S, Cazier JBG-SM, Youhanna S, Badro DA, Kamatani Y, Hager J, et al. Large scale association analysis identifies three susceptibility loci for coronary artery disease. PLoS One. 2011;6:e29427.
31. Della Gaspera B, Armand AS, Lecolle S, Charbonnier F, Chanoine C. Mef2d acts upstream of muscle identity genes and couples lateral myogenesis to dermomyotome formation in Xenopus laevis. PLoS One. 2012;7:e52359.
32. Davies OR, Lin CY, Radzisheuskaya A, Zhou X, Taube J, Blin G, et al. Tcf15 primes pluripotent cells for differentiation. Cell Rep. 2013;3:472–84.
33. Paulus WJ, Tschöpe C, Sanderson JE, Rusconi C, Flachskampf FA, Rademakers FE, et al. How to diagnose heart failure: a consensus statement on the diagnosis of heart failure with normal left ventricular ejection fraction by the Heart Failure and Echocardiography Associations of the European Society of Cardiology. Eur Heart J. 2007;28:2539–50.
34. Zile MR, Brutsaert DL. New concepts in diastolic dysfunction and diastolic heart failure: part I. Diagnosis, prognosis, and measurements of diastolic function. Circulation. 2002;105:1387–93.
35. Nyholt DR. Genetic case–control association studies - correcting for multiple testing. Hum Genet. 2001;109:564–5.
36. Hill AB. The environment and disease: association or causation? Proc R Soc Med. 1965;58:295–300.
37. Thanassoulis G, Peolso GM, Pencina MJ, Hoffmann U, Fox CS, Cupples LA, et al. A genetic risk score is associated with incident cardiovascular disease and coronary artery calcium: the Framingham Heart Study. Circ Cardiovasc Genet. 2012;5:113–21.
38. de Vries PS, Kavousi M, Lightart S, Uitterlinden AG, Hofman A, Franco OH, et al. Incremental predictive value of 152 single nucleotide polymorphisms in the 10-year risk prediction of incident coronary heart disease: the Rotterdam Study. Int J Epidemiol. 2015;44:682–8.

2

MC1R diversity in Northern Island Melanesia has not been constrained by strong purifying selection and cannot explain pigmentation phenotype variation in the region

Heather L. Norton[1*], Elizabeth Werren[2] and Jonathan Friedlaender[3]

Abstract

Background: Variation in human skin pigmentation evolved in response to the selective pressure of ultra-violet radiation (UVR). Selection to maintain darker skin in high UVR environments is expected to constrain pigmentation phenotype and variation in pigmentation loci. Consistent with this hypothesis, the gene *MC1R* exhibits reduced diversity in African populations from high UVR regions compared to low-UVR non-African populations. However, *MC1R* diversity in non-African populations that have evolved under high-UVR conditions is not well characterized.

Methods: In order to test the hypothesis that MC1R variation has been constrained in Melanesians the coding region of the MC1R gene was sequenced in 188 individuals from Northern Island Melanesia. The role of purifying selection was assessed using a modified McDonald Kreitman's test. Pairwise F_{ST} was calculated between Melanesian populations and populations from the 1000 Genomes Project. The SNP rs2228479 was genotyped in a larger sample (n = 635) of Melanesians and tested for associations with skin and hair pigmentation.

Results: We observe three nonsynonymous and two synonymous mutations. A modified McDonald Kreitman's test failed to detect a significant signal of purifying selection. Pairwise F_{ST} values calculated between the four islands sampled here indicate little regional substructure in *MC1R*. When compared to African, European, East and South Asian populations, Melanesians do not exhibit reduced population divergence (measured as F_{ST}) or a high proportion of haplotype sharing with Africans, as one might expect if ancestral haplotypes were conserved across high UVR populations in and out of Africa.

The only common nonsynonymous polymorphism observed, rs2228479, is not significantly associated with skin or hair pigmentation in a larger sample of Melanesians.

Conclusions: The pattern of sequence diversity here does not support a model of strong selective constraint on *MC1R* in Northern Island Melanesia This absence of strong constraint, as well as the recent population history of the region, may explain the observed frequencies of the derived rs2228479 allele. These results emphasize the complex genetic architecture of pigmentation phenotypes, which are controlled by multiple, possibly interacting loci. They also highlight the role that population history can play in influencing phenotypic diversity in the absence of strong natural selection.

Keywords: Pigmentation phenotype, Natural selection, Island Melanesia

* Correspondence: heather.norton@uc.edu
[1]Department of Anthropology, University of Cincinnati, 481 Braunstein Hall, PO Box 210380, Cincinnati, OH 45221, USA
Full list of author information is available at the end of the article

Background

Human skin and hair pigmentation are highly variable traits that are controlled by multiple genetic loci [1–14]. Skin pigmentation in humans is tightly correlated with the intensity of ultra-violet radiation (UVR) [15]; darker pigmentation is commonly observed in populations originating from regions of higher UVR, while lighter skin color is common to populations in lower UVR regions. The geographic structure of human skin pigmentation variation strongly supports a model in which pigmentation phenotype has been influenced by natural selection [15–17]. It has been proposed that darker skin color in hominins evolved with the loss of body fur and hair, becoming established some time around 1.2 million years ago (mya) [18], presumably to protect against UV-induced damage to DNA and folic acid photolysis [19, 20]. Ultimately alleles causing lighter skin increased in frequency in populations expanding into lower UVR regions, possibly to increase the potential for vitamin D synthesis [15, 21].

Skin pigmentation is primarily determined by the amount, type, and distribution of melanin, one of the primary chromophores of the skin. Melanin (particularly the alkali-insoluble brown-black eumelanin) acts as both a barrier to and filter of UVR—it scatters UVR and limits its penetration of the epidermis [22, 23]. This photoprotective property takes on particular evolutionary significance in regions where UVR is high, as both the long (UVA) and short (UVB) wavelength radiation that reaches the earth's surface have cytotoxic and mutagenic effects with potentially significant effects on fitness. UVB is absorbed by DNA, resulting in mutations such as cyclobutane pyrimidine dimers (CPDs) and pyrimidine (6–4) pyrimidine photoproducts [24, 25], both of which play a role in photocarcinogenesis [26]. The ability of melanin, particularly eumelanin, to minimize the potential cancer-causing properties of UVR has led some to speculate that darker skin pigmentation should be favored by natural selection in regions where UVR is high [20], although others argue that this is at best a weak selective force given the late age of onset of many fatal skin cancers [15, 16, 27, 28]. A perhaps more relevant evolutionary argument for the evolution and maintenance of a highly melanized skin in regions of high UVR is the need to minimize UVA mediated photolysis of the B-vitamin folate [19]. Folic acid is involved in DNA synthesis and repair as well as spermatogenesis [29, 30], and folic acid deficiencies in reproductive-age females have been linked to an increased risk of neural tube birth defects [31, 32]. As with the photoprotection hypothesis, the folic acid hypothesis predicts that darker skin color should be maintained by purifying selection in high UVR regions [15, 16, 33].

A genetic locus known to influence human skin and hair pigmentation is the *MC1R* gene, which encodes the melanocortin-1 receptor, a 7-pass transmembrane G-protein coupled receptor found on the surface of melanocyte cells. When the hormone α-MSH binds to the MC1R, activation of adenyl cyclase results in increased levels of cAMP. This leads to an increase in the activity of tyrosinase, the rate-limiting enzyme of melanogenesis, as well as increased levels of tyrosinase-related proteins (TRP)-1 and –2 [34, 35], ultimately resulting in eumelanin synthesis. However, if instead the antagonist agouti-signaling protein binds to the MC1R, pheomelanin production results. Despite its small size (951 bp), the *MC1R* gene harbors a high number of polymorphisms, including several loss-of-function mutations that are associated with reduced skin color, melanoma, freckling, and red or blond hair [7, 36–39]. *MC1R* mutations are also associated with a reduced DNA-repair capacity, possibly explaining the link between *MC1R* and melanoma risk [40, 41].

MC1R has received particular attention in studies of human evolution because of its unusual levels and patterns of sequence diversity. Unlike many loci, which exhibit higher levels of diversity in African populations, *MC1R* diversity is highest in Eurasian populations [42, 43]. The lower sequence diversity observed in Africans is commonly attributed to purifying selection, while the higher diversity in Eurasian populations has been alternatively interpreted as being due to either relaxed functional constraint or positive selection for lighter skin color [42, 43]. It has been suggested that the higher diversity observed at *MC1R* can be attributed in part to a high mutation rate related to the elevated CpG content of the region [37]. This makes the reduced diversity in high-UVR African populations all the more notable.

Evidence for purifying selection on *MC1R* in African and other high-UVR populations rests primarily on low levels of nucleotide diversity and the ratios of nonsynonymous to synonymous polymorphisms and divergent sites (assessed using McDonald-Kreitman and HKA tests) [42–44]. The higher sequence diversity observed at *MC1R* in non-African populations has been argued to be a signal of diversifying selection [43], although this is not supported by McDonald Kreitman and HKA tests [42]. Tests of natural selection that rely on the site frequency spectrum, inter-population divergence, and extended haplotype homozygosity are also in conflict as to the nature of selection acting on *MC1R* in non-African populations [10, 37, 42, 45, 46].

Because of its association with several phenotypic traits in European (and to a lesser extent East Asian) populations, *MC1R* has been extensively sequenced in populations living in low UVR regions. However, far less is known of *MC1R* sequence diversity in high UVR populations outside of Africa, including whether or not polymorphisms segregating in such populations may be responsible for variation in skin or hair pigmentation phenotype in these regions. Because *MC1R* is commonly associated with mutations that lead to a *decrease* in skin pigmentation, it is generally assumed that *MC1R* variation is tightly constrained by purifying selection in high UVR regions. However, it is possible that mutations leading to an *increase* in the synthesis of eumelanin may have been favored by positive selection. While such mutations in *MC1R* are believe to have played an important role in adaptation following the loss of body hair in hominins 1.2 mya [18], there is little evidence to date for more recent mutations with this effect occurring in humans. As such, most investigations of selection acting on *MC1R* in high-UVR populations focus on the role of purifying, rather than positive, selection.

Early reports of sequence variation in *MC1R* indicated that the gene was under strong functional constraint in populations from Papua New Guinea as well as Africa, as one might expect if purifying selection has acted to remove nonsynonymous variants resulting in lighter skin color [42, 43] across high UVR regions. However, those studies sampled only a small number of Melanesians (32 chromosomes) [42], resulting in a limited picture of *MC1R* diversity in high-UVR non-African populations. In order to accurately assess the extent of variation at *MC1R* in Melanesia, a broader sample is critical, due to the complex population history of the region [47–51]. Archaeological evidence indicates that modern humans reached Near Oceania (Sahul, the New Guinea and Australia landmasses, the Bismarck Archipelago, and much of the Solomon islands) by 49,000 YBP [52, 53], spreading as far east as the island of Buka in the Solomons by 29,000 YBP [54]. The region later saw a major influx of migrants speaking languages belonging to the Proto-Oceanic Austronesian language family around 4 KYA [55, 56] originating from a homeland in Taiwan. Thus, while *MC1R* variation is expected to have been tightly constrained during much of the first ~30,000–40,000 years of human habitation in Melanesia, the migration of Austronesian speakers into the region and their subsequent admixture with resident populations may have introduced nonsynonymous mutations more commonly found in low-UVR regions. If not quickly removed by strong purifying selection such variants may contribute to observed variation in pigmentation phenotype.

Populations from Island Melanesia are darkly pigmented compared to populations of northern Europe and Asia [57], as expected if pigmentation phenotype reflects adaptation to UVR intensity [15]. However, despite experiencing high levels of UVR, Northern Island Melanesian populations exhibit a striking amount of variation in skin pigmentation [58]. Perhaps even more unusually, some Melanesians also exhibit a characteristic "blond hair" phenotype that is observed from Northern

Island Melanesia throughout the Solomon Islands, which can be partially explained by a nonsynonymous variant in the *TYRP1* gene [58, 59]. Variation in skin and hair pigmentation is particularly pronounced between different islands of Northern Island Melanesia, including the island of Bougainville (located at the northwest tip of the Solomon Islands chain) and islands in the Bismarck Archipelago. It is highly unlikely that this variation is due to very fine scale adaptation to UVR differences in the region [58]. Instead, this variation suggests that even in a high-UVR environment pigmentation phenotype may vary so long as it is maintained above a protective melanin "threshold" [58, 60]. *MC1R*, known to influence skin and hair color in European and East Asian populations [1, 36, 38, 39, 61, 62] is a possible candidate to explain a portion of this observed variation.

Here we survey sequence variation in the coding region of the *MC1R* gene in 188 Island Melanesians from four different islands: New Hanover, New Britain, New Ireland, and Bougainville. To our knowledge this is the most extensive survey of *MC1R* sequence variation in the region to date. We use these data to address three questions pertaining to the evolution of *MC1R* sequence variation in Northern Island Melanesian populations and the role of *MC1R* in shaping phenotypic diversity in the region. Specifically we set out to test the hypothesis that variation in *MC1R* has been constrained in these populations by *purifying* selection. We also characterized regional (inter-island) and global levels of variation at *MC1R* in order to assess the roles of selection and population history in shaping variation at *MC1R* in Northern Island Melanesia. Finally, we tested for associations between *MC1R* polymorphisms and quantitatively measured skin and hair pigmentation phenotype to evaluate the role of *MC1R* in shaping local pigmentation phenotype.

Methods
Sample collection
The pigmentation measurements reported here were originally collected as part of a larger study examining phenotypic variation and population history in Island Melanesia by H.L.N. and J.S.F conducted in 2000 and 2003. Individuals were sampled from islands throughout Northern Island Melanesia, with an emphasis on the islands of New Britain, New Hanover, New Ireland, and Bougainville (Fig. 1). Individuals speaking languages belonging to the Oceanic division of the Austronesian (AN) phylum as well as speakers of unrelated non-Austronesian (Papuan) languages were recruited [50]. In total 1135 adults were sampled, of which 188 were used here in DNA sequencing and an additional 444 in the genotyping of the rs2228479 SNP. DNA samples were chosen for sequencing and genotyping based on the

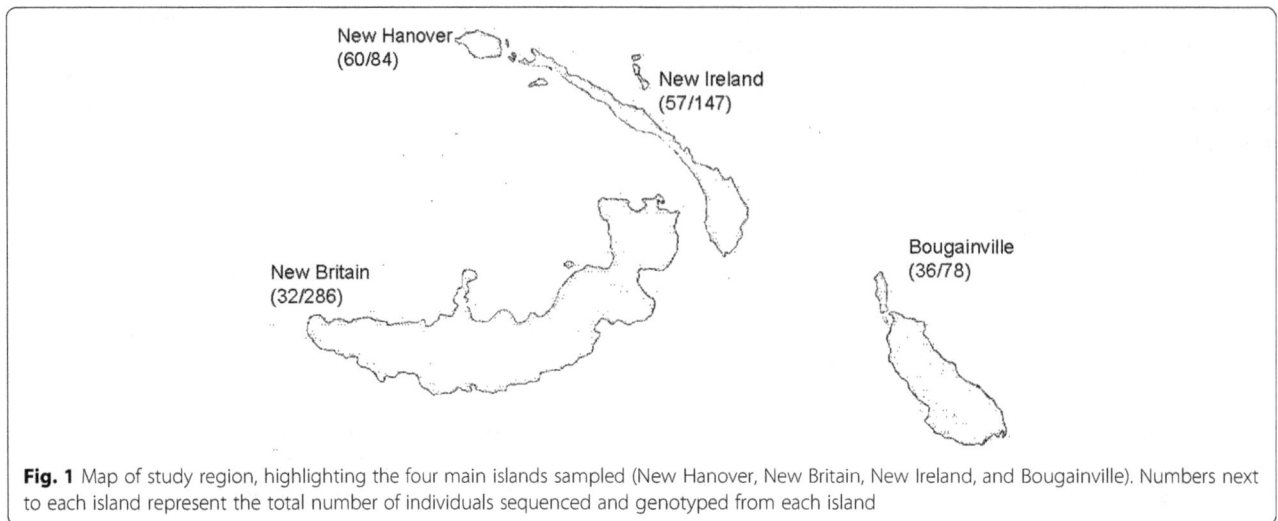

Fig. 1 Map of study region, highlighting the four main islands sampled (New Hanover, New Britain, New Ireland, and Bougainville). Numbers next to each island represent the total number of individuals sequenced and genotyped from each island

following factors: the quantity and quality of available DNA, representation of all four major islands and two linguistic phyla in the region, and, where possible, an attempt to focus heavily on 1–3 neighborhoods or subpopulations within an island (in an effort to minimize intra-island substructure).

Individuals were assigned to categories according to sex, island, neighborhood, and linguistic phylum. In order to be assigned to a particular island, neighborhood, or linguistic phylum an individual and both of his or her parents needed to also be from that island/neighborhood or speak a language belonging to the same phylum. All individuals gave their informed consent to participate in the study, including the measurement of pigmentation data and the examination and publication of genetic variation. Institutional Review Board (IRB) approval for data collection and analyses were obtained from Temple University (IRB 99–226), The Pennsylvania State University (IRB 00 M558-2), and the Papua New Guinea Medical Research Advisory Committee. Approval was granted for all data collection sites. This was a collaborative project with the PNG Institute of Medical Research.

Pigmentation measurement

Quantitative measurements of skin and hair pigmentation were taken using the DermaSpectrometer (Cortex Technology, Hadsund, Denmark) following the practices described in Norton et al. [58]. The DermaSpectrometer estimates the concentrations of the two primary chromophores of the skin, hemoglobin and melanin, and reports this as the melanin (M) index. The M index provides a quantitative measure of skin color that is due primarily to the effects of melanin alone [63].

MC1R sequencing and genotyping

The coding region of the $MC1R$ gene (951 bp) was amplified in the 191 Melanesian samples from the islands of Bougainville ($N = 36$), New Hanover ($n = 60$), New Britain ($n = 32$), and New Ireland ($n = 57$). Sequence was also obtained from three individuals who could not be clearly assigned to any one of these four islands (see *Sample Collection*). Product amplification was verified using gel electrophoresis, and amplified products were purified for sequencing using the GeneJet Purification kit (Life Technologies). Sequence data were aligned to the human reference sequence using the Geneious 6.0 software package [64]. Variants were identified using the "Find Heterozygotes" feature in Geneious and confirmed by visual inspection. Haplotypes were computationally phased using the program Phase 2.1.1 [65], and all phases were correctly inferred ($p > 0.85$) Melanesian sequences were later aligned with the chimpanzee (*Pan troglodytes*) $MC1R$ sequence and African (LWK, YRI), East Asian (CHB, CHS, JPT), European (CEU, FIN, GBR, IBS, and TSI), and South Asian (BEB, GIH, ITU, PJL, STU) samples sequenced as part of the 1000 Genomes Project (http://www.1000Genomes.org, accessed on August 28, 2014).

In order to test for the effect of the derived allele at rs2228479 on skin and hair pigmentation in a larger sample of Melanesians we utilized published genotype and phenotype data from these individuals [66]. To obtain genotype data on additional individuals we designed separate amplification primers (available from the authors on request) to amplify a 695 bp region around rs2228479. The restriction enzyme NspI, which recognizes the derived allele, was used to digest these amplified fragments under standard conditions. Digested products were visualized using gel electrophoresis.

Statistical analyses

Summary statistics for *MC1R* sequence diversity and polymorphism and divergence were calculated in DNASp v5.10.1 [67, 68]. Two measures of nucleotide diversity were calculated: π [69], which is based on the average number of nucleotide differences between two sequences randomly drawn from a sample; and θ, which is based on the proportion of segregating sites in the sample [70]. Tajima's D, a summary of the allele frequency spectrum that tests the null hypothesis of mutation-drift equilibrium and constant population size (under which π and θ are approximately equal) was also calculated [71]. A negative value of Tajima's D indicates an excess of rare alleles, which may occur when a population is growing or when a gene is targeted by purifying selection. Positive values may be indicative of a population bottleneck or balancing selection. We assessed statistical significance of observed Tajima's D values by comparing them to values obtained from 10,000 simulations under a standard neutral model. Because demographic history (e.g. a population bottleneck) can mimic the effects of selection, we also compared observed values in the Melanesian sample those obtained under 10,000 simulations of a simple bottleneck model marking the divergence of Melanesian populations from an ancestral Eurasian population. As no well-defined demographic model has yet been developed to characterize the population history of the five populations used here (the four 1000 Genome populations and the Melanesians), we use bottleneck and divergence parameters estimates for a New Guinea Highlands population estimated by Wollstein et al. [72]. This model is designed primarily to capture the reduction in diversity associated with the initial colonization of Melanesia.

In order to test the hypothesis that *MC1R* has been shaped by purifying selection in Island Melanesian populations we use a modification of the McDonald-Kreitman test, which compares the ratio of nonsynonymous to synonymous polymorphisms to the ratio of nonsynonymous to synonymous fixed differences [73]. An advantage of using the McDonald-Kreitman test over other site-frequency spectrum based tests (such as Tajima's D and Fay and Wu's H) is that it is relatively insensitive to variation in demographic history [74, 75]. Due to the small number of segregating sites that we observe in the Melanesian sample (five), we follow Harding et al. [42] and use an exact test described by Sokal and Rohlf (1969). In this test, we use the number of fixed nonsynonymous (ten) to synonymous (six) changes between humans and chimpanzee, and assume that new nonsynonymous and synonymous mutations will occur with a binomial probability of 0.625 and 0.375, respectively [42]. The sum of probabilities for the observed ratio of nonsynonymous ($n = 3$) to synonymous ($n = 2$) polymorphisms in Melanesians as well as for all other

equally or less likely probabilities provides a test of this null hypothesis.

Allele frequency distinctions between islands and between linguistic phyla were tested for using a χ^2 test implemented in DNASp [67]. Tests of Hardy Weinberg Equilibrium for each Melanesian island were performed in PLINK [76]. Pairwise F_{ST} values for the coding region of *MC1R* were calculated using DNASp between each of the four Melanesian islands, between each island and each population from the 1000 Genomes Project, and between the Melanesian region as a whole 1000 Genomes African, East Asian, European, and South Asian regional groupings. Median-joining haplotype networks showing the relationship of *MC1R* haplotypes among Melanesian islands and between Melanesians and the 1000 Genomes samples were constructed using Network 4.613 [77].

Associations between the only common (frequency > 0.05) *MC1R* nonsynonymous polymorphism, rs2228479, and skin and hair pigmentation in the Melanesian population were tested for in PLINK [76] using linear regression with an additive model. We tested for associations in the full Melanesian sample using island as a covariate in an effort to control for inter-island substructure in the region [48]. We also tested for associations on each island separately. Due to the skewed distribution of the pigmentation measurements, skin and hair M index values were first transformed using a Box-Cox procedure (skin $M \lambda = -0.788$, hair $M \lambda = 2.93$). The distribution of untransformed and transformed skin and hair M index values are depicted in Additional files 1 and 2.

Results
MC1R sequence variation

In our sample of 188 sequenced Melanesians, we observed a total of five segregating sites: three of these were synonymous mutations, two were nonsynonymous. Of these five mutations, one (a nonsynonymous Ile → Leu mutation at position 89986456) is novel and has not been previously reported. However, this mutation is rare in the sample, occurring in three individuals (all from the island of New Hanover). The most common *MC1R* mutations observed here are rs2228479 (V92M), rs885479 (R163G), and rs2228478 (T314T), which occur at frequencies of 15.4, 4.5, and 22.3 % in the resequenced sample, respectively. *MC1R* haplotypes and their frequencies in the Melanesian sample are reported in Table 1.

We calculated average nucleotide diversity (measured as π and θ) in the total Melanesian sample, as well as for each island and linguistic phylum separately. These are displayed in Table 2, with values for Africans, East Asians, Europeans, and South Asians sequenced by the 1000 Genomes Project included for comparison. π in the full Melanesian sample (0.00075) was lower than that

Table 1 Melanesian *MC1R* haplotypes and total number of observed haplotypes for each island and linguistic phylum. Note—because some individuals could not be assigned to an island and/or phylum, the sum of the haplotypes for a given island/phylum may not equal the total number of observed haplotypes in the total Melanesian sample

HAPLOTYPE #	rs2228479 89985940	Rs372929572 89986083	rs885479 89986154	Ile → Leu 89986456	rs2228478 89986608	Island				Languages		Total Melanesian
						Bougainville	New Hanover	New Britain	New Ireland	Austronesian	Papuan	
1	G	C	A	A	A	0	5	6	6	16	1	17
2 (human consensus)	G	C	G	A	A	57	69	45	82	155	99	258
3	A	C	G	A	A	0	9	3	2	12	2	14
4	A	C	G	A	G	8	22	4	9	38	5	43
5	G	C	G	A	G	7	13	6	12	31	8	39
6	G	T	G	A	A	0	0	0	0	0	0	2
7	G	C	G	C	G	0	1	0	1	2	0	2
8	G	C	G	C	A	0	1	0	0	1	0	1

Table 2 Summary statistics for *MC1R* in Melanesian sample and for 1000 Genomes populations (AFR = LWK, YRI; EUR = CEU, FIN, GBR, IBS, TSI; EAS = CHB, CHD, JPT; SAS = BEB, GIH, ITU, PJL, STU)

Population	2 N	bp	S	# of haplotypes	π	θ	Tajima's D	FWH
Melanesia	376	954	5	8	0.00075	0.00081	−0.117	−0.543
Bougainville	72	954	2	3	0.00056	0.00043	0.486	−0.761
New Hanover	122	954	5	7	0.00099	0.00097	0.036	−0.191
New Britain	64	954	3	5	0.00067	0.00067	0.012	−0.850
New Ireland	112	954	4	6	0.00065	0.00079	−0.361	−0.715
Austronesian speaking	246	954	4	7	0.00087	0.00069	0.445	−0.362
Non-Austronesian speaking	124	954	3	5	0.00050	0.00058	−0.249	−0.934
AFR	370	954	10	13	0.00086	0.00162	−1.019	0.160
LWK	194	954	9	10	0.00082	0.00161	−1.12	0.123
YRI	176	954	7	9	0.00090	0.00128	−0.631	0.193
EUR	758	954	17	18	0.00104	0.00247	−1.338	−2.749
CEU	170	954	9	10	0.00101	0.00165	−0.902	−2.866
FIN	186	954	11	11	0.00107	0.00199	−1.109	−0.863
GBR	178	954	10	11	0.00122	0.00182	−0.778	−0.437
IBS	28	954	4	5	0.00054	0.00108	−1.30	−1.481
TSI	196	954	14	15	0.00087	0.00251	−1.65	−0.898
EAS	572	954	8	10	0.00138	0.00121	0.269	−0.802
CHB	194	954	6	6	0.00142	0.00108	0.657	−0.787
CHD	200	954	5	5	0.00157	0.00089	1.451	−0.285
JPT	178	954	6	8	0.00098	0.00109	−0.220	−1.697
SAS	978	954	23	24	0.00061	0.00323	−1.939	−2.662
BEB	172	954	7	8	0.00070	0.00128	−0.989	−0.611
GIH	206	954	13	14	0.00059	0.00231	−1.833	−0.635
ITU	204	954	12	13	0.00064	0.00213	−1.700	−2.672
PJL	192	954	10	11	0.00060	0.00180	−1.562	−0.677
STU	204	954	9	10	0.00056	0.00160	−1.479	−0.741

observed for the AFR, EAS, and EUR regional samples. However, it was higher than the observed value for the SAS samples (0.00061). The Melanesian θ value (0.00081) was less than the reported value for all four 1000 Genomes population samples. Thinking that the lower diversity levels observed in the Melanesian sample might reflect the smaller sample size of the Melanesians (2 *N* = 376) compared to the larger regional 1000 Genome populations (2 N range: 370–758), we randomly subsampled all regional populations (MEL, AFR, EAS, EUR, and SAS) to obtain 370 haplotypes (the total number in the smallest, AFR, sample) from each and recalculated summary statistics. This process was repeated 100 times to obtain a range of π and θ values for each subsampled population (see Additional file 3: Table S1). As expected, subsampling leads to a reduction in diversity for for both π and θ in all five populations, with Melanesian diversity

at both π and θ remaining low. Mean π is lower in the subsampled Melanesian sample than in the subsampled South Asian sample, and mean θ in the subsampled Melanesians is the lowest of all the subsampled population values.

Mean levels of diversity and haplotype distribution across the four Melanesian islands sampled here can be found in Tables 1 and 2. Of the four islands, diversity is lowest on Bougainville (π = 0.00059, θ = 0.00043), where only two segregating sites and three haplotypes are present. Diversity levels are highest on New Hanover (π = 0.00099, θ = 0.00097), where the highest number segregating sites (five) and haplotypes (seven) are observed. Diversity values for each of the four Melanesian islands also indicate relatively low levels of variation compared to the individual 1000 Genomes populations. Notably π values in non-Austronesian speakers (π = 0.00050) are roughly one-half of those reported for Austronesian speakers ($\pi = 0.00087$), indicating a sharp reduction in diversity among non-Austronesians.

Evidence for purifying selection

Given the ten fixed nonsynonymous and six fixed synonymous changes between humans and chimpanzees, we estimate the likelihood of observing three nonsynonymous polymorphisms out of five total segregating sites to be 0.275. These data do not support a model in which variation in *MC1R* has been tightly constrained by strong purifying selection in Melanesian populations. Although site-frequency-spectrum based statistics such as Tajima's D and Fay and Wu's H are sensitive to demographic history, we also calculated values for these statistics in the full Melanesian sample, as well as for each individual Melanesian population (Table 2). Tajima's D is slightly negative in the population samples from New Ireland and in non-Austronesian speakers, while it is slightly positive on the remaining islands and in Austronesian speakers Tajima's D in the full Melanesian sample was –0.117. None of these values fall outside the 95 % confidence intervals of Tajima's D values obtained from simulations under either the standard neutral (–1.412 – 1.696) or bottleneck model (–0.911 – 2.932). Fay and Wu's H is slightly negative in all Melanesian populations sampled (–0.850 – -0.191). As with Tajima's D, none of these values are outside the 95 % confidence limits estimated for either the standard neutral (–2.857 – 1.349) or bottleneck models (–2.986 – 1.014).

Inter-population variation

Among the four Melanesian islands sampled here, pairwise F_{ST} values at *MC1R* are generally low (0.000–0.042), with the highest values reported for comparisons between New Hanover and New Britain (0.042), and the lowest between New Ireland and Bougainville (0.000) and between New Ireland and New Britain (0.000). By comparison, pairwise F_{ST} values at *MC1R* among the five European, two African, three East Asian, and five South Asian populations in the 1000 Genomes Project ranged from 0.001–0.065 (European) 0.000 (African), 0.022–0.089 (East Asian), and 0.000–0.004 (South Asian) (see Additional file 3: Table S2). Among the eight *MC1R* haplotypes observed in the Melanesian sample, three of the five common haplotypes (haplotype frequency > 1 %) can be found on each of the 4 islands (Table 1, Fig. 2), also indicating relatively little regional substructure at *MC1R*.

Pairwise F_{ST} values between the full Melanesian sample and the 1000 Genomes AFR, EUR, EAS, and SAS regional population samples are found in Table 3. Surprisingly, Melanesians exhibit the lowest pairwise F_{ST} values with South Asians ($F_{ST(MEL-SAS)} = 0.027$) and the highest with East Asians ($F_{ST(MEL-EAS)} = 0.270$). However, this high divergence between Melanesians and East Asians is characteristic of generally high levels of divergence between East Asians and the other 3 populations at *MC1R* (F_{ST} values range between 0.255 and 0.326 for all East Asian-specific F_{ST} comparisons). There is no evidence of a particularly strong affinity between Melanesians and the high-UVR African populations of

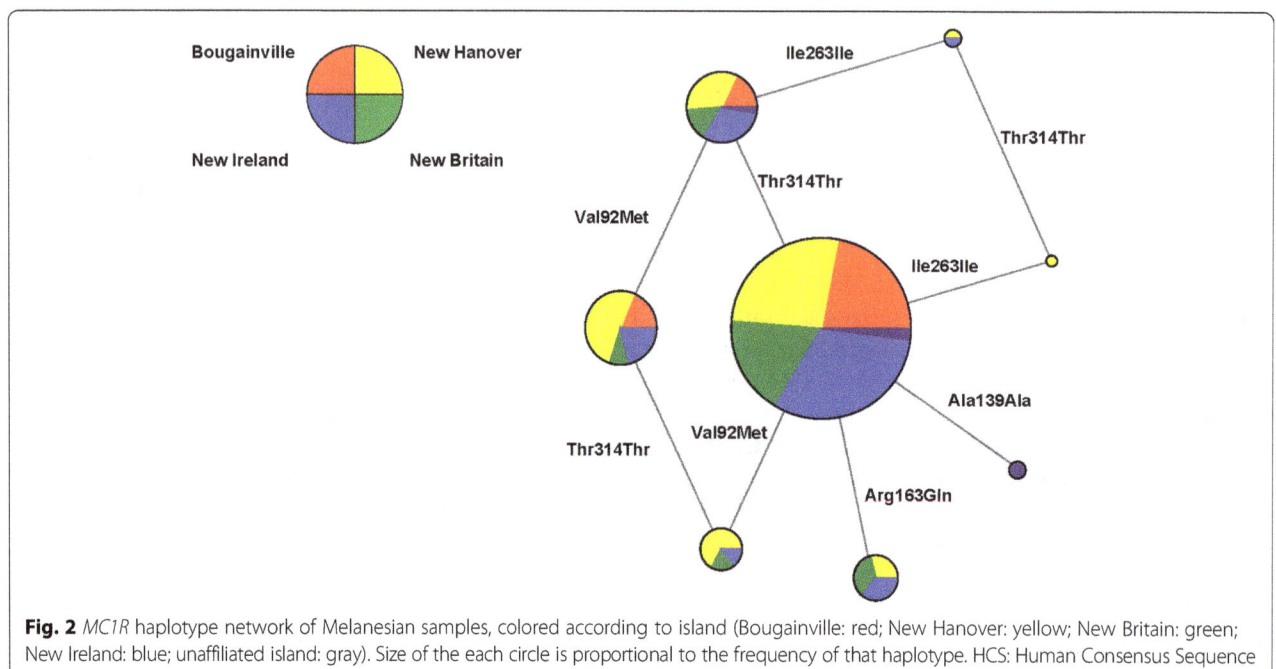

Fig. 2 *MC1R* haplotype network of Melanesian samples, colored according to island (Bougainville: red; New Hanover: yellow; New Britain: green; New Ireland: blue; unaffiliated island: gray). Size of the each circle is proportional to the frequency of that haplotype. HCS: Human Consensus Sequence

Table 3 Pairwse F_{ST} of *MC1R* between Melanesian and 1000 genomes regional populations

AFR	0.088			
EUR	0.054	0.148		
EAS	0.270	0.326	0.255	
SAS	0.027	0.075	0.048	0.306
	MEL	AFR	EUR	EAS

the 1000 Genomes Project (Table 3, Fig. 3). The majority of common (>1 %) Melanesian *MC1R* haplotypes are found in Melanesians as well as Africans, East Asians, Europeans, and South Asians Of the eight haplotypes observed in the Melanesian sample, none are shared exclusively with any of these groups (Fig. 3). When pairwise F_{ST} is calculated between each Melanesian island and the individual 1000 Genome populations there is

little evidence to suggest particularly low levels of divergence between a specific island and any individual 1000 Genomes population (see Additional file 3: Table S2).

rs228479 frequency and associations with phenotype across Melanesia

The only common nonsynonymous SNP observed in the Melanesian sample was the rs2228479 polymorphism. To better assess the frequency of the derived allele at this site across Northern Island Melanesia we genotyped this allele in 444 additional individuals from New Britain, New Hanover, New Ireland, and Bougainville. The frequency of the derived allele at this locus in each Melanesian island and population are reported in Table 4. The frequency of the derived allele at rs2228479 was significantly different between New Hanover and the 3 other islands sampled (Chi-squared test, all $p < 0.001$).

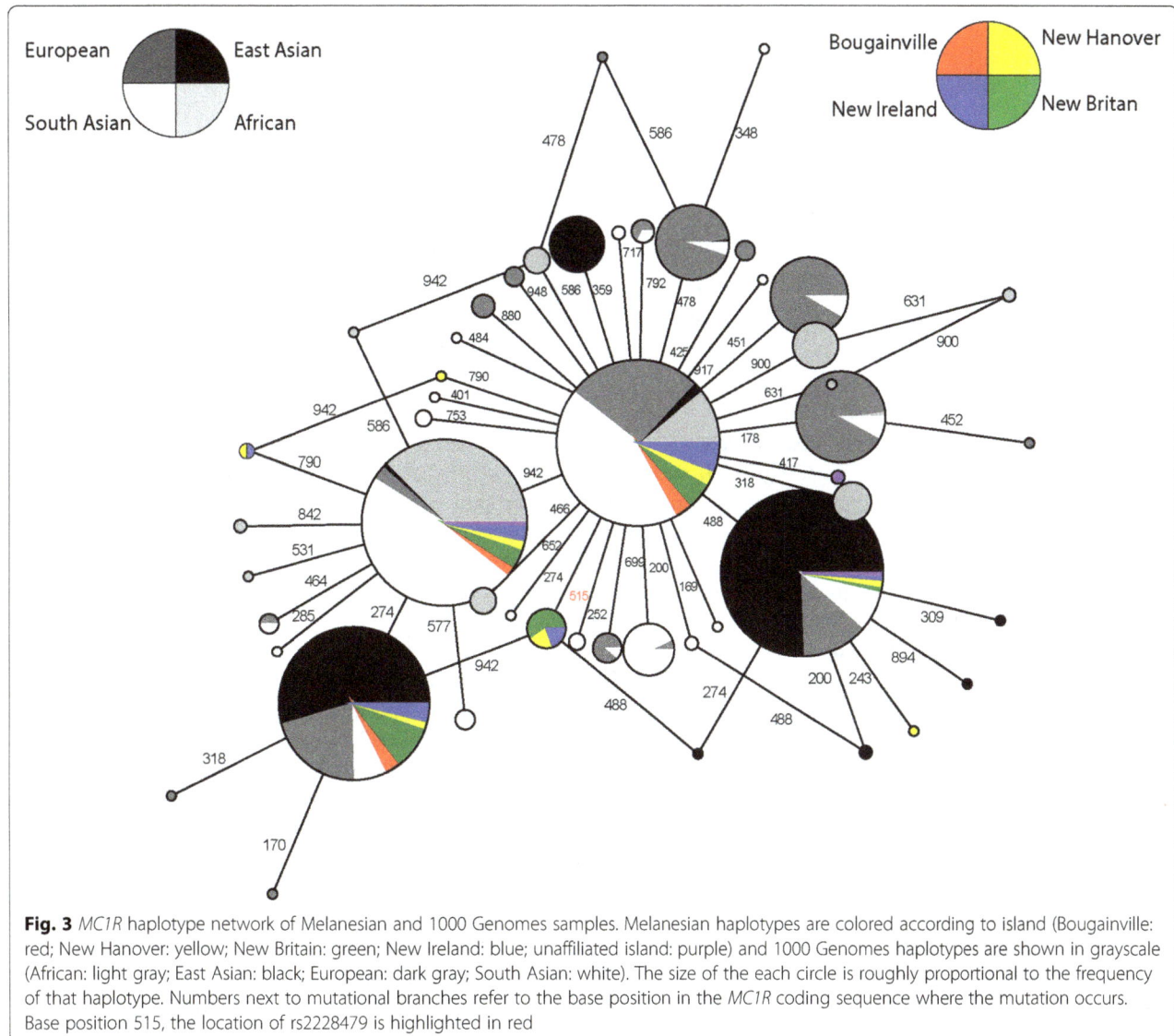

Fig. 3 *MC1R* haplotype network of Melanesian and 1000 Genomes samples. Melanesian haplotypes are colored according to island (Bougainville: red; New Hanover: yellow; New Britain: green; New Ireland: blue; unaffiliated island: purple) and 1000 Genomes haplotypes are shown in grayscale (African: light gray; East Asian: black; European: dark gray; South Asian: white). The size of the each circle is roughly proportional to the frequency of that haplotype. Numbers next to mutational branches refer to the base position in the *MC1R* coding sequence where the mutation occurs. Base position 515, the location of rs2228479 is highlighted in red

Table 4 Raw mean (and S.D.) skin and hair M index values and frequency of rs2228479 derived allele in islands and neighborhoods sampled here

Island/Neighborhood	N	Mean skin M (SD)	Mean hair M	rs2228479 A allele frequency
Bougainville	78	89.8 (9.7)	147.6 (14.8)	0.10
Kukuavo	**32**	**91.9 (8.5)**	**138.6 (8.9)**	**0.12**
Saposa Island	16	82.6 (8.3)	150.1 (7.7)	0.16
New Britain	286			0.15
Arimegi Island	44	65.2 (5.3)	157.5 (11.1)	0.33
Kariai (Anem)	**16**	**71.3 (4.3)**	**149.0 (8.8)**	**0.09**
Pureling (Anem)	**12**	**67.0 (6.9)**	**146.5 (12.2)**	**0.08**
Kisiluvi (Mamousi)	29	68.8 (6.0)	155.0 (17.0)	0.05
Lingite (Mamousi)	13	69.3 (7.7)	164.4 (17.4)	0.04
"other" Mamousi	20	67.5 (4.6)	159.0 (15.7)	0.07
Loso (Nakanai)	10	70.9 (6.8)	164.3 (28.4)	0.05
Uasilau (Ata)	**38**	**67.4 (5.3)**	**150.6 (15.1)**	**0.09**
Lugei (Ata)	**11**	**67.0 (5.6)**	**173.9 (21.1)**	**0.04**
Bileki (Nakanai)	18	66.7 (6.3)	143.3 (15.4)	0.19
Ubili (Nakanai)	27	67.6 (6.5)	166.1 (20.1)	0.19
New Ireland	147	73.9 (8.4)	151.5 (22.4)	0.17
Tigak	25	72.9 (8.6)	156.0 (16.5)	0.32
Nailik	19	72.8 (6.7)	151.1 (25.1)	0.21
Kabil (Kuot)	**32**	**76.1 (8.3)**	**151.1 (17.0)**	**0.13**
Notsi	16	74.1 (9.9)	161.1 (21.2)	0.09
Madak	23	77.0 (8.3)	145.6 (28.3)	0.07
New Hanover	84	76.7 (7.4)	154.5 (16.2)	0.26
North Lavongai	62	76.7 (6.9)	155.2 (16.3)	0.30
South Lavongai	13	77.3 (9.9)	150.3 (14.2)	0.15

Populations in boldface type speak non-Austronesian languages

The frequency of the derived allele also varied significantly between the Austronesian speaking (0.19) and non-Austronesian speaking (0.10) population sample ($\chi^2 = 17.06$, $p < 0.0001$).

We tested for an association between genotype at rs2228479 and quantitatively measured skin and hair color using this expanded sample. Mean skin and hair values (standardized and normalized as well as raw values) are found in Table 4. Because the observed rs2228479 genotypes differed significantly from expectations under Hardy Weinberg equilibrium in the full Northern Island Melanesian sample ($p = 3.494 \times 10^{-5}$), possibly indicating significant population substructure, and because of previous reports of substructure in the region [48], we tested for associations in the full sample using island as a covariate as well as on each island separately. In the full sample rs2228479 is not significantly associated with variation in hair pigmentation ($p = 0.7201$). Associations with skin pigmentation are suggestive (0.0635), but not significant. These should also be interpreted with caution, since this analysis does not account for potential substructure within island groupings [48]. rs2228479 is not significantly associated with lighter skin color on any of the four islands tested, nor is it significantly associated with hair color ($p > 0.05$ for all tests).

Discussion

The populations sampled here are all from Northern Island Melanesia, a region where UVR is high and variation in skin phenotype and at pigmentation loci is expected to be constrained by purifying selection. While pigmentation is generally dark, notable variation in both skin and hair pigmentation is observed [58], indicating the likely presence of coding or regulatory mutations in pigmentation genes. Recently an allele associated with lighter hair color in populations from the Solomon Islands was identified [59], the distribution of which is restricted to the Solomons and parts of the Bismarck Archipelago [59, 78]. This suggests that at least some of the observed variation in skin and hair phenotype in the region may be caused by population-specific alleles not currently reported in large public databases. In this paper, we set out to characterize variation in a well-known pigmentation candidate gene, *MC1R*, in a large sample of individuals from Northern Island Melanesia in order to test the following hypotheses: a) that *MC1R* variation is constrained by purifying selection in Melanesia, as it is in other high UVR populations; b) that, given this selective constraint, Melanesians should appear more closely related to other high-UVR populations (e.g., Africans) than to populations with which they share a more recent common history (other non-Africans, and specifically East Asians), and c) that nonsynonymous variants in *MC1R* are significantly associated with skin and/or hair pigmentation phenotype in Melanesian populations.

Purifying selection
Given the high levels of UVR in Melanesia, we expected that variation in *MC1R* would be constrained in Melanesian populations, particularly in non-Austronesian speakers, as the ancestors of modern non-Austronesian speakers have been resident in the region for ~ 30 kya. Surprisingly, out of the five segregating sites that we observed in this Melanesian sample, three are nonsynonymous polymorphisms. One of these is rare (occurring at a frequency of < 1 % in the sample), while the two others, rs2228479 and rs885479, occur at frequencies of 15.4 and 4.5 % in the sequenced sample. A modified McDonald's Kreitman's test (suitable for tests involving a small number of segregating sites) does not support a model of purifying selection. Tajima's D and Fay and Wu's H values, while slightly negative, also do not indicate a significant departure from neutrality, suggesting that *MC1R* in Melanesians has not been subject to either purifying or positive selection. This absence of strong purifying selection is surprising given the high UVR levels in the region [58]. However, while we find little evidence to support strong purifying selection acting on *MC1R* in this Melanesian population sample it is likely that purifying selection has influenced variation at other pigmentation loci.

Interestingly, when we apply this modified McDonald-Kreitman's test to the African 1000 Genomes data we also fail to reject a model of neutrality. Ten segregating sites are observed among the LWK and YRI samples, 4 of which are nonsynonymous and 6 of which are synonymous. The likelihood of observing such a pattern is $p = 0.553$. This differs from observations of Harding et al. [42], who observed only five *MC1R* haplotypes in Africans, and zero nonsynonymous polymorphisms. Using those data, the authors estimated that the probability of observing zero nonsynonymous polymorphisms out of five total polymorphisms was 0.0198, causing them to reject a model of neutral evolution in Africans. Similarly low levels of variation were observed in Africans by Rana et al. [43]. However, a subsequent study by John et al. [44] reported the presence of three nonsynonymous polymorphisms in Africans, including one (rs3212366) that occurred at a frequency of 0.11. Despite the presence of these nonsynonymous polymorphisms, however, a McDonald-Kreitman's test indicated that variation in these African samples was constrained by purifying selection. However, an HKA test failed to reject the neutral model, and while site-frequency-spectrum based tests (e.g. Tajima's D, Fay and Wu's H) indicated a significant departure from neutrality, the authors noted that this could be confounded by demographic history [44]. These results demonstrate that nonsynonymous polymorphisms are present (and tolerated) in African populations (and presumably other high UVR populations). The 1000 Genome data set used here is roughly seven times the size of the dataset used in these earlier studies, which may explain the higher proportion of polymorphisms (both nonsynonymous and synonymous) that we observe. Notably, though, none of these reach a frequency greater than 0.025 in the African sample (frequency range is 0.005–0.022). This is in contrast to the nonsynonymous polymorphisms observed in the Melanesian sample, which range in frequency from 0.003 to 0.157.

Inter-population variation
Melanesia is a region known for its extensive diversity, often characterized by high levels of inter-island phenotypic and genotypic variation [48, 57, 78, 79]. This variation can be attributed, at least in part, to the complex population history of the region [47–51]. Given these background levels of high inter-island diversity, as well as known differences in skin and hair pigmentation across islands in the region, one might expect to observe high levels of inter-island divergence at *MC1R*. However, levels of pairwise F_{ST} between islands are relatively low (0.000–0.031) suggesting that *MC1R* is not significantly differentiated among islands. This is also visualized in the Melanesian *MC1R* haplotype networks (Fig. 2), which indicate that the majority of common Melanesian *MC1R* haplotypes are shared across islands.

Constraint on pigmentation phenotype is expected to lead to lower levels of inter-population divergence at pigmentation loci between high UVR populations. In our dataset this could be visualized as reduced population differentiation (measured as F_{ST}) and high levels of haplotype sharing between Melanesian and African populations. Here we observe an F_{ST} value of 0.088 between our sequenced Melanesian sample and the African populations (LWK and YRI) of the 1000 Genomes Project. Without comparisons to F_{ST} values for other genes sequenced in the same population samples it is difficult to assess whether this is unusually high or low compared to average levels of divergence between these specific populations. However, previous studies have reported the mean F_{ST} between a single Melanesian (Nasioi) and African (Mende) population to be 0.182, based on allele frequencies at ~11,000 SNPs distributed throughout the genome [11]. Other published estimates based on sequence data from non-coding autosomal regions indicate that Melanesian-African F_{ST} values range from 0.199–0.283 [80]. These suggest that while still high, the reported F_{ST} values here between Melanesians and Africans may be lower than average levels of genome-wide divergence between populations from these two regions. However, if *MC1R* function were highly conserved across high-UVR populations we would expect to observe much lower values, as well as a high degree of exclusive Melanesian-African haplotype sharing, which is not evident (see Fig. 3).

The phenotypic role and evolutionary history of rs2228479

The derived rs2228479 allele is associated with red hair and fair skin color in European populations [36], and has been associated with lighter skin color in South Asians [13] and freckling and solar letinges in a Japanese population [61, 62]. rs2228479 also has a significant gene-gene interaction effect with the *OCA2* locus that may influence pigmentation in Tibetan populations [1]. When we typed this SNP (the only nonsynonymous mutation to reach a frequency > 5 % in our sequenced sample) in an expanded sample from across Northern Island Melanesia (total rs2228479 sample size = 635 individuals) we observe the derived allele occurring at a frequency of 16.1 % in the region. In comparison, this allele occurs at a frequency of 11-33 % in the East Asian 1000 Genomes populations, although it is generally rare in the European, African, and South Asian samples (0–12 %).

Previous studies have demonstrated that this allele results in impaired *MC1R* function relative to the human consensus sequence [81], consistent with reported associations with pigmentation phenotype. However, we find no association between rs2228479 and skin or hair color in the full sample or on any of the individual four islands that make up the bulk of the Melanesian sample used here. We have tested for associations on each island separately to avoid confounding effects of population substructure across the region (which can lead to false-positive associations). However, this results in smaller sample sizes and a loss of power to detect associations. A more robust method to control for population substructure in the larger region would be to would be to utilize genotype data from a large number of markers across the genome and conduct a PCA analyses to identify potential substructure [82, 83]. Significant PCs can then be included as covariates in association analyses.

A second possible explanation for this lack of association is that the effects of the rs2228479 are modified by epistatic interactions with other pigmentation loci [84]. This might explain the moderate frequencies of the derived rs2228479 allele in our Melanesian sample: despite the reduction in MC1R function caused by this allele, its effects are masked by other pigmentation loci and it is treated as effectively neutral. These results also indicate that polymorphisms at loci other than *MC1R* are necessary to explain the observed variation in Melanesian skin and hair pigmentation. One locus known to influence hair pigmentation in the region, rs387907171 SNP in the *TYRP1* gene [59], can explain a small proportion of hair pigmentation on these islands [78], but other, yet-to-be-identified loci also clearly play a role in shaping Melanesian pigmentation phenotype.

While epistatic effects between *MC1R* and other pigmentation loci may partially explain the frequency of the derived rs2228479 allele observed here, the demographic history of Melanesian populations may also be a contributing factor. In particular, we note the differences in the frequency of the derived rs2228479 allele between Austronesian and non-Austronesian speakers, and suggest that this allele may have been introduced to the region via Austronesian-mediated migration. Supporting this, some of the highest reported frequencies of this allele are found in East Asia, specifically in aboriginal populations from Taiwan, the putative Austronesian homeland [85], As further evidence, we note the distribution of the derived allele in Indonesian populations to the west, where it is absent from multiple non-Austronesian speaking populations, but is observed at a frequency of 0.27 in the Austronesian-speaking Biak islanders. Also relevant are the sharply decreased diversity levels of nucleotide diversity observed among the non-Austronesian individuals in our sample, which suggest that non-Austronesian *MC1R* diversity may reflect a history of long-term constraint, while the higher diversity among Austronesian speakers is indicative of a recent history outside of the region. However, most phenotypic and genetic studies indicate that sufficient admixture between the later arriving Austronesians and the descendants of the earliest inhabitants of the region has occurred to blur clear-cut distinctions between Austronesian and non-Austronesian speakers [47, 48, 58, 78, 86], and we caution against interpreting patterns of *MC1R* diversity in this sample strictly along linguistic lines. Consistent with the picture of long-term admixture, we note that haplotypes carrying the derived rs2228479 allele are observed on all islands, and in both Austronesian and non-Austronesian-speaking individuals.

Conclusion

In this study we sequenced the *MC1R* gene in a large sample ($n = 188$) of individuals from four islands in

Northern Island Melanesia, and find little indication of strong purifying selection acting on this gene. We also find little evidence of high levels of inter-island diversity in the region, contrary to other loci [48]. While F_{ST} values between our Melanesian sample and African populations sequenced in the 1000 Genomes Project indicate lower than average levels of divergence, we do not find strong evidence that ancestral *MC1R* haplotypes have been conserved in populations from these two high UVR regions. The only common nonsynonymous *MC1R* mutation observed in the Melanesian sample, rs2228479, is not associated with pigmentation phenotype in the region, possibly due to epistatic effects at other loci. These epistatic effects along with recent migrations of East Asian individuals via the Austronesian expansion can explain the observed patterns of *MC1R* diversity across the region. Pigmentation is a complex phenotype that exhibits extensive variation within and between populations, and is controlled by multiple interacting genetic loci. While *MC1R* has a strong influence on pigmentation in several non-African populations, it does not play a significant role in shaping variation in populations from Northern Island Melanesia. Clearly a better understanding of pigmentation candidate gene diversity in Melanesian populations will be necessary in order to better characterize the genetic architecture of the pigmentation phenotype in this region.

Availability of supporting data
The DNA sequences supporting the material analyzed here are available in the GenBank repository, with accession numbers KT863240- KT863426.

Competing interests
The authors declare that they have no competing interests.

Authors' contributions
JSF and HLN collected the Melanesian samples and phenotypic data used in the study. HLN designed the experiments. EW and HLN conducted the sequencing and genotyping. HLN analyzed the genetic and phenotypic data. EW, JSF, and HLN revised the manuscript. All authors read and approved the final manuscript.

Acknowledgements
The authors are grateful to the people of Papua New Guinea for their enthusiastic participation in this work. This work was supported by the University of Cincinnati.

Author details
[1]Department of Anthropology, University of Cincinnati, 481 Braunstein Hall, PO Box 210380, Cincinnati, OH 45221, USA. [2]Department of Anthropology, 101 West Hall, University of Michigan, 1085 South University Ave, Ann Arbor, MI 48109, USA. [3]Department of Anthropology, Temple University, Gladfelter Hall, 1115 West Berks Street, Philadelphia, PA 19122, USA.

References
1. Akey JM, Wang H, Xiong M, Wu H, Liu W, Shriver MD, et al. Interaction between the melanocortin-1 receptor and P genes contributes to inter-individual variation in skin pigmentation phenotypes in a Tibetan population. Hum Genet. 2001;108(6):516–20.
2. Ang KC, Ngu MS, Reid KP, Teh MS, Aida ZS, Koh DX, et al. Skin color variation in Orang Asli tribes of Peninsular Malaysia. PLoS One. 2012;7(8):e42752.
3. Basu Mallick C, Iliescu FM, Mols M, Hill S, Tamang R, Chaubey G, et al. The light skin allele of SLC24A5 in South Asians and Europeans shares identity by descent. PLoS Genet. 2013;9(11), e1003912.
4. Beleza S, Johnson NA, Candille SI, Absher DM, Coram MA, Lopes J, et al. Genetic architecture of skin and eye color in an African-European admixed population. PLoS Genet. 2013;9(3), e1003372.
5. Bonilla C, Boxill LA, Donald SA, Williams T, Sylvester N, Parra EJ, et al. The 8818G allele of the agouti signaling protein (ASIP) gene is ancestral and is associated with darker skin color in African Americans. Hum Genet. 2005;116(5):402–6.
6. Edwards M, Bigham A, Tan J, Li S, Gozdzik A, Ross K, et al. Association of the OCA2 polymorphism His615Arg with melanin content in east Asian populations: further evidence of convergent evolution of skin pigmentation. PLoS Genet. 2010;6(3), e1000867.
7. Flanagan N, Healy E, Ray A, Philips S, Todd C, Jackson IJ, et al. Pleiotropic effects of the melanocortin 1 receptor (MC1R) gene on human pigmentation. Hum Mol Genet. 2000;9(17):2531–7.
8. Guenther CA, Tasic B, Luo L, Bedell MA, Kingsley DM. A molecular basis for classic blond hair color in Europeans. Nat Genet. 2014;46(7):748–52.
9. Lamason RL, Mohideen MA, Mest JR, Wong AC, Norton HL, Aros MC, et al. SLC24A5, a putative cation exchanger, affects pigmentation in zebrafish and humans. Science. 2005;310(5755):1782–6.
10. Myles S, Somel M, Tang K, Kelso J, Stoneking M. Identifying genes underlying skin pigmentation differences among human populations. Hum Genet. 2007;120(5):613–21.
11. Norton HL, Kittles RA, Parra E, McKeigue P, Mao X, Cheng K, et al. Genetic evidence for the convergent evolution of light skin in Europeans and East Asians. Mol Biol Evol. 2007;24(3):710–22.
12. Shriver MD, Parra EJ, Dios S, Bonilla C, Norton H, Jovel C, et al. Skin pigmentation, biogeographical ancestry and admixture mapping. Hum Genet. 2003;112(4):387–99.
13. Stokowski RP, Pant PV, Dadd T, Fereday A, Hinds DA, Jarman C, et al. A genomewide association study of skin pigmentation in a South Asian population. Am J Hum Genet. 2007;81(6):1119–32.
14. Valenzuela RK, Henderson MS, Walsh MH, Garrison NA, Kelch JT, Cohen-Barak O, et al. Predicting phenotype from genotype: normal pigmentation. J Forensic Sci. 2010;55(2):315–22.
15. Jablonski NG, Chaplin G. The evolution of human skin coloration. J Hum Evol. 2000;39(1):57–106.
16. Jablonski NG, Chaplin G. Colloquium paper: human skin pigmentation as an adaptation to UV radiation. Proc Natl Acad Sci U S A. 2010;107 Suppl 2:8962–8.
17. Relethford JH. Hemispheric difference in human skin color. Am J Phys Anthropol. 1997;104(4):449–57.
18. Rogers A, Iltis D, Wooding S. Genetic variation at the MC1R locus and the time since loss of human body hair. Curr Anthropol. 2004;45(1):105–8.
19. Branda RF, Eaton JW. Skin color and nutrient photolysis: an evolutionary hypothesis. Science. 1978;201(4356):625–6.
20. Robbins A. Biological perspectives on human pigmentation. Cambridge, NY: Cambridge University Press; 1991.
21. Loomis WF. Skin-pigment regulation of vitamin-D biosynthesis in man. Science. 1967;157(3788):501–6.
22. Kaidbey KH, Agin PP, Sayre RM, Kligman AM. Photoprotection by melanin–a comparison of black and Caucasian skin. J Am Acad Dermatol. 1979;1(3):249–60.

23. Brenner M, Hearing VJ. The protective role of melanin against UV damage in human skin. Photochem Photobiol. 2008;84(3):539–49.

24. Sage E. Distribution and repair of photolesions in DNA: genetic consequences and the role of sequence context. Photochem Photobiol. 1993;57(1):163–74.

25. Tadokoro T, Kobayashi N, Zmudzka BZ, Ito S, Wakamatsu K, Yamaguchi Y, et al. UV-induced DNA damage and melanin content in human skin differing in racial/ethnic origin. FASEB J. 2003;17(9):1177–9.

26. Vink AA, Roza L. Biological consequences of cyclobutane pyrimidine dimers. J Photochem Photobiol B. 2001;65(2–3):101–4.

27. Blum HF. Does the melanin pigment of human skin have adaptive value? An essay in human skin have adaptive value? An essay in human ecology and the evolution of race. Q Rev Biol. 1961;36:50–63.

28. Jablonski NG, Chaplin G. Skin cancer was not a potent selective force in the evolution of protective pigmentation in early hominins. Proc Biol Sci Royal Soc. 2014;281(1789):20140517.

29. Fenech M. Folate (vitamin B9) and vitamin B12 and their function in the maintenance of nuclear and mitochondrial genome integrity. Mutat Res. 2012;733(1–2):21–33.

30. Mathur U, Datta SL, Mathur BB. The effect of aminopterin-induced folic acid deficiency on spermatogenesis. Fertil Steril. 1977;28(12):1356–60.

31. Lucock M, Daskalakis I, Briggs D, Yates Z, Levene M. Altered folate metabolism and disposition in mothers affected by a spina bifida pregnancy: influence of 677c – > t methylenetetrahydrofolate reductase and 2756a – > g methionine synthase genotypes. Mol Genet Metab. 2000;70(1):27–44.

32. Off MK, Steindal AE, Porojnicu AC, Juzeniene A, Vorobey A, Johnsson A, et al. Ultraviolet photodegradation of folic acid. J Photochem Photobiol B. 2005;80(1):47–55.

33. Jablonski NG. The evolution of human skin and skin color. Annu Rev Anthropol. 2004;33:585–623.

34. Slominski A, Tobin DJ, Shibahara S, Wortsman J. Melanin pigmentation in mammalian skin and its hormonal regulation, vol. 84. 2004.

35. Abdel-Malek Z, Swope VB, Suzuki I, Akcali C, Harriger MD, Boyce ST, et al. Mitogenic and melanogenic stimulation of normal human melanocytes by melanotropic peptides. Proc Natl Acad Sci U S A. 1995;92(5):1789–93.

36. Valverde P, Healy E, Jackson I, Rees JL, Thody AJ. Variants of the melanocyte-stimulating hormone receptor gene are associated with red hair and fair skin in humans. Nat Genet. 1995;11(3):328–30.

37. Martinez-Cadenas C, Lopez S, Ribas G, Flores C, Garcia O, Sevilla A, et al. Simultaneous purifying selection on the ancestral MC1R allele and positive selection on the melanoma-risk allele V60L in south Europeans. Mol Biol Evol. 2013;30(12):2654–65.

38. Box NF, Wyeth JR, O'Gorman LE, Martin NG, Sturm RA. Characterization of melanocyte stimulating hormone receptor variant alleles in twins with red hair. Hum Mol Genet. 1997;6(11):1891–7.

39. Rees JL. The genetics of sun sensitivity in humans. Am J Hum Genet. 2004;75(5):739–51.

40. Kadekaro AL, Kavanagh R, Kanto H, Terzieva S, Hauser J, Kobayashi N, et al. alpha-Melanocortin and endothelin-1 activate antiapoptotic pathways and reduce DNA damage in human melanocytes. Cancer Res. 2005;65(10):4292–9.

41. Song X, Mosby N, Yang J, Xu A, Abdel-Malek Z, Kadekaro AL. α-MSH activates immediate defense responses to UV-induced oxidative stress in human melanocytes. Pigment Cell Melanoma Res. 2009;22(6):809–18.

42. Harding RM, Healy E, Ray AJ, Ellis NS, Flanagan N, Todd C, et al. Evidence for variable selective pressures at MC1R. Am J Hum Genet. 2000;66(4):1351–61.

43. Rana BK, Hewett-Emmett D, Jin L, Chang BH, Sambuughin N, Lin M, et al. High polymorphism at the human melanocortin 1 receptor locus. Genetics. 1999;151(4):1547–57.

44. John PR, Makova K, Li WH, Jenkins T, Ramsay M. DNA polymorphism and selection at the melanocortin-1 receptor gene in normally pigmented southern African individuals. Ann N Y Acad Sci. 2003;994:299–306.

45. Harris EE, Meyer D. The molecular signature of selection underlying human adaptations. Am J Phys Anthropol. 2006;Suppl 43:89–130.

46. Hider JL, Gittelman RM, Shah T, Edwards M, Rosenbloom A, Akey JM, et al. Exploring signatures of positive selection in pigmentation candidate genes in populations of East Asian ancestry. BMC Evol Biol. 2013;13:150.

47. Delfin F, Myles S, Choi Y, Hughes D, Illek R, van Oven M, et al. Bridging near and remote Oceania: mtDNA and NRY variation in the Solomon Islands. Mol Biol Evol. 2012;29(2):545–64.

48. Friedlaender JS, Friedlaender FR, Reed FA, Kidd KK, Kidd JR, Chambers GK, et al. The genetic structure of Pacific Islanders. PLoS Genet. 2008;4(1):e19.

49. Friedlaender JS, Gentz F, Green K, Merriwether DA. A cautionary tale on ancient migration detection: mitochondrial DNA variation in Santa Cruz Islands, Solomon Islands. Hum Biol. 2002;74(3):453–71.

50. Lindström E, Terrill A, Reesnick G, Dunn M. The languages of island Melanesia. In: Friedlaender J, editor. Genes, language, and culture history in the Southwest Pacific. New York: Oxford University Press; 2007. p. 118–39.

51. Summerhayes G. Island Melanesian pasts: a view from archaeology. In: Friedlaender JS, editor. Genes, language, and culture history in the Southwest Pacific. New York: Oxford University Press; 2007. p. 10–35.

52. Groube L, Chappell J, Muke J, Price D. A 40,000 year-old human occupation site at Huon Peninsula, Papua New Guinea. Nature. 1986;324(6096):453–5.

53. Summerhayes GR, Leavesley M, Fairbairn A, Mandui H, Field J, Ford A, et al. Human adaptation and plant use in highland New Guinea 49,000 to 44,000 years ago. Science. 2010;330(6000):78–81.

54. Wickler S, Spriggs M. Pleistocene human occupation of the Solomon Islands, Melanesia. Antiquity. 1988;62:703–6.

55. Bellwood P, Dizon E. The batanes archaeological project and the "Out of Taiwan" hypothesis for austronesian dispersal. J Austronesian Stud. 2005;1:1–31.

56. Kirch P. Peopling of the pacific: a holistic anthropological perspective. Annu Rev Anthropol. 2010;39:131–48.

57. Norton H, Koki G, Friedlaender J. Pigmentation and candidate gene variation in Northern Island Melanesia. In: Population genetics, linguistics, and culture history in the southwest pacific: a synthesis. 2007. p. 96–112.

58. Norton HL, Friedlaender JS, Merriwether DA, Koki G, Mgone CS, Shriver MD. Skin and hair pigmentation variation in Island Melanesia. Am J Phys Anthropol. 2006;130(2):254–68.

59. Kenny EE, Timpson NJ, Sikora M, Yee MC, Moreno-Estrada A, Eng C, et al. Melanesian blond hair is caused by an amino acid change in TYRP1. Science. 2012;336(6081):554.

60. Chaplin G. Geographic distribution of environmental factors influencing human skin coloration. Am J Phys Anthropol. 2004;125(3):292–302.

61. Motokawa T, Kato T, Hashimoto Y, Katagiri T. Effect of Val92Met and Arg163Gln variants of the MC1R gene on freckles and solar lentigines in Japanese. Pigment Cell Res. 2007;20(2):140–3.

62. Yamaguchi K, Watanabe C, Kawaguchi A, Sato T, Naka I, Shindo M, et al. Association of melanocortin 1 receptor gene (MC1R) polymorphisms with skin reflectance and freckles in Japanese. J Hum Genet. 2012;57(11):700–8.

63. Diffey BL, Farr PM, Ive FA. The establishment and clinical value of a dermatological photobiology service in a district general hospital. Br J Dermatol. 1984;110(2):187–94.

64. Kearse M, Moir R, Wilson A, Stones-Havas S, Cheung M, Sturrock S, et al. Geneious Basic: an integrated and extendable desktop software platform for the organization and analysis of sequence data. Bioinformatics. 2012;28(12):1647–9.

65. Stephens M, Smith NJ, Donnelly P. A new statistical method for haplotype reconstruction from population data. Am J Hum Genet. 2001;68(4):978–89.

66. Norton H, Koki G, Friedlaender J. Pigmentation and candidate gene variation in Northern Island Melanesia. In: Friedlaender J, editor. Genes, language, and culture history in the Southwest Pacific. New York: Oxford University Press; 2007. p. 96–112.

67. Rozas J. DNA sequence polymorphism analysis using DnaSP. Methods Mol Biol. 2009;537:337–50.

68. Librado P, Rozas J. DnaSP v5: a software for comprehensive analysis of DNA polymorphism data. Bioinformatics. 2009;25(11):1451–2.

69. Nei M, Li WH. Mathematical model for studying genetic variation in terms of restriction endonucleases. Proc Natl Acad Sci U S A. 1979;76(10):5269–73.

70. Watterson GA. On the number of segregating sites in genetical models without recombination. Theor Popul Biol. 1975;7(2):256–76.

71. Tajima F. Statistical method for testing the neutral mutation hypothesis by DNA polymorphism. Genetics. 1989;123(3):585–95.

72. Wollstein A, Lao O, Becker C, Brauer S, Trent RJ, Nurnberg P, et al. Demographic history of Oceania inferred from genome-wide data. Curr Biol. 2010;20(22):1983–92.

73. McDonald JH, Kreitman M. Adaptive protein evolution at the Adh locus in Drosophila. Nature. 1991;351(6328):652–4.

74. Nielsen R. Molecular signatures of natural selection. Annu Rev Genet. 2005;39:197–218.

75. Nielsen R. Statistical tests of selective neutrality in the age of genomics. Heredity (Edinb). 2001;86(Pt 6):641–7.

76. Purcell S, Neale B, Todd-Brown K, Thomas L, Ferreira MA, Bender D, et al. PLINK: a tool set for whole-genome association and population-based linkage analyses. Am J Hum Genet. 2007;81(3):559–75.

77. Bandelt HJ, Forster P, Rohl A. Median-joining networks for inferring intraspecific phylogenies. Mol Biol Evol. 1999;16(1):37–48.

78. Norton HL, Correa EA, Koki G, Friedlaender JS. Distribution of an allele associated with blond hair color across Northern Island Melanesia. Am J Phys Anthropol. 2014;153(4):653–62.

79. Scheinfeldt L, Friedlaender F, Friedlaender J, Latham K, Koki G, Karafet T, et al. Unexpected NRY chromosome variation in Northern Island Melanesia. Mol Biol Evol. 2006;23(8):1628–41.

80. Wall JD, Cox MP, Mendez FL, Woerner A, Severson T, Hammer MF. A novel DNA sequence database for analyzing human demographic history. Genome Res. 2008;18(8):1354–61.

81. Xu X, Thornwall M, Lundin LG, Chhajlani V. Val92Met variant of the melanocyte stimulating hormone receptor gene. Nat Genet. 1996;14(4):384.

82. Patterson N, Price AL, Reich D. Population structure and eigenanalysis. PLoS Genet. 2006;2(12), e190.

83. Price AL, Patterson NJ, Plenge RM, Weinblatt ME, Shadick NA, Reich D. Principal components analysis corrects for stratification in genome-wide association studies. Nat Genet. 2006;38(8):904–9.

84. Pospiech E, Wojas-Pelc A, Walsh S, Liu F, Maeda H, Ishikawa T, et al. The common occurrence of epistasis in the determination of human pigmentation and its impact on DNA-based pigmentation phenotype prediction. Forensic Sci Int Genet. 2014;11:64–72.

85. Ding Q, Hu Y, Xu S, Wang CC, Li H, Zhang R, et al. Neanderthal origin of the haplotypes carrying the functional variant Val92Met in the MC1R in modern humans. Mol Biol Evol. 2014;31(8):1994–2003.

86. Duggan AT, Evans B, Friedlaender FR, Friedlaender JS, Koki G, Merriwether DA, et al. Maternal history of Oceania from complete mtDNA genomes: contrasting ancient diversity with recent homogenization due to the Austronesian expansion. Am J Hum Genet. 2014;94(5):721–33.

3

Genetic variants in the upstream region of activin receptor IIA are associated with female fertility in Japanese Black cattle

Shinji Sasaki[1*], Takayuki Ibi[2], Tamako Matsuhashi[3], Kenji Takeda[3], Shogo Ikeda[4], Mayumi Sugimoto[1] and Yoshikazu Sugimoto[5]

Abstract

Background: Female fertility, a fundamental trait required for animal reproduction, has gradually declined in the last 2 decades in Japanese Black cattle. To identify associated genetic variants in Japanese Black cattle, we evaluated female fertility as a metric to describe the average inverse of the number of artificial inseminations required for conception from the first through the fourth parity ($ANAI_4$) and conducted a genome-wide association study (GWAS) using 430 animals with extreme $ANAI_4$ values from 10,399 animals.

Results: We found that 2 variants, namely a single-nucleotide polymorphisms (SNP; g.48476925C > T) and a 3-bp indel (g.48476943_48476946insGGC), in the upstream region of the activin receptor IIA gene (*ACVR2A*) were associated with $ANAI_4$. *ACVR2A* transcripts from Japanese Black cattle of the *Q* haplotype, defined by the SNP and the 3-bp indel, with increased $ANAI_4$ were 1.29–1.32-fold more abundant than *q*-derived transcripts. In agreement, reporter assay results revealed that the activity of the *ACVR2A* promoter was higher in reporter constructs with the *Q* haplotype than in those with the *q* haplotype by approximately 1.2 fold. Expression of exogenous *ACVR2A* induced dose-dependent increases of reporter activity from the follicle-stimulating hormone, beta polypeptide (*FSHB*) promoter in response to activin A in a pituitary gonadotrophic cell line. The findings suggested that sequence variations in the upstream region of *ACVR2A* with the *Q* haplotype increased *ACVR2A* transcription, which in turn induced *FSHB* expression. This association was replicated using a sample population size of 1,433 animals; the frequency of the *Q* haplotype was 0.39, and *Q*-to-*q* haplotype substitution resulted in an increase of 0.02 in terms of $ANAI_4$.

Conclusions: This GWAS identified variants in the upstream region of *ACVR2A*, which were associated with female fertility in Japanese Black cattle. The variants affected the level of *ACVR2A* mRNA expression, which could lead to an allelic imbalance. This association was replicated with a sample population of 1,433 animals. Thus, the results suggest that the *Q* haplotype could serve as a useful marker to select Japanese Black cattle with superior female fertility.

Keywords: Female fertility, Reproductive efficiency, Genome-wide association study, Activin receptor IIA (*ACVR2A*), Japanese Black cattle, Beef cattle

---

* Correspondence: sasakis@siag.or.jp
[1]National Livestock Breeding Center, Odakura, Nishigo, Fukushima 961-8511, Japan
Full list of author information is available at the end of the article

Background

Fertile female cattle show clear estrous in a timely manner and become pregnant within a minimum number of artificial inseminations (AIs). However, over the past 2 decades, female fertility in AI breeding programs has been gradually declining in Japanese Black cattle; e.g., the first-AI conception rate decreased from 67.4 % to 56 % between 1992 and 2012 in Japan [1]. The trend has been also observed in dairy cattle [1, 2]. Additional AIs increase costs related to semen, hormonal treatments, and AI technician fees, as well as feeding until the next AI. Therefore, farmers and breeders pay close attention to genetic factors related to improving female fertility for greater reproductive performance and profitability.

Recently, using a high-density single-nucleotide polymorphism (SNP) array [3, 4], genome-wide association studies (GWAS) have enabled researchers to scan the entire genome for related genetic factors and have identified a quantitative trait locus (QTL) for fertility-related traits in various cattle breeds [5–13] (reviewed in [14, 15]); however, QTLs for fertility-related traits in Japanese Black cattle have remained fully unknown.

Japanese Black cattle are highly valued owing to the abundant marbling of meat caused by intramuscular fat depositions [16]. Strict selection for marbling under a closed breeding system in Japan [17] has made Japanese Black cattle genetically distinct from European cattle breeds [18]. Therefore, genome-wide QTL screening for female fertility needs to be applied to this breed.

The genetic parameters for calving interval-related traits have been evaluated until animals reach approximately 4 to 5 years of age in Japanese Black cattle in Japan [19]. Thus, the AI records of 10,399 animals at each parity, from the first through the third or fourth parity (see Methods section) were available. Multiple records from single animals are important for accurately evaluating fertility performances. However, the number of cows decreases as the age of cows increases, reflecting the culling of animals with lower fertility. Thus, in this study, to identify variants associated with female fertility in Japanese Black cattle, we evaluated female fertility as a metric to describe the average inverse of the number of AIs required for conception [20] from first through fourth parity ($ANAI_4$) and conducted a GWAS for this trait. The current GWAS identified associated variants in the upstream region of the activin receptor IIA gene (*ACVR2A*), which serve as a key regulator of follicular growth in the ovaries by controlling follicle-stimulating hormone (*FSH*) expression.

Results and discussion

A QTL for $ANAI_4$ was identified on bovine chromosome 2 (BTA2) in Japanese Black cattle

The heritability of $ANAI_4$ was estimated to be 0.02 using the numerator-relationship matrix among 10,399 animals based on pedigree information, consistent with previous studies reporting the inverse of the number of inseminations required for conception [20] and female conception-related traits in cattle (reviewed in [15, 21]). As shown in Additional file 1, the distribution was sufficiently wide to discriminate between the upper- and lower-performance groups. Selective genotyping using animals with phenotypic values that deviate from the population mean is an effective method for reducing the sample sizes required to detect common SNPs associated with traits [22]. We selected 256 cows from the upper extreme (85th percentile, average $ANAI_4$ was 0.927) of the distribution and 174 cows from the lower extreme (15th percentile, average $ANAI_4$ is 0.399) among 10,399 cows with $ANAI_4$ records. These samples were genotyped using the BovineSNP50K BeadChip, comprising probes for 54,001 SNPs. A total of 33,303 autosomal SNPs that passed our quality control criteria (call rate > 99 %, minor allele frequency > 0.01, Hardy–Weinberg equilibrium $P > 0.001$, and inclusion of the SNP on the BovineHD BeadChip) [23] were used for the association study. Analysis was performed using EMMAX software [24], which is based on a linear-mixed model approach using a genetic-relationship matrix estimated by SNP genotypes to model the correlation between the phenotypes of the sample subjects. The genomic-inflation factor (λ GC) in this analysis was 1.058, indicating that the sample was appropriate for an association study. A quantile–quantile (Q–Q) plot showed that 2 SNPs deviated from the distribution under the null hypothesis (Additional file 2). Two SNPs on BTA2 reached the Bonferroni-corrected threshold for genome-wide significance ($P < 1.5 \times 10^{-6}$; Fig. 1a, Table 1). The 2 SNPs, Hapmap43862-BTA-47538 and Hapmap43863-BTA-47554, were located within a 200-kbp window from 48,240,577 bp to 48,440,885 bp on BTA2 and were in linkage disequilibrium (LD) with each other ($P < 6.52 \times 10^{-7}$–8.32×10^{-7}, $D' = 1$, $r^2 = 0.99$), as shown in Table 1. This QTL has not been previously reported for reproductive-related traits in cattle (reviewed in [14, 15]; for a cattle QTLdb, see [25]).

To characterize the region on BTA2 in more detail—in particular, the extent of LD in the QTL region, the genotypes of 430 animals for 33,303 SNPs were imputed using BEAGLE software [26, 27] with phased haplotype data inferred from 586,812 SNPs (BovineHD Beadchip) in 1,041 Japanese Black cattle served as the reference [23]. We previously estimated the imputation accuracy by comparing the true genotypes to the imputed genotypes, indicating that such imputation was highly accurate [23]. Subsequently, 4 SNPs associated with $ANAI_4$ were detected within the 218-kbp from 48,225,372 bp to 48,443,632 bp on BTA2 using EMMAX ($D' = 1$, r^2 ranging from 0.99 to 1.00; Fig. 1b, Table 1). Except for the 4 SNPs on BTA2, we did not detect any imputed SNPs

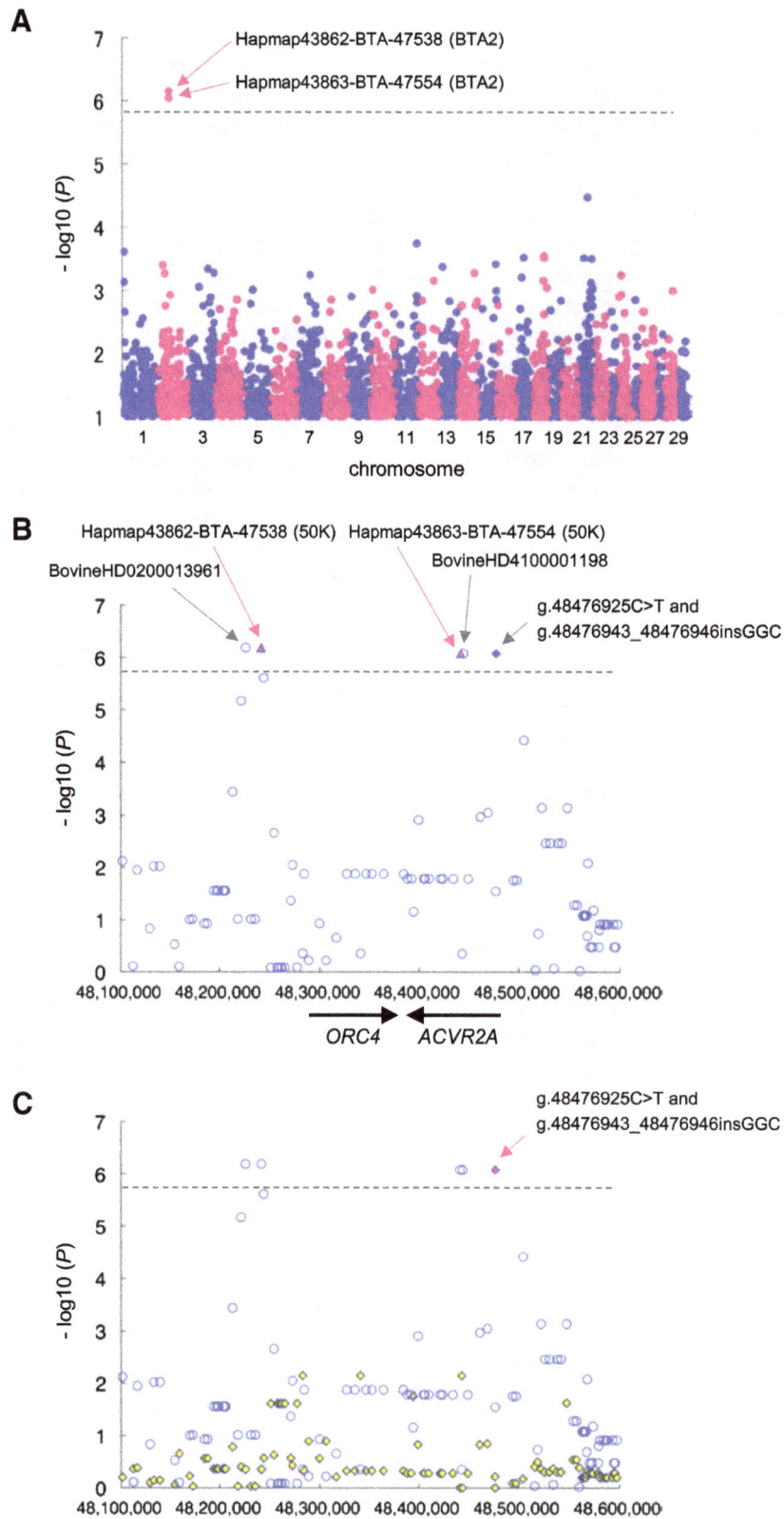

Fig. 1 (See legend on next page.)

Fig. 1 Association of SNPs and a indel with ANAI$_4$ in 430 Japanese Black cattle. **a** Manhattan plot of the association of 33,303 SNPs (BovineSNP50K BeadChip) with ANAI$_4$ in 430 Japanese Black cattle. The chromosomes are distinguished with alternating colors (blue, odd numbers; red, even numbers). The chromosome number is indicated on the X-axis. The dashed line is the Bonferroni-corrected threshold for genome-wide significance ($-\log10$ (P) = 5.823). The vertical axis is broken for P values below -log10 (P) = 1. **b** Regional plot of the locus on BTA2 associated with ANAI$_4$. SNPs from the BovineSNP50K BeadChip are shown as red triangles. The imputed SNPs are shown as unfilled blue circles. g.48476925C > T SNP and the g.48476943_48476946insGGC indel are shown as filled blue diamonds. Genes and their directions of transcription are noted at the bottom of the plot. **c** A conditioned analysis was performed by including the haplotype, defined by g.48476925C > T SNP and the g.48476943_48476946insGGC indel, as covariates in the model. The 2 red, filled diamonds indicate the g.48476925C > T and g.48476943_48476946insGGC (arrow) variants. The blue, unfilled circles and the yellow, filled diamonds represent P values on a $-\log10$ scale before and after conditioning, respectively. The positions shown are based on the UMD3.1 assembly of the bovine genome

that were significantly associated with ANAI$_4$ ($P < 0.05$, Bonferroni-corrected).

Variants in the upstream region of *ACVR2A* were associated with ANAI$_4$

The LD region harbors 2 genes: origin recognition complex subunit 4 (*ORC4*) and *ACVR2A* (Fig. 1b, Additional file 3). Out of 4 associated SNPs, 2 SNPs were located in the intronic region of *ACVR2A* and 2 SNPs were located on centromeric side at a distance of 42 to 57 kbp from *ORC4*, respectively (Fig. 1b). To detect associated polymorphisms in *ORC4* and *ACVR2A*, we sequenced all exons and upstream regions, beginning 2,131 bp upstream of the start codon, of each gene in 3 animals with homozygous *Q* and *q* haplotypes that were defined by the genotypes of Hapmap43862-BTA-47538 (48,240,577 bp) and Hapmap43863-BTA-47554 (48,440,885 bp) (Fig. 1a, B, Table 1). In the region of *ACVR2A*, we found a 21-bp indel (g.48476691_484766711delGAGCTCGCGGCGGT GGCGGCC) in the 5′-untranslated region (UTR) and a SNP in 3′-UTR (g.48381943A > G) in exons 1 and 11, respectively. We also found a SNP (g.48476925C > T) and a 3-bp indel (g.48476943_48476946insGGC) in the upstream region of *ACVR2A* (716 bp and 737–740 bp upstream of the start codon, respectively). We did not found any variants in the exons or in the region upstream of the *ORC4* transcription start site. To ascertain whether the *ACVR2A* variants were associated with ANAI$_4$, we genotyped the variants in 430 animals used in the GWAS and analyzed the association with ANAI$_4$, using EMMAX software and a genetic-relationship matrix for the animals. The SNP (g.48476925C > T) and the 3-bp indel (g.48476943_48476946insGGC) in the upstream region of *ACVR2A* produced a highly significant signal ($P = 8.32 \times 10^{-7}$; Fig. 1b, Table 1), whereas the 21-bp indel in the 5′-UTR and the SNP in the 3′-UTR of *ACVR2A* were not associated with ANAI$_4$ ($P = 0.028$ and $P = 0.51$, respectively). Subsequently, 5 SNPs and 1 indel associated with ANAI$_4$ were detected within the 251. 5-kbp region spanning 48,225,372 bp to 48,476,946 bp on BTA2 ($D′ = 1$, r^2 ranging from 0.99 to 1.00; Fig. 1b, Table 1).

We then performed conditioned analysis to ascertain whether there were any other significantly associated SNPs in the region. The *Q* haplotype, defined by the SNP (g.48476925C > T) and the 3-bp indel (g.48476943_48476 946insGGC) in the upstream region of *ACVR2A*, was individually included as a covariate in the linear-mixed model. After conditioning, the associations of other SNPs were no longer evident (Fig. 1c), indicating that the region contained a single QTL.

ACVR2A is a type-II transforming growth factor-beta (TGF-beta) receptor with serine/threonine kinase activity, which is involved in initial activin binding. Such binding leads to the recruitment and phosphorylation of type-I TGF-beta receptors and activates transcription of specific target genes, such as the FSH, polypeptide beta gene (*FSHB*) (reviewed in [28]). FSH is a heterodimer, consisting of an alpha and beta subunit. FSH is produced in gonadotropes in the anterior pituitary gland in response to activin by autocrine and paracrine mechanisms, and stimulates the growth and recruitment of immature follicles in the ovary [29]. Given these facts, *ACVR2A* could be a reasonable candidate gene for the fertility trait.

Variants in the 5′-upstream region of *ACVR2A* were involved in allelic imbalances of *ACVR2A* mRNA expression

Variants in the upstream region of the *ACVR2A* gene could potentially affect the activity of the promoter and, thus, they may contribute to an allelic imbalance in *ACVR2A* mRNA expression. To compare the relative abundance of *Q*- versus *q*-derived transcripts of *ACVR2A*, we performed allelic-imbalance testing with heterozygous samples [30]. The exonic SNPs in 5′ and 3′-UTRs of *ACVR2A* were not associated with ANAI$_4$. The intronic SNPs were transcribed as primary mRNAs in the nucleus; thus, we amplified an intronic *ACVR2A* SNP, BovineHD4100001198 (48,443,632 bp, Table 1), which is in perfect LD with the *Q* haplotype that was defined by the g.48476925C > T SNP and the 3-bp g.48476943_48476946insGGC indel genotypes ($r^2 = 1$). We observed that *ACVR2A* was ubiquitously expressed in female cow tissues including primary dermal fibroblasts

Table 1 SNPs and a indel with genome-wide significant associations with ANAI$_4$ on BTA2

BTA	SNP and indel-ID	Reference SNP-ID_number[c]	position (bp)_UMD3.1[d]	allele_1[e]	Minor allele frequency (upper extrame)	Minor allele frequency (lower extrame)	allele_2[f]	odds ratio	P-value
2	BovineHD0200013961[b]	rs110523739	48225372	G	0.45	0.29	A	1.98	6.52E-07
2	Hapmap43862-BTA-47538[a]	rs41636186	48240577	A	0.45	0.29	G	1.98	6.52E-07
2	Hapmap43863-BTA-47554[a]	rs41636197	48440885	G	0.45	0.29	A	1.96	8.32E-07
2	BovineHD4100001198[b]	rs41636199	48443632	A	0.45	0.29	G	1.96	8.32E-07
2	g.48476925C>T		48476925	T	0.45	0.29	C	1.96	8.32E-07
2	g.48476943_48476946insGGC		48476943_48476946	GGC	0.45	0.29	(-)	1.96	8.32E-07

[a]SNPs included in the Illumina Bovine SNP50K BeadChip; [b]Imputed SNPs from the Illumina Bovine HD BeadChip; [c]Reference SNP ID numbers (rs) were obtained using the SNPchiMp v.3 database (http:// bioinformatics.tecnoparco.org/SNPchimp/). [d]The positions are based on the UMD3.1 assembly of the bovine genome. [e]Minor allele; [f]Major allele. The upper and lower extremes correspond to ANAI$_4$ values above the 85th percentile and below the 15th percentile, respectively

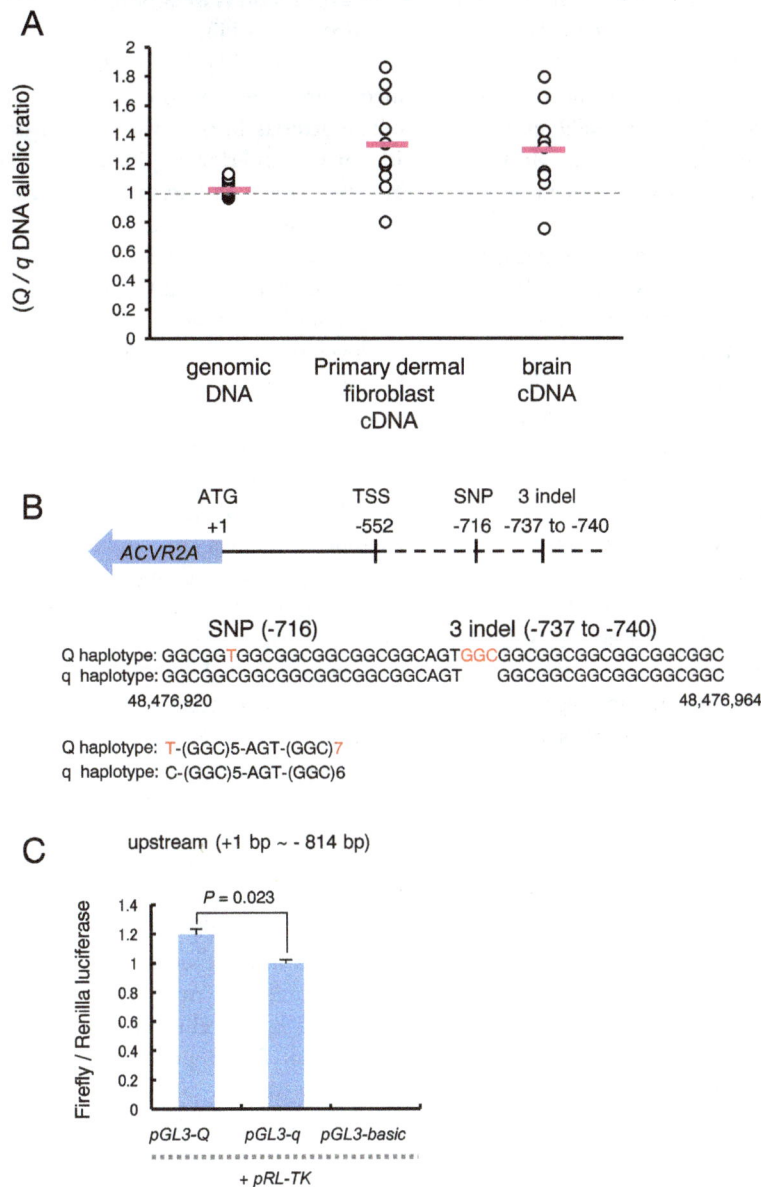

Fig. 2 The upstream region of *ACVR2A* affects *ACVR2A* expression. **a** Allelic-imbalance test for the level of *ACVR2A* mRNA expression in heterozygotes. cDNA from primary dermal fibroblasts and brain, and gDNA from heterozygous animals were amplified using primers to BovineHD4100001198 (48,443,632 bp), which is located in an intron of *ACVR2A* (Table 1). The PCR product was directly sequenced. Peak heights at the SNP site were quantified using PeakPicker 2 software [30]. The Y-axis shows the ratio of the peak height of the *Q* allele to that of the *q* allele in gDNA ($n = 16$, mean = 1.035), cDNA from primary dermal fibroblasts ($n = 13$, mean = 1.32), or cDNA from brains ($n = 9$, mean = 1.29). Red bars show the mean. The *P* values for the difference between cDNA and gDNA were 0.009 in primary dermal fibroblasts and 0.044 in brain, respectively, as determined by performing *t* tests. **b** Schematic representation of the positions of variants in the 5′ upstream region of *ACVR2A*. "SNP (−716)" and "3 indel (−737 to −740)" represent g.48476925C > T and g.48476943_48476946insGGC, respectively. "ATG (+1)" and "TSS (−522)" represent the start codon and the predicted transcriptional start site, respectively. A *Q* and *q* haplotype sequence alignment of the upstream region of *ACVR2A* encompassing 48,476,920 and 48,476,964 bp is shown. **c** Luciferase reporter assays for the 5′ upstream region of *ACVR2A*-derived *Q* and *q* haplotypes. The 5′ upstream region (814 bp upstream from the start codon) derived from the *Q* and *q* haplotypes was cloned into the firefly luciferase pGL3-*Q* and pGL3-*q* plasmids, respectively. The Firefly-to-Renilla luminescence ratios observed after cotransfecting HeLa cells were measured to evaluate the effects of the 5′ upstream region. Bars represent the mean ± SEM obtained in triplicate from 3 independent experiments. *P* values determined by *t* tests are shown

and brain tissues (Additional file 4). Therefore, we compared the relative abundances of *Q*- versus *q*-derived *ACVR2A* transcripts in primary dermal fibroblast ($n = 13$)

and brain tissues ($n = 9$) from heterozygotes (Fig. 2a). We isolated samples of genomic DNA (gDNA) and complementary DNA (cDNA) derived from total RNA of whole-cell lysates and then compared their allelic ratios using PeakPicker2 software [30]. The results showed that the Q-derived *ACVR2A* transcripts were 1.32-fold and 1.29-fold more abundant than q-derived transcripts in primary dermal fibroblasts and in brain (t-test, $P = 0.009$ and $P = 0.044$, respectively). In contrast, Q-derived *ACVR2A* gDNA was detected equally well as q-derived gDNA (Fig. 2a).

To determine whether the imbalance in the *ACVR2A* transcript ratio between the haplotypes was attributable to the variants in the upstream region, we cloned the upstream region, beginning 814-bp upstream of the start codon, which included the g.48476925C > T SNP and the 3-bp indel g.48476943_48476946insGGC from both the Q and q haplotypes, into luciferase-reporter constructs (Fig. 2b). We then transfected HeLa cells with these constructs and measured the resulting luciferase activities at 24 h post-transfection. The activity was higher for the Q constructs than for the q constructs, with a difference of approximately 1.2 fold (t-test, $P = 0.023$) (Fig. 2c). These results suggest that the variants including the SNP (g.48476925C > T) and the 3-bp indel (g.48476943_48476946insGGC) in the upstream region of *ACVR2A* affected the level of *ACVR2A* mRNA expression, which could lead to an allelic imbalance in *ACVR2A* mRNA expression.

We did not observe that the SNP (g.48476925C > T) or the 3-bp indel (g.48476943_48476946insGGC) resided in a transcription factor-binding site in the upstream region of *ACVR2A* in either the Q or q haplotypes, using the TRANSFAC Professional database [31]. The SNP (g.48476925C > T) is a T(Q) to C(q) SNP located 716 bp upstream of the start codon of *ACVR2A*, and the 3-bp indel (g.48476943_48476946insGGC) is a (GGC)n trinucleotide repeat with either 7(Q) or 6 (q) copies located 737–740 bp upstream of the start codon of *ACVR2A* (Fig. 2b). Recently, Karim et al. reported that 2 causative variants for bovine stature QTL in the upstream region of *PLAG1* influence the promoter activity and reflect differential binding of nuclear factors [32]. The causative variants were a (CCG)n trinucleotide repeat with either 11(Q) or 9(q) copies located immediately upstream of the presumed *PLAG1* transcription start site and a G(Q) to A(q) SNP located 12-bp upstream of the (CCG)n trinucleotide repeat [32]. The repetitive, GC-rich triplet composition and the location were similar to that observed in the upstream region of *ACVR2A* (Fig. 2b). Therefore, it is possible that the variants in the upstream region of *ACVR2A* could also affect promoter activity and cause differential binding of transcriptional factors between the Q and q haplotype, which in turn may have caused the imbalance observed in *ACVR2A* transcription between animals with Q and q haplotypes.

FSHB expression is dependent on ACVR2A expression in a gonadotrope cell line

Matzuk et al. [31] found that female *Acvr2a*-knockout mice were infertile and showed evidence of multiple follicle atresias in the ovaries. Consistently, FSHB expression in the anterior pituitary was suppressed in *Acvr2a*-knockout mice, and serum FSH was decreased in female homozygous *Acvr2a*-knockout mice (35.2 ± 5.2 ng/ml) compared to female wild-type mice (83.4 ± 20.6 ng/ml), indicating that ACVR2A affects folliculogenesis by regulating *FSHB* expression. However, it is unknown whether the level of ACVR2A influences *FSHB* expression in a dose-dependent manner.

The mouse gonadotrope LβT2 cell line [33, 34] shows *FSHB* expression and FSH secretion that is induced in response to activin A through ACVR2A [35–37]. Therefore, this cell type serves as a good model for analyzing whether the level of *ACVR2A* expression influences *FSHB* expression, using an *FSHB* promoter-reporter plasmid [36]. Although the transfection efficiency in LβT2 cells was very low (8.8 ± 1.62 % of 3,368 cells; Additional file 5: Figure S5A) [36], the co-transfection efficiency was approximately 100 % (100 % of 3,368 cells; Additional file 5: Figure S5B–D). These findings indicated that *FSHB* promoter-reporter activity was arguably detected in cells co-transfected cells with the *ACVR2A* expression plasmid. In addition, we confirmed that the level of ACVR2A was dependent on amount of the *ACVR2A* expression plasmid used for the transfections (Additional file 6: Figure S6).

To determine whether the level of *ACVR2A* affects *FSHB* expression in response to activin A, we co-transfected LβT2 cells with an *ACVR2A*-expression plasmid and an *FSHB* promoter-reporter plasmid. The results showed that activin A-induced *FSHB* promoter activity was dependent on the amount of *ACVR2A* expression plasmid used (Fig. 3; Tukey–Kramer post-hoc test, $P < 0.01$). These finding suggested that the level of *ACVR2A* affected the level of *FSHB* expression in response to activin A.

Taken together, these results suggested that the variants in the upstream region of *ACVR2A* in the Q-haplotype animals induced an increase in *ACVR2A* transcription relative to that observed in the q animals, which in turn could induce *FSHB* expression. Cattle are polyestrous animals and display estrous behavior approximately every 21 days. Normally, 2–3 waves of follicular growth occur in the ovaries during each estrous cycle, which are induced by a transient increase of FSH concentration [38–40]. Subsequently, a single dominant follicle is selected in a manner that is dependent on decline of the FSH concentration [41], followed by ovulation of the single dominant follicle or atresia. Although cattle are mono-ovulatory species, FSH concentrations influence the emergence of co-

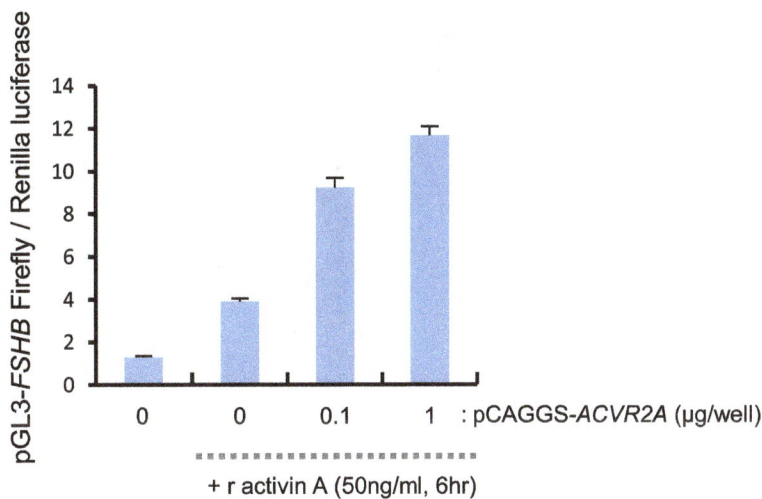

Fig. 3 *FSHB* promoter activity is dependent on *ACVR2A* expression in response to activin A in LβT2 cells. The pGL3-Basic-*FSHB* (chr2:107,059, 651–107,061,641 on the GRCm38/mm10 mouse genome assembly) and pCAGGS-*ACVR2A* plasmids were cotransfected with pRL-TK into LβT2 cells. At 18 h post-transfection, recombinant activin A was added for 6 h, and the Firefly-to-Renilla luminescence ratios were measured. Bars represent the mean luminescence ± SEM obtained in triplicate from 3 independent experiments. *P* values were determined by Tukey–Kramer post-hoc test (*P* < 0.01)

dominant follicles [42]. Moreover, administration of exogenous recombinant FSH in cattle induces multiple follicles growth during super-ovulation [43]. In this manner, follicular growth at several stages is closely associated with circulating FSH concentrations. Therefore, although quantitative differences in FSHB expression between Q- and q-haplotype animals may be subtle, it is possible that these differences govern folliculogenesis. Thus, further analyses regarding differences of serum FSH levels and follicle size at each stage between Q and q-haplotype animals could help to elucidate the mechanisms underlying the variants of *ACVR2A* that influence female fertility.

Of note, mutation of TGF-beta family ligand and its receptor BMP15 [44] and the type-I receptor ALK6 [45, 46] were identified as causative variants in the ovulation rate in sheep. Inactivation of 1 copy of *BMP15* increased the ovulation rate, whereas inactivation of both copies of *BMP15* showed blocked follicular growth. In contrast, *ALK6* mutant alleles increased the ovulation rate in an additive manner, indicating that the TGF-beta family ligand and its receptor functions in folliculogenesis in a dosage-sensitive manner. Similarly, ACVR2A, which is a TGF-beta family type II receptor, may also function in female fertility in cattle in a dose-dependent manner.

Replication of the associated haplotype in a sample of 1,433 animals

To validate the GWAS results and estimate the effective size of the QTL and the allele frequencies, we examined whether the Q haplotype was associated with $ANAI_4$ in a sample of 1,433 animals. We genotyped 1,433 animals randomly selected from the remainder of the cohort

from the same farm used for the GWAS (Additional file 7). The results showed that the Q haplotype was significantly associated with $ANAI_4$ compared to the q haplotype (Tukey–Kramer post-hoc test, $P = 0.017$; Table 2). The Q haplotype frequency was 0.39, indicating that the haplotype is common in Japanese Black cattle. We fitted a linear-mixed model to the $ANAI_4$ values in the additive model and used restricted maximum likelihood (REML) analysis to estimate the variance explained by the haplotype. We estimated the proportion of total genetic variance attributable to the Q haplotype as 0.1 (Table 2). The Q-to-q haplotype substitution effect on $ANAI_4$ was 0.02 (Table 2).

Conclusion

This GWAS identified variants in the upstream region of *ACVR2A*, which were associated with female fertility in Japanese Black cattle. The variants affected the level of *ACVR2A* mRNA expression, which led to an allelic imbalance. Expression of exogenous *ACVR2A* induced dose-dependent increases of *FSHB* expression in response to activin A. Finally, we replicated this association and estimated the effect in a sample of 1,433 animals. Thus, the results suggest that Q haplotype could serve as a useful marker in Japanese Black cattle to select animals with superior female fertility in Japanese Black cattle.

Methods

Ethics statement

All animal experiments were performed according to the guidelines for care and use of laboratory animals of

Table 2 Proportion of genetic variance attributable to haplotypes associated with ANAI$_4$

BTA	BTA SNP and indel-id	positon (bp) _UMD3,1[a]	Q/q haplotype	Number of animals genotyped for the haplotype	Minor allele frequency	Heritability	Haplotype effect on total genetic variance[b]	Haplotype substitution effect[c] P-value[d]	P-value[d]
2	g.48476925C>T and g.48476943_48476946insGGC	48476925 and 48476943. 48476946	T and GCC/ C and (-)	1433	0.39	0.02	0.1	0.02	0.017

[a]The positions shown are based on the UMD3.1 assembly of the bovine genome.

[b]The effects of the haplotype were estimated as the least-square means from GLM analysis. The statistical model for GLM analysis included the fixed variables of the farm, birth year, and haplotype. The genetic variance explained by the haplotype was calculated based on estimates of the haplotype effect and the haplotype frequency [55]. Total genetic variance was estimated using the MTDF-REML programs. The effect size of haplotype was estimated as the proportion of genetic variance explained by the haplotype.

[c]The average ANAI$_4$ values for QQ and qq were 0.69 and 0.65, respectively.

[d]Results were tested by a 1-way ANOVA, followed by the Tukey–Kramer test for multiple comparisons

Shirakawa Institute of Animal Genetics, and this research project was approved by the Shirakawa Institute of Animal Genetics Committee on Animal Research (H21-1). We have obtained the written agreement from the cattle owners to use the samples and data.

Collection of phenotype data

Data were collected from cattle farms, and the data management system for Japanese Black cattle was described in a previous study [47, 48]. The original data included 63,775 records for reproductive females born from 1992 to 2006. The data were selected using 9 selection criteria: 1) data were not missing for the cow from first to fourth parity, 2) the cow did not have twins in parturition, 3) the cow did not receive any embryo transfers, 4) the cow did not have any abortions, 5) the length of all gestations ranged from 261 to 310 days, 6) the calving interval ranged from 276 to 730 days, 7) the age of the cow at the first calving was be less than 1,128 days, 8) the cow was reared in a single farm, and 9) each breeding farm had more than 10 records from each birth year. After applying these selection criteria, the final dataset contained 10,399 records. In this study, female fertility was evaluated as a metric to describe the average inverse of the number of artificial inseminations required for conception [20] from the first parity through the fourth parity. For example, if conception is achieved at the first, second, third, or fourth AI, ANAI values are 1, 0.5, 0.33, and 0.25, respectively. ANAI$_4$ values were corrected for the effects of the individual farm and birth year. The heritability and variance components of phenotypic variance were estimated with the REML procedure using the MTDF-REML programs [49]. We fitted single-traits animal models with random effects and fixed effects:

$$y_{ij} = \mu + year_i + farm_j + u_{ij} + e_{ij}$$

where y_{ij} is the observation of ij for the traits, μ is the total mean, $year_i$ is the fixed effect of birth year i (15 classes, 1992 to 2006), $farm_j$ is the fixed effect of farm j (174 classes), and u_{ij} is the infinitesimal genetic effect of animal ij, which is distributed as $N(0, \mathbf{A}\sigma_u^2)$, where $\mathbf{A}$ is the numerator relationship matrix, and e_{ij} is the residual effect. Pedigrees of the base population animals were traced back for 2 generations to create the numerator relationship matrix, and 10,399 animals were included in the pedigree analysis.

Selection of samples for the GWAS and DNA sample collection

Samples were selected from the upper extreme (85th percentile, 1,560 animals) and the lower extreme (15th percentile, 1,560 animals) for ANAI$_4$. To reduce population stratification, we selected less than 5 cows derived from a single sire in each extreme, resulting in 256 cows for the upper extreme and 174 cows for the lower extreme. Whole blood was collected from each cow, and gDNA was isolated using the Easy-DNA gDNA Purification Kit (Invitrogen, Cat. #K1800-01).

GWAS for ANAI$_4$

A total of 430 DNA samples were genotyped using the BovineSNP50K BeadChip (version 1, Illumina), which comprises probes for 54,001 SNPs. The UMD3.1 assembly [50] was used to map the positions of the SNPs. The data were analyzed using PLINK software, v1.07 [51]. A total of 33,303 autosomal SNPs that passed our quality control criteria (call rate > 99 %, minor allele frequency > 0.01, Hardy–Weinberg equilibrium P > 0.001, and inclusion of the SNP on the Illumina Bovine HD BeadChip) [23] were used for the association study. We performed a GWAS using the trait as a binary variable, as is commonly done in case–control studies. To analyze the upper- and lower-performance phenotypes, we used a linear mixed model with a genetic relationship matrix for the binary phenotypes using the EMMAX software [24].

Linkage disequilibrium and diplotype analysis

Haploview 4.2 software [52] was used to analyze linkage disequilibria between the SNPs. The diplotypes of the GWAS samples were estimated using BEAGLE3.3.2 software [26, 27].

Imputation of SNPs

The genotypes of 33,303 SNPs were imputed using BEAGLE 3.3.2 software, with phased haplotype data inferred from 586,812 SNPs in 1,041 Japanese Black cattle as the reference [23].

Expression analysis

For real-time quantitative PCR, we extracted total RNA from cow tissues, primary bovine dermal fibroblasts, and bovine endometrial epithelial cells (Cell Application, Inc., Cat. #B932-05) using RNeasy Mini Kits (Qiagen, Cat. #74104), after which the total RNA was treated with DNase I. cDNA was synthesized from 50 ng total RNA using the ReverTra Ace-α Kit (Toyoba, Cat. #FSK-101) with random primers, according to the manufacturer's instructions. The *ACVR2A* gene was detected with the following primers and probe: forward, 5′-catgggatt agtcctgtgggaac-3′; reverse, 5′-cctcaaatggcagcatgtattca-3′; and probe, 5′-tacaggtccatctgcagcagtacagcga-3′. Real-time PCR was performed on a 7900HT Real-Time PCR System (Applied Biosystems) using the comparative Ct method with glyceraldehyde-3-phosphate dehydrogenase mRNA serving as the internal control.

Allelic imbalance test

To quantify the allelic imbalance of *ACVR2A* transcripts, we designed PCR primers to BovineHD4100001198 (48,443,632 bp) on BTA2, located in an intron of *ACVR2A*. The forward primer was 5'-aacctagaaaccgtag aaagacga-3', and the reverse primer was 5'-gatggc atctcttggctcat-3'. We used 50 ng of template cDNA from primary bovine dermal fibroblasts or bovine brain tissues (medulla oblongata), or 10 ng of gDNA from heterozygous animals for PCR amplification with TaKaRa Ex Taq HS DNA Polymerase (TaKaRa, Cat. #RR006). The PCR product was directly sequenced and purified with the CleanSEQ system (Agencourt, Cat. #A29154). Peak heights at polymorphic sites were quantified using PeakPicker 2 software [30]. Allelic imbalances were estimated as the ratio of the peak height of the Q allele to that of the q allele in cDNA and in gDNA. Calibration curves were generated using data obtained by mixing varying amounts of gDNA from Q and q homozygotes.

Luciferase reporter assays

To measure the effects of the 5'-upstream region of *ACVR2A* on *ACVR2A* expression, the 814-bp fragment upstream of the start codon of *ACVR2A* of each haplotype (Q and q haplotype) including a SNP (g.48476925C > T) and a 3-bp indel (g.48476943_48476946insGGC), was PCR amplified using PrimeSTAR Max DNA Polymerase (Takara, Cat. #R045A). PCR was performed using gDNA, a forward primer (5'-GGGGTACCcaaatctcctcgcgctcac-3'; uppercase letters indicate the *Kpn*I linker), and a reverse primer (5'-TCCCCCGGGactttgcagcagctcccatt-3'; uppercase letters indicate the *Sma*I linker). The PCR products were cloned into the *Kpn*I and *Sma*I sites of the pGL3-Basic Vector (Promega, Cat. #E1751). The sequence and orientation of the insert were confirmed by sequencing. The pGL3-Basic Vector was used for mock transfections. For cell culture, HeLa S3 cells were maintained in Dulbecco's modified Eagle's medium (DMEM; Sigma, Cat. #D5796) with 10 % fetal calf serum (FCS; Sigma, Cat. #F-2442) supplemented with 2 mM L-glutamine (Gibco, Cat. #25030-081) and 100 units/ml penicillin and 100 μg/ml streptomycin (Gibco, Cat. #15140-122). Using Lipofectamine 2000 (Invitrogen, Cat. #11668-019), we transfected 5×10^4 cells per well in a 24-well plate with a mixture of 750 ng of the reporter vector and 10 ng of the pRL-TK Renilla vector (Promega, Cat. #E2241) to calibrate transfection efficiency. Luciferase assays were performed 24 h post-transfection using the Dual Luciferase Reporter Assay System (Promega, Cat. #E1910) and the GloMax (Promega, Cat. #E6521).

FSHB promoter-reporter assay

To measure the *FSHB* promoter activity, the promoter region of *FSHB* (chr2:107,059,651–107,061,641 on the

GRCm38/mm10 mouse genome assembly) [36] was PCR amplified from gDNA of a C57BL/6NJ mouse using a forward primer (5'-GGGGTACCcctgttcattaaccactgagct-3'; uppercase letters indicate the *Kpn*I linker) and a reverse primer (5'-CCGCTCGAGcactgagtcaagttacacctca-3'; uppercase letters indicate the *Xho*I linker). The PCR products were cloned into the *Kpn*I and *Xho*I sites of pGL3-Basic. The sequence and orientation of the insert were confirmed by sequencing. To express the ACVR2A protein, the coding region of *ACVR2A* (NM_007396, 664–2,205 bp) was PCR amplified from cDNA from LβT2 cells using a forward primer (5'-GCTCTAGAatgggagctgctgc aaagttggc-3'; uppercase letters indicate the *Xba*I linker) and a reverse primer (5'-CGGAATTCctaagcgtaatcagg aacgtcgtaagggtatagactagattctttgggaggaaagtc-3'; uppercase letters indicate the *Eco*RI linker, and underlined letters indicate the C-terminal hemagglutinin [HA] tag for ACVR2A, respectively). The PCR product was cloned into the *Xba*I and *Eco*RI sites of the pCAGGS vector [53]. The sequence and orientation of the insert were confirmed by sequencing. The expression of ACVR2A was confirmed by western blotting with an anti-HA antibody 3 F10 (Roche, Cat. #11 867 423 001, 100 ng/ml). Immunoreactivity was detected with a horseradish peroxidase-conjugated donkey anti-rat IgG antibody (Jackson ImmunoResearch, Cat. #712-035-153) and the ECL Prime Western Blotting Detection Reagent (GE Healthcare, Cat. #RPN2232). Chemiluminescence was detected with an ImageQuant LAS 4000 (GE Healthcare) and quantified using the ImageQuant TL Analysis Toolbox. LβT2 cells [33, 34] were maintained in DMEM with 10 % charcoal-stripped FCS (Gibco, Cat. #12676-029) supplemented with 2 mM L-glutamine, 20 nM dexamethasone (Sigma, Cat. #D4902), 0.1 mM non-essential amino acids (Gibco, Cat. #11140-050), and 100 units/ml penicillin and 100 μg/ml streptomycin.

To determine whether the expression level of *ACVR2A* affected *FSHB* reporter activity in response to activin A, we transfected 2×10^5 cells per well in a 24-well plate with a mixture of 200 ng of the *FSHB* promoter-reporter vector, pCAGGS-*ACVR2A* (using the amounts of plasmid indicated in Fig. 3) and 10 ng of pRL-TK Renilla to determine transfection efficiencies. After 18 h, 50 ng/ml recombinant activin A (R&D, Cat. #338-AC; 50 μg/ml stock in 0.1 % bovine serum albumin/phosphate-buffered saline) was added to the culture medium, and luciferase assays were performed at 6 h post-activin A stimulation using the Dual Luciferase Reporter Assay System and the GloMax.

To examine co-transfection efficiencies in LβT2 cells, we used the pCAGGS-EGFP (Clontech, Cat. #6085-1) and pCAGGS-mCherry (Clontech, Cat. #632522) vectors. At 24 h post-transfection, cells were examined with a confocal microscope (FV1000, Olympus Optical) and fluorescence-positive cells were counted using ImageJ software, version 1.46 [54].

Replication study

For the replication study, we used 1,433 samples from the remainder of the cohort from the same farm selected for the GWAS (Additional file 7). The g.48476925C > T SNP was genotyped by directly sequencing PCR products using a forward primer (5′-acaatctcctcgcgctcac-3′) and a reverse primer (5′-caagttctggtccaggctct-3′). PCR products were sequenced using the reverse primer and the BigDye Terminator v.3.1 Cycle Sequencing Kit (Applied Biosystems), followed by electrophoresis using an ABI 3730 sequencer (Applied Biosystems) and genotyping using SeqScape software, V2.5 (Applied Biosystems). The 3-bp indel (g.48476943_48476946insGGC) was PCR-amplified using a forward primer (5′-acaatctcctcg cgctcac-3′) and a reverse primer (5′-fluorescein amidite-caagttctggtccaggctct-3′). PCR products were electrophoresed using an ABI 3730 sequencer (Applied Biosystems) and genotyped using GeneScan analysis software (Applied Biosystems) and GeneMapper software v3.7 (Applied Biosystems).

Estimation of the genetic variance explained by the haplotype and effect size of haplotype

The effects of the haplotype were estimated as the least square means in generalized linear model (GLM) analyses. The statistical model for GLM analysis included fixed variables, such as the farm, birth year, and haplotype. The genetic variance explained by the haplotype was calculated based on estimates of the haplotype effect and the haplotype frequency [55]. Total genetic variance was estimated using the MTDF-REML program. The effect size of haplotype was estimated as the proportion of genetic variance explained by the haplotype.

Availability of supporting data

The data sets supporting the results of this article are included within the article and its additional file.

Additional files

Additional file 1: Distribution of ANAI$_4$ in 10,399 cows in this study. (PDF 31 kb)

Additional file 2: Quantile-quantile plots of the GWAS results for ANAI$_4$. The red dots represent the observed − log10 P values, and the straight line represents the expected − log10 P values under the null hypothesis. (PPTX 68 kb)

Additional file 3: Genes on BTA2 in the ANAI$_4$-associated region. The positions are based on the UMD3.1 assembly of the bovine genome. (XLSX 36 kb)

Additional file 4: Relative expression of ACVR2 in cow tissues and cells. Relative ACVR2 expression levels in tissues and cells are indicated on the Y-axis. Total RNA was extracted from tissues (1–16), primary

dermal fibroblasts (17) from 2 female Japanese Black cattle, or from bovine primary endometrial epithelial cells (18). Relative gene expression levels in the different tissues are shown as mean quantities relative to the value observed in ovarian tissue (dotted line). (PPTX 50 kb)

Additional file 5: Transfection and co-transfection efficiencies of LβT2 cells. (A) To examine the transfection efficiency of LβT2 cells, we used the pCAGGS-EGFP and pCAGGS-mCherry vectors. At 24 h post-transfection, double fluorescence-positive cells were counted using ImageJ software. The data shown represent the transfection efficiencies of LβT2 cells in 5 experiments. The average transfection efficiency was 8.8 ± 1.62 % (3,368 cells). (B–D) To examine co-transfection efficiency in LβT2 cells, we used the pCAGGS-EGFP and pCAGGS-mCherry vectors. At 24 h post-transfection, double fluorescence-positive cells (B, magenta) were counted using ImageJ software. All GFP positive cells (C, green) were mCherry-positive (D, red), representing 100 % of 3,368 cells. (PPTX 6646 kb)

Additional file 6: Expression of pCAGGS-ACVR2A in LβT2 cells. (A) To determine whether the pCAGGS-ACVR2A plasmid was expressed in LβT2 cells, we transfected 2×10^5 cells per well in a 24-well plate with a mixture of 200 ng of the FSHB promoter-reporter plasmid, pCAGGS-ACVR2A (the amount of each plasmid is indicated in the lanes) and 10 ng of pRL-TK Renilla. HA-ACVR2A expression was confirmed by western blot analysis using an anti-HA antibody. Unstained Precision Plus protein standards were used as a marker (BioRad, Cat. #161-0363). HA-ACVR2A expression was detected with multiple bands (~57.8 kDa) because TGF-beta type-II receptors are normally modified by phosphorylation and glycosylation, causing then to migrate heterogeneously in sodium dodecyl sulfate-polyacrylamide gel electrophoresis gels. (B) Relative ACVR2A band intensities were measured using the ImageQuant TL Analysis Toolbox. The amounts of loaded proteins were calibrated by Coomassie Brilliant Blue-stained bands between 37 and 75 kDa. The bars represent the mean ± SEM observed in triplicate from 3 independent experiments. (PPTX 706 kb)

Additional file 7: Distribution of ANAI$_4$ in 1,433 cows for the replication study. A sample population ($n = 1,433$) was derived from the remainder of the cohort from the same farm used for the GWAS. (PPTX 47 kb)

Abbreviations

ANAI$_4$: Average of the inverse of the number of artificially inseminations required for conception from first through fourth parity; BTA: Bovine (Bos taurus) chromosome; GWAS: Genome-wide association study; LD: Linkage disequilibrium; PCR: Polymerase chain reaction; QTL: Quantitative trait locus or loci; SNP: Single nucleotide polymorphism.

Competing interests

The authors declare no conflict of interest.

Authors' contributions

SS and TI conceived and designed the study. SS, YS performed SNP genotyping. SS performed the GWAS analysis. TI collected and analyzed the data. SS and TI performed replication studies. SS, TM, KT, SI, and MS performed the expression and functional analyses. SS, TI, and YS wrote the manuscript. All authors read and approved the final manuscript.

Acknowledgements

We would like to thank Toshio Watanabe helping with the data analysis, as well as Takatoshi Kojima and other lab members for providing generous support and valuable suggestions. We are grateful to Pamela Mellon for the LβT2 cells and Jun-ichi Miyazaki for the pCAGGS plasmid. We are grateful to Emiko Watanabe for technical assistance. This work was supported by funding from the KIEIKAI Research Foundation (H24-KIEIKAI and H25-KIEIKAI) to SS, the JSPS KAKENHI (Grant Number 26450384) to SS, and the Japan Racing and Livestock Promotion to YS.

Author details

[1]National Livestock Breeding Center, Odakura, Nishigo, Fukushima 961-8511, Japan. [2]Graduate School of Environmental and Life Science, Okayama University, Tsushima-naka, Okayama 700-8530, Japan. [3]Gifu Prefectural

Livestock Research Institute, Kiyomi, Takayama, Gifu 506-0101, Japan. [4]Cattle Breeding Development Institute of Kagoshima Prefecture, Osumi, So, Kagoshima 899-8212, Japan. [5]Shirakawa Institute of Animal Genetics, Japan Livestock Technology Association, Odakura, Nishigo, Fukushima 961-8061, Japan.

References

1. Annual report of conceptional rate in Japan, Livestock Improvment Association of Japan. [http://liaj.or.jp/giken/gijutsubu/seieki/jyutai.htm]

2. Lucy MC. Reproductive loss in high-producing dairy cattle: where will it end? J Dairy Sci. 2001;84(6):1277–93.

3. Matukumalli LK, Lawley CT, Schnabel RD, Taylor JF, Allan MF, Heaton MP, et al. Development and characterization of a high density SNP genotyping assay for cattle. PLoS One. 2009;4(4):e5350.

4. Rincon G, Weber KL, Eenennaam AL, Golden BL, Medrano JF. Hot topic: performance of bovine high-density genotyping platforms in Holsteins and Jerseys. J Dairy Sci. 2011;94(12):6116–21.

5. Kim ES, Berger PJ, Kirkpatrick BW. Genome-wide scan for bovine twinning rate QTL using linkage disequilibrium. Anim Genet. 2009;40(3):300–7.

6. Huang W, Kirkpatrick BW, Rosa GJ, Khatib H. A genome-wide association study using selective DNA pooling identifies candidate markers for fertility in Holstein cattle. Anim Genet. 2010;41(6):570–8.

7. Sahana G, Guldbrandtsen B, Bendixen C, Lund MS. Genome-wide association mapping for female fertility traits in Danish and Swedish Holstein cattle. Anim Genet. 2010;41(6):579–88.

8. Olsen HG, Hayes BJ, Kent MP, Nome T, Svendsen M, Lien S. A genome wide association study for QTL affecting direct and maternal effects of stillbirth and dystocia in cattle. Anim Genet. 2010;41(3):273–80.

9. Olsen HG, Hayes BJ, Kent MP, Nome T, Svendsen M, Larsgard AG, et al. Genome-wide association mapping in Norwegian Red cattle identifies quantitative trait loci for fertility and milk production on BTA12. Anim Genet. 2011;42(5):466–74.

10. Sahana G, Guldbrandtsen B, Lund MS. Genome-wide association study for calving traits in Danish and Swedish Holstein cattle. J Dairy Sci. 2011;94(1):479–86.

11. Schulman NF, Sahana G, Iso-Touru T, McKay SD, Schnabel RD, Lund MS, et al. Mapping of fertility traits in Finnish Ayrshire by genome-wide association analysis. Anim Genet. 2011;42(3):263–9.

12. Hawken RJ, Zhang YD, Fortes MR, Collis E, Barris WC, Corbet NJ, et al. Genome-wide association studies of female reproduction in tropically adapted beef cattle. J Anim Sci. 2012;90(5):1398–410.

13. Sugimoto M, Sasaki S, Gotoh Y, Nakamura Y, Aoyagi Y, Kawahara T, et al. Genetic Variants Related to Gap Junctions and Hormone Secretion Influence Conception Rates in Cows. Proc Natl Acad Sci U S A. 2013;110(48):19495–500.

14. Fortes MR, Deatley KL, Lehnert SA, Burns BM, Reverter A, Hawken RJ, et al. Genomic regions associated with fertility traits in male and female cattle: advances from microsatellites to high-density chips and beyond. Anim Reprod Sci. 2013;141(1–2):1–19.

15. Kirkpatrick BW. Genetics of Reproduction in Cattle. In: Garrick DJ, Ruvinsky A, editors. The Genetics of Cattle. 2nd ed. Boston: CABI Publishing; 2014: 260–283

16. Cameron PJ, Zembayashi M, Lunt DK, Mitsuhashi T, Mitsumoto M, Ozawa S, et al. Relationship between Japanese beef marbling standard and intramuscular lipid in the M. longissimus thoracis of Japanese Black and American Wagyu Cattle. Meat Sci. 1994;38(2):361–4.

17. Namikawa K. Japanese Beef Cattle - Historical Breeding Processes of Japanese Beef Cattle and Preservation of Genetic Resources as Economic Farm Animal (in Japanese): Wagyu Registry Association. Wagyu. 1992.

18. McKay SD, Schnabel RD, Murdoch BM, Matukumalli LK, Aerts J, Coppieters W, et al. An assessment of population structure in eight breeds of cattle using a whole genome SNP panel. BMC Genet. 2008;9:37.

19. Wagyu Registry Association. The breeding value estimation for the number of calves produced at a specific age (in Japanese). Wagyu. 2010;252(61):40–8.

20. Weller JI, Ezra E. Genetic analysis of somatic cell score and female fertility of Israeli Holsteins with an individual animal model. J Dairy Sci. 1997;80(3):586–93.

21. Kirkpatrick BW. Genetics and Biology of Reproduction in Cattle. In: Fries R, Ruvinsky A, editors. The Genetics of Cattle. Boston: CABI Publishing; 1999: 391–410.

22. Xing C, Xing G. Power of selective genotyping in genome-wide association studies of quantitative traits. BMC Proc. 2009;3 Suppl 7:S23.

23. Sasaki S, Ibi T, Watanabe T, Matsuhashi T, Ikeda S, Sugimoto Y. Variants in the 3' UTR of General Transcription Factor IIF, polypeptide 2 affect female calving efficiency in Japanese Black cattle. BMC Genet. 2013;14:41.

24. Kang HM, Sul JH, Service SK, Zaitlen NA, Kong SY, Freimer NB, et al. Variance component model to account for sample structure in genome-wide association studies. Nat Genet. 2010;42(4):348–54.

25. CattleQTLdb. [http://www.animalgenome.org/cgi-bin/QTLdb/BT/index]

26. Browning SR, Browning BL. Rapid and accurate haplotype phasing and missing-data inference for whole-genome association studies by use of localized haplotype clustering. Am J Hum Genet. 2007;81(5):1084–97.

27. Browning BL, Browning SR. A unified approach to genotype imputation and haplotype-phase inference for large data sets of trios and unrelated individuals. Am J Hum Genet. 2009;84(2):210–23.

28. Bernard DJ, Tran S. Mechanisms of activin-stimulated FSH synthesis: the story of a pig and a FOX. Biol Reprod. 2013;88(3):78.

29. Kumar TR, Wang Y, Lu N, Matzuk MM. Follicle stimulating hormone is required for ovarian follicle maturation but not male fertility. Nat Genet. 1997;15(2):201–4.

30. Ge B, Gurd S, Gaudin T, Dore C, Lepage P, Harmsen E, et al. Survey of allelic expression using EST mining. Genome Res. 2005;15(11):1584–91.

31. Matys V, Kel-Margoulis OV, Fricke E, Liebich I, Land S, Barre-Dirrie A, et al. TRANSFAC and its module TRANSCompel: transcriptional gene regulation in eukaryotes. Nucleic Acids Res. 2006;34(Database issue):D108–110.

32. Karim L, Takeda H, Lin L, Druet T, Arias JA, Baurain D, et al. Variants modulating the expression of a chromosome domain encompassing PLAG1 influence bovine stature. Nat Genet. 2011;43(5):405–13.

33. Alarid ET, Windle JJ, Whyte DB, Mellon PL. Immortalization of pituitary cells at discrete stages of development by directed oncogenesis in transgenic mice. Development. 1996;122(10):3319–29.

34. Thomas P, Mellon PL, Turgeon J, Waring DW. The L beta T2 clonal gonadotrope: a model for single cell studies of endocrine cell secretion. Endocrinology. 1996;137(7):2979–89.

35. Graham KE, Nusser KD, Low MJ. LbetaT2 gonadotroph cells secrete follicle stimulating hormone (FSH) in response to active A. J Endocrinol. 1999;162(3):R1–5.

36. Bernard DJ. Both SMAD2 and SMAD3 mediate activin-stimulated expression of the follicle-stimulating hormone beta subunit in mouse gonadotrope cells. Mol Endocrinol. 2004;18(3):606–23.

37. Rejon CA, Hancock MA, Li YN, Thompson TB, Hebert TE, Bernard DJ. Activins bind and signal via bone morphogenetic protein receptor type II (BMPR2) in immortalized gonadotrope-like cells. Cell Signal. 2013;25(12):2717–26.

38. Pierson RA, Ginther OJ. Ultrasonographic appearance of the bovine uterus during the estrous cycle. J Am Vet Med Assoc. 1987;190(8):995–1001.

39. Sirois J, Fortune JE. Ovarian follicular dynamics during the estrous cycle in heifers monitored by real-time ultrasonography. Biol Reprod. 1988;39(2):308–17.

40. Adams GP, Matteri RL, Kastelic JP, Ko JC, Ginther OJ. Association between surges of follicle-stimulating hormone and the emergence of follicular waves in heifers. J Reprod Fertil. 1992;94(1):177–88.

41. Ginther OJ, Bergfelt DR, Kulick LJ, Kot K. Pulsatility of systemic FSH and LH concentrations during follicular-wave development in cattle. Theriogenology. 1998;50(4):507–19.

42. Lopez H, Sartori R, Wiltbank MC. Reproductive hormones and follicular growth during development of one or multiple dominant follicles in cattle. Biol Reprod. 2005;72(4):788–95.

43. Jainudeen MR, Wahid H, Hafez E, S E. Ovulation Induction, Embryo Production and Transfer. In: Hafez B, Hafez E, S E, editors. Reproduction in farm animals. 7th ed. Oxford: Blackwell Publishing; 2000: 405–430.

44. Galloway SM, McNatty KP, Cambridge LM, Laitinen MP, Juengel JL, Jokiranta TS, et al. Mutations in an oocyte-derived growth factor gene (BMP15) cause increased ovulation rate and infertility in a dosage-sensitive manner. Nat Genet. 2000;25(3):279–83.

45. Wilson T, Wu XY, Juengel JL, Ross IK, Lumsden JM, Lord EA, et al. Highly prolific Booroola sheep have a mutation in the intracellular kinase domain of bone morphogenetic protein IB receptor (ALK-6) that is expressed in both oocytes and granulosa cells. Biol Reprod. 2001;64(4):1225–35.

46. Mulsant P, Lecerf F, Fabre S, Schibler L, Monget P, Lanneluc I, et al. Mutation in bone morphogenetic protein receptor-IB is associated with increased ovulation rate in Booroola Merino ewes. Proc Natl Acad Sci U S A. 2001;98(9):5104–9.

47. Ibi T, Hirooka H, Kahi AK, Sasae Y, Sasaki Y. Genotype x environment interaction effects on carcass traits in Japanese Black cattle. J Anim Sci. 2005;83(7):1503–10.

48. Ibi T, Kahi AK, Hirooka H. Effect of carcass price fluctuations on genetic and economic evaluation of carcass traits in Japanese Black cattle. J Anim Sci. 2006;84(12):3204–11.

49. Boldman KG, Kriese LA, Van Vleck LD, Kachman SD. A Manual for Use of MTDFREML. A Set of Programs to Obtain Estimates of Variances and Covariances [Draft]. USDA-ARS, Washington. DC. 1993.

50. Center for Bioinformatics and Computational Biology at University of Maryland. [ftp://ftp.ccb.jhu.edu/pub/data/assembly/Bos_taurus/ Bos_taurus_UMD_3.1/]

51. Purcell S, Neale B, Todd-Brown K, Thomas L, Ferreira MA, Bender D, et al. PLINK: a tool set for whole-genome association and population-based linkage analyses. Am J Hum Genet. 2007;81(3):559–75.

52. Barrett JC, Fry B, Maller J, Daly MJ. Haploview: analysis and visualization of LD and haplotype maps. Bioinformatics. 2005;21(2):263–5.

53. Niwa H, Yamamura K, Miyazaki J. Efficient selection for high-expression transfectants with a novel eukaryotic vector. Gene. 1991;108(2):193–9.

54. Image J. [http://imagej.nih.gov/ij/]

55. Falconer DS, Mackay TFC. Introduction to quantitative genetics. 4th ed. Essex: Longman Group Ltd; 1996.

4

MicroRNA profiles reveal female allotetraploid hybrid fertility

Rong Zhou, Yanhong Wu, Min Tao, Chun Zhang and Shaojun Liu[*]

Abstract

Background: The bisexual fertile tetraploid fish is important in biological evolution. Tetraploid fish fertility is the key factor for stable inheritance. Therefore, elucidating tetraploid fish fertility at the molecular level is essential. MicroRNAs regulate gene expression and are involved in many aspects of gonad development.

Methods: Total RNA was isolated using TRIzol, followed by constructing small RNA libraries. And then, the qualified libraries were sequenced with the HiSeq 2500 SE50 system. The obtained clean reads were analyzed to identify conserved and novel miRNAs, and evaluate the expression, and also predict the target genes. The differential expressions of miRNAs were confirmed by RT-PCR.

Results: In this study, allotetraploid hybrid fish (4nAT) and diploid red crucian carp (RCC) ovaries were used to compare miRNA profiles. The results indicated that most of the highly expressed miRNAs were closely correlated with ovary maturation, and displayed no significant differences in expression. Moreover, 34 up-regulated and nine down-regulated miRNAs were found in 4nAT. The differentially expressed miRNAs were primarily involved in metabolism, defense mechanisms, and cytoskeleton production.

Conclusions: This is the first study to provide new epigenetic evidences for tetraploid fish fertility and phenotypic changes as a result of increased ploidy.

Keywords: MicroRNA, Fertility, Polyploidy, High throughout sequencing

Background

Polyploidization is one of important features of species evolution [1, 2]. Duplication of the whole genome was proved to be occurred during long-term evolution of many eukaryotes [3]. The polyploidy phenomenon, arose when a rare meiotic or mitotic event resulted in the formation of gametes with more than one set of chromosomes, is common in plants but rare in animals, especially vertebrates [4, 5]. Thus, polyploid organisms contain at least three or more complete chromosome sets, with four being the most common (tetraploidy) [5]. Polyploidy is divided into autopolyploidy and allopolyploidy. Allopolyploidy involves distant hybridization and duplication of divergent parental genomes, this induces more rapid genome evolution than autopolyploidy [6]. For survival, the allopolyploid individual, with more than two differential genomes in the same nucleus, must experience challenges

and balance the extra biochemical diversity and gene expression of multiple genomes [7]. For inheritance, the allotetraploid individuals must be bisexual fertile to maintain the genetic characteristics from one generation to the next. Therefore, it is essential to prove the fertility of polyploid individuals at the morphological and molecular levels before studying their special roles in biological evolution.

The allotetraploid hybrid (4nAT) was derived from the distant hybridization of female red crucian carp (RCC) and male common carp (CC), which is the first bisexual fertile allotetraploid fish, even vertebrate reported so far [8]. In the F_2 hybrid lineage, the hybrid offspring could produce unreduced diploid gametes [9]. Thus, self-crossing of F_2 resulted in the appearance of allotetraploid hybrids in F_3, which is then stably inherited through to F_{24} [10]. The cytogenetic studies revealed that 4nAT possessed 200 chromosomes, which form 100 bivalents during meiosis I to produce stable diploid gametes [11, 12]. Observations of gonad tissue sections have revealed that 4nAT was matured at

* Correspondence: lsj@hunnu.edu.cn
Key Laboratory of Protein Chemistry and Developmental Biology of the State Education Ministry of China, College of Life Sciences, Hunan Normal University, Changsha, Hunan 410081, People's Republic of China

1 year and experienced a similar gonad developmental process as diploid RCC [10, 11]. Molecular evidences have shown that regulation of reproductive endocrine in *4n*AT is normal [13–15]. All these studies have shown that *4n*AT is bisexual fertile. However, the epigenetic regulation of reproductive characteristics (e.g., fertility and gamete ploidy) between allotetraploid and diploid fishes has not been studied to date.

MicroRNA is a kind of highly conserved endogenous noncoding small RNA approximately 22 nt in length, first discovered in the early 1990s by Lee and Wightman in *C. elegans* in which it regulates heterochronic gene *lin-14* expression to mediate temporal pattern formation [16, 17]. Mature miRNAs are formed through three sequential steps: (1) the longer nascent transcripts (termed pri-miRNAs) are transcribed by RNA polymerase II in the nucleus [18]; (2) the ~70 nt long precursors of miRNAs (termed pre-miRNAs) are generated in the nucleus by Drosha and then transported to the cytoplasm [19]; (3) the ~22 nt long mature miRNAs are formed via processing by Dicer in the cytoplasm [20]. MiRNA, as an important epigenetic modification, regulates gene expression by recognizing and binding 3'-untranslated regions of target mRNAs to either block gene translation or induce mRNA cleavage [21]. Target deletion of Dicer 1 in mouse ovaries provided the first empirical evidence that miRNAs are critical for the normal development of the female reproductive system and fertility [22, 23]. Moreover, a series of gonad-specific expressed miRNAs have been identified through the comparison of gonad and other tissues, or different gonad developmental stages both in mammals and teleost fishes [24–26].

In this study, we compared the miRNA profiles of diploid RCC and *4n*AT ovaries, and found several differentially expressed miRNAs, including 34 up-regulated and nine down-regulated in *4n*AT. Target gene prediction and functional annotation analysis revealed that genes targeted by differentially expressed miRNAs were primarily involved in the metabolic, cytoskeleton, and defense systems. However, miRNAs related to ovary maturation were abundant and exhibited similar expression levels in both groups of samples. This study provides epigenetic evidences for female *4n*AT fertility and phenotypic changes resulting from increased ploidy.

Results

General description of miRNA libraries

In this study, we chose diploid RCC and *4n*AT ovaries to analyze gene expression and regulation coupled with species evolution driven by polyploidy. In each group, three individuals were used as biological replicates. Thus, six miRNA libraries were constructed in total, followed by high throughput sequencing. General bioinformatic analysis revealed that on average 15.521 M and 15.28 M raw reads were obtained in the two groups, respectively. After

filtering by sequence length (18–30 nt) and removing the low quality reads including those containing "N", 11.819 M and 12.414 M clean reads were obtained for the subsequent analysis. The detailed read numbers of each sample were shown in Additional file 1: Table S1.

The zebrafish genome was selected as the reference genome. Blasted against the GenBank and Rfam databases, ~12 % clean reads in diploid RCC and 14 % in *4n*AT were mapped to the genome with the standard of permitting no more than one base pair mismatch. Percentages of other types of RNAs (including rRNA, snRNA, snoRNA, and tRNA) and unmapped sequences were shown in Additional file 1: Table S2.

The length distribution of clean reads mainly concentrated in two regions, 21–23 nt and 26–29 nt (Fig. 1a). However, the 21–23 nt reads were most likely to be the miRNAs (Fig. 1b). Statistical analysis of the base sequences indicated that the first base is biased to "U", the tenth base biased to "A", and the probability of "U" at the second to forth bases was very low (Fig. 2).

Identification of distinct miRNAs

To identify the conserved and novel miRNAs, we used the miRDeep2 software for mapping the small RNA sequences to the reference genome. The results revealed that there were 441 mature miRNAs in diploid RCC, including 414 conserved and 30 novel miRNAs; 418 mature miRNAs were identified in *4n*AT, including 377 conserved and 41 novel miRNAs.

The composition and expression patterns of miRNAs are closely correlated with the biological characteristics of tissues. High throughput sequencing evaluated the frequency of different miRNAs in different samples. In both diploid RCC and *4n*AT ovaries, miR-143-3p exhibited the highest expression with approximately 145,643 reads on average; followed by miR-22-3p, miR-51-5p, miR-202-5p and let-7-5p with over 70,000 reads (Additional file 1: Table S3). In contrast, some miRNAs expressed less than five reads on average, such as miR-683, miR-4257, miR-4030-3p, miR-211-3p, miR-3243, and miR-3920 (Additional file 1: Table S3). This result revealed that the expression levels of different miRNAs were varied greatly.

Differential expressed miRNAs in 4nAT as compared to diploid RCC

To investigate the differential expression of miRNAs in the ovaries of different ploidy fishes, correlations among the biological replicates in each group were first analyzed. Triplicate diploid RCC were closely correlated with $R^2 > 0.92$. In the three *4n*AT replicates, samples four and five were closely correlated with $R^2 = 0.95$, while sample six was not closely correlated with the other two $R^2 = 0.78$ and 0.69, respectively (Additional file 1: Table S4, Additional file 2: Figure S1). This might be the

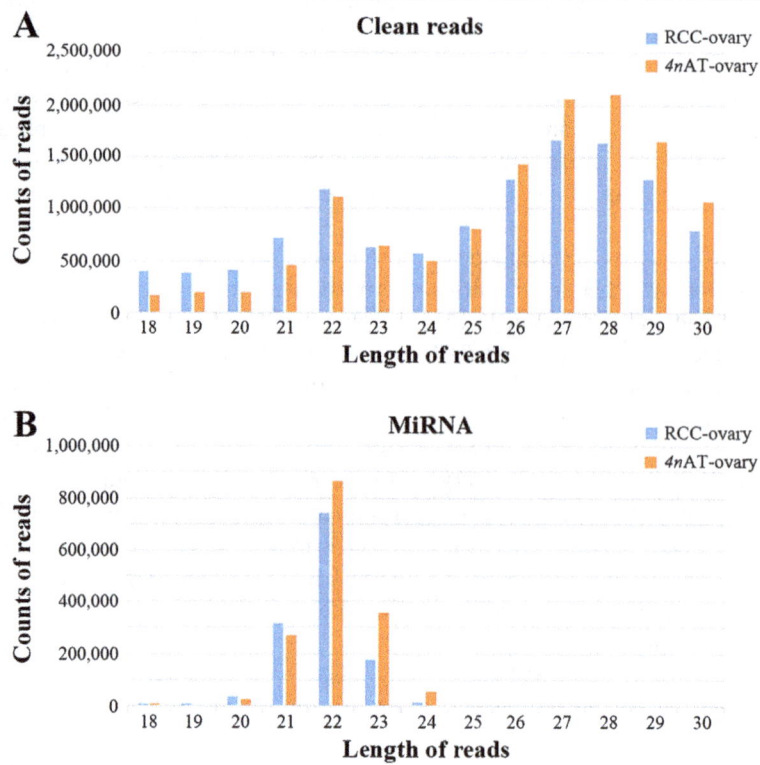

Fig. 1 Length distribution of clean reads (**a**) and miRNAs (**b**)

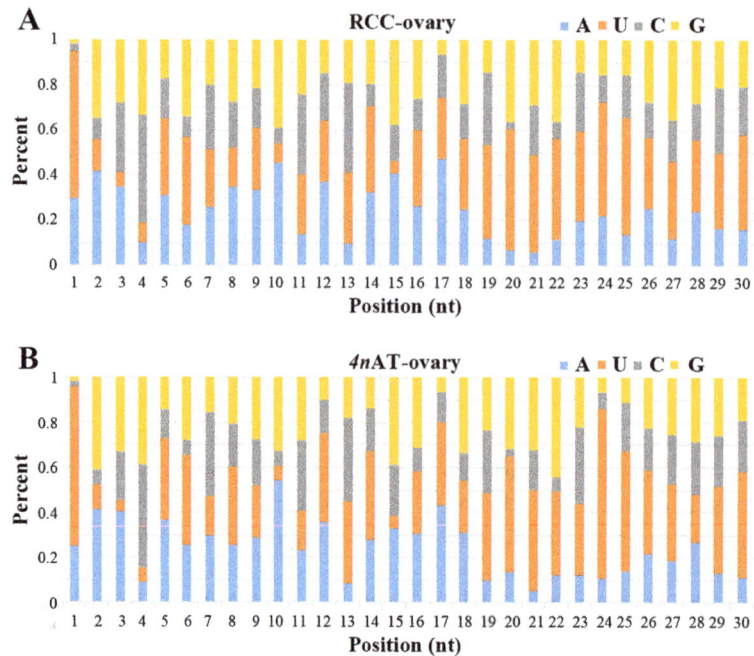

Fig. 2 Base bias of small RNA at each position in diploid RCC (**a**) and 4nAT (**b**)

result of individual differences. Thus, sample six of the *4n*AT group was excluded from the subsequent analysis.

The main objective of this study is to illustrate the differential expression of miRNAs in the ovaries of different ploidy fishes. The relative expression level of each miRNA could be calculated according to the deep sequencing results. The differential expressions of miRNAs were screened with the standard of false discovery rate (FDR) <0.01 and fold change (FC) >2. We found that 9 conserved miRNAs were down-regulated, whereas 30 conserved and 4 non-conserved miRNAs were up-regulated in *4n*AT (Fig. 3, Additional file 1: Table S5). Among the down-regulated miRNAs, miR-6843-3p exhibited the greatest fold change, followed by miR-1738 and miR-3526. Whereas among the up-regulated miRNAs, miR-2285q exhibited the greatest fold change, followed by miR-503-5p, miR-3183, miR-301a-5p, miR-2492-3p, miR-63 k-3p, miR-1421 m-5p, and miR-3086-5p (Additional file 1: Table S5).

To validate the relative differential expressions based on high throughput sequencing, five differentially expressed miRNAs (miR-6843-3p, miR-81b-p, miR-42-5p, miR-21-5p, and miR-2368-3p) were analyzed for relative expression levels by quantitative real-time PCR in the ovaries of different ploidy fishes (Fig. 4). The real-time PCR results were consistent with those of the high throughput sequencing.

Target gene prediction and function annotation

To illustrate the biological processes and physiological functions in which the differentially expressed miRNAs involved, the target gene sequences were predicted by the miRanda database. Thirty-nine unigenes targeted by up-regulated miRNAs were identified, but no gene information corresponding to down-regulated miRNAs were found (Additional file 1: Table S6). The function annotation and module classification of target genes were carried out by BLAST searches against the NR, Swiss-Prot, GO, COG, and KEGG databases.

Annotation with the COG database revealed that the differentially expressed target genes were mainly assigned to macromolecular transport and metabolism (e.g., amino acids, nucleotides, carbohydrates, and coenzymes), transcription, defense mechanisms, cytoskeleton production, and other general functions (Fig. 5).

GO analysis indicated that the differentially expressed target genes were distributed in the following three classifications (Fig. 6). As for the cellular component, there were more differentially expressed target genes in the intracellular parts (e.g., cell, organelles, membranes, macromolecular complex, and membrane-enclosed lumen) and extracellular matrix. Regarding molecular function, there were more differentially expressed target genes involved in in binding, catalytic activity, transporter activity, molecular transducer activity, receptor activity, and nucleic acid binding

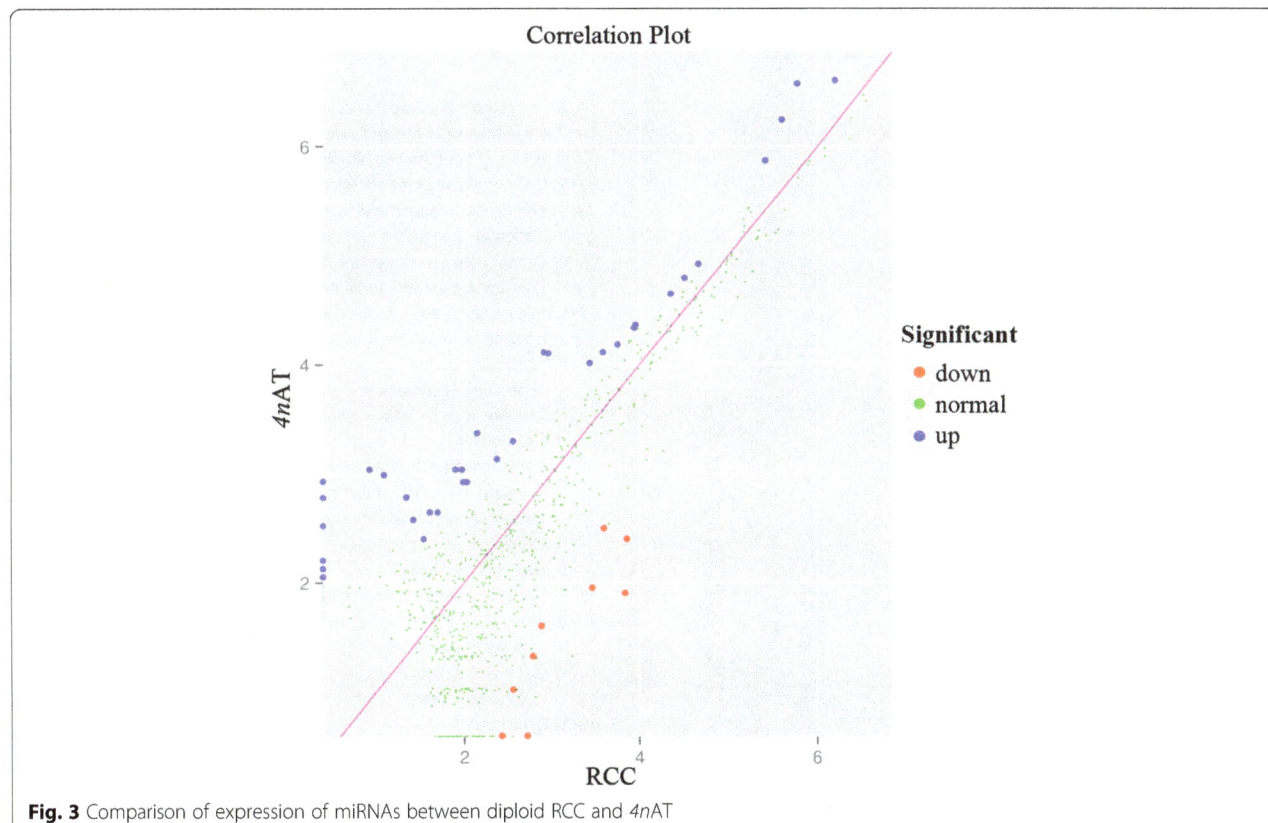

Fig. 3 Comparison of expression of miRNAs between diploid RCC and *4n*AT

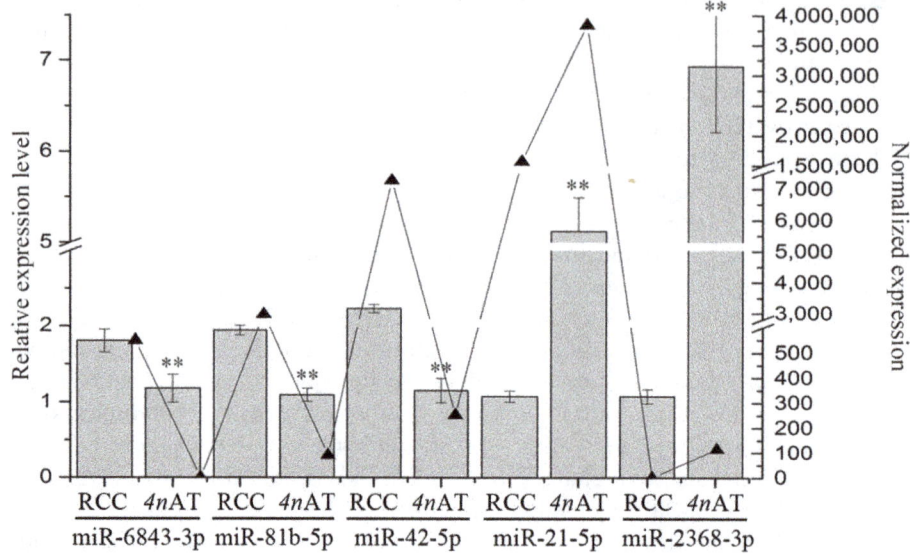

Fig. 4 The relationship between relative expression levels of miRNA validated by RT-PCR and normalized expression derived from sequencing. Column charts depicted the expression levels of miRNAs measured by RT-PCR. The triangles represented the expression after normalization from sequencing

transcription factor activity. As for the biological and cellular processes, biological regulation, metabolic processes, developmental processes, response to stimulus, multicellular organismal processes, localization, cellular component organization or biogenesis, signaling, establishment of localization, and immune system processes had higher numbers of differentially expressed target genes.

KEGG pathway analysis revealed that the differentially expressed target genes mainly focused on 11 signaling pathways, such as metabolic pathways (glycine, serine, threonine, galactose, starch, sucrose, amino sugar, nucleotide sugar, porphyrin, and chlorophyll metabolism), ABC transporters, one carbon pool by folate, pentose and glucoronate interconversions, lysosome, neuroactive

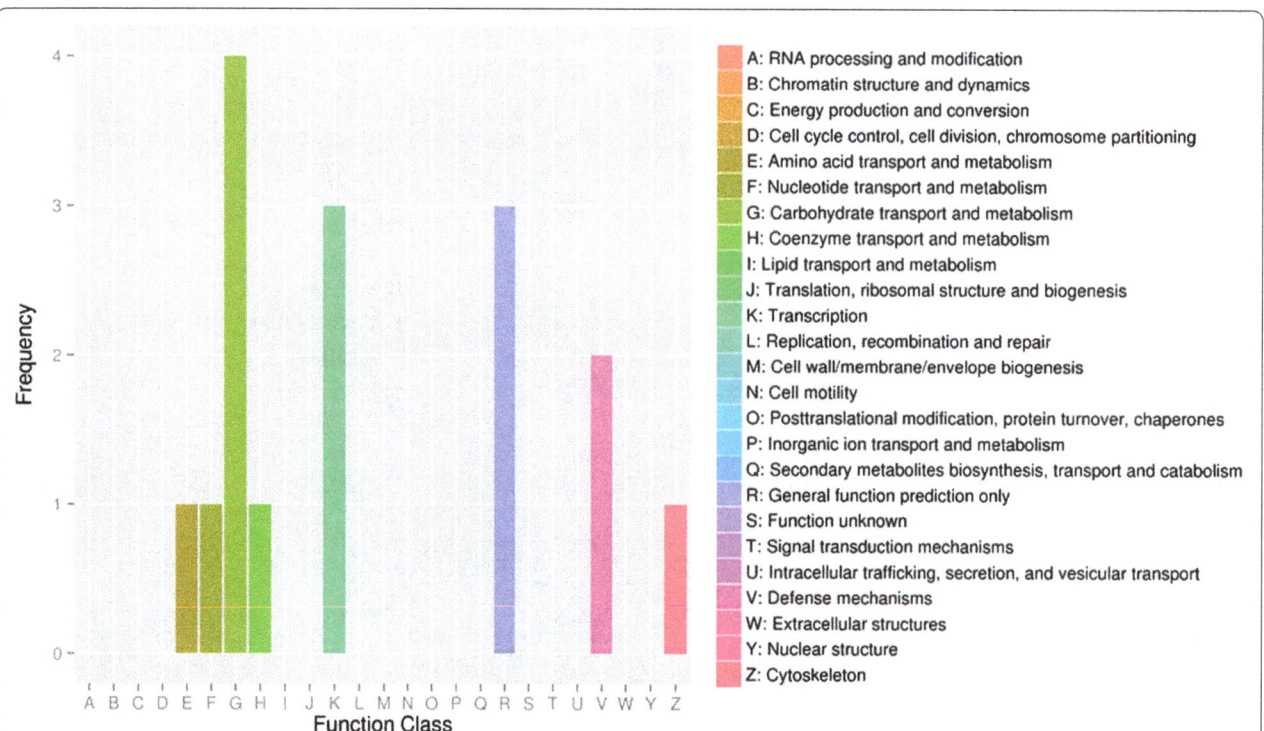

Fig. 5 Function annotation of genes targeted by differential expressed miRNAs with COG database

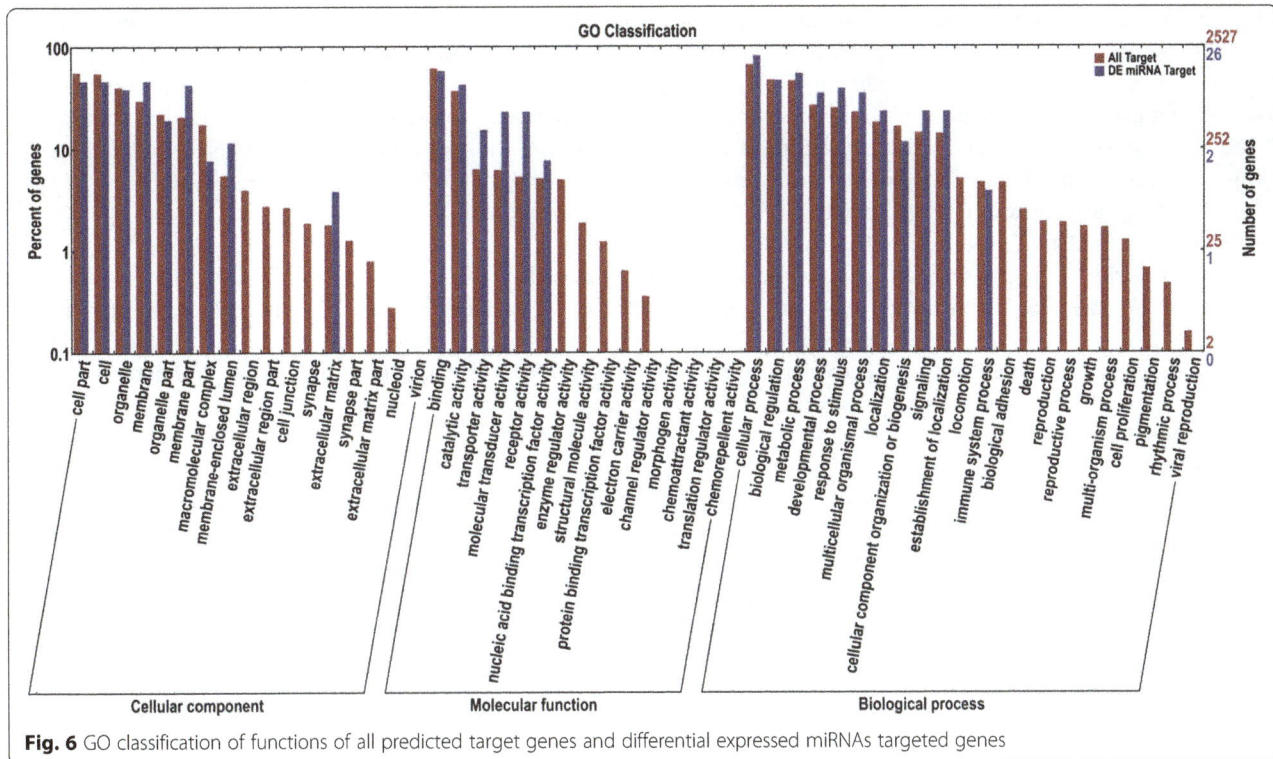

Fig. 6 GO classification of functions of all predicted target genes and differential expressed miRNAs targeted genes

ligand-receptor interaction, and the calcium signaling pathway (Additional file 1: Table S7).

Discussion

Polyploidization is one of important features of species evolution. Distant hybridization is a useful strategy for transferring the genome of one species to another, thus resulting in hybrid vigor (phenotype) and polyploidy formation (genotype) in the offspring [10]. Through combining distant hybridization, genetic breeding, and self-crossing, the allotetraploid hybrid line has been established from F_3 to F_{24}, these artificially cultured polyploid fish with stable inheritance are the first bisexual fertility allotetraploid fish, even vertebrate [8]. Molecular genetic marker analysis, including AFLP and ISSR, between 4nAT and their parents revealed that the genetic characteristics of the 4nAT population is stable and represent a bias to the maternal progenitor [27]. The formation of the 4nAT population was not only important for studying biological evolution, but was also the important step in producing sterile triploid fish by inter-ploid hybridization [10]. The sterile triploid fish exhibited strong disease resistance, fast growth rate, and reduced environmental risk for its sterility, which acted as a perfect aquatic species being widely cultured in China, and obtained great economic benefits [10]. However, the premise of the above described advantages is clarifying the reproductive characteristics of 4nAT at the morphological and molecular levels.

In this study, we compared the ovarian miRNA profiles of RCC and 4nAT, and found that the expressions of different miRNAs were varied from 5 to 150,000 reads. The highly expressed miRNAs, i.e., >70,000 reads on average, were closely correlated with ovary maturation. For example, miR-202-5p, which is abundant in both groups of samples, has been reported as a gonad-specific miRNA in frogs [28], Atlantic halibut [29], humans, mice [27], rats [30, 31], and rainbow trout [25]. Analysis of miRNA profiles in the complete ovary development and maturation process in rainbow trout revealed that miR-202-5p expression is much higher in the stage prior to germinal vesicle breakdown compared to any other stage during the process [25]. Other miRNAs, including miR-143a-3p, miR-miR-126-3p, and miR-101-3p, are also abundant in the final oocyte maturation in rainbow trout [25, 32]. Moreover, the second most abundant miRNA in this study, miR-22-3p, has a complex interplay with steroids (estrogen and androgen) [33]. Differential expression analysis revealed that the expression of the gonad development-related miRNAs listed above did not differ significantly between the two groups of samples. These results are consistent with the reproductive characteristics of 4nAT. Both of the fishes are sexually matured at the age of 1 year and have a similar ovary development process [8].

Non-additive gene expression patterns in polyploid plants or animals have been widely studied for response to genomic shock coupled with the balancing of more than two sets differential chromosomes in one nucleus [5, 7]. Global analysis of miRNA transcriptomes in *Squalius alburnoides*, comparing diploid and triploid hybrids, indicated that the diploid and triploid hybrids share most of their miRNA expression profiles, and the triploid hybrids tend to regulate gene expression to a diploid state [34]. Comparative analysis of miRNAs profiles in this research revealed that majority of miRNAs, obtained from sequencing, displayed no significant differential expression between diploid RCC and 4nAT. These results might be correlated with the non-additive gene expression patterns in polyploids. Among the differentially expressed miRNAs, the majority were up-regulated. For the miRNA is transcribed based on the DNA templates, this phenomenon might be caused by the double size of the DNA alleles in the allotetraploid hybrid. DNA methylation and miRNA are both important negative epigenetic regulators of gene expression. Xiao et al. [35] have reported that, compared with the parents, the genomic DNA in allotetraploid hybrids displayed hypermethylation levels. Studies in other polyploid species, such as *Brassica* [36], wheat [37], and *Arabidopsis* [38] demonstrated a similar situation in genomic DNA methylation. For instance, the differentially expressed miRNAs might be correlated with gene expression, which coupled with the change of phenotype, in the different ploidy animals.

Cacalier-Smith et al. found that genome size is generally correlated with cell volume and nuclear volume [39]. The surface area of the nuclear envelope available for nucleocytoplasmic transport of RNA is crucial for determining the cell growth rate and metabolism [39]. During evolution, the balance between DNA content and cell volume has been adjusted to allow reasonable growth rate. The cell geometry in polyploidies, large cells tending to have smaller surface area to volume ratios, would reduce the available nuclear envelope for nucleocytoplasmic transport, thus limiting the metabolism and growth rate. However, these effects of polyploidization at the phenotypic level are not exact and universal, which also depend on the environment [40]. The allotetraploid hybrids used in this study displayed a smaller body size and slower growth rate, although the volume of blood cell nuclei and the size of gametes were larger than in the diploid parental fish [8, 10]. Annotation of genes targeted by differentially expressed miRNAs with public databases indicated that the function of target genes mainly focused on macromolecular metabolism and transport, cellular skeleton, defense mechanisms, and transcription. Variations in morphological traits are governed by complex and well-balanced programs of gene activation and silencing [41]. The functions of target genes of differentially expression miRNAs were consistent with the morphological trait variations

in the allotetraploid hybrids and may have a molecular basis for these stable phenotypical changes caused by polyploidy. Moreover, global DNA methylation analysis in 4nAT also revealed that sequences involved in metabolism and disease resistance displayed DNA methylation variation, which is consistent with the miRNA profile results and phenotypic change [35].

Conclusions

This is the first study to describe the expression profiles and involvement of miRNAs in gene regulation coupled with polyploidy in ovary tissues. Molecular evidence revealed that the ovarian development process is similar in diploid RCC and 4nAT, whereas the differential expressions of miRNAs and mRNAs are mainly caused by ploidy change. Therefore, our results provide strong epigenetic evidences for the fertility of female allotetraploid fish and phenotypical changes caused by polyploidy.

Methods
Ethics statement

All experiments were approved by Animal Care Committee of Hunan Normal University and carried out according to the Care and Use of Agricultural Animals in Agricultural Research and Teaching Guidelines, approved by the Science and Technology Bureau of China. The fish were deeply anesthetized with 100 mg/L MS-222 (Sigma, St. Louis, MO, USA) prior to dissection.

Animals and sample collecting

RCC and 4nAT used in this experiment were cultured in ponds at the Protection Station of Polyploidy Fish, Hunan Normal University, and fed with artificial feed once per day. One-year-old fish were sampled in April. The ploidy of fish was determined by flow cytometry analysis as previously described. The fish were then anesthetized and dissected. The ovaries were removed from the fish, frozen in liquid nitrogen, and stored at −80 °C prior to RNA isolation. Three fish of each type were collected and analyzed.

Total RNA isolation

Total RNA isolation was performed with TRIzol® reagent (Invitrogen, Carlsbad, CA, USA) according to the manufacturer's instructions. In brief, 100 mg of ovary tissue was homogenized with 1 ml TRIzol with a glass homogenizer, and then centrifuged at $12,000 \times g$ for 10 min at 4 °C to discard the precipitate. Purification with phenol/chloroform (BBI, Shanghai, China) was then carried out followed by precipitation with isopropyl alcohol (BBI, Shanghai, China)
and dissolution in RNase-free water. The quality and concentration of total RNA were verified by 1.5 % agarose electrophoresis and detecting the ratio of 260/280 by

spectrophotometry (Eppendorf, Westbury, NY, USA). The integrity of total RNA was evaluated with an Agilent 2100 Bioanalyzer (Agilent Technologies, Palo Alto, Calif., Germany).

MiRNA library construction and sequencing

A small RNA library was constructed with a TruSeq Small RNA Sample Preparation Kit (Illumina, San Diego, CA, USA) according to the manufacturer's instructions based on the special structural characteristics of small RNAs, with a phosphate group in the 5' left and a hydroxyl in the 3' left. Using total RNA as the initial sample, the small RNA was ligated to a 3' adapter at first, then a 5' adapter following the addition of the reverse transcription primer to create cDNA constructs. After PCR amplification with forward and reverse primers complementary to the 5' and 3' adapters, the products were purified with 6 % PAGE and selected with the proper size to construct the small RNA library. The effective concentration of the small RNA library was verified by Qubit 2.0 (Invitrogen, Carlsbad, CA, USA) and Q-PCR. The insert size of the library was evaluated with an Agilent 2100 Bioanalyzer (Agilent Technologies, Palo Alto, Calif., Germany). The qualified small RNA libraries were sequenced with the HiSeq 2500 SE50 system (Illumina, San Diego, CA, USA).

Bioinformatic analysis

Clean reads were obtained by filtering low quality reads and selecting 18–30 nt reads from the raw reads for the further analysis. The clean reads were searched against the GenBank, Rfam, and ZFIN databases to discard the annotated non-coding RNA reads and find the reads mapped to the reference genome for miRNA identification. The remaining sequences were subsequently used to identify the conserved and novel miRNAs with miR-Deep2 software [42]. To analyze the differential expression of miRNAs in RCC and $4n$AT ovaries, miRNA reads were then normalized to obtain the miRNA expression using the DEGseq software (http://www.bioconductor.org/packages/release/bioc/html/DEGseq.html). To understand the function of differentially expressed miRNAs, potential target genes for miRNAs were predicted by the miRanda database, then annotated and the functional modules identified by BLAST searches against the NR, Swiss-Prot, GO, COG, and KEGG databases. The correlation analysis of biological replicates was assessed with the Pearson correlation coefficient (R^2) with pairwise comparisons >0.92.

Quantitative real-time PCR

To validate the differential expression of miRNA identified from the sequencing results, the relative expressions of miRNAs were evaluated by quantifying the miRNA

Table 1 Specific primers used in Q-PCR

Primer	Sequence
miR-6843-3p-RT	CTCAACTGGTGTCGTGGAGTCGGCAATTCAGTTGAG CAGCAT
miR-6843-3p-F	ACACTCCAGCTGGGTTGGTCTCTGTA
miR-81b-5p-RT	CTCAACTGGTGTCGTGGAGTCGGCAATTCAGTTGAG ACTGTGA
miR-81b-5p-F	ACACTCCAGCTGGGCCGGGTGTGTGTT
miR-42-5p-RT	CTCAACTGGTGTCGTGGAGTCGGCAATTCAGTTGAG ACTGTGA
miR-42-5p-F	ACACTCCAGCTGGGCTGGGTGTGTGCT
miR-21-5p-RT	CTCAACTGGTGTCGTGGAGTCGGCAATTCAGTTGAGG CCAACAC
miR-21-5p-F	ACACTCCAGCTGGGTAGCTTATCAGACT
miR-2368-3p-RT	CTCAACTGGTGTCGTGGAGTCGGCAATTCAGTTGAGA AAAAGCC
miR-2368-3p-F	ACACTCCAGCTGGGGCTGTCAGAAAGGG
Actin-RTF	TCTACAACGAGCTGCGTGTTG
Actin-RTR	CCTGTTGGCTTTGGGATTGA

stem-loop. The total RNA was used to obtain cDNA with a reverse transcription kit (Invitrogen, Carlsbad, CA, USA) with specific primers for each miRNA. The PCR reaction was performed using ABI SYBRGreen PCR master mix (Applied Biosystems, Foster City, CA, USA) on the ABI 7500 PCR system (Applied Biosystems, Foster City, CA, USA) with specific primers (Table 1). β-actin was used as the internal control. For each sample, three independent repetitions were tested.

Statistical analysis

The quantitative real-time PCR data were expressed as means ± SD and the significant differences were confirmed by t-tests in SPSS 13.0 software. P-values < 0.05 and < 0.01 were taken to indicate a statistical difference.

Availability of supporting data

All the short read sequences were available in the NCBI Sequence Read Archive (http://www.ncbi.nlm.nih.gov/sra) with a study number SRX1233598 and SRX1239635.

Abbreviations
RCC: Red crucian carp; CC: Common carp; 4nAT: The allotetraploid hybrids.

Competing interests
The authors declare that they have no competing interests.

Authors' contributions
SJL, RZ and MT designed the experiments; RZ, YHW and CZ performed the experiments; SJL and RZ performed the statistical analysis and wrote the manuscript. All authors read and approved the final manuscript.

Acknowledgements
This work was supported by the China Postdoctoral Science Foundation (Grant No. 2014 M560645 and No. 2015 T80874), the National Natural Science Foundation of China (Grant No. 31430088), Major international cooperation projects of the National Natural Science Foundation of China (Grant No. 31210103918), the Doctoral Fund of Ministry of Education of China (Grant No. 20114306130001), the Cooperative Innovation Center of Engineering and New Products for Developmental Biology of Hunan Province (20134486), the construct program of the key discipline in Hunan province and China.

References

1. Otto SP, Whitton J. Polyploid incidence and evolution. Annu Rev Genet. 2000;34(1):401–37.
2. Van de Peer Y, Maere S, Meyer A. The evolutionary significance of ancient genome duplications. Nat Rev Genet. 2009;10(10):725–32.
3. Soltis DE, Soltis PS. Polyploidy: recurrent formation and genome evolution. Trends Ecol Evol. 1999;14(9):348–52.
4. Song C, Liu S, Xiao J, He W, Zhou Y, Qin Q, et al. Polyploid organisms. Sci China Life Sci. 2012;55(4):301–11.
5. Comai L. The advantages and disadvantages of being polyploid. Nat Rev Genet. 2005;6(11):836–46.
6. Ozkan H, Levy AA, Feldman M. Allopolyploidy-induced rapid genome evolution in the wheat (Aegilops-Triticum) group. Plant Cell Online. 2001;13(8):1735–47.
7. Yoo M-J, Liu X, Pires JC, Soltis PS, Soltis DE. Nonadditive gene expression in polyploids. Annu Rev Genet. 2014;48:485–517.
8. Liu S, Liu Y, Zhou G, Zhang X, Luo C, Feng H, et al. The formation of tetraploid stocks of red crucian carp × common carp hybrids as an effect of interspecific hybridization. Aquaculture. 2001;192(2):171–86.
9. Shao-Jun L, Yuan-Dong S, Kai-Kun L, Yun L. Evidence of different ploidy eggs produced by diploid F 2 hybrids of Carassius auratus (♀) × Cyprinus carpio (♂). Acta Genet Sin. 2006;33(4):304–11.
10. Liu S. Distant hybridization leads to different ploidy fishes. Sci China Life Sci. 2010;53(4):416–25.
11. Sun Y-D, Liu S-J, Zhang C, Li J-Z, Huang W-R, Zhang J, et al. The chromosome number and gonadal structure of F9-F11 allotetraploid crucian-carp. Yi Chuan Xue Bao. 2003;30(5):414–8.
12. Zhang C, He X, Liu S, Sun Y, Liu Y. Chromosome pairing in meiosis I in allotetraploid hybrids and allotriploid crucian carp. Acta Zool Sin. 2004;51(1):89–94.
13. Long Y, Liu S, Huang W, Zhang J, Sun Y, Zhang C, et al. Comparative studies on histological and ultra-structure of the pituitary of different ploidy level fishes. Sci China C Life Sci. 2006;49(5):446–53.
14. Long Y, Tao M, Liu S, Zhong H, Chen L, Tao S, et al. Differential expression of Gnrh2, Gthβ, and Gthr genes in sterile triploids and fertile tetraploids. Cell Tissue Res. 2009;338(1):151–9.
15. Long Y, Zhong H, Liu S, Tao M, Chen L, Xiao J, et al. Molecular characterization and genetic analysis of Gnrh2 and Gthβ in different ploidy level fishes. Proc Natl Acad Sci U S A. 2009;19(11):1569–79.
16. Lee RC, Feinbaum RL, Ambros V. The C. elegans heterochronic gene lin-4 encodes small RNAs with antisense complementarity to lin-14. Cell. 1993;75(5):843–54.
17. Wightman B, Ha I, Ruvkun G. Posttranscriptional regulation of the heterochronic gene lin-14 by lin-4 mediates temporal pattern formation in C. elegans. Cell. 1993;75(5):855–62.
18. Lee Y, Kim M, Han J, Yeom KH, Lee S, Baek SH, et al. MicroRNA genes are transcribed by RNA polymerase II. EMBO J. 2004;23(20):4051–60.
19. Lee Y, Ahn C, Han J, Choi H, Kim J, Yim J, et al. The nuclear RNase III Drosha initiates microRNA processing. Nature. 2003;425(6956):415–9.
20. Lee Y, Jeon K, Lee JT, Kim S, Kim VN. MicroRNA maturation: stepwise processing and subcellular localization. EMBO J. 2002;21(17):4663–70.
21. Krol J, Loedige I, Filipowicz W. The widespread regulation of microRNA biogenesis, function and decay. Nat Rev Genet. 2010;11(9):597–610.
22. Hong X, Luense LJ, McGinnis LK, Nothnick WB, Christenson LK. Dicer1 is essential for female fertility and normal development of the female reproductive system. Endocrinology. 2008;149(12):6207–12.
23. Gonzalez G, Behringer RR. Dicer is required for female reproductive tract development and fertility in the mouse. Mol Reprod Dev. 2009;76(7):678–88.
24. Ahn HW, Morin RD, Zhao H, Harris RA, Coarfa C, Chen Z-J, et al. MicroRNA transcriptome in the newborn mouse ovaries determined by massive parallel sequencing. Mol Hum Reprod. 2010;16(7):463–71.
25. Juanchich A, Le Cam A, Montfort J, Guiguen Y, Bobe J. Identification of differentially expressed miRNAs and their potential targets during fish ovarian development. Biol Reprod. 2013;88(5):128.
26. Xiao J, Zhong H, Zhou Y, Yu F, Gao Y, Luo Y, et al. Identification and characterization of microRNAs in ovary and testis of Nile tilapia (Oreochromis niloticus) by using solexa sequencing technology. PLoS One. 2014;9(1):e86821.
27. Liu L, Yan J, Liu S, Liu D, You C, Zhong H, et al. Evolutionary analysis of allotetraploid hybrids of red crucian carp × common carp, based on ISSR, AFLP molecular markers and cloning of cyclins genes. Chin Sci Bull. 2009;54(16):2849–61.
28. Armisen J, Gilchrist MJ, Wilczynska A, Standart N, Miska EA. Abundant and dynamically expressed miRNAs, piRNAs, and other small RNAs in the vertebrate Xenopus tropicalis. Genome Res. 2009;19(10):1766–75.
29. Schmid M. Sex-biased miRNA expression in Atlantic halibut (Hippoglossus hippoglossus) brain and gonads. Sex Dev. 2012;6:257–66.
30. Landgraf P, Rusu M, Sheridan R, Sewer A, Iovino N, Aravin A, et al. A mammalian microRNA expression atlas based on small RNA library sequencing. Cell. 2007;129(7):1401–14.
31. Ro S, Song R, Park C, Zheng H, Sanders KM, Yan W. Cloning and expression profiling of small RNAs expressed in the mouse ovary. RNA. 2007;13(12):2366–80.
32. Ma H, Hostuttler M, Wei H, Rexroad III CE, Yao J. Characterization of the rainbow trout egg microRNA transcriptome. PLoS One. 2012;7(6):e39649.
33. Cochrane DR, Cittelly DM, Richer JK. Steroid receptors and microRNAs: relationships revealed. Steroids. 2011;76(1):1–10.
34. Inácio A, Pinho J, Pereira PM, Comai L, Coelho MM. Global analysis of the small RNA transcriptome in different ploidies and genomic combinations of a vertebrate complex–the Squalius alburnoides. PLoS One. 2012;7(7):e41158.
35. Xiao J, Song C, Liu S, Tao M, Hu J, Wang J, et al. DNA methylation analysis of allotetraploid hybrids of red crucian carp (Carassius auratus red var.) and common carp (Cyprinus carpio L.). PLoS One. 2013;8(2):e56409.
36. Chen ZJ, Pikaard CS. Epigenetic silencing of RNA polymerase I transcription: a role for DNA methylation and histone modification in nucleolar dominance. Genes Dev. 1997;11(16):2124–36.
37. Kashkush K, Feldman M, Levy AA. Gene loss, silencing and activation in a newly synthesized wheat allotetraploid. Genetics. 2002;160(4):1651–9.
38. Madlung A, Masuelli RW, Watson B, Reynolds SH, Davison J, Comai L. Remodeling of DNA methylation and phenotypic and transcriptional changes in synthetic Arabidopsis allotetraploids. Plant Physiol. 2002;129(2):733–46.
39. Cavalier-Smith T. Nuclear volume control by nucleoskeletal DNA, selection for cell volume and cell growth rate, and the solution of the DNA C-value paradox. J Cell Sci. 1978;34(1):247–78.
40. Otto SP. The evolutionary consequences of polyploidy. Cell. 2007;131(3):452–62.
41. Huminiecki L, Conant GC. Polyploidy and the evolution of complex traits. Int J Evol Biol. 2012;2012:292068.
42. Friedländer MR, Mackowiak SD, Li N, Chen W, Rajewsky N. miRDeep2 accurately identifies known and hundreds of novel microRNA genes in seven animal clades. Nucleic Acids Res. 2012;40(1):37–52.

5

Spatially differentiated expression of quadruplicated green-sensitive RH2 opsin genes in zebrafish is determined by proximal regulatory regions and gene order to the locus control region

Taro Tsujimura[1,2]* (ID), Ryoko Masuda[1], Ryuichi Ashino[1] and Shoji Kawamura[1]*

Abstract

Background: Fish are remarkably diverse in repertoires of visual opsins by gene duplications. Differentiation of their spatiotemporal expression patterns and absorption spectra enables fine-tuning of feature detection in spectrally distinct regions of the visual field during ontogeny. Zebrafish have quadruplicated green-sensitive (RH2) opsin genes in tandem (*RH2-1, −2, −3, −4*), which are expressed in the short member of the double cones (SDC). The shortest wavelength RH2 subtype (*RH2-1*) is expressed in the central to dorsal area of the adult retina. The second shortest wave subtype (*RH2-2*) is expressed overlapping with *RH2-1* but extending outside of it. The second longest wave subtype (*RH2-3*) is expressed surrounding the *RH2–2* area, and the longest wave subtype (*RH2-4*) is expressed outside of the *RH2-3* area broadly occupying the ventral area. Expression of the four RH2 genes in SDC requires a single enhancer (RH2-LCR), but the mechanism of their spatial differentiation remains elusive.

Results: Functional comparison of the RH2-LCR with its counterpart in medaka revealed that the regulatory role of the RH2-LCR in SDC-specific expression is evolutionarily conserved. By combining the RH2-LCR and the proximal upstream region of each RH2 gene with fluorescent protein reporters, we show that the RH2-LCR and the *RH2-3* proximal regulatory region confer no spatial selectivity of expression in the retina. But those of *RH2-1, −2* and *−4* are capable of inducing spatial differentiation of expression. Furthermore, by analyzing transgenic fish with a series of arrays consisting of the RH2-LCR and multiple upstream regions of the RH2 genes in different orders, we show that a gene expression pattern related to an upstream region is greatly influenced by another flanking upstream region in a relative position-dependent manner.

Conclusions: The zebrafish RH2 genes except *RH2-3* acquired differential *cis*-elements in the proximal upstream regions to specify the differential expression patterns. The input from these proximal elements collectively dictates the actual gene expression pattern of the locus, context-dependently. Importantly, competition for the RH2-LCR activity among the replicates is critical in this collective regulation, facilitating differentiation of expression among them. This combination of specificity and generality enables seemingly complicated spatial differentiation of duplicated opsin genes characteristic in fish.

Keywords: Zebrafish, opsin, RH2, Gene duplication, Subfunctionalization, Expression, Gene regulation, RH2-LCR, Gene order

* Correspondence: t-tsujimura@umin.ac.jp; kawamura@k.u-tokyo.ac.jp
[1]Department of Integrated Biosciences, Graduate School of Frontier Sciences, the University of Tokyo, Kashiwanoha 5-1-5, Kashiwa 277-8562 Chiba, Japan
Full list of author information is available at the end of the article

Background

In vertebrates, visual opsins are classified into five phylogenetic types that originated in their common ancestor. RH1 is the rod opsin or rhodopsin responsible for dim-light vision. The other four are cone opsins for color vision: they are SWS1, SWS2, RH2 and M/LWS, and mainly sensitive to UV, blue, green and red light, respectively [1]. These different types of opsin genes are generally expressed in distinct types of photoreceptor cells that are arrayed in the retina to assure color discrimination [2].

Among vertebrates, fish have experienced gain and loss of visual opsins repeatedly by gene duplications and deletions. For example, zebrafish (*Danio rerio*) have ten visual opsin genes: a tandem array of spectrally distinct two LWS genes (*LWS-1* and *LWS-2*) and that of four RH2 genes (*RH2-1*, *RH2-2*, *RH2-3* and *RH2-4*), single-copy SWS1 and SWS2 genes, and two RH1 genes [3, 4]. Medaka (*Oryzias latipes*) [5] and cichlid (Nile tilapia, *Oreochromis niloticus*) [6, 7] have nine and eight opsin genes, respectively.

Newly replicated daughter genes are identical; hence typically only one is likely to be preserved and the others become pseudogenes due to functional redundancy. Nevertheless, if the replicates undergo a process of subfunctionalization, i.e. taking a different part of original function which the ancestral gene had, both genes are more likely to be preserved [8]. The subtype opsin genes in fish indeed achieved the subfunctionalization by differential spatial and temporal expression patterns within the retina as well as divergent absorption spectra of the encoding photopigments [3, 5, 9–14].

In the case of zebrafish, all the four subtypes of RH2 are expressed in the short (or accessory) member of double cones (SDCs) and both two LWS subtypes are expressed in the long (or principal) member of double cones (LDCs) [13, 15, 16]. However, they are differentiated in the expression pattern in the retina [13]. Fish eyes continue to grow through the lifetime by adding new cells to the peripheral zones [17]. Concomitantly, early-expressed subtypes are located centrally in the adult retina. The shortest wavelength RH2 subtype (*RH2-1*) is expressed earliest and in the central to the dorsal area of the adult retina. The second shortest wave subtype (*RH2-2*) is expressed subsequently overlapping with *RH2-1* but extending outside of it. The longer wave RH2 subtype (*RH2-3*) is expressed later and in a region surrounding the *RH2–2* area, and the longest wave RH2 subtype (*RH2-4*) is also expressed later and outside of the *RH2-3* area, broadly occupying the ventral area. Similarly, the shorter wave LWS subtype (*LWS-2*) is expressed earlier and in the central-to-dorsal area in the adult retina, and the longer wave LWS subtype (*LWS-1*) is expressed later in the development and confined

peripherally with largely occupying the ventro-nasal area of the adult retina [13]. Thus, in zebrafish, each replicated opsin gene is expressed in a portion of the expression area of their hypothetical ancestral gene, which is presumed to have been expressed throughout the retina, while maintaining the cell-type specificity. As a result, retinal regions that detect spectrally distinct portions of the visual field in the water acquired different spectral sensitivity and presumably different color vision [13].

Medaka and cichlids also show differential expression of tandemly-arrayed opsin genes, which were replicated independently from zebrafish [5, 9–12]. Hence, fish appear to have established regulatory mechanisms for the differential expression of subtype opsins repeatedly. We wondered how fish could accomplish such seemingly complicated regulation for the differential expression in parallel.

We previously showed that, in zebrafish, a single enhancer, named RH2-LCR, was located at the 15-kb upstream of the RH2 gene cluster and was necessary and sufficient for the SDC-specific expression of all the four RH2 genes [18]. In larvae, it was shown that the relative distance from the RH2-LCR to the genes affects their expression levels whereas *RH2-4* is relatively insensitive to the distance effect [18]. We also showed that the two LWS genes of zebrafish are regulated by a single enhancer (LAR) [19]. In the case of LWS genes, the closer gene to LAR, *LWS-1*, is expressed later and more peripherally than *LWS-2* [13]. The proximal upstream region of *LWS-1* appears to have an active role in specifying the spatial expression to the peripheral retina while that of *LWS-2* allows expression throughout the retina in conjunction with LAR. In the presence of the *LWS-1* upstream region, however, the gene expression from the *LWS-2* promoter is excluded from the area where *LWS-1* is expressed [19]. From these observations we hypothesize that the tandemly arrayed genes compete for their interaction with the RH2-LCR/LAR and that the relative distance influences the likeliness, e.g. a closer gene can have a greater chance to interact with it, while a proximal regulatory region can alter this stereotype pattern [19].

In this study, we test the hypothesis for the zebrafish RH2 genes. We first investigate the evolutionary and functional conservation of the RH2-LCR. We further make a series of transgenic fish carrying reporter genes under variously reordered proximal upstream regions from the RH2 genes together with the RH2-LCR.

Results

The RH2-LCR regulatory function is evolutionarily conserved

The genomic sequence of the RH2 locus was compared between two superorders Ostariophysi (e.g. zebrafish) and Acanthopterygii (e.g. medaka and *Tetraodon*). Although the RH2 gene duplications are known to have occurred independently in the two superorders [3, 5], a portion of the RH2-LCR was highly conserved among them (Fig. 1a, b). This is consistent with the RH2 opsin genes being expressed in the SDC in both zebrafish and medaka [13, 15, 16, 20].

To test if the sequence conservation reflects its functional importance, we introduced five deletions to the RH2-LCR in the *RH2-1*/GFP-PAC. The *RH2-1*/GFP-PAC is a PAC-vectored clone, modified from the RH2-PAC containing all the four RH2 genes to replace the exon 1 of *RH2-1* with a GFP (green fluorescent protein) reporter gene [18]. The reporter expression in the zebrafish retina was lost only when the central 100 bp of the RH2-LCR was deleted, which corresponds to the region of the highest sequence similarity between species (Additional file 1: Figure S1).

To further test its functional conservation, we introduced the orthologous sequence of medaka to zebrafish for its regulatory activity. We used a BAC-vectored clone from medaka [5] encompassing the orthologous RH2-LCR and all the medaka RH2 genes (*RH2-A, –B, –C*) in which the exon 1 of *RH2-A* was replaced with GFP. We observed GFP expression in the SDCs in zebrafish (Fig. 1c, d). The removal of the LCR abolished the GFP expression in the retina (Fig. 1c). Consistently, when the medaka RH2-LCR was injected together with the 3-kb upstream region of *RH2-A* conjugated to a GFP reporter, the GFP expression was also observed in the SDCs in zebrafish (Fig. 1e). Thus, the RH2-LCR is an evolutionarily conserved regulatory region that has been present prior to the independent gene duplication events in zebrafish and medaka lineages, and has maintained the regulatory function to drive SDC-specific gene expression.

When we coupled the RH2-LCR with the proximal upstream region of *keratin 8* as a basal promoter, which presumably has no spatial specificity of expression in the retina [18, 21, 22], the GFP reporter was expressed in all SDCs throughout the retina of the transgenic fish (Fig. 2a). Thus, the RH2-LCR confers no spatial selectivity in the retina on the expression regulation, reflecting its presumed ancestral state prior to gene duplications.

Roles of proximal regulatory regions in area specificity

We further modified the RH2-PAC to create the *RH2-1*/GFP-*RH2-2*/RFP-PAC and the *RH2-3*/GFP-*RH2-4*/RFP-PAC in which the exon 1 of *RH2-1* or *RH2-3* was replaced with GFP and that of *RH2-2* or *RH2-4* was replaced with RFP (red fluorescent protein), respectively, so that we could visualize the expression pattern of two RH2 genes simultaneously (Additional file 1: Figures S2 and S3). We confirmed that these transgenic zebrafish lines indeed recapitulated the corresponding RH2 genes'

expression as we previously showed using the single-gene replacement constructs (*RH2-1*/GFP-PAC, *RH2-2*/GFP-PAC, *RH2-3*/GFP-PAC, and *RH2-4*/GFP-PAC) [18]. This further confirmed that the RH2-PAC contains the complete set of *cis*-regulatory regions.

We next established transgenic zebrafish lines using only the upstream regions of *RH2-1*, *RH2-2*, *RH2-3* and *RH2-4* coupled respectively with the RH2-LCR and the GFP reporter (Fig. 2b): the upstream sequence of *RH2-1* (4.2 kb) and the entire intergenic regions upstream of *RH2-2* (3.0 kb), of *RH2-3* (2.6 kb), and of *RH2-4* (7.4 kb). The transgenic lines showed that the *RH2-1* and *RH2-2* constructs drove GFP expression confined to the central-to-dorsal area of the retina and the *RH2-4* construct to the ventral area, largely recapitulating the respective expression pattern of these genes (Fig. 2c, see also Additional file 1: Figures S2 and S3) [13]. By contrast, the *RH2-3* construct drove GFP expression in all SDCs of the retina, markedly different from its native narrow expression pattern surrounding the native *RH2-2* expression area (Fig. 2c, see also Additional file 1: Figure S3C, D) [13]. Taken together, the proximal upstream regions of *RH2-1, –2* and *–4* are capable of specifying expression area in the retina whereas that of *RH2-3* is not. It was also noted that the transgenic fish with the *RH2-3* construct showed considerable ectopic expression of GFP in non-SDC photoreceptors throughout the retina, while those with the *RH2-1, –2* and *–4* constructs did not (Fig. 2d).

Effect of relative distance to the RH2-LCR among proximal regulatory regions on the spatial expression pattern in the retina

We then combined two genes' proximal regulatory regions, each coupled with either GFP or RFP reporter gene, under the RH2-LCR (Fig. 3), and established transgenic zebrafish lines with these constructs. This was to test the previously inferred effect of relative distance to the RH2-LCR among genes [18].

When the *RH2-1* upstream region was placed closer to the RH2-LCR than the other and was combined with either of the *RH2-3*, *RH2-4* or *keratin 8* upstream region (Fig. 3a, b, and Additional file 1: Figure S4A, B), GFP expression driven by the *RH2-1* upstream region was similar to its native pattern, confined in the central-to-dorsal region of the retina. On the other hand, expression pattern of the second gene, represented by RFP expression, varied. The *RH2-3* upstream region drove RFP expression in a narrow area outside the *RH2-1*/GFP area in one transgenic line, similar to its native pattern (Fig. 3a). In another transgenic line with the same construct, the RFP expression was evident in the central-to-dorsal area but weaker in the ventral area (Additional file 1: Figure S4A). Interestingly, the RFP and GFP expression showed

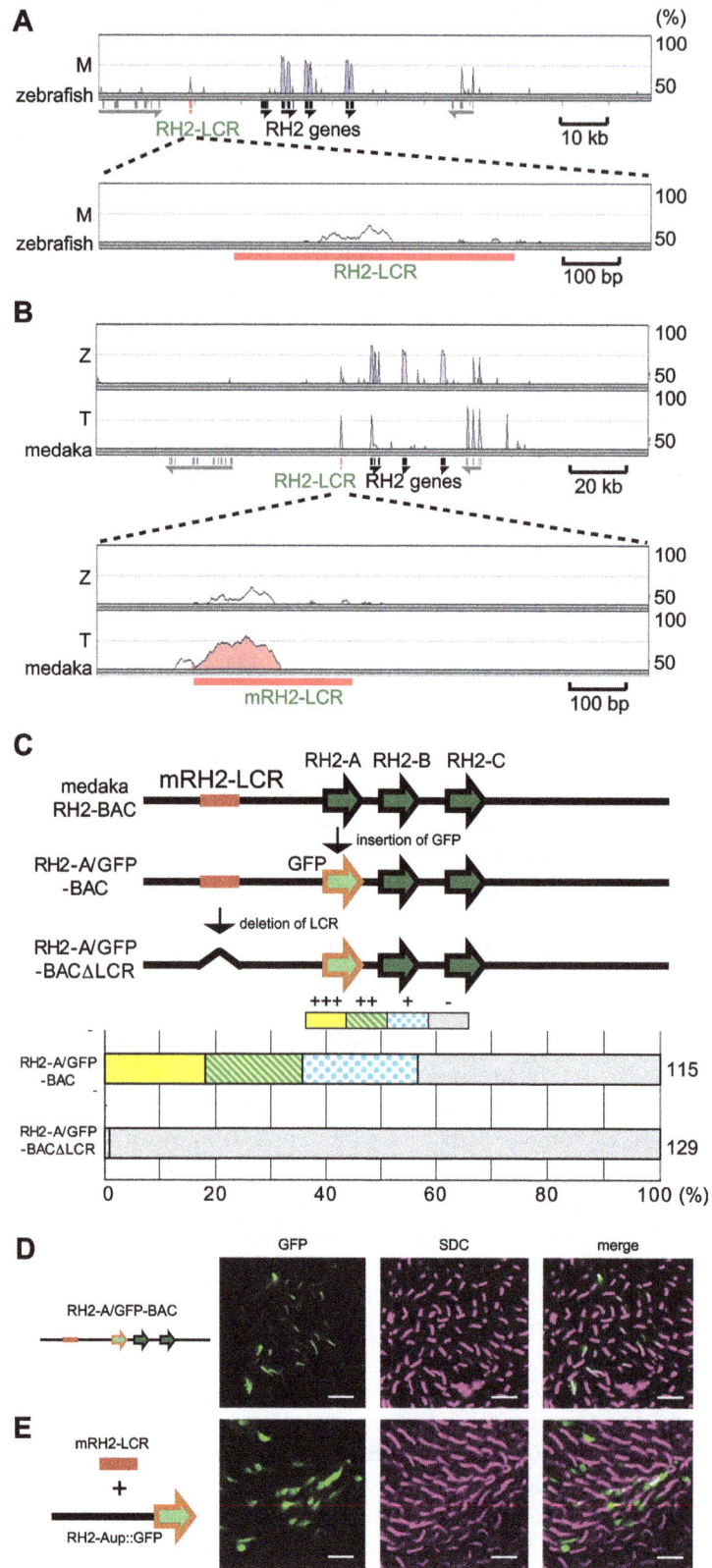

Fig. 1 (See legend on next page.)

(See figure on previous page.)
Fig. 1 The RH2-LCR is an evolutionarily conserved enhancer in teleosts. **a** Sequence comparison of the RH2 locus between zebrafish and medaka (M) is shown using mVISTA program. The baseline corresponds to the zebrafish RH2 region. An enlarged illustration around the RH2-LCR is also shown. Black and gray bars under the chart are the exons of the RH2 genes and the other genes, respectively. The red bar indicates the RH2-LCR. The sequence homology is indicated to the right of the chart. **b** Sequence comparisons of the RH2 locus of medaka with zebrafish (Z) and *Tetraodon* (T) are shown as in (**a**). Homology regions colored with gray correspond to coding regions of genes and those colored with pink correspond to conserved non-coding sequences. Note that *Tetraodon* has higher homology at the RH2-LCR than zebrafish, reflecting their closer relation with medaka. **c** Construction of the medaka *RH2-A*/GFP-BAC clones (upper panel) and expression levels at 5 dpf of the GFP reporter in zebrafish injected with the BAC clones above (lower panel). The histogram shows the percentage of eyes graded into four levels (+++, ++, + and -) according to the number of GFP-expressing cells in the retina. The names to the left of the histogram indicate the constructs injected. The numbers to the right of the histogram show the total number of eyes examined. **d, e** Whole mount retinas of 7-dpf zebrafish embryos injected with the *RH2-A*/GFP-BAC (**d**) and with mixture of the medaka RH2-LCR fragment and the GFP reporter under the 3-kb upstream region of *RH2-A*. GFP fluorescent signals appear as green (*left*) and immunostaining signals of SDCs by the anti-RH2 antibody appear as magenta (*middle*). Overlap of the two signals appears as white. The right panels are the overlays of the left and middle panels. Scale bars = 10 μm

reciprocal gradients in intensity, though expression was not totally exclusive (Additional file 1: Figure S4A). Thus, in these two lines, the presence of the *RH2-1* upstream region in front seems to restrict the induction by the *RH2-3* upstream region to the cells in the central-to-dorsal retina that do not or only weakly express GFP. It is of note that the placement of the *RH2-3* upstream region at the second position appears to enhance the expression of the first gene (Fig. 3a). The RFP expression by the *RH2-4* upstream region was similar to its native pattern confined to the ventral area of the retina (Fig. 3b). The RFP expression by the *keratin 8* upstream region was not clearly detected (Additional file 1: Figure S4B).

When the *RH2-4* upstream region was placed after *keratin 8* upstream region, the reporter expression by the *RH2-4* upstream region was again similar to its native pattern confined to the ventral retina while the reporter expression by the *keratin 8* upstream region was in the entire retina (Fig. 3c). When the *RH2-4* upstream region was placed in front of the *RH2-3* upstream region (Fig. 3d), the reporter from the *RH2-4* upstream was expressed strongly in the ventral retina, though also detected elsewhere in the retina. The GFP expression by the *RH2-3* upstream region was weak overall, being almost absent in the ventral area. Thus, the two reporters show reciprocal gradients of expression in this transgenic line (Fig. 3d).

In the above cases, the reporter expression from the *RH2-3* upstream was reshaped by another upstream region located closer to the RH2-LCR (Fig. 3a, d, Additional file 1: Figure S4A). However, when the *RH2-3* upstream region was placed closer to the RH2-LCR than the other (*RH2-2* or *RH2-4* upstream region) (Fig. 3e, f), the reporter expression by the *RH2-3* upstream region was in the entire retina. On the other hand, the reporter gene expression by the *RH2-2* and *RH2-4* upstream regions was not clearly detected (Fig. 3e, f). We speculate, in these cases, the upstream region of *RH2-3* outcompeted those

of the other RH2 genes and blocked the access of the RH2-LCR to the downstream.

These results indicate that the RH2 genes' expression pattern in the retina is not only governed by their own proximal upstream regions but also by those of the other RH2 genes nearby on the array. Importantly, their relative distance (or arrayed order) to the RH2-LCR greatly matters to the effect of the latter.

To further test the effect of relative distance with each other in such context-dependent regulation of the RH2 locus, we established transgenic zebrafish lines of modified RH2-PAC clones, in which the RH2-LCR was translocated from the 15-kb upstream of the gene array to the region immediately downstream of *RH2-3* in the RH2-PAC (Fig. 4a). We previously showed that this configuration increased the expression level of *RH2-3* and decreased those of *RH2-1* and *RH2-2* in larvae by transient transgenic assays with GFP reporters [18]. Consistently, in the adult transgenic fish carrying the RH2-PAC clone with GFP replacing *RH2-3*, the expression of *RH2-3*/GFP was clearly extended towards the dorsal area of the retina (Fig. 4b *middle*). By contrast, the expression of *RH2-2*/GFP was abolished (Fig. 4b *left*) and that of *RH2-4*/GFP was maintained in the ventral area and the dorsal tip of the retina in this configuration (Fig. 4b *right*). Thus, upon the translocation, the closer *RH2-3* became the target of the RH2-LCR in the central to dorsal area of the retina. The LCR activity promoting the more distant *RH2-1*/*RH2-2* was reduced. In the ventral retina, *RH2-4* was still predominantly activated, excluding the activation of *RH2-3*.

Discussion

The present study shows that (i) the regulatory role of the RH2-LCR in SDC-specific expression is evolutionarily conserved; (ii) the RH2-LCR and the *RH2-3* proximal regulatory region provide no spatial selectivity of expression in the retina; (iii) the proximal regulatory regions of *RH2-1*, −2 and −4 are capable of inducing spatial

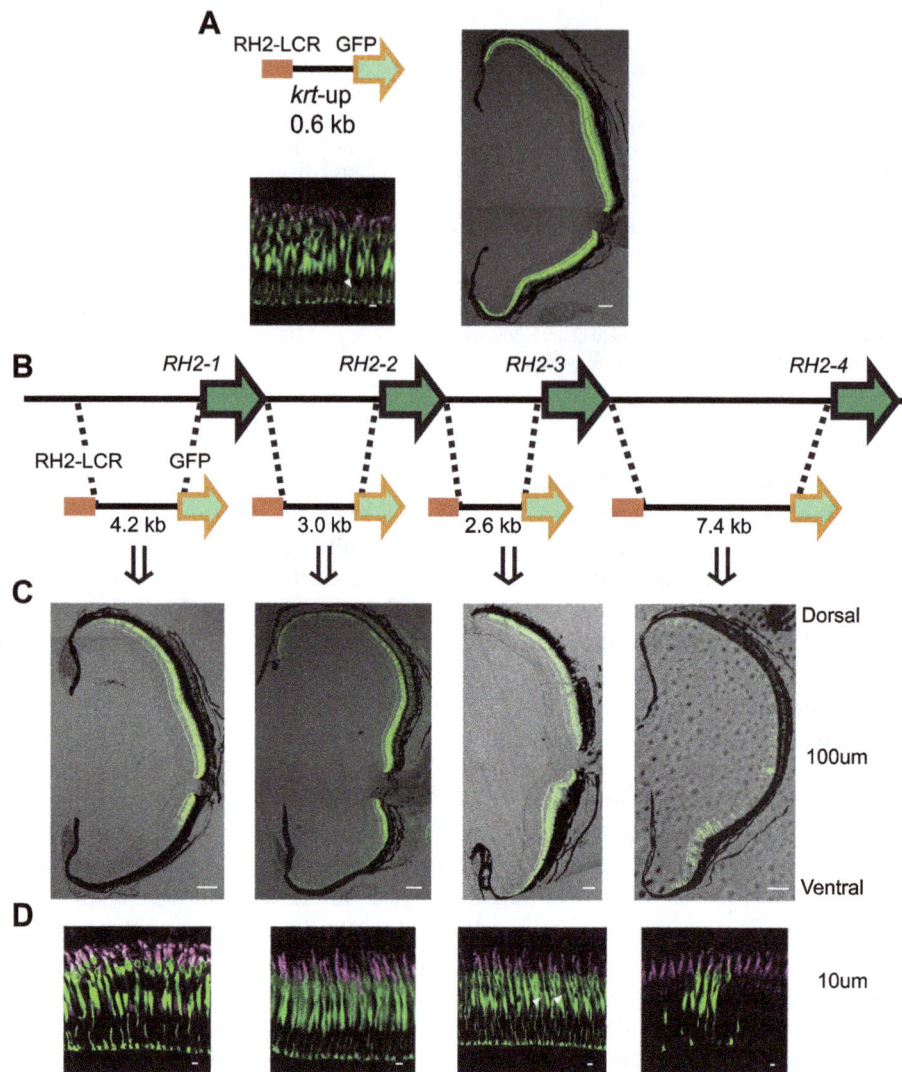

Fig. 2 GFP expression patterns specified by the RH2 upstream sequences. **a** The promoter of *keratin 8* attached with the RH2-LCR was linked to a GFP reporter (*top left*). The GFP was mostly expressed in the SDCs with some ectopic expression as indicated by a white arrowhead (*bottom left*). The transverse sections of the retinas of the adult transgenic zebrafish showed GFP expression in the entire region from the dorsal to the ventral retina (*right*). Scale bars = 10 μm (bottom left), 100 μm (*right*). **b** Schematic representation of the RH2 upstream constructs with the RH2-LCR. **c** Images of the transverse sections of the retinas of the adult transgenic zebrafish possessing the respective constructs. The dorsal side is at the top and the ventral side is at the bottom. **d** Vertical sections of the photoreceptor layer in the same adult retinas as in (**c**). Immunostaining signals of SDCs by the anti-RH2 antibody appear as magenta, GFP fluorescent signals appear as green, and overlap of the two signals appears as white. Note the weak ectopic expression by the *keratin 8* and *RH2-3* upstream construct in non-SDC photoreceptor cells as indicated by white arrowheads (**a**, **d**). Scale bars = 100 μm (**c**) and 10 μm (**d**)

differentiation of expression; (iv) these regulatory regions influence with each other to modulate the actual output in a position-dependent manner, which most strikingly determines the spatial pattern of *RH2-3* expression. Since the ancestral RH2 gene was likely expressed in the entire area of the retina, we suggest that the upstream of *RH2-3* maintained the ancestral regulatory feature while those of the others achieved functional modification.

The following lines of evidence support that competition for the interaction with the RH2-LCR among the replicated genes underlies the position-dependent regulation.

First, we often observed reciprocal gradients of expression between two reporter genes in our transgenic fish including those in which one is completely repressed (Fig. 3, Additional file 1: Figure S4). Furthermore, genes closer to the RH2-LCR, which should have a higher chance to interact with the enhancer, were more likely and broadly activated, while activation of other genes located further from the LCR was often interrupted (Figs. 3 and 4, Additional file 1: Figure S4).

However, the competition does not seem to be the only mechanism governing the context-dependent regulation

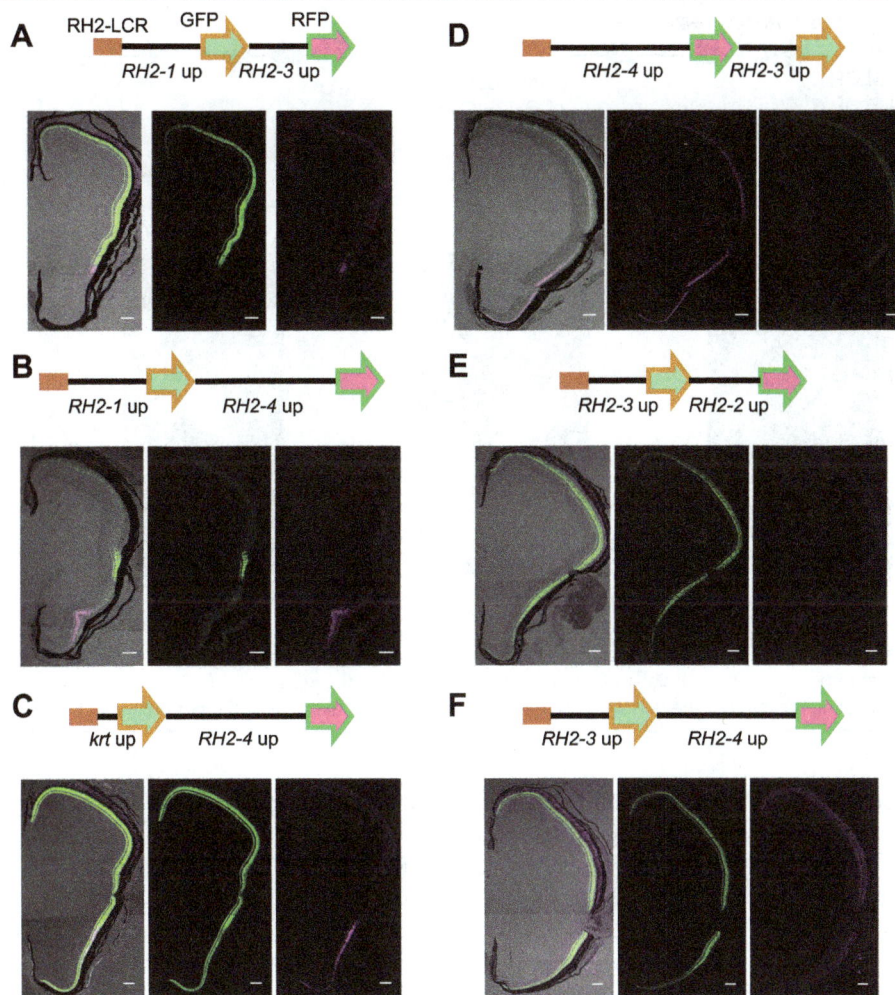

Fig. 3 Presence of a competitive promoter modulates reporter expression by the RH2-LCR and RH2 upstream elements. **a-f** Schematic representations of constructs with the RH2-LCR and double promoter-reporter sets are depicted at the top. RH2-LCR is represented as a red rectangle. The GFP and RFP reporters are represented as green and magenta arrows, respectively. The upstream sequences used to drive the reporters are indicated below. The lower panels are transverse sections of retinas from the adult transgenic fish with the respective constructs. The middle and right panels are fluorescence from the first and second reporters, respectively. The left is the overlay of the middle and right panels with DIC images of the same retina. The GFP signals appear as green and the RFP signals appear as magenta. Note that the signal in the right panel of (**f**) is autofluorescence from the retinal pigment epithelium as evident in the overlaid image. The dorsal side is at the top, and the ventral side is at the bottom. Scale bars = 100 μm

of the RH2 locus. The absence of RFP expression from the *RH2-3* upstream sequence in the ventral retina (Fig. 3a and Additional file 1: Figure S4A) might indicate that the regulatory region upstream of *RH2-1*, which was located between the RH2-LCR and the *RH2-3* upstream in the transgenic lines, represses gene expression in the ventral zone not only of itself but also of *RH2-3* (Fig. 5). It should also be noted that in the double reporter constructs in this study, any promoters including that of *RH2-4* failed to induce gene expression when located downstream of the *RH2-3* promoter due to its blocking activity (Fig. 3e, f). Nevertheless the expression of endogenous *RH2-4*, located downstream of *RH2-3*, is robustly induced in the ventral area (see Additional file 1: Figure S3). This fact implies a mechanism that should interfere the blocking activity by

RH2-3 over *RH2-4*. Perhaps, the repressive action by the upstream of *RH2-1* (and also of *RH2-2*) towards *RH2-3* in the ventral retina might play a role in it (Fig. 5). On the other hand, the translocation of the RH2-LCR to the downstream of *RH2-3* diminished the activation of *RH2-1* [18] and *RH2-2*, while extending the expression of *RH2-3* (Fig. 4). Taken together, these results show that the appropriate positioning of the *cis*-regulatory elements within the locus is crucial for the collective regulation of the quadruplicates.

It should also be noted that the reporter expression induced by the *RH2-3* promoter with the RH2-LCR was not only in the SDCs but also in other types of the photoreceptors. Although such ectopic expression was not clear in the upstream sequences of the other genes,

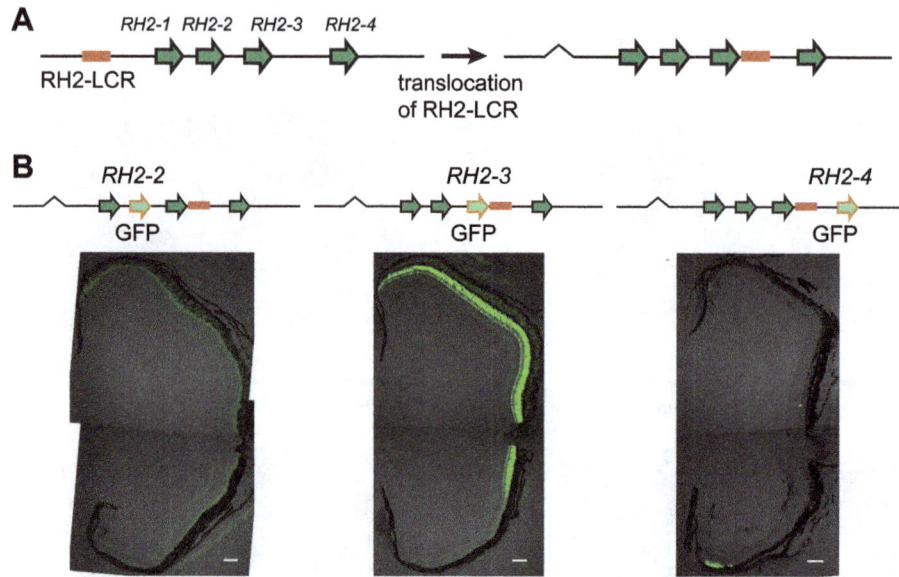

Fig. 4 Translocation of the RH2-LCR revealed gene-order dependent competition for the RH2 regulation. **a** Translocation of the RH2-LCR. The RH2-LCR was removed from the original position and then re-inserted into the immediate downstream of *RH2-3* in the RH2-PAC clones [18]. **b** Transverse sections of the adult transgenic zebrafish retinas of the PAC constructs where *RH2-2* (*left*), *RH2-3* (*middle*) and *RH2-4* (*right*) are replaced with the GFP reporter respectively. The GFP signals appear as green. Note that in the left panel the green signal is saturated and only the autofluorescence from the retina is captured. The dorsal side is at the top, and the ventral side is at the bottom. Scale bars = 100 μm

previous studies, in fact, reported that the RH2-LCR sometimes induced weak expression in long single cones, where SWS2 is specifically expressed normally [18, 23]. When reporter genes are integrated in the RH2-PAC clone, however, we have never seen their expression ectopically in non-SDC photoreceptor cells. Therefore, there should be a *cis*-regulatory mechanism that involves not only the RH2-LCR but also other *cis*-elements within the locus to strictly specify the RH2 expression to the SDCs. It can be further speculated that this collective regulation might depend on the genomic context, as was suggested in the *Hoxd* cluster and the *Fgf8* locus, where multiple enhancers cooperatively defines the expression patterns of the target genes [24, 25]. Recently, *sine oculis* homeobox homolog 7 (Six7) was implicated in regulation of the RH2 genes in zebrafish [26]. To deepen our understanding of the cell type-specific regulation, roles of such *trans*-regulatory factors should be studied in parallel with *cis*-regulatory mechanisms.

We asked in this study how the ceaseless duplications and differentiations of the opsin genes in teleosts are accompanied by elaborate building of regulatory mechanism

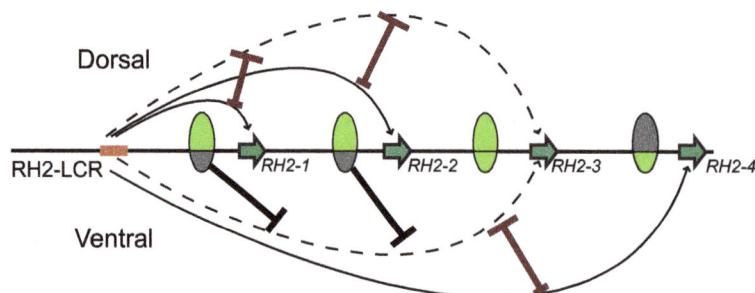

Fig. 5 The proposed model of the collective regulation of the RH2 locus in zebrafish. The regulatory activity of the proximal regions in the dorsal and ventral retina is depicted by the top and bottom part of the divided ovals, respectively, at each position. Light green and gray indicate the active and repressive states, respectively. Basically, the RH2-LCR induces gene expression if the upstream region of the target is active as depicted by the arched arrows. However, competition with other members of the locus (whiskered brown bars) and the repressive regulation by the upstream regions of *RH2-1* and *RH2-2* in the ventral retina (black bars) inhibits the interaction of *RH2-3* with the RH2-LCR as indicated by the dashed arrows, restricting its expression into the narrow banded area between the central and ventral retina. Importantly, the dominance in the competitive regulation (indicated by the sizes of the whiskers) greatly depends on the relative position to the RH2-LCR.

to have the replicates differentially expressed with each other. We found that the competitive regulation between the replicates by a single enhancer plays an important role in the differentiation of the RH2 genes. Remarkably, the LWS in zebrafish utilizes a similar mechanism for their differential expression [19]. We propose that such competitive regulation is advantageous to preserve the replicated opsin genes from pseudogenization in fish, since the competition can intrinsically differentiate and subfunctionalize the replicates by assigning them to a distinct set of the photoreceptor cells. At the same time it precludes a void space in the retina that expresses none of the replicates: without competitive regulation, it might be possible that all lose cis-elements necessary to be expressed in some part of the retina through accumulation of mutations in their regulatory regions during the process of differentiation of expression patterns. In fact, it was shown that tandem duplication is the exclusively predominant event in the expansion of the opsin repertoires in fish rather than the whole genome duplication or retroposon-mediated duplications, which do not allow such coordinate regulation in cis [27].

In addition, the similarity between the RH2 and LWS in the expression pattern (i.e.. central-to-dorsal vs. ventral) as well as in the cis-regulatory mechanisms strongly indicates that the both systems utilize shared trans-regulatory components that distinguish different areas of the retina, which might have helped their convergent differentiation. In fact, a recent study revealed that retinoic acid signaling regulates the differential expression of the two LWS genes [28]. Such an extracellular signal might also impact on the regulation of the RH2 genes, though its involvement is still elusive. On the other hand, it should be emphasized that zebrafish also acquired new mutations in the upstream sequences of some subtypes (RH2-1, RH2-2, RH2-4 and LWS-1) independently for several times to have the spatially polarized differentiation patterns (this study and [19]). This seems to be in a sharp contrast with the duplicated red/green opsin genes in catarrhine primates including human, which have seemingly more or less equivalent promoters with each other, and, as a result, are randomly expressed in the retina in a mutually exclusive way, underlying the trichromacy [29].

The β-globin cluster in human is a well-known example of spatiotemporally-patterned regulation among tandem replicates. The developmental switching of the gene expression along the tandemly clustered genes is also regulated by stage-specific cis-elements associated with some of the early-expressed genes together with the competitive regulation that is a function of the gene order [30, 31]. On the other hand, random competition among duplicated genes with apparently equivalent promoters takes place in regulation of olfactory receptor genes

to improve the dimension of odor recognition [32–34]. An artificially induced duplication of Protocadherin-α cluster also resulted in stochastic expression of replicates, but not patterned one [35]. These cases might indicate that the only competition among duplicated genes tends to result in stochastic regulation by shared enhancers, and that differential stereotyped spatiotemporal expression further requires additional cis-regulatory elements associated with some, but not necessarily all, of the replicates to dictate stage or tissue specificity.

Conclusions

Our study highlights the differential cis-elements embedded in the upstream regions of the RH2 genes as well as their relative distance to the RH2-LCR as fundamental cis-regulatory features that collectively shape the differential expression of the quadruplicated opsin genes in zebrafish (Fig. 5). Other fish such as medaka, cichlids, guppy (Poecilia reticulata), four-eyed fish (Anableps anableps) and barfin flounder (Verasper moseri) are also known to have opsin genes duplicated and differentiated in the spatiotemporal expression patterns to adjust their visual sensitivity to heterogeneity in their ontogeny and environment with fine-tuned absorption spectra of the visual pigments [5, 9–12, 27]. Therefore it is anticipated that these differentiated opsins also adopted proximal cis-regulatory mutations to differentiate the expression patterns, probably based on concerted regulation through enhancer-sharing generated upon duplications. In order to deepen our understanding of the unique expansion of the opsin repertoires in fish, future studies should further clarify the evolutionary steps of the gene regulatory mechanisms in different lineages.

Methods

Sequence comparison of RH2 locus among zebrafish, medaka and Tetraodon

The sequence surrounding the zebrafish RH2 locus was obtained from Ensembl genome database of zebrafish, and corresponds to the nucleotide position 32265746–32385745 (120 kb) of chromosome 6 in zebrafish assembly version 6. The sequence surrounding the medaka RH2 locus was obtained from medaka UT genome browser and corresponds to the nucleotide position 1505256–1705255 (200 kb) of scaffold84 of version 1.0. The sequence surrounding the Tetraodon RH2 locus was obtained from Ensembl genome database of Tetraodon, and corresponds to the nucleotide position 5110001–5140000 (30 kb) of chromosome 11 in Tetraodon (Tetraodon nigroviridis) assembly version 7 in a reverse orientation. The Sequence alignment between zebrafish, medaka and Tetraodon was made with the mVISTA program [36, 37] using the AVID algorithm [38]. Window size was set as 100 bp.

Usage of zebrafish

All animal protocols were approved by the University of Tokyo animal care and use committee (Approval numbers C-09-02 and C-09-03). The strains of zebrafish (*Danio rerio*) used in the present study were WIK [39] and TL [40], each for microinjection and for mating with transgenic fish of WIK, respectively. They were maintained at 28.5 °C in a 14-h light/10-h dark cycle as described by [41].

Modification of zebrafish RH2-PAC and medaka RH2-BAC clones

The RH2-PAC and the medaka RH2-BAC (33O2) clones were obtained in [18] and [5], respectively. The insertion of reporter genes and the removal and translocation of the RH2-LCR was all done by the recombineering technique in EL250 [42]. We inserted two I-SceI recognition sites into the vector backbones as described in [19] and [18] in order to facilitate integration of the construct into the genome with the meganuclease [43]. An additional file describes the details of the construction (Additional file 1: Document S1).

Construction of reporter-expression plasmids

We used pT2GFP-TKPA [19], a derivative from the plasmid clone pT2AL200R150G, which contains the Tol2 transposase recognizing sequences, L200 and R150 [44], as a basal vector backbone for the construction of the GFP expression constructs of the upstream regions of *RH2-1, –2, –3, –4* and *keratin 8*, attached with the RH2-LCR to integrate the transgene via Tol2 transposon-mediated transgenesis. The double reporter constructs with the RH2-LCR were also made in the same vector backbone. An additional file describes the details of the construction (Additional file 1: Document S1).

Microinjection of DNA constructs into zebrafish embryos

We utilized three different methods of microinjection for transgensis of reporter constructs. We linearized the GFP reporter plasmid of the 3-kb upstream sequence of medaka *RH2-A* by restriction digest with Eco47III to perform co-injection with the linear fragment of medaka RH2-LCR that was amplified by PCR. The co-injection protocol is described in [18, 19]. The plasmid DNAs derived from pT2GFP-TKPA were prepared at final concentration of 25 ng/μl using Plasmid Mini Kit (QIAGEN) or Plasmid Midi Kit (QIAGEN), and were resuspended in 0.1 M KCl and tetramethyl-rhodamin dextran added as a tracer. They were co-injected into the cytoplasm of embryos at the one-cell stage with mRNA of Tol2 transposase of 27 ng/μl that was prepared through *in vitro* transcription from pCS-TP using mMASSAGE mMACHINE kit (Ambion) [44, 45]. The RH2-PAC-derived constructs (20 ng/μl) were injected with I-SceI

meganuclease (0.5 units/μl) (New England Biolabs, Beverly, MA) in the solution of 0.5 X commercial meganuclease buffer with tetramethyl-rhodamin dextran tracer [43].

Establishment of transgenic zebrafish

For the generation of transgenic lines, the injected embryos were grown to sexual maturity and crossed with non-injected fish in a pair-wise fashion. The genomic DNA extracted from a pool of the resulting embryos was examined for the presence of the transgene by PCR amplification of the GFP DNA segment as described in [46]. Importantly, the screening did not rely on presence of the fluorescence in the eyes to avoid biased selection of founder lines. Fish of the subsequent generations were screened again for the presence of the transgene by PCR amplification of the GFP from genomic DNA extracted from the fins. The spatial expression patterns of the reporters in the retina were analyzed in the generation of the offspring from the injected fish (F1) or later.

Transient assay of GFP expression levels in zebrafish embryos

The injected embryos were grown in 0.003 % 1-phenyl-2-thiourea after 12–24 h post fertilization (hpf) to disrupt pigment formation. Then their eyes were examined at 5 dpf (days post fertilization) for GFP fluorescence under a dissecting fluorescent microscope, and the number of eyes expressing GFP was determined as described by [47]. The eyes were scored as "+++", "++", "+", and "-" when GFP was expressed in more than 50 cells, in 5–50 cells, in 1–4 cells, and in no cells per eye, respectively.

Immunohistochemistry and image capture

Immunostaining was carried out against the retina sections from adult fish or embryonic whole-mount eye cups, following the protocol described before [47]. Images of GFP, RFP and Cy3 fluorescence were captured using a Zeiss 510 laser-scanning confocal microscope (Carl Zeiss). In case the entire part of the retina sections could not be captured by one image, two overlapping images were collected and then aligned manually to represent the whole retina as a single image.

Availability of data and materials

All the supporting data are included as additional files.

Abbreviations

RH1: Rhodopsin; SWS1: Short-wave-sensitive-1 cone-opsin; SWS2: Short-wave-sensitive-2 cone-opsin; RH2: Rod-opsin-like cone-opsin; LWS: Long-wave-sensitive cone-opsin; SDC: Short member of double cones; LDC: Long member of double cones; RH2-LCR: RH2-locus control region; LAR: LWS activating region; PAC: P1-derived artificial chromosome; BAC: Bacterial artificial chromosome; GFP: Green fluorescent protein; RFP: Red fluorescent protein; Hpf: Hours post fertilization; Dpf: Days post fertilization.

Competing interests

The authors declare no conflict of interests.

Authors' contributions

T.T. and S.K. conceived the study. T.T. performed most of the experiments with help from R.M. and R.A.. T.T. and S.K. wrote the manuscript. All authors have read and approved the final version of the manuscript.

Acknowledgements

We are grateful to Dr. N. G. Copeland for the *E. coli* strains EL250, to Drs. T. Vihtelic and D. Hyde for the antibody against the zebrafish green opsin, to Dr. V. Korzh for the construct of the *krt8* promoter, and to Dr. K. Kawakami for the plasmid vectors of the Tol2 transposon and transposase. We thank Dr. Deborah L. Stenkamp for proofreading and commenting on the manuscript. T.T. was supported by a Grant-in-Aid for JSPS Fellows (19–4820). S.K. was supported by Grants-in-Aid for Scientific Research (B) (16405015), (A) (19207018), and (A) (22247036) from the Japan Society for the Promotion of Science and Grants-in-Aid for Scientific Research on Priority Areas "Comparative Genomics" (20017008) and "Cellular Sensor" (21026007) from the Ministry of Education, Culture, Sports, Science, and Technology of Japan.

Author details

¹Department of Integrated Biosciences, Graduate School of Frontier Sciences, the University of Tokyo, Kashiwanoha 5-1-5, Kashiwa 277-8562 Chiba, Japan. ²Department of Advanced Nephrology and Regenerative Medicine, Division of Tissue Engineering, the University of Tokyo Hospital, Hongo 7-3-1, Bunkyo-ku 113-8655Tokyo, Japan.

References

1. Yokoyama S. Molecular evolution of vertebrate visual pigments. Prog Retin Eye Res. 2000;19(4):385–419.
2. Kawamura S. Evolutionary diversification of visual opsin genes in fish and primates. In: Inoue-Murayama M, Kawamura S, Weiss A, editors. From genes to animal behavior: Social structures, personalities, communication by color. Tokyo: Springer; 2011. p. 329–49.
3. Chinen A, Hamaoka T, Yamada Y, Kawamura S. Gene duplication and spectral diversification of cone visual pigments of zebrafish. Genetics. 2003;163(2):663–75.
4. Morrow JM, Lazic S, Chang BS. A novel rhodopsin-like gene expressed in zebrafish retina. Vis Neurosci. 2011;28(4):325–35.
5. Matsumoto Y, Fukamachi S, Mitani H, Kawamura S. Functional characterization of visual opsin repertoire in Medaka (*Oryzias latipes*). Gene. 2006;371(2):268–78.
6. Carleton KL, Kocher TD. Cone opsin genes of African cichlid fishes: Tuning spectral sensitivity by differential gene expression. Mol Biol Evol. 2001;18(8):1540–50.
7. Spady TC, Parry JW, Robinson PR, Hunt DM, Bowmaker JK, Carleton KL. Evolution of the cichlid visual palette through ontogenetic subfunctionalization of the opsin gene arrays. Mol Biol Evol. 2006;23(8):1538–47.
8. Zhang J. Evolution by gene duplication: an update. Trends Ecol Evol. 2003;18(6):292–8.
9. Carleton KL, Spady TC, Streelman JT, Kidd MR, McFarland WN, Loew ER. Visual sensitivities tuned by heterochronic shifts in opsin gene expression. BMC Biol. 2008;6:22.
10. Hofmann CM, Carleton KL. Gene duplication and differential gene expression play an important role in the diversification of visual pigments in fish. Integr Comp Biol. 2009;49(6):630–43.
11. Owens GL, Rennison DJ, Allison WT, Taylor JS. In the four-eyed fish (*Anableps anableps*), the regions of the retina exposed to aquatic and aerial light do not express the same set of opsin genes. Biol Lett. 2012;8(1):86–9.
12. Rennison DJ, Owens GL, Allison WT, Taylor JS. Intra-retinal variation of opsin gene expression in the guppy (*Poecilia reticulata*). J Exp Biol. 2011;214(Pt 19):3248–54.
13. Takechi M, Kawamura S. Temporal and spatial changes in the expression pattern of multiple red and green subtype opsin genes during zebrafish development. J Exp Biol. 2005;208(Pt 7):1337–45.
14. Davies WI, Collin SP, Hunt DM. Molecular ecology and adaptation of visual photopigments in craniates. Mol Ecol. 2012;21(13):3121–58.
15. Raymond PA, Barthel LK, Rounsifer ME, Sullivan SA, Knight JK. Expression of rod and cone visual pigments in goldfish and zebrafish: a rhodopsin-like gene is expressed in cones. Neuron. 1993;10(6):1161–74.
16. Vihtelic TS, Doro CJ, Hyde DR. Cloning and characterization of six zebrafish photoreceptor opsin cDNAs and immunolocalization of their corresponding proteins. Vis Neurosci. 1999;16(3):571–85.
17. Stenkamp DL. Neurogenesis in the fish retina. Int Rev Cytol. 2007;259:173–224.
18. Tsujimura T, Chinen A, Kawamura S. Identification of a locus control region for quadruplicated green-sensitive opsin genes in zebrafish. Proc Natl Acad Sci U S A. 2007;104(31):12813–8.
19. Tsujimura T, Hosoya T, Kawamura S. A single enhancer regulating the differential expression of duplicated red-sensitive opsin genes in zebrafish. PLoS Genet. 2010;6(12):e1001245.
20. Hisatomi O, Satoh T, Tokunaga F. The primary structure and distribution of killifish visual pigments. Vision Res. 1997;37(22):3089–96.
21. Gong Z, Ju B, Wang X, He J, Wan H, Sudha PM. Green fluorescent protein expression in germ-line transmitted transgenic zebrafish under a stratified epithelial promoter from *keratin8*. Dev Dyn. 2002;223(2):204–15.
22. Parinov S, Kondrichin I, Korzh V, Emelyanov A. *Tol2* transposon-mediated enhancer trap to identify developmentally regulated zebrafish genes in vivo. Dev Dyn. 2004;231(2):449–59.
23. Fang W, Bonaffini S, Zou J, Wang X, Zhang C, Tsujimura T, et al. Characterization of transgenic zebrafish lines that express GFP in the retina, pineal gland, olfactory bulb, hatching gland, and optic tectum. Gene Expr Patterns. 2013;13(5–6):150–9.
24. Montavon T, Soshnikova N, Mascrez B, Joye E, Thevenet L, Splinter E, et al. A regulatory archipelago controls Hox genes transcription in digits. Cell. 2011;147(5):1132–45.
25. Marinic M, Aktas T, Ruf S, Spitz F. An integrated holo-enhancer unit defines tissue and gene specificity of the Fgf8 regulatory landscape. Dev Cell. 2013;24(5):530–42.
26. Ogawa Y, Shiraki T, Kojima D, Fukada Y. Homeobox transcription factor Six7 governs expression of green opsin genes in zebrafish. Proc Biol Sci. 2015;282(1812):20150659.
27. Rennison DJ, Owens GL, Taylor JS. Opsin gene duplication and divergence in ray-finned fish. Mol Phylogenet Evol. 2012;62(3):986–1008.
28. Mitchell DM, Stevens CB, Frey RA, Hunter SS, Ashino R, Kawamura S, et al. Retinoic Acid signaling regulates differential expression of the tandemly-duplicated long wavelength-sensitive cone opsin genes in zebrafish. PLoS Genet. 2015;11(8):e1005483.
29. Smallwood PM, Wang Y, Nathans J. Role of a locus control region in the mutually exclusive expression of human red and green cone pigment genes. Proc Natl Acad Sci U S A. 2002;99(2):1008–11.
30. Hanscombe O, Whyatt D, Fraser P, Yannoutsos N, Greaves D, Dillon N, et al. Importance of globin gene order for correct developmental expression. Genes Dev. 1991;5(8):1387–94.
31. Tanimoto K, Liu Q, Bungert J, Engel JD. Effects of altered gene order or orientation of the locus control region on human beta-globin gene expression in mice. Nature. 1999;398(6725):344–8.
32. Fuss SH, Omura M, Mombaerts P. Local and cis effects of the H element on expression of odorant receptor genes in mouse. Cell. 2007;130:373–84.
33. Nishizumi H, Kumasaka K, Inoue N, Nakashima A, Sakano H. Deletion of the core-H region in mice abolishes the expression of three proximal odorant receptor genes in cis. Proc Natl Acad Sci U S A. 2007;104:20067–72.

34. Serizawa S, Miyamichi K, Nakatani H, Suzuki M, Saito M, Yoshihara Y, et al. Negative feedback regulation ensures the one receptor-one olfactory neuron rule in mouse. Science. 2003;302(5653):2088–94.

35. Kaneko R, Abe M, Hirabayashi T, Uchimura A, Sakimura K, Yanagawa Y, et al. Expansion of stochastic expression repertoire by tandem duplication in mouse Protocadherin-alpha cluster. Sci Rep. 2014;4:6263.

36. Frazer KA, Pachter L, Poliakov A, Rubin EM, Dubchak I. VISTA: Computational tools for comparative genomics. Nucleic Acids Res. 2004;32 (Web Server issue):W273–279.

37. Mayor C, Brudno M, Schwartz JR, Poliakov A, Rubin EM, Frazer KA, et al. VISTA: Visualizing global DNA sequence alignments of arbitrary length. Bioinformatics. 2000;16(11):1046–7.

38. Bray N, Dubchak I, Pachter L. AVID: A global alignment program. Genome Res. 2003;13(1):97–102.

39. Rauch G-J, Granato, M., and Haffter, P.: A polymorphic zebrafish line for genetic mapping using SSLPs on high-percentage agarose gels. Tech Tips Online 1997, T01208.

40. Haffter P, Odenthal J, Mullins MC, Lin S, Farrell MJ, Vogelsang E, et al. Mutations affecting pigmentation and shape of the adult zebrafish. Dev Genes Evol. 1996;6(4):260–76.

41. Westerfield M. The zebrafish book: A guide for the laboratory use of zebrafish (*Danio Rerio*). Eugene: University of Oregon Press; 1995.

42. Lee EC, Yu D, Martinez de Velasco J, Tessarollo L, Swing DA, Court DL, et al. A highly efficient *Escherichia coli*-based chromosome engineering system adapted for recombinogenic targeting and subcloning of BAC DNA. Genomics. 2001;73(1):56–65.

43. Thermes V, Grabher C, Ristoratore F, Bourrat F, Choulika A, Wittbrodt J, et al. *I-SceI* meganuclease mediates highly efficient transgenesis in fish. Mech Dev. 2002;118(1–2):91–8.

44. Urasaki A, Morvan G, Kawakami K. Functional dissection of the *Tol2* transposable element identified the minimal *cis*-sequence and a highly repetitive sequence in the subterminal region essential for transposition. Genetics. 2006;174(2):639–49.

45. Kawakami K, Takeda H, Kawakami N, Kobayashi M, Matsuda N, Mishina M. A transposon-mediated gene trap approach identifies developmentally regulated genes in zebrafish. Dev Cell. 2004;7(1):133–44.

46. Hamaoka T, Takechi M, Chinen A, Nishiwaki Y, Kawamura S. Visualization of rod photoreceptor development using GFP-transgenic zebrafish. Genesis. 2002;34(3):215–20.

47. Luo W, Williams J, Smallwood PM, Touchman JW, Roman LM, Nathans J. Proximal and distal sequences control UV cone pigment gene expression in transgenic zebrafish. J Biol Chem. 2004;279(18):19286–93.

6

A study of vertebra number in pigs confirms the association of vertnin and reveals additional QTL

Gary A. Rohrer[*], Dan J. Nonneman, Ralph T. Wiedmann and James F. Schneider

Abstract

Background: Formation of the vertebral column is a critical developmental stage in mammals. The strict control of this process has resulted in little variation in number of vertebrae across mammalian species and no variation within most mammalian species. The pig is quite unique as considerable variation exists in number of thoracic vertebrae as well as number of lumbar vertebrae. At least two genes have been identified that affect number of vertebrae in pigs yet considerable genetic variation still exists. Therefore, a genome-wide association (GWA) analysis was conducted to identify additional genomic regions that affect this trait.

Results: A total of 1883 animals were phenotyped for the number of ribs and thoracolumbar vertebrae as well as successfully genotyped with the Illumina Porcine SNP60 BeadChip. After data editing, 41,148 SNP markers were included in the GWA analysis. These animals were also phenotyped for kyphosis. Fifty-three 1 Mb windows each explained at least 1.0 % of the genomic variation for vertebrae counts while 16 regions were significant for kyphosis. Vertnin genotype significantly affected vertebral counts as well. The region with the largest effect for number of lumbar vertebrae and thoracolumbar vertebrae were located over the Hox B gene cluster and the largest association for thoracic vertebrae number was over the Hox A gene cluster. Genetic markers in significant regions accounted for approximately 50 % of the genomic variation. Less genomic variation for kyphosis was described by QTL regions and no region was associated with kyphosis and vertebra counts.

Conclusions: The importance of the Hox gene families in vertebral development was highlighted as significant associations were detected over the A, B and C families. Further evaluation of these regions and characterization of variants within these genes will expand our knowledge on vertebral development using natural genetic variants segregating in commercial swine.

Keywords: Rib, Thoracic, Lumbar, Vertebra, Kyphosis, GWAS, Pig, Swine

Background

Number of specific vertebrae in most mammalian species is quite conserved. The degree of conservation is location dependent with greater conservation towards the anterior end of the animal. For example, the number of cervical vertebrae in most mammalian species is fixed at seven whereas the number of caudal vertebrae among the 133 mammalian species observed ranged from 2 to 37 [1]. Variation in the number of thoracic and lumbar vertebrae across species does exist, but Narita and Kuratani [2]

noted that the summation of thoracic and lumbar vertebrae (called thoracolumbar) was more conserved as most species were fixed at 19. Variation within species in thoracic, lumbar and thoracolumbar numbers has been observed in a limited number of mammalian species. The domestic pig is one such species where variation exists for vertebrae numbers. Surprisingly, a dramatic amount of variation exists in swine for a trait where most species exhibit no variation. Wild boars and Meishan pigs appear to be similar to most mammalian species with a thoracolumbar number of 19. However, commercial pigs of European origin typically range from 21 to 23 thoracolumbar vertebrae [3]. While it is often speculated that variation in pigs exists due to selection for

* Correspondence: Gary.Rohrer@ars.usda.gov
United States Department of Agriculture, Agricultural Research Service,, U.S. Meat Animal Research Center, Clay Center, NE 68933, USA

increased body size, similar or greater changes in body size has occurred due to selection in other domesticated species (for example: cattle, sheep and dog) and yet variation within these species has not been widely reported.

Number of thoracolumbar vertebrae can affect carcass length, which is an economically important trait in pig production; however, vertebrae number only accounts for 10–15 % of the phenotypic variation in carcass length [3, 4]. Early reports indicated that pigs with more vertebrae suffered more lameness and structural problems [5, 6] as well as gave birth to more stillborn piglets [7]. These reports attempted to link disruption/modification of vertebral development with other abnormalities. Within the USMARC population a back curvature defect characterized as kyphosis has been observed and determined to have a genetic basis [8]. While this defect is not strictly associated with number of vertebrae, there is a tendency for increased incidence with greater number of ribs.

A number of genome scans have been conducted for carcass length and vertebra numbers in pigs. The first studies evaluated crosses between commercial breeds and either Meishan or Wild Boar pigs [9, 10]. Two major QTL were discovered in these studies and causative genetic variation has been identified in NR6A1 [11] and Vertnin [12]. Despite these discoveries, considerable unexplained genetic variation affecting thoracic, lumbar and thoracolumbar vertebra numbers still exists in commercial swine. Therefore, the objective of this study was to search for additional loci that affect vertebra numbers in a commercial composite swine population containing Duroc, Landrace and Yorkshire germplasm and determine if vertebra number is associated with animal performance or kyphosis.

Methods
Data collection
Care and handling of all animals included in this study was according to procedures outlined in *Guide for Care and Use of Agricultural Animals in Agricultural Research and Teaching* [13] and approved by the USMARC Animal Care and Use Committee. Animals were from a closed commercial-type composite population created by mating Duroc and Landrace boars to Yorkshire-Landrace composite females. Subsequently Duroc-sired animals were mated to Landrace-sired animals resulting in a population of ½ Landrace, ¼ Duroc and ¼ Yorkshire. Phenotyped animals were born in the fourth through eighth generations after composite formation (between May 2005 and November 2009). Animals were delivered to the USMARC abattoir and harvested using standard humane methods. The abattoir used electrical stunning followed by exsanguination under the supervision of USDA-FSIS inspectors. The age at slaughter ranged from 5 to 30 months. The number of ribs (RIB) and number of thoracolumbar (TLV) vertebrae were counted on the right

side of the carcass. Number of ribs was considered the number of thoracic vertebrae and number of lumbar vertebrae (LVN) was computed by subtracting the number of ribs from the number of thoracolumbar vertebrae (LVN = TLV − RIB). Only animals with complete vertebra records were included in analyses.

Back curvature scores were also collected at the same time as vertebra counts based on the scoring system reported by Holl et al. [7]. The scoring was considered a continuous variable with values ranging from 0 (normal back curvature) to 3 (carcass exhibited severe kyphosis). All carcasses were scored by the same person.

Genomic DNA was extracted from tail tissue of all animals as described by Schneider et al. [14] and applied to Illumina Porcine SNP60 BeadChips [15] (Illumina Inc., San Diego, CA) per manufacturer's instructions. Beadchips were read at USDA-ARS-BARC and data were interpreted with BeadStudio software (Illumina Inc; San Diego, CA). Markers with call rates < 80 %, minor allele frequencies of < 0.05 or significant deviation from Hardy-Weinberg equilibrium ($P < 1.0$ E −06) were eliminated from the analysis as well as markers not uniquely assigned to a position in build 10.2 of the porcine genome. Vertnin genotypes were determined for phenotyped animals by genotyping NV024 and NV090 [12] using the MassARRAY system (Agena BioSciences; San Diego, CA) based on manufacturer's protocols. Pigs whose genotype was not determined with the MassARRAY system were imputed based on their genotype for MARC0038565 from the SNP60 BeadChip and linkage phase among these markers in the animal's parents. A total of 726 phenotyped pigs as well as 188 parents of phenotyped pigs were genotyped for NV024 and NV090, and the remaining 1166 phenotyped animals had their genotypes imputed. The imputed genotypes of 96 animals were verified by PCR amplification of a fragment spanning the PRE-1 insertion site, reported as NV123 [12]. Vertnin genotypes were reported as the number of copies of the B allele for NV090.

The final data set included genotypic data from 41,148 markers. There were 1883 animals with vertebrae and genotypic data. For the back curvature, 2027 animals had both phenotypic and genotypic data available.

Statistical analyses
Bayesian genome-wide association analyses were conducted using GenSel version 4.61R (www.animalgenome.org/community/projects/BIGS/). A covariate for vertnin genotype, expressed as number of copies of the B allele for NV090 [12], was fit in the model for all vertebra number traits. For kyphosis, only birth contemporary group was fitted as preliminary analyses indicated vertnin genotype did not affect this trait. Marker effects were fitted as additive effects based on the number of copies of the B allele (as defined by Illumina's BeadStudio software).

A 1 Mb window analysis was conducted as described by Wolc et al. [16] and implemented in GenSel version 4.61R. This approach was used as linkage disequilibrium and marker density would tend to distribute the effect of a causative variant across multiple SNP markers in the analysis. Therefore, the genome was divided into non-overlapping 1 Mb windows and estimates of genetic variation attributed to each marker that had a non-zero effect within the window was summed for every 40th iteration of the chain [16]. Under an infinitesimal model, each window would be expected to explain 0.04 % of the genomic variation. Thus, we concluded that windows explaining in excess of 10 fold the expectation (>0.4 %) should be considered a region of interest. Windows that accounted for more than 1 % of the genomic variation were considered significant, as all windows explaining >1.0 % genomic variation were determined to be highly significant ($P < 0.001$) after bootstrap analysis for other traits with similar heritability [17]. Within significant windows, the SNP marker with the largest estimated effect was reported.

Candidate genes within significant 1 Mb windows were selected after manually inspecting the regions using the UCSC Goldenpath genome browser (http://www.genome.ucsc.edu/), build 10.2 of the porcine genome. Annotated genes within QTL regions were considered candidates if they were involved in embryonic development or bone formation.

Association of number of vertebrae with performance was conducted using PROC GLM in SAS version 9.2 (Cary, NC). Weight and ultrasound estimated backfat thickness was recorded at 22 weeks of age on 3182 pigs and the model fit contemporary group and sex as fixed effects, age as a covariate and vertebra number analyzed as a covariate and as a categorical variable in separate analyses. Only one vertebra trait was fit at a time. Also, 1755 animals with vertebra counts had given birth to at least one litter. The percentage of stillborn piglets was analyzed with vertebra counts fitted similarly to weight and backfat analyses.

Results

Descriptive statistics for the traits measured are presented in Table 1. Average values for RIB, LVN and TLV were 15.4, 6.1 and 21.6, respectively. The range observed for RIB was 14 to 17, LVN was 4 to 8 and TLV was 19

to 23. All SNP markers explained 16.1, 12.0, and 24.1 % of the total phenotypic variation of RIB, LVN, and TLV, respectively. The number of 1 Mb windows explaining more than 1 % of the genomic variation detected was 19 for RIB, 16 for LVN, and 18 for TLV. The percentage of genetic variation explained by these windows was 59 % for RIB, 48 % for LVN, and 52 % for TLV. The regression coefficients for vertnin genotype were significantly different from zero for RIB (–0.76), LVN (0.21), and TLV (–0.55).

Table 2 displays the genomic regions significantly associated with at least one trait as well as additional associations of interest for these 1 MB windows. In total, loci affecting number of vertebra spanned 15 chromosomes and resided in 49 unique 1 Mb windows; however, eight pairs of adjacent 1 Mb windows were present and may actually represent only one causative genetic variant. If the adjacent regions were associated with the same vertebra trait, then the associations are likely due to a single causative gene based on strong linkage disequilibrium values ($r^2 > 0.99$) between SNP markers with the largest estimated effects. This was evident for the adjacent regions of SSC 5: 70 and 71 Mb (RIB) and SSC 6:102 and 103 Mb (TLV). Chromosome 16 had three regions associated with LUM. Strong linkage disequilibrium was observed for the regions SSC 16:30 and 31 Mb, but less disequilibrium was observed between SSC 16:29 and 30 Mb ($r^2 = 0.56$). The r^2 values among the SNP markers with the largest estimated effects for the other four adjacent regions did not exceed 0.50.

There were no significant associations of any vertebra trait with first rib, last rib or last lumbar fat thickness ($P > 0.50$). Animals with more TLV grew faster ($P < 0.03$), but no other associations with growth were observed. Furthermore, there was no association between RIB, LVN or TLV of a sow and the number of stillborn piglets delivered in her first litter ($P > 0.40$).

The frequency of animals with some level of kyphosis was 24.7 %; however, most animals (17.4 %) were only mildly affected. The analysis of kyphosis indicated that genetic markers accounted for 16.6 % of the phenotypic variance. In total, 16 windows 1 Mb in size each explained more than 1 % of the genomic variation (Table 3). Cumulatively, these 16 windows explained 39.4 % of the genomic variation. These regions resided

Table 1 Descriptive statistics for phenotypic data analyzed in the study. Genomic variation is the amount of phenotypic variation associated with genotypic data and genomic heritability is the ratio of genomic to phenotypic variation

Trait	Mean	Range	Genomic Variation	Phenotypic variation	Genomic heritability
Thoracic vertebrae (RIB)	15.42	14 to 17	0.0203	0.1258	0.1610
Lumbar vertebrae (LVN)	6.12	4 to 8	0.0200	0.1656	0.1202
Thoracolumbar vertebrae (TLV)	21.55	19 to 23	0.0405	0.1677	0.2412
Kyphosis	0.33	0 to 3	0.0699	0.4222	0.1655

Table 2 Results from GWAS for vertebral traits including chromosome, one megabase window and percent of genomic variation associated with the one megabase window for all significant associations (>1.0 %). Significant regions which were also suggestive (>0.4 %) for other traits are also listed. Potential candidate genes are presented in the last column

Chromosome	Mb	Thoracic variation	Lumbar variation	Thoracolumbar variation	Potential candidate gene symbol
4	114	1.11			MAB21L3
5	1	2.22			WNT7B
5	19		1.46		HOXC
5	70	1.80		0.40	TULP3
5	71	1.96			WNT5B
5	102	1.03			ALX1
6	81	2.35			MATN1
6	83	1.03			COL16A1
6	93		0.54	1.30	ARHGAP28
6	96			1.40	MYOM1
6	98			4.29	ADCYAP1
6	99	8.99		2.38	GATA6
6	102			1.13	ZNF521
6	103	0.59		2.68	ZNF521
6	146	2.86			LRP8
7	5			1.20	BMP6
7	54	1.04			ANKRD34C
7	103		1.20		VRTN
7	119			1.73	C14orf159
8	93			1.01	MAML3
8	98			1.06	PCDH10
9	14		2.30		TENM4
9	121		1.00		EZH2
9	124		6.03		TPK1
10	3		1.22		RGS18
11	25			1.49	AKAP11
12	19	1.28	1.25		MEOX1
12	24			10.26	HOXB
12	26		8.59	4.76	COL1A1
12	27			3.66	CHAD
12	34			6.21	MSI2
14	49	1.18			KREMEN1
14	75	1.38			SIRT1
14	80			1.23	UNC5B
15	30		1.18		CNTNAP5
16	18		4.80		CDH6
16	19			3.80	C5orf22
16	29		8.26		FGF10
16	30		1.08		FGF10
16	31		3.27		HCN1
16	45		1.20		RNF180
17	49		0.50	1.99	PTPRT

Table 2 Results from GWAS for vertebral traits including chromosome, one megabase window and percent of genomic variation associated with the one megabase window for all significant associations (>1.0 %). Significant regions which were also suggestive (>0.4 %) for other traits are also listed. Potential candidate genes are presented in the last column *(Continued)*

18	8	2.01		CLEC5A
18	48		2.36	FKBP14
18	50	17.35	2.70	HOXA
18	51	2.50		HOXA
18	54	2.80		RAMP3
X	41	4.72		CASK
X	140	1.58		MTMR1

on 10 chromosomes and 13 unique regions separated by at least 10 Mb.

Additional file 1: Table S1 contains information on all 1 MB regions that exceeded 0.4 % of the genomic variation along with relevant data on the SNP within the window that had the largest estimated effect on the phenotype.

Discussion

Validation of Vertnin for vertebrae number

Variation in vertnin is clearly associated with the number of ribs and thoracolumbar vertebra in pigs. Our estimates of vertnin's effect (0.76 ribs per copy of the mutant allele) on vertebra development concur with previous studies [12, 18]. The current study is the first genome wide association study that has fit variation in vertnin as a fixed effect. By accurately adjusting for variation in vertnin, we were able to detect additional loci affecting vertebra development in pigs

that would have been masked without adjusting for vertnin (data not shown). In addition, the adjustment appeared to account for all phenotypic variation associated with this region for RIB and TLV. An association with SSC 7:103 Mb with LVN was detected. The most significant marker (DIAS0001088) exhibited a very low level of linkage disequilibrium (Additional file 2: Figure S1) with vertnin and was approximately 500 kb from vertnin, so it is possible a different gene is responsible for this association. Despite the large effect of vertnin, there are a number of other loci that affect the development of the vertebral column in pigs that can be exploited via selection to alter the number and type of vertebra in pigs. By utilizing this approach, 18 novel 1 Mb windows were discovered (based on QTLdb; http://www.animalgenome.org/cgi-bin/QTLdb/SS/index), and several other regions were only represented by QTL spanning 20 or more Mb.

Table 3 Results from GWAS for kyphosis including chromosome, one megabase window and percent of genomic variation associated with the one megabase window for all significant associations (>1.0 %). Potential candidate genes are presented in the last column

Chromosome	Mb	Percent genomic variation	Potential candidate gene symbol
1	287	1.83	TNC
2	1	2.7	SHANK2
2	7	4.01	FLRT1
2	12	2.02	GLYAT
5	72	1.93	CECR2
5	73	2.28	CPNE8
6	60	1.07	CCDC27
6	105	1.55	
7	125	4.04	EMX2/VRK1
8	96	2.27	
13	145	1.87	ITGB5
15	143	1.39	
16	63	1.25	TENM2
X	5	2.27	
X	39	3.47	MID1IP1
X	136	5.45	SLITRK2

Comparison with other studies in swine for vertebrae number

Only four direct overlaps were observed among studies counting vertebrae. Ren and coworkers [19] found a QTL affecting both total vertebrae and thoracic vertebrae number on SSC 12 spanning 0.1-54.4 Mb in a population containing Duroc and Erhualian germplasm. Within this broad range, our study found seven different significant associations in five different 1 Mb windows (19, 24, 26, 27 and 34 Mb). All but one (19 Mb) significantly affected number of thoracolumbar vertebra. The window SSC 12:19 Mb affected both RIB and LUM while SSC 12:26 Mb was associated with LUM as well as TLV. Ren and coworkers [19] also reported a QTL spanning SSC 7:103 Mb for LUM. In a Duroc-Pietrain population, Edwards et al. [20] reported a QTL for number of ribs spanning SSC 7:54 Mb where we report a QTL for thoracic vertebra number. Harmegnies et al. [21] found QTL for rib number at SSC 18:24–52.2 Mb in a commercial population overlapping QTL for thoracic vertebra number at SSC 18:48–54 Mb in the present study. Surprisingly, the region of SSC 16:30–35 Mb was identified as controlling number of ribs in 2 different studies using commercial pigs [21, 22] but this region was associated with number of lumbar vertebrae in the current population.

Most published QTL that overlap these results measured carcass length rather than actual vertebra numbers. Several of these studies utilized F2 populations containing Chinese germplasm and overlap QTL detected in the current study on chromosomes 4 [23], 6 [24], 7 [25], 9 [24, 26], 14 [27], 17 [28], 18 [21] and X [9, 29]. Studies evaluating carcass length in commercial pigs confirmed QTL located on chromosome 6 [30–32], 12 [33] and 18 [33].

While carcass length is an economically important trait in pig production, vertebra numbers only account for a small proportion of the variation in carcass length [4]. Vertebra numbers have been associated with leaner carcasses [4] but data from our population was unable to substantiate this association ($P > 0.50$ for all analyses). Growth rate was associated ($P < 0.03$) with TLV but neither of the other two vertebra counts. Animals with more thoracolumbar vertebrae grew faster. Research reported nearly 50 years ago [5, 6] indicated that animals with more vertebrae suffered from lameness and movement problems. We do not monitor locomotion in our population, but one would presume lameness would result in reduced growth. If this presumption is correct, then animals with more thoracolumbar vertebra were not suffering from issues with mobility. A second production trait that has been associated with number of vertebrae is rate of stillbirths. Rees Evans [7] found that sows with more vertebrae had a higher incidence of stillborn piglets. Fredeen and Newman [34] contradicted the earlier report and our data also did not find any association

with number of stillborn piglets and sow vertebra numbers ($P > 0.40$).

Four QTL regions were associated with more than one trait. The window of SSC 6:99 Mb was associated with coordinated changes in number of thoracic and thoracolumbar vertebrae while SSC 12:26 Mb was associated with coordinated changes in number of lumbar and thoracolumbar vertebrae numbers. Therefore these two regions increased the total number of vertebra but only affected one type of vertebra. Contrarily, SSC 18:50 Mb had a large effect on number of thoracic vertebrae and an opposite effect on number of lumbar vertebrae resulting in no significant change in number of thoracolumbar vertebra numbers. Finally, SSC 12:19 Mb had small effects on both thoracic and lumbar vertebra numbers with an undetectable effect on thoracolumbar vertebrae numbers. These associations were only marginally over the 1.0 % genomic variation threshold and the SNP markers with the largest effect on both traits were separated by 146 kb and had very different allele frequencies so the effects appear to be independent of each other.

To date, only two genes affecting vertebra numbers in pigs have been identified. Mikawa et al. [11] discovered a mutation in NR6A1 using crosses of Asian and European pigs , located on SSC 1:299 Mb; however, variation in this gene has been fixed in commercial (European descent) pigs based on genotypes and evidenced by a strong selective sweep [35]. No evidence of additional genetic variation in NR6A1 was found in the current population studied. Vertnin was discovered by the same group of scientists [12] and since its discovery it has been associated with vertebra numbers in multiple populations and within most germplasm [8,12,18,19]. Estimates of vertnin's effect have been consistent at approximately 0.5 to 0.6 additional ribs (or thoracolumbar vertebrae) per copy of the mutant allele.

The genome wide association study discovered QTL for a surprisingly large percentage of the estimated genomic variation (48 to 59 %). The power to detect QTL in this study is quite high as it has the most phenotyped animals within a single population and utilized over 40,000 genetic markers. However, the genetic architecture of vertebral development should also be considered. Development of the vertebral column is a critically important process for survival, so it is highly conserved and strictly regulated [2]. Few species exhibit phenotypic variation for these traits and likely do not possess genetic variation. Therefore, it should not be surprising that the phenotypic variation observed in the pig may be due to a small set of genes. This was evident by our estimates of π in the Bayes Cπ analyses as estimates were > 0.999, indicating approximately 40 QTL should be expected. All of these facts together indicate that variation in vertebra numbers in pigs does not fit the infinitesimal model as well as it does a model where a few genes with moderate to large effects regulate this phenotype.

Further investigation of the QTL regions discovered in this study is necessary to determine the gene(s) associated with variation in vertebra numbers. This information will permit the use of marker-assisted selection decisions for producers wanting to change vertebra numbers in their populations as well as provide insight into the genetic mechanisms controlling this important developmental process. Determining the genes responsible for variation in vertebra numbers in pigs may also provide insight into why this developmental process is much less conserved in pigs than most other mammalian species.

Identification of candidate genes for vertebrae number

Spatial cues which will result in segmental structures, such as vertebrae, result from the combinatorial expression of Hox genes within specific somites of the developing embryo [36, 37]. Manipulation of Hox genes can cause transformations of the vertebral column in mice [38, 39] and expression patterns of various Hox genes correlates with anatomical position across species [2]. Therefore, it shouldn't be surprising that three of the QTL discovered are located where a cluster of Hox genes are located. The largest QTL for TLV (SSC 12:24) overlaps the *HoxB* cluster. This QTL tends to affect the number of total vertebrae or segments either independently of vertebra type or possibly by increasing number of lumbar vertebra (association detected at 24 Mb). The region associated with the most genomic variation for RIB was located over the *HoxA* (SSC 18:50 Mb) gene cluster. As previously noted this region appears to convert vertebrae from thoracic to lumbar or *vice versa*. Finally, the *HoxC* cluster (SSC 5:19 Mb) appears to have a small effect on number of lumbar vertebrae. Along with Hox genes, the WNT gene family also is a key regulator of embryonic development. Two RIB QTL colocate with the family members *WNT5B* (SSC 5:71.09 Mb) and *WNT7B* (SSC 5:0.66 Mb).

A few mutations have been studied in mice. One spontaneous recessive mutation caused split and/or fused ribs and was named rib-vertebrae [40]. Evaluation of this mutation implicated the Notch signaling pathway [41] and a discovery of a regulatory mutation in the gene *TBX6* was discovered [42]. While we did not find any associations near where *TBX6* should map, a family member, *TBX4*, does reside on SSC 12:38.23 Mb which is close to a QTL affecting TLV. Interestingly, *TBX4* is known to interact with *FGF10* which resides on SSC 16:30.24 Mb, which is central to LVN QTL on SSC16 at 29, 30 and 31 Mb. Another candidate gene associated with NOTCH signaling is MAML3 (SSC 8:92.37 Mb) located near a TLV QTL at SSC 8:93 Mb.

McPerron et al. [43] showed that knockout of *GDF11* resulted in anterior transformation of the vertebral column

and increased the number of ribs in a dose dependent (additive) manner. *GDF11* in pigs is located on SSC 5:22.7 Mb only 3 Mb from a QTL associated with number of lumbar vertebrae, but the *HoxC* cluster is located directly in the QTL region and is a more likely candidate gene. *NODAL* is a TGF-beta superfamily member involved in early embryogenesis located at SSC 14:79.27 Mb, less than 1 Mb from a QTL for TLV (SSC 14:80) and a region of interest for RIB (Additional file 1: Table S1); thus, this gene may be responsible for the variation observed in the current study. Another candidate in this region at 80 Mb is the netrin-1 receptor *UNC5B*, which is a regulator of osteoclast differentiation [44]. Finally, teneurin-4 (*TENM4*) on SSC 9:14.33 Mb has been shown to be a critically important developmental gene necessary for the appropriate development of somites and thus the skeleton [45]. In addition, Lossie et al. [45] reported that mutations in the coding region of *TENM4* resulted in fusion of thoracic and/or lumbar vertebrae. *TENM4's* location within a LVN QTL region and reported phenotypic effects of mutants in mice makes it a prime candidate for additional studies.

Several candidate genes involved in bone formation, osteoblast differentiation or chondrocyte differentiation include *ZNF521* on SSC 6:102.99 Mb [46], *BMP6* on SSC 7:5.16 Mb [47, 48], and *CHAD* on SSC 12:26.83 Mb [49] were all associated with TLV while *CLEC5A* on SSC 18:8.31 Mb and *RAMP3* on SSC 18:55.16 Mb were associated with RIB. While none of these genes have been reported to affect number of vertebra, their involvement in bone formation is intriguing and worthy of further investigation.

Several of these QTL overlap regions predicted to contain copy number variation based on analyses of SNP60 BeadChip data of the current population [50]. While many of the CNV in Wiedmann et al. [50] were predicted to segregate in only a few animals, QTL located on SSC 12:19 Mb (RIB and TLV), 16:19 Mb (TLV) and 16:29 Mb (LVN) were variable in over 20 families. However, predicted number of copies for specific animals by Wiedmann et al. [50] did not correlate well with phenotypic data adjusted for VRTN genotype.

Genetic basis of Kyphosis

Kyphosis has been reported in commercial pigs. However, the kyphosis originally described by Holl et al. [7] within the USMARC and a Duroc-Landrace F2 population had not been previously reported. Lindholm-Perry et al. [51] conducted a genome scan on a subset of the animals used in the current study with a much lower marker density. They discovered several nominally significant associations in the USMARC and the Duroc-Landrace F2 population, but no associations were consistent across populations. Only a couple of the associations reported by Lindholm-Perry et al. [51] were near QTL reported in the current study. The QTL in the present study at SSC 5:72 and 73 Mb is near the association with marker 2928_3 (rs# 45434241), the QTL at SSC 6:60 Mb is only 5 Mb from PLOD1 which was associated with kyphosis in the linkage analysis and finally the QTL at SSC 6:105 Mb is near the microsatellite marker APR18. However, the largest QTL from the present study (SSC 2:7 Mb, 7:125 Mb and SSC X:136 Mb) were not detected by Lindholm-Perry et al. [51]. The difference in results can be explained by the nearly two-fold increase in phenotyped animals, a 200-fold increase in marker density and utilizing a Bayesian analysis.

While there were no direct overlapping associations for kyphosis and any vertebra trait, there were a number of QTL located within 2 Mb of each other. The large kyphosis QTL at SSC X:39 Mb (3.5 % genomic variation) was near the RIB QTL at SSC X:41 Mb (4.7 % genomic variation) while the kyphosis QTL at SSC 5:72–73 Mb (collectively accounting for 4.2 % of the genomic variation) was only 2 Mb away from a RIB QTL at SSC 5:70–71 Mb (collectively accounting for 3.8 % of the genomic variation). Whether these colocalizations are due to pleiotropic effects or merely linked QTL it may help explain the trend seen where for each additional rib, the frequency of mildly affected pigs increases by 3 %. The kyphosis QTL at SSC 6:105 Mb and 8:96 Mb were near TLV QTL, but both of these QTL had smaller estimated effects.

Conclusion

Regulation of vertebral column development in pigs is controlled by numerous genes. In most species, genetic variation in this regulatory system is not tolerated. The pig is quite unique in the variability present and the amount of phenotypic variation displayed as a species. These results highlight the importance of the HOX gene families in embryonic development. Candidate genes in TGF-beta superfamily members were also detected. Further studies of vertebra numbers in pigs will provide insight into this developmental mechanism and provide natural genetic variants for future basic research. In addition, modification of vertebra numbers in commercial swine is possible and can be fortified by use of genetic markers.

Abbreviations

BMP6: bone morphogenetic protein 6; *CHAD*: chondroadherin; CLEL5A: C-type lectin domain family 5 member A; CNV: copy number variation; FASS: Federation of Animal Science Societies; FGF10: fibroblast growth factor 10; FSIS: Food Safety Inspection Service; GDF11: growth/differentiation factor 11; GWA: genome-wide association; HOX: Homeo box; LVN: number of lumbar vertebrae; Mb: megabase; NODAL: nodal homolog precursor; NR6A1: nuclear receptor subfamily 6 group A member 1; PLOD1: procollagen-lysine,2-oxoglutarate 5-dioxygenase 1; QTL: quantitative trait locus (loci); RAMP3: receptor activity-modifying protein 3 precursor; RIB: number of ribs; SNP: single nucleotide polymorphism; SSC: sus scrofa; TBX6: T-box transcription factor 6; *TENM4*: teneurin-4; TLV: number of thoracolumbar; UNC5B: netrin receptor UNC5B precursor; USDA: United States Department of Agriculture; USMARC: U.S. Meat Animal Research Center; VRTN: vertnin; ZNF521: zinc finger protein 521.

Competing interests

The authors declare that they have no competing interests.

Authors' contributions

GAR collected phenotypic and genotypic data and wrote the manuscript. DJN contributed to interpretation of results and identification of candidate genes. RTW processed genotypic data and mapped markers to sus scrofa build 10.2. JFS developed the statistical analysis process and conducted genome wide association analyses. All authors contributed to the design of this study, read, and approved the final manuscript.

Acknowledgements

The authors would like to acknowledge the expert technical assistance of K. Simmerman and L. Parnell.

Mention of trade names or commercial products in this publication is solely for the purpose of providing specific information and does not imply recommendation or endorsement by the U.S. Department of Agriculture. The U.S. Department of Agriculture (USDA) prohibits discrimination in all its programs and activities on the basis of race, color, national origin, age, disability, and where applicable, sex, marital status, familial status, parental status, religion, sexual orientation, genetic information, political beliefs, reprisal, or because all or part of an individual's income is derived from any public assistance program. (Not all prohibited bases apply to all programs.) Persons with disabilities who require alternative means for communication of program information (Braille, large print, audiotape, etc.) should contact USDA's TARGET Center at (202) 720–2600 (voice and TDD). To file a complaint of discrimination, write to USDA, Director, Office of Civil Rights, 1400 Independence Avenue, S.W., Washington, D.C. 20250–9410, or call (800) 795–3272 (voice) or (202) 720–6382 (TDD). USDA is an equal opportunity provider and employer.

References

1. Owen R. Descriptive catalogue of the osteological series contained in the museum of the Royal College of Surgeons of England. London: Royal College of Surgeons; 1853.
2. Narita Y, Kuratani S. Evolution of the vertebral formulae in mammals: A perspective on developmental constraints. J Exp Zool (Mol Dev Evol). 2005;304B:91–106.
3. King JWB, Roberts RC. Carcass length in the bacon pig; its association with vertebrae numbers and prediction from radiographs of the young pig. Anim Prod. 1960;2:59–65.
4. Borchers N, Reinsch N, Kalm E. The number of ribs and vertebrae in a Piétrain cross: variation, heritability and effects on performance traits. J Anim Breed Genet. 2004;121:392–403.
5. Duckworth JE, Holmes W. Selection for carcass length in Large White pigs. Anim Prod. 1968;10:359–72.
6. Meyer H, Lindfeld A. [Number and length of vertebrae in the improved German country-pig] (In German). Dtsch Tieraztl Wochenschr. 1969;76:448–53.
7. Rees Evans ET. Investigations on the vertebral column of the Welsh pig. Proc Br Soc Anim Prod. 1954; pp. 65–73.

8. Holl JW, Rohrer GA, Shackelford SD, Wheeler TL, Koohmaraie M. Estimates of genetic parameters for kyphosis in two crossbred swine populations. J Anim Sci. 2008;86:1765–9.

9. Rohrer GA, Keele JW. Identification of quantitative trait loci affecting carcass composition in swine: II. Muscling and wholesale product yield traits. J Anim Sci. 1998;76:2255–62.

10. Mikawa S, Hayashi T, Nii M, Shimanuki S, Morozumi T, Awata T. Two quantitative trait loci on Sus scrofa chromosomes 1 and 7 affecting the number of vertebrae. J Anim Sci. 2005;83:2247–54.

11. Mikawa S, Morozumi T, Shimanuki S-I, Hayashi T, Uenishi H, Domukai M, et al. Fine mapping of a swine quantitative trait locus for number of vertebrae and analysis of an orphan nuclear receptor, germ cell nuclear factor (NR6A1). Genome Res. 2007;17:586–93.

12. Mikawa S, Sato S, Nii M, Morozumi T, Yoshioka G, Imaeda N, et al. Identification of a second gene associated with variation in vertebral number in domestic pigs. BMC Genet. 2011;12:5.

13. FASS. (Federation of Animal Science Societies). Guide for the Care and Use of Agricultural Animals in Research and Teaching, Third Edition. Champaign, IL: Federation of Animal Science Societies; 2010.

14. Schneider JF, Rempel LA, Snelling WM, Wiedmann RT, Nonneman DJ, Rohrer GA. Genome-wide association study of swine farrowing traits. Part II: Bayesian analysis of marker data. J Anim Sci. 2012;90:3360–7.

15. Ramos AM, Crooijmans RPMA, Affara NA, Amaral AJ, Archibald AL, Beever JE, et al. Design of a high density SNP genotyping assay in the pig using SNPs identified and characterized by next generation sequencing technology. PLoS ONE. 2009;4, e6524.

16. Wolc A, Arango J, Settar P, Fulton JE, O'Sullivan NP, Preisinger, et al. Genome-wide association analysis and genetic architecture of egg weight and egg uniformity in layer chickens. Anim Genet. 2012;43 Suppl 1:87–96.

17. Schneider JF, Nonneman DJ, Wiedmann RT, Vallet JL, Rohrer GA. Genomewide association and identification of candidate genes for ovulation rate in swine. J Anim Sci. 2014;92:3792–803.

18. Fan Y, Xing Y, Zhang Z, Ai H, Ouyang Z, Ouyang J, et al. A further look at porcine chromosome 7 reveals VRTN variants associated with vertebral number in Chinese and Western pigs. PLoS ONE. 2013;8, e62534.

19. Ren DR, Ren J, Ruan GF, Guo YM, Wu LH, Yang GC, et al. Mapping and fine mapping of quantitative trait loci for the number of vertebrae in a White Duroc x Chinese Erhualian intercross resource population. Anim Genet. 2012;43:545–51.

20. Edwards DB, Ernst CW, Raney NE, Doumit ME, Hoge MD, Bates RO. Quantitative trait locus mapping in an F2 Duroc x Pietrain resource population: II. Carcass and meat quality traits. J Anim Sci. 2008;86:254–66.

21. Harmegnies N, Davin F, De Smet S, Buys N, Georges M, Coppieters W. Results of a whole-genome quantitative trait locus scan for growth, carcass composition and meat quality in a porcine four-way cross. Anim Genet. 2006;37:543–53.

22. Soma Y, Uemoto Y, Sato S, Shibata T, Kadowaki H, Kobayashi E, et al. Genome-wide mapping and identification of new quantitative trait loci affecting meat production, meat quality, and carcass traits within a Duroc purebred population. J Anim Sci. 2011;89:601–8.

23. Knott SA, Nyström PE, Andersson-Eklund L, Stern S, Marklund L, Andersson L, et al. Approaches to interval mapping of QTL in a multigenerational pedigree: the example of porcine chromosome 4. Anim Genet. 2002;33:26–32.

24. Paszek AA, Wilkie PJ, Flickinger GH, Miller LM, Louis CF, Rohrer GA, et al. Interval mapping of carcass and meat quality traits in a divergent swine cross. Anim Biotechnol. 2001;12:155–65.

25. Yue G, Stratil A, Cepica S, Schröffel Jr J, Schröffelova D, Fontanesi L, et al. Linkage and QTL mapping for Sus Scrofa chromosome 7. J Anim Br Genet. 2003;120 Suppl 1:56–65.

26. Wei WH, Duan Y, Haley CS, Ren J, de Koning DJ, Huang LS. High throughput analyses of epistasis for swine body dimensions and organ weights. Anim Genet. 2011;42:15–21.

27. Dragos-Wendrich M, Sternstein I, Brunsch C, Moser G, Bartenschlager H, Reiner G, et al. Linkage and QTL mapping for Sus scrofa chromosome 14. J Anim Breed Genet. 2003;120:111–8.

28. Pierzchala M, Cieslak D, Reiner G, Bartenschlager H, Moser G, Geldermann H. Linkage and QTL mapping for Sus scrofa chromosome 17. J Anim Breed Genet. 2003;120:132–7.

29. Ma J, Ren J, Guo Y, Duan Y, Ding N, Zhou L, et al. Genome-wide identification of quantitative trait loci for carcass composition and meat quality in a large-scale White Duroc x Chinese Erhualian resource population. Anim Genet. 2009;40:637–47.

30. Evans GJ, Giuffra E, Sanchez A, Kerje S, Davalos G, Vidal O, et al. Identification of quantitative trait loci for production traits in commercial pig populations. Genetics. 2003;164:621–7.

31. Kim J-J, Rothschild MF, Beever J, Rodriguez-Zas S, Dekkers JCM. Joint analysis of two breed cross populations in pigs to improve detection and characterization of quantitative trait loci. J Anim Sci. 2005;83:1229–40.

32. Choi I, Steibel JP, Bates RO, Raney NE, Rumph JM, Ernst CW. Identification of carcass and meat quality QTL in an F$_2$ Duroc x Pietrain pig resource population using different least-squares analysis models. Frontiers in Genetics. 2011;2:18.

33. Liu G, Jennen DGJ, Tholen E, Juengst H, Kleinwächter T, Hölker M, et al. A genome scan reveals QTL for growth, fatness, leanness and meat quality in a Duroc-Pietrain resource population. Anim Genet. 2007;38:241–52.

34. Fredeen HT, Newman JA. Rib and vertebral numbers in swine. I. Variation observed in a large population. Can J Anim Sci. 1962;42:232–9.

35. Yang G, Ren J, Zhang Z, Huang L. Genetic evidence for the introgression of Western NR6A1 haplotype into Chinese Licha breed associated with increased vertebral number. Anim Genet. 2009;40:247–50.

36. Kessel M, Gruss P. Homeotic transformations of murine vertebrae and concomitant alteration of Hox codes induced by retinoic acid. Cell. 1991;67:89–104.

37. Oh SP, Yeo C-Y, Lee Y, Schrewe H, Whitman M, Li E. Activin type IIA and IIB receptors mediate Gdf11 signaling in axial vertebral patterning. Genes Dev. 2002;16:2749–54.

38. Carrasco AE, López SL. The puzzle of Hox genes. Int J Dev Biol. 1994;38:557–64.

39. Maconochie M, Nonchev S, Morrison A, Krumlauf R. Paralogous Hox genes: Function and regulation. Annu Rev Genet. 1996;30:529–56.

40. Theiler K, Varnum DS. Development of rib-vertebrae: a new mutation in the house mouse with accessory caudal duplications. Anat Embryol Berl. 1985;173:111–6.

41. Beckers J, Schlautmann N, Gossler A. The mouse rib-vertebrae mutation disrupts anterior-posterior somite patterning and genetically interacts with a Delta1 null allele. Mech Dev. 2000;95:35–46.

42. Watabe-Rudolph M, Schlautmann N, Papaioannou VE, Gossler A. The mouse rib-vertebrae mutation is a hypomorphic Tbx6 allele. Mech Dev. 2002;119:251–6.

43. McPherron AC, Lawler AM, Lee S-J. Regulation of anterior/posterior patterning of the axial skeleton by growth/differentiation factor 11. Nat Genet. 1999;22:260–4.

44. Mediero A, Ramkhelawon B, Perez-Aso M, Moore KJ, Cronstein BN. Netrin-1 is a critical autocrine/paracrine factor for osteoclast differentiation. J Bone Miner Res. 2015;30:837–54.

45. Lossie AC, Nakamura H, Thomas SE, Justice MJ. Mutation of l7Rn3 shows that Odz4 is required for mouse gastrulation. Genetics. 2005;169:285–99.

46. Addison WN, Fu MM, Yang HX, Lin Z, Nagano K, Gori F, et al. Direct transcriptional repression of Zfp423 by Zfp521 mediates a bone morphogenic protein-dependent osteoblast versus adipocyte lineage commitment switch. Mol Cell Biol. 2014;34:3076–85.

47. Perry MJ, McDougall KE, Hou S, Tobias JH. Impaired growth plate function in bmp-6 null mice. Bone. 2008;42:216–25.

48. Solloway MJ, Dudley AT, Bikoff EK, Lyons KM, Hogan BLM, Robertson EJ. Mice lacking Bmp6 function. Dev Genet. 1998;22:321–39.

49. Hessle L, Stordalen GA, Wenglén C, Petzold C, Tanner EK, Brorson S-H, et al. The skeletal phenotype of chondroadherin deficient mice. PLoS ONE. 2013;8:e63080. Erratum in: PLoS One. 2013;8(7).

50. Wiedmann RT, Nonneman DJ, Rohrer GA. Genome-wide copy number variations using SNP genotyping in a mixed breed swine population. PLoS ONE. 2015;10, e0133529.

51. Lindholm-Perry AK, Rohrer GA, Kuehn LA, Keele JW, Holl JW, Shackelford SD, et al. Genomic regions associated with kyphosis in swine. BMC Genet. 2010;11:112.

7

A test of somatic mosaicism in the androgen receptor gene of Canada lynx (*Lynx canadensis*)

Melanie B. Prentice[1*], Jeff Bowman[2] and Paul J. Wilson[3]

Abstract

Background: The *androgen receptor*, an X-linked gene, has been widely studied in human populations because it contains highly polymorphic trinucleotide repeat motifs that have been associated with a number of adverse human health and behavioral effects. A previous study on the *androgen receptor* gene in carnivores reported somatic mosaicism in the tissues of a number of species including Eurasian lynx (*Lynx lynx*). We investigated this claim in a closely related species, Canada lynx (*Lynx canadensis*). The presence of somatic mosaicism in lynx tissues could have implications for the future study of exonic trinucleotide repeats in landscape genomic studies, in which the accurate reporting of genotypes would be highly problematic.

Methods: To determine whether mosaicism occurs in Canada lynx, two lynx individuals were sampled for a variety of tissue types (lynx 1) and tissue locations (lynx 1 and 2), and 1,672 individuals of known sex were genotyped to further rule out mosaicism.

Results: We found no evidence of mosaicism in tissues from the two necropsied individuals, or any of our genotyped samples.

Conclusions: Our results indicate that mosaicism does not manifest in Canada lynx. Therefore, the use of hide samples for further work involving trinucleotide repeat polymorphisms in Canada lynx is warranted.

Keywords: Somatic mosaicism, Androgen receptor, Canada lynx, Trinucleotide repeats

Background

The X-linked *androgen receptor* (*AR*) gene codes for a transcription factor that controls the binding of androgens in different tissue types [1–3]. The organization and location of the *AR* gene on the X-chromosome has been conserved for both male and female placental, marsupial and monotreme mammals [3, 4]. Androgenic hormones including testosterone and dihydrotestosterone are integral in a number of bodily processes, most notably sexual differentiation and development [5]. The wide range of functions that the *AR* gene encompasses has concurrently lead to a range of disease-associated phenotypes, which have been linked to variable tandem trinucleotide repeats occurring in the first codon of the

AR gene coding sequence [6]. Trinucleotide repeats are repeat structures that consist of units that are 3 nucleotides long, caused by the selection against frame-shift mutations which would alter the reading frame of the transcribed protein [7]. The natural variation of these repeats within humans indicates that these motifs have a critical role in "normal" protein function and evolutionary adaptation [8, 9]. More specifically, trinucleotide repeats are known to affect phenotype, such that disease in humans has been attributed to frequency of repeats exceeding a certain threshold, beyond which, the transcriptional activity of the *AR* gene is affected [10, 11]. For this reason, trinucleotide repeat fragments of the *AR* gene have been extensively studied in humans for their potential role in infertility [12, 13], aggressive or dominant behavior [14–16], criminal activity [17, 18], personality disorders [19, 20], and the development of some cancers and other diseases [21–25].

* Correspondence: melanieprenti@trentu.ca
[1]Department of Environmental & Life Sciences, Trent University, 1600 West Bank Drive, Peterborough K9J 7B8, ON, Canada
Full list of author information is available at the end of the article

Studies of the *AR* gene in wildlife are rare but are likely to become more frequent in the future as the role of trinucleotide markers in mediating adaptive evolution in contemporarily short time-frames becomes more clear [26]. While it is well understood that climate change will have profound effects on wildlife [27], we are currently unable to predict whether species will be able to adapt and evolve new strategies to cope with the increasing environmental change. The characterization of exonic standing genetic variability will therefore allow for a better understanding of the adaptive capacities of populations to be resilient to the effects of stressful events including climate change. As a result, there is a recognized need to identify and characterize the genetic variability of fitness-related traits [28] and the response of genes to environmental change [29, 30]. Trinucleotide repeats are particularly desirable candidates for studies of the genomics of adaptation because they occur in as many as 20 % of human genes, have relatively higher rates of mutation than single nucleotide polymorphisms (SNPs), and can show consistently high levels of within-population variation [6, 26]. Importantly, such high rates of mutation may facilitate adaptation to stressors (e.g., climate change) in contemporarily short timeframes. Recently, several studies have demonstrated the potential evolutionary and adaptive importance of trinucleotide repeats within clock genes in both birds [31] and fish [32]. Thus, the study of trinucleotide repeat structures in a range of other vertebrate species [8, 26, 33, 34] offers the potential to use the properties of microsatellite repeats [35] to understand the genomics of rapid adaptation.

Historically, the characterization of the *AR* gene has been affected by biological and technical issues, with implications for accurate genotyping. More specifically, somatic mutations and allele peak morphology issues have been encountered upon scoring size separated alleles differing in the number of exonic trinucleotide repeats [36–38]. Mosaicism in biological systems can be defined as "the presence of more than one genetically distinct cell line in a single organism" in which tissue-to-tissue genetic variations occur that may not follow Mendelian rules of inheritance ([39]; p. 748). More recently, Köhler et al. (2005) [p. 106] describe somatic mosaicism as "different proportions of cells containing either mutant or wild-type proteins that are present in various tissues of the same individual [22]". Telenius et al. (1994) provided the first report of heterogenic somatic mosaicism of CAG repeats in tissues [40]. Since then, several studies have detected tissue-specific somatic mosaicism of CAG repeats in the *AR* gene in both the neural and non-neural tissues of individuals with Huntington's disease, spinal bulbar muscular atrophy, spinocerebellar ataxia type 1, denatorubural-pallidoluysian atrophy and

Machado-Joseph disease [21]. For individuals with androgen insensitivity syndrome, genotype-phenotype discrepancies have been traced to somatic mosaicism of the *AR* gene itself [36, 37].

Much of the research conducted on the *AR* gene to date has involved the study of human disease. Trinucleotide repeats in the *AR* gene have yet to be correlated with transcriptional activity in species other than humans, and the limited number of studies that have been conducted on other species suggests lower levels of variability than in humans [41, 42]. Of particular interest is a study by Wang et al. (2012) who examined the variability of *AR* trinucleotide repeat in carnivores through sequencing of the first exon in the *AR* gene (containing three trinucleotide repeat tracts) [42]. The authors reported a change in CAG repeat number in the same tissues of a number of carnivore species, indicating tissue-specific mosaicism patterns in the *AR* gene of studied species. In their study, somatic mosaicism was evident in all three poly-glutamine tracts within exon 1of the *AR* gene, with a maximum extent of five alleles in several carnivore species. The authors concluded that the higher frequency of tissue-specific mosaicism in the *AR* gene of carnivores compared to other studied taxa implies that carnivores tend to exhibit mosaicism [42].

The objective of our study was to test for somatic mosaicism in a carnivore, the Canada lynx (*Lynx canadensis*). Canada lynx are closely related to the Eurasian lynx (*Lynx lynx*), one of the species shown by Wang et al. (2012) to exhibit somatic mosaicism. We consider it important to evaluate the potential for somatic mosaicism in Canada lynx before conducting further research on the *AR* gene. If allelic patterns of mosaicism are revealed, simple genotyping of individuals may not provide conclusive results with respect to genetic variability of individuals at this gene, which could complicate high throughput genotyping of individuals at the *AR* gene. Further, if mosaicism in this gene is caused by trinucleotide repeat instabilities, there will be important consequences for future studies that wish to examine trinucleotide repeat variability in wildlife species at any gene. This makes the investigation of potential somatic mutations a worthwhile goal as somatic mosaicism could significantly confound the use of trinucleotide repeat markers in the study of the adaptive genomics of wildlife. In such a case, we will need to begin considering the more dynamic nature of genes within genomes when designing studies, in particular those containing trinucleotide repeats.

We test the hypothesis that somatic mosaicism occurs in the androgen receptor gene in Canada lynx. We report *AR* genotypes for multiple samples taken from two necropsied lynx, as well as hide samples from lynx sampled at multiple locations across Canada.

Methods

To address the question of whether or not Canada lynx exhibit mosaicism at the *AR* gene, we designed a study that was composed of two levels of analysis. First, we conducted necropsies and tissue sampling of two lynx individuals (one full carcass and one hide), which allowed for multiple samples of various tissue types to be taken from one individual and a variety of sampling locations spanning the entire lynx carcasses in both individuals. Second, as we recognize that the sample size from the necropsies alone is limited, we genotyped additional samples collected across the Canada lynx range to verify our findings on a broader scale. Canada lynx are currently listed as not at risk by the Committee on the Status of Endangered Wildlife in Canada (COSEWIC), and are legally harvested annually. Thus, we obtained our additional samples either through licensed, commercial fur harvest, or under the authority of the Ontario Ministry of Natural Resources and Forestry (OMNRF). While sequence data would provide additional information about repeat purity (i.e., perfect vs. imperfect repeat structures) and the potential for SNPs within the flanking regions of the repeats, we conducted microsatellite genotyping on all of our samples as mosaicism can very easily be detected as size based variants. Mosaicism was evident in [42] largely based on size, indicating that if mosaicism is present in our study species, we should be able to detect it given our study used the same primers as [42] in addition to our large sample size.

Necropsy sampling

To test the hypothesis that somatic mosaicism exists in Canada lynx tissues, a necropsy was conducted for strategic sampling of two lynx individuals. The first individual (lynx 1) consisted of an entire carcass and the second (lynx 2) was a hide. The lynx carcass was a road-killed individual that was collected by the Ontario Ministry of Natural Resources and Forestry in 2010 and stored frozen until tissue sampling was conducted to ensure optimal preservation of high-quality tissues for DNA extraction. The lynx hide was collected in 2006 from a fur harvester in Ontario, Canada. It was important for the purpose of assessing the influence of the *AR* gene in different tissues, to obtain and analyze the genetic profile of a large number of different cell types. A total of 87 hide, muscle, liver and brain samples were taken from the two individuals. The liver we sampled had five lobes; two main lobes rested on top of three smaller lobes.

DNA extraction, quantification and amplification

DNA extraction and quantification was solely performed on the necropsy samples. DNA for the remaining 1,672 lynx samples (979 males and 693 females) was previously extracted from hide tissue according to the protocols outlined in [43], and was available in working concentration for PCR amplification. The availability of hide tissues from both museum specimens and fur auction houses makes this tissue type highly accessible for the genetic surveying of Canada lynx and other furbearer populations (e.g., [44–47]). The hide samples in our study represent individuals trapped in Yukon, British Columbia, Alberta, Manitoba, Ontario, and Quebec, Canada, as well as Alaska, USA.

Tissues were prepared for extraction by mincing approximately 1 mm X 1 mm pieces of tissue and placing it in 500ul of 1X lysis buffer [4 M Urea, 0.2 M NaCl, 0.5 % n-lauroyl sarcosine, 10 mM 1,2-cyclohexanediaminetetraacatic acid (CDTA), 0.1 M Tris–HCl (pH 8) and 600 U/ml proteinase K (Roche Applied Science, Laval QC)]. DNA from tissues was extracted by a modified version of the MagneSil® (Promega) manufacturers protocol, in which 200ul of the prepared tissues was substituted for the suggested 60ul of whole blood, and the number of wash steps was reduced [48]. All liquid handling was carried out by a JANUS® Automated Workstation from Perkin Elmer. Extracted DNA was quantified by PicoGreen® (Invitrogen) method according to the manufacturers protocols [49, 50].

From quantification, samples were normalized to a working concentration of 2.5 ng/ul and amplified with the primers developed by [42], which capture a ~700 bp region of exon 1 containing three trinucleotide repeat tracts. Amplification was conducted in a 10ul reaction containing deionized water (Invitrogen), 1X PCR Reaction Buffer (Invitrogen), 2 mM $MgCl_2$ (Invitrogen), 0.2 mM dNTP solution (Invitrogen), 0.2 mg/mL BSA, 0.4uM forward and reverse primers (forward primer labeled with the fluorescent dye HEX) (Integrated DNA Technologies), 0.025U Invitrogen Platinum Taq DNA Polymerase, and 5 ng of DNA. The PCR reaction was run in a Bio-Rad DNA Engine Dyad and Dyad Disciple thermocycler under the following conditions: 95 °C for 10 min; followed by 29 cycles of 94 °C for 30 s, 58 °C for 1 min, and 72 °C for 1 min, and completed with a step of 65 °C for 15 min.

Difficulties and biases in PCR amplification have been previously reported for the *AR* gene (e.g., [38]), most likely due to the high GC content in many exonic trinucleotide repeat fragments including *AR*. Many researchers have since obtained successful amplification and improved results by substituting Invitrogen Platinum Taq DNA Polymerase for the standard Invitrogen Taq DNA Polymerase (e.g., [51]). Such improvements were also evident in our study (Fig. 1).

Sexing of lynx necropsy individuals

The knowledge of sex for each individual allowed for the development of a search image for detecting mosaicism.

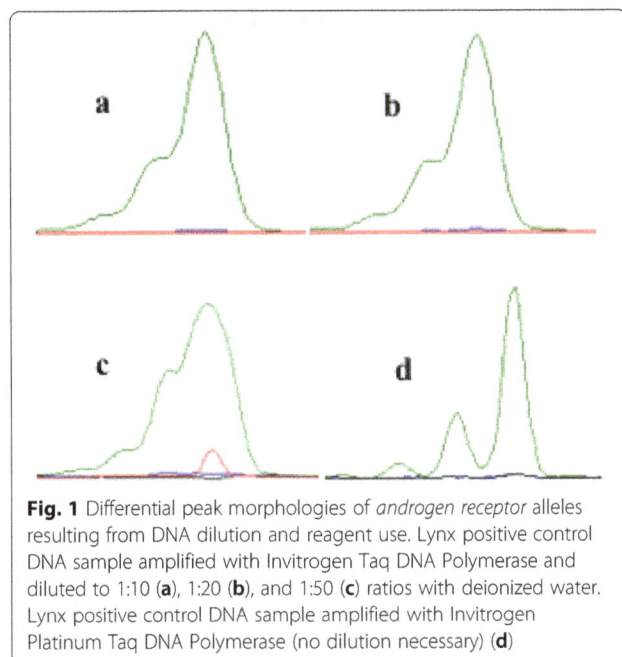

Fig. 1 Differential peak morphologies of *androgen receptor* alleles resulting from DNA dilution and reagent use. Lynx positive control DNA sample amplified with Invitrogen Taq DNA Polymerase and diluted to 1:10 (**a**), 1:20 (**b**), and 1:50 (**c**) ratios with deionized water. Lynx positive control DNA sample amplified with Invitrogen Platinum Taq DNA Polymerase (no dilution necessary) (**d**)

For male lynx tissues, a homozygous genotype is expected as the *AR* gene is X-linked, and males should therefore only inherit a single copy of the gene. In our study, any heterozygote male individual is a candidate for exhibiting mosaicism. Female lynx can be homozygous or heterozygous at the *AR* gene naturally, however, the allelic diversity of lynx at this locus predicts three allele patterns should be observed if mosaicism is occurring. If mosaicism were detected in female individuals with three alleles, the extent of mosaicism in females would still be an underestimate given that heterozygous females could be undetected somatic homozygous individuals. In the necropsy analysis, mosaicism would be suggested if more than the expected number of alleles were discovered across multiple samples from the same individual (i.e., more than one allele for males and two alleles for females across all samples).

To confirm sex of necropsied lynx, two samples from each individual (one hide and one muscle from Lynx 1 and two hide samples from Lynx 2) were amplified at two sex loci. The first primer pair, SRY-Y53-3D-F and SRY-Y53-3C-R amplified a ~218 bp region of the SRY genetic marker [52]. The second locus, a ~447 bp region of the ZFX/ZFY genetic marker, was amplified with the primer pair ZFX-P3-3EZ-F and ZFX-P3-5EZ-R [53]. Amplification was conducted in a 10ul reaction containing deionized water (Invitrogen), 10X PCR Reaction Buffer (Invitrogen), 50 mM $MgCl_2$ (Invitrogen), 100 mM dNTP solution (Invitrogen), 3 mg/mL BSA, 40uM forward and reverse primers (Integrated DNA Technologies) mentioned above (forward primers labeled with the fluorescent dye HEX), 0.0375U Invitrogen Taq DNA Polymerase, and

5 ng of DNA. The PCR reaction was run in a Bio-Rad DNA Engine Dyad and Dyad Disciple thermocycler under the following conditions: 94 °C for 15 min; followed by 29 cycles of 94 °C for 30 s, 52 °C for 1 min 30 s, and 72 °C for 1 min 30 s, and completed with a step of 60 °C for 45 min. Amplified samples were run on an 80 mL, 1.5 % agarose gel stained with ethidium bromide at 90 volts for 45 min, and visualized under ultraviolet light and to determine sex. Female individuals were identified by the presence of two bands, and males, by the presence of three bands on the gel. Controls of a known male and female lynx were included to rule out technological errors and strengthen conclusions.

Genotyping

For genotyping, 5ul of MapMarker 1000 X-Rhodamine (MM-1000-Rox) size standard (BioVentures) was mixed into 1 mL of deionized HiDi Formamide (Applied Biosystems), and 9.5ul of this product was added to 0.5ul of each amplified sample. Genotyping was performed on the Applied Biosystems 3730 DNA Analyzer. Genotypes were scored with SoftGenetics LLC GeneMarker AFLP/Genotyping Software Version 1.91. We used GenAlEx version 6.5 (Peakall & Smouse 2006, 2012) to calculate allele and genotype frequencies for both males and females.

Results & discussion

We observed ten different alleles across all genotypes samples, ranging between sizes 711–744 bp (including flanking sequence). The smallest three alleles observed were only found in a single female individual each, and

Table 1 Allele frequencies of the trinucleotide repeat tracts within exon 1 of the *androgen receptor* (*AR*) gene in Canada lynx (*Lynx canadensis*)

Allele	Frequency (Males only)	Frequency (Females only)	Frequency (All samples)
711	0.000	0.001	0.000
714	0.000	0.001	0.000
720	0.000	0.001	0.000
726	0.109	0.078	0.096
729	**0.401**	**0.379**	**0.392**
732	**0.289**	**0.335**	**0.313**
735	0.100	0.119	0.108
738	0.077	0.075	0.076
741	0.007	0.009	0.008
744	0.008	0.003	0.006

Frequencies are shown for male samples only (N = 979), female samples only (N = 693), and both males and females combined (all samples; N = 1672). As the *AR* gene is X-linked, and all males are therefore homozygous, allele frequencies are equivalent to genotype frequencies for males. No individuals were observed with alleles 717 or 723. The two most common alleles are in bold

Table 2 Genotype frequencies of the trinucleotide repeat tract within exon 1 of the *androgen receptor* (*AR*) gene in Canada lynx (*Lynx canadensis*). Frequencies are shown for female samples only ($N = 693$)

First allele/Second allele	711	714	720	726	729	732	735	738	741	744
711	-	-	-	-	-	-	-	-	-	-
714	0.001	-	-	-	-	-	-	-	-	-
720	-	-	-	-	-	-	-	-	-	-
726	-	-	-	0.009	-	-	-	-	-	-
729	-	-	-	0.059	0.160	-	-	-	-	-
732	-	-	-	0.039	0.253	0.123	-	-	-	-
735	-	-	-	0.019	0.078	0.081	0.020	-	-	-
738	-	-	0.001	0.017	0.040	0.048	0.017	0.012	-	-
741	-	-	-	0.004	0.006	0.004	0.001	0.001	-	-
744	-	-	-	-	0.001	0.001	0.001	0.001	-	-

no individuals with alleles 717 or 723 within the allelic range were found. The most common alleles were observed in the middle of the allelic range (Tables 1 and 2).

Sex identification indicated that the necropsied lynx represented one female (lynx 1) and one male (lynx 2) specimen. Of the tissues analyzed at the *AR* gene from these individuals (62 from lynx 1 and 25 from lynx 2), all resulted in a single clear genotype for each individual (a consistent homozygote and heterozygote genotype across all tissue samples for the male and female, respectively).

Additional genotyping of the 1,672 lynx samples did not detect somatic mosaicism in any of our male or female Canada lynx samples, although a single sample was removed from the data set due to contamination (see Additional file 1). All other samples fell within our search image of what is expected in a typical individual not exhibiting mosaicism (all males were homozygotes and no females exhibited more than two alleles). The absence of any evidence of mosaicism in Canada lynx does not provide conclusive evidence that it is not present in other, unanalyzed individuals, however, given the high allelic diversity of the *AR* gene in Canada lynx, if undetected, mosaicism would still only be present at a negligible level due to the large sample size we surveyed. For the purposes of our study, the overall lack of detection, coupled with our large sample size indicates that mosaic events do not pose a high risk of confounding large-scale analyses and genotyping in this study system, nor is an important biological mechanism within Canada lynx.

Our findings are inconsistent with those of Wang et al. (2012) who found *AR* mosaicism in multiple carnivore species [42]. It is possible that expression of the somatic mutation causing *AR* mosaicism is absent in Canada lynx in particular, but does manifest in Eurasian lynx and other carnivore tissues at a higher rate. As we evaluated a large sample of lynx hides, we suggest that lynx hide tissue can be used to study the *AR* gene in Canada lynx without the risk of issues caused by mosaicism.

Conclusions

The implications of somatic mosaicism within exonic trinucleotide repeat polymorphisms can have important influences on the accurate reporting and use of genotypes in studies of landscape genomics. This potential issue, however, is rarely considered in research outside of human disease studies. As the role of exonic repeat fragments in mediating adaptive evolution becomes clearer, it is likely that the prevalence of their use in wildlife genomic studies will increase. This makes the evaluation of somatic mosaicism in these repeat fragments imperative. In this study, we report no evidence of mosaicism in our two necropsied lynx individuals, or our larger screening of Canada lynx hide tissue. All males were homozygous for a single allele, and there was no evidence of more than two alleles in females, which would have been predicted if mosaicism was present given the allelic diversity of the gene in lynx. Our results indicate that even if mosaicism is present in this species, its prevalence is low given our inability to detect mosaicism in our large sample size. Therefore, the use of hide samples for further work involving trinucleotide repeat polymorphisms in Canada lynx is warranted, given that the *AR* gene appears to follow typical patterns of a X-linked gene in this species.

Availability of data

Genotypic data supporting the findings of this study can be found on the Dryad Digital Repository: http://dx/doi.org/10.5061/dryad.h43c1.

Abbreviations
AR: Androgen receptor; SNPs: Single nucleotide polymorphisms;
COSEWIC: Committee on the Status of Endangered Wildlife in Canada;
OMNRF: Ontario Ministry of Natural Resources and Forestry.

Competing interests
The authors declare that they have no conflict of interest.

Authors' contributions
PJW and JB participated in the design and coordination of the study, the
interpretation of the results and helped in editing the draft manuscript.
MBP carried out the necropsy sampling, carried out the molecular genetic
analyses and drafted the manuscript. All authors read and approved the
final manuscript.

Acknowledgements
The authors would like to acknowledge the North American Fur Auctions
(NAFA) for the contribution of all Canada lynx hide samples, and the Ontario
Ministry of Natural Resources and Forestry (OMNRF) for the contribution
of the two lynx carcasses utilized in this study. We would also like to
acknowledge Carrie Sadowski for her help in conducting the lynx necropsy
sampling, and Marina Kerr and Cornelya Klutsch for help with the analytical
troubleshooting of the data.
This study was funded by the Natural Sciences and Engineering Research
Council of Canada (grant number STPGP 391719–10) and the Ontario
Ministry of Natural Resources and Forestry.

Author details
[1]Department of Environmental & Life Sciences, Trent University, 1600 West
Bank Drive, Peterborough K9J 7B8, ON, Canada. [2]Wildlife Research and
Monitoring Section, Ontario Ministry of Natural Resources and Forestry, 2140
East Bank Drive, Peterborough K9J 7B8, ON, Canada. [3]Biology Department,
Trent University, 1600 West Bank Drive, Peterborough K9J 7B8, ON, Canada.

References
1. Lubahn DB, Joseph DR, Sar M, Tan J, Higgs HN, Larson RE, et al. The human androgen receptor: complementary deoxyribonucleic acid cloning, sequence analysis and gene expression in prostate. Mol Endocrinol. 1988;2:1265–75.
2. Colvard DS, Eriksent EF, Keetingt PE, Wilsont EM, Lubahnt DB, Frencht FS, et al. Identification of androgen receptors in normal human osteoblast-like cells. Proc Natl Acad Sci U S A. 1989;86:854–7.
3. Gelmann EP. Molecular Biology of the Androgen Receptor. J Clin Oncol. 2002;20:3001–15.
4. Spencer JA, Watson JM, Lubahn DB, Joseph DR, French FS, Wilson EM, et al. The androgen receptor gene is located on a highly conserved region of the X chromosomes of marsupial and monotreme as well as eutherian mammals. J Hered. 1991;82:134–9.
5. Traish AM, Goldstein I, Kim NN. Testosterone and erectile function: from basic research to a new clinical paradigm for managing men with androgen insufficiency and erectile dysfunction. Eur Urol. 2008;52:54–70.
6. Ryan CP, Crespi BJ. Androgen receptor polyglutamine repeat number: models of selection and disease susceptibility. Evol Appl. 2013;6:180–96.
7. Duitama JA, Zablotskaya R, Gemayel A, Jansen S, Belet JR, Vermeech KJ, et al. Large-scale analysis of tandem repeat variability in the human genome. Nucleic Acids Res. 2014;42:5728–41.
8. Haerty W, Golding BG. Low-complexity sequences and single amino acid repeats: not just "junk" peptide sequences. Genome. 2010;53:753–62.
9. King DG, Hannan AJ. Evolution of simple sequence repeats as mutable sites. In: Tandem Repeat Polymorphisms: Genetic Plasticity, Neural Diversity and Disease. New York: Landes Biosciences, Texas & Springer Science & Business Media; 2012. p. 10–23.
10. Bhandari R, Brahmachari SK. Analysis of CAG/CTG triplet repeats in the human genome: Implication in transcription factor gene regulation. J Biosci. 1995;20:613–27.
11. Buchanan G, Yang M, Cheong A, Harris JM, Irvine RA, Lambert PF, et al. Structural and functional consequences of glutamine tract variation in the androgen receptor. Hum Mol Genet. 2004;13:1677–92.
12. Dowsing AT, Yong EL, Clark M, Mclachlan RI, de DM K, Trounson AO. Linkage between male infertility and trinucleotide repeat expansion in the androgen-receptor gene. Lancet. 1999;354:640–3.
13. Mifsud A, Sim CKS, Boettger-Tong H, Moreira S, Lamb DJ, Lipshultz LI, et al. Trinucleotide (CAG) repeat polymorphisms in the androgen receptor gene: molecular markers of risk for male infertility. Fertil Steril. 2001;75:275–81.
14. Archer J. The influence of testosterone on human aggression. Br J Psychol. 1991;82:1–28.
15. Archer J. Testosterone and human aggression: an evaluation of the challenge hypothesis. Neurosci Biobehav Rev. 2006;30:319–45.
16. Scordalakes EM, Rissman EF. Aggression and arginine vasopressin immunoreactivity regulation by androgen receptor and estrogen receptor alpha. Genes Brain Behav. 2004;3:20–6.
17. Cheng D, Hong C-J, Liao D-L, Tsai S-J. Association study of androgen receptor CAG repeat polymorphism and male violent criminal activity. Psychoneuroendocrinology. 2006;31:548–52.
18. Rajender S, Pandu G, Sharma JD, Gandhi KPC, Singh L, Thangaraj K. Reduced CAG repeats length in androgen receptor gene is associated with violent criminal behavior. Int J Legal Med. 2008;122:367–72.
19. Jönsson EG, von Gertten C, Gustavsson JP, Yuan Q-P, Lindblad-Toh K, Forslund K, et al. Androgen receptor trinucleotide repeat polymorphism and personality traits. Psychiatr Genet. 2001;11:19–23.
20. Seidman SN, Araujo AB, Roose SP, McKinlay JB. Testosterone level, androgen receptor polymorphism, and depressive symptoms in middle-aged men. Biol Psychiatry. 2001;50:371–6.
21. Ito Y, Tanaka F, Yamamoto M, Doyu M, Nagamatsu M, Riku S, et al. Somatic mosaicism of the expanded CAG trinucleotide repeat in mRNAs for the responsible gene of Machado-Joseph disease (MJD), dentatorubral-pallidoluysian atrophy (DRPLA), and spinal and bulbar muscular atrophy (SBMA). Neurochem Res. 1998;23:25–32.
22. Köhler B, Lumbroso S, Leger J, Audran F, Grau ES, Kurtz F, et al. Androgen insensitivity syndrome: somatic mosaicism of the androgen receptor in seven families and consequences for sex assignment and genetic counseling. J Clin Endocrinol Metab. 2005;90:106–11.
23. Song Y-N, Geng J-S, Liu T, Zhong Z-B, Liu Y, Xia B-S, et al. Long CAG repeat sequence and protein expression of androgen receptor considered as prognostic indicators in male breast carcinoma. PLoS One. 2012;7:e52271.
24. Summers K, Crespi B. The androgen receptor and prostate cancer: a role for sexual selection and sexual conflict? Med Hypotheses. 2008;70:435–43.
25. Tanaka F, Reeves MF, Ito Y, Matsumoto M, Li M, Miwa S, et al. Tissue-specific somatic mosaicism in spinal and bulbar muscular atrophy is dependent on CAG-repeat length and androgen receptor-gene expression level. Am J Hum Genet. 1999;65:966–73.
26. Gemayel R, Cho J, Boeynaems S, Verstrepen KJ. Beyond Junk-Variable Tandem Repeats as Facilitators of Rapid Evolution of Regulatory and Coding Sequences. Genes (Basel). 2012;3:461–80.
27. Bellard C, Berteksmeier C, Leadley P, Thuiller W, Courchamp F. Impacts of climate change on the future of biodiversity. Ecol Lett. 2012;15:365–77.
28. Berteaux D, Réale D, McAdam AG, Boutin S. Keeping pace with fast climate change: Can arctic life count on evolution? Integr Comp Biol. 2004;44:140–51.
29. Dawson TP, Jackson ST, House JI, Prentice IC, Mace GM. Beyond predictions: biodiversity conservation in a changing climate. Science. 2011;332:53–8.
30. Franks S, Hoffmann A. Genetics of climate change adaptation. Annu Rev Genet. 2012;46:185–208.
31. Johnsen A, Fidler AE, Kuhn S, Carter KL, Hoffmann A, Barr IR, et al. Avian Clock gene polymorphism: evidence for a latitudinal cline in allele frequencies. Mol Ecol. 2007;16:4867–80.

32. O'Malley KG, Ford MJ, Hard JJ. Clock polymorphisms in Pacific salmon: evidence for variable selection along a latitudinal gradient. Proc R Soc B Biol Sci. 2010;277:3703–14.

33. Laidlaw J, Gelfand Y, Ng KW, Garner HR, Ranganathan R, Benson G, et al. Elevated basal slippage mutation rates among the Canidae. J Hered. 2007;98:452–60.

34. Gemayel R, Vinces MD, Legendre M, Verstrepen KJ. Variable tandem repeats accelerate evolution of coding and regulatory sequences. Annu Rev Genet. 2010;44:445–77.

35. Press MO, Carlson KD, Queitsch C. The overdue promise of short tandem repeat variation for heritability. Trends Genet. 2014;11:504–12.

36. Holterhus P-M, Brüggenwirth HT, Hiort O, Kleinkauf-Houcken A, Kruse K, Sinnecker GHG, et al. Mosaicism due to a somatic mutation of the androgen receptor gene determines phenotype in androgen insensitivity syndrome. J Clin Endocrinol Metab. 1997;82:3584–9.

37. Gottlieb B, Beitel LK, Trifiro MA. Somatic mosaicism and variable expressivity. Trends Genet. 2001;17:79–82.

38. Mutter GL, Boynton KA. PCR bias in amplification of androgen receptor alleles, a trinucleotide repeat marker used in clonality studies. Nucleic Acids Res. 1995;23:1411–18.

39. Youssoufian H, Pyeritz RE. Mechanisms and consequences of somatic mosaicism in humans. Nat Rev Genet. 2002;3:748–58.

40. Telenius H, Kremer B, Goldberg YP, Theilmann J, Andrew SE, Zeisler J, et al. Somatic and gonadal mosaicism of the Huntington disease gene CAG repeat in brain and sperm. Nat Genet. 1994;6:409–14.

41. Mubiru JN, Cavazos N, Hemmat P, Garcia-Forey M, Shade RE, Rogers J. Androgen receptor CAG repeat polymorphism in males of six non-human primate species. J Med Primatol. 2012;41:67–70.

42. Wang Q, Zhang X, Wang X, Zeng B, Jia X, Hou R, et al. Polymorphism of CAG repeats in androgen receptor of carnivores. Mol Biol Rep. 2012;39:2297–303.

43. Row JR, Gomez C, Koen EL, Bowman J, Murray DL, Wilson PJ. Dispersal promotes high gene flow among Canada lynx populations across mainland North America. Conserv Genet. 2012;13:1259–68.

44. Beauclerc KB, Bowman J, Schulte-Hostedde AI. Assessing the cryptic invasion of a domestic conspecific: American mink in their native range. Ecol Evol. 2013;3:2296–309.

45. Zigouris J, Schaefer JA, Fortin C, Kyle CJ. Phylogeography and post-glacial recolonization in wolverines (Gulo gulo) from across their circumpolar distribution. PLoS One. 2013;8:e83837.

46. Koen EL, Bowman J, Lalor JL, Wilson PJ. Continental-scale assessment of the hybrid zone between bobcat and Canada lynx. Biol Conserv. 2014;178:107–15.

47. Koen EL, Bowman J, Wilson PJ. Isolation of peripheral populations of Canada lynx (Lynx canadensis). Can J Zoolog. 2015;93:521–30.

48. Promega: MagneSil® ONE, Fixed Yield Blood Genomic System. 2012:1–8.

49. Invitrogen: Quant-iT™ PicoGreen® dsDNA Reagent and Kits. 2008:1–7.

50. Ahn SJ, Costa J, Emanuel JR. PicoGreen quantitation of DNA: effective evaluation of samples pre- or post-PCR. Nucleic Acids Res. 1996;24:2623–5.

51. Gustafson DR, Wen MJ, Koppanati BM. Androgen receptor gene repeats and indices of obesity in older adults. Int J Obes. 2003;27:75–81.

52. Fain S, LeMay J. Gender identification of humans and mammalian wildlife species from PCR amplified sex linked genes. Proc Am Adacemy Forensic Sci. 1995;1:34.

53. Aasen E, Medrano JF. Amplification of the Zfy and Zfx genes for sex identification in humans, cattle, sheep and goats. Nat Biotechnol. 1990;8:1279–81.

8

Leveraging local ancestry to detect gene-gene interactions in genome-wide data

Hugues Aschard[1*], Alexander Gusev[1], Robert Brown[2] and Bogdan Pasaniuc[2,3,4]

Abstract

Background: Although genome-wide association studies have successfully identified thousands of variants associated to complex traits, these variants only explain a small amount of the entire heritability of the trait. Gene-gene interactions have been proposed as a source to explain a significant percentage of the missing heritability. However, detecting gene-gene interactions has proven to be very difficult due to computational and statistical challenges. The vast number of possible interactions that can be tested induces very stringent multiple hypotheses corrections that limit the power of detection. These issues have been mostly highlighted for the identification of pairwise effects and are even more challenging when addressing higher order interaction effects. In this work we explore the use of local ancestry in recently admixed individuals to find signals of gene-gene interaction on human traits and diseases.

Results: We introduce statistical methods that leverage the correlation between local ancestry and the hidden unknown causal variants to find distant gene-gene interactions. We show that the power of this test increases with the number of causal variants per locus and the degree of differentiation of these variants between the ancestral populations. Overall, our simulations confirm that local ancestry can be used to detect gene-gene interactions, solving the computational bottleneck. When compared to a single nucleotide polymorphism (SNP)-based interaction screening of the same sample size, the power of our test was lower on all settings we considered. However, accounting for the dramatic increase in sample size that can be achieve when genotyping only a set of ancestry informative markers instead of the whole genome, we observe substantial gain in power in several scenarios.

Conclusion: Local ancestry-based interaction tests offer a new path to the detection of gene-gene interaction effects. It would be particularly useful in scenarios where multiple differentiated variants at the interacting loci act in a synergistic manner.

Keywords: Gene-gene interaction, GWAS, Local ancestry, Statistical genetics

Background

Advances in high-throughput genotyping technologies have enabled large-scale studies of genetic variation, from genome-wide association studies (GWAS) to inference of population history. The most notable use of high-throughput genotyping has been in GWAS where researchers have reproducibly identified thousands of genetic variants associated with many complex traits and common diseases. Despite the great success in identifying variants that contribute risk to disease, the majority of the genetic component of human traits and diseases remains unexplained. A potential source for this missing heritability is gene-gene interactions that alter disease risk in a coordinated fashion, for example when several genes are acting synergistically on a trait. Although of potential great interest, robust identification of gene-gene interactions has largely remained elusive, and despite numerous studies only a few interaction effects have been detected in human data [1–3]. Most genetic association studies of gene-gene interaction have focused on the joint effect on pairs of single nucleotide polymorphisms (SNPs) and used brute force approaches to evaluate a large number of pairs on homogenous populations (e.g. individuals of European ancestry only), while alternative strategies using heterogeneous populations have been seldom considered [4, 5].

* Correspondence: haschard@hsph.harvard.edu
[1]Department of Epidemiology, Harvard School of Public Health, Boston, MA, USA
Full list of author information is available at the end of the article

The development of accurate methods for discerning population structure have allowed for studies across different ethnicities including admixed populations (i.e. populations with recent ancestry from more than one continent such as African Americans). In addition to the standard linkage disequilibrium (LD) between nearby markers (used by GWAS to tag hidden causal variants) admixed populations exhibit another form of correlation among variants at a coarser scale due to chromosomal segments of distinct ancestry that is commonly referred to as admixture-LD [6]. This enables admixture mapping to be an effective approach for identifying disease loci that differ in frequency across populations [7–11]. A key component of such studies is the inference of ancestry at each locus in the genome. Several computational and statistical tools, including HAPMIX [12], LAMP-LD [13], EILA [14], and LANC-CSVs [15] can now be used to reliably call local ancestry. Although local ancestry has been traditionally used in admixture mapping, recent works use analyses of local ancestry to yield novel insights into the dynamics of recombination rate across the genome, to make demographic inferences from genetic data of admixed populations, as well as to understand the genetic basis of complex traits [16–18].

In this work we explore the use of local ancestry in recently admixed populations to find signals of gene-gene interaction that affect disease risk. We introduce an approach that leverages the correlation between local ancestry and the hidden unknown causal variants to find distant –e.g. on different chromosomes– gene-gene interactions. Our proposed approach uses multiple linear regression to model the interaction effect between pairs of local ancestry segments. Hence, as opposed to the standard approaches that test all pairs of SNPs assayed in GWAS (e.g. on the order of 10^{12} pairs for a standard GWAS of 2.5 million SNPs), we propose to test interaction only between pairs of local ancestries (on the order of 5×10^5 pairs for recent admixtures). By performing a much smaller number of statistical tests, our approach solves the computational bottleneck and reduces the multiple testing correction burden. We derive the analytical formulation for our test assuming a single causal variant for each interacting locus and investigate its performance across a wide range of parameter values. Motivated by recent works that show ever-increasing evidence for multiple causal variants per locus [19–21], we extend our approach to allow for multiple causal variants at each interacting locus. We find that local ancestry can be used to find gene-gene interactions, with power increasing with the number of causal variants per locus and the degree of differentiation in the frequency of the causal variants between the ancestral populations. Assuming equal sample size, the test based on pairwise genotyped SNPs appears to be more powerful than the ancestry-based interaction test in most scenarios. However, when accounting for the increase in sample size that can be achieved for a fixed budget when measuring local ancestry only (e.g. based on ancestry informative markers, AIMs), we observed a substantial increase in power under various scenarios.

Results and discussion
Overview of the approach
A standard approach for finding pairwise SNP interactions is to test for non-zero effect size of the product term of the two SNPs considered. The underlying assumption is that the SNPs tested in the model are either the interacting causal variants or correlated to the actual causal variants through LD. Indeed, only a finite number of SNPs are assayed in GWAS (today's genotyping arrays assay a few million SNPs), with true biologically causal variants likely remaining untyped. While a number of additional SNPs can be imputed on a genome-wide scale, the presence of the causal variants in the data can only be assumed for whole-genome sequence data. It is likely the causal variants will only be tagged by the SNPs analyzed. In admixed populations, correlation between SNPs also exists at a coarse level due to the segments of recent ancestry (admixture-LD). Similar to the pairwise SNP interaction screening, we can tag the hidden causal variants using admixture-LD and test for the presence of interaction at hidden causal variants by testing for interaction at the level of local ancestry.

Testing for interaction under a single causal variant per locus assumption
We first considered a scenario where two common SNPs are located on two physically distant segments in the genome, thus independent from each other, and have an interactive effect on a quantitative phenotype, while all other SNPs at the locus harboring these two causals have no effect (Fig. 1a). We derived the performances of three interaction tests, based on full sequence data (S_S), genotypic data from a 1 M (1 million) SNPs chip (S_G) and local ancestry (S_L). Figure 2 presents the sample size required for each of the three interaction tests to achieve a significance level of 5 % with 80 % power after correction for multiple testing. The sample sizes are plotted for a range of correlation levels between the causal variant and the tagging SNP or tagging ancestry segment. More specifically, we refer to ρ_{GC} for the correlation between the true interaction term and the interaction term derived from the best tag from the genotyping chip, and to ρ_{LC} for the correlation between the true interaction term and the local ancestry interaction term between the two segments harboring the causals. For simplicity we assumed here that local ancestry is inferred with high accuracy (r^2 between true ancestry and

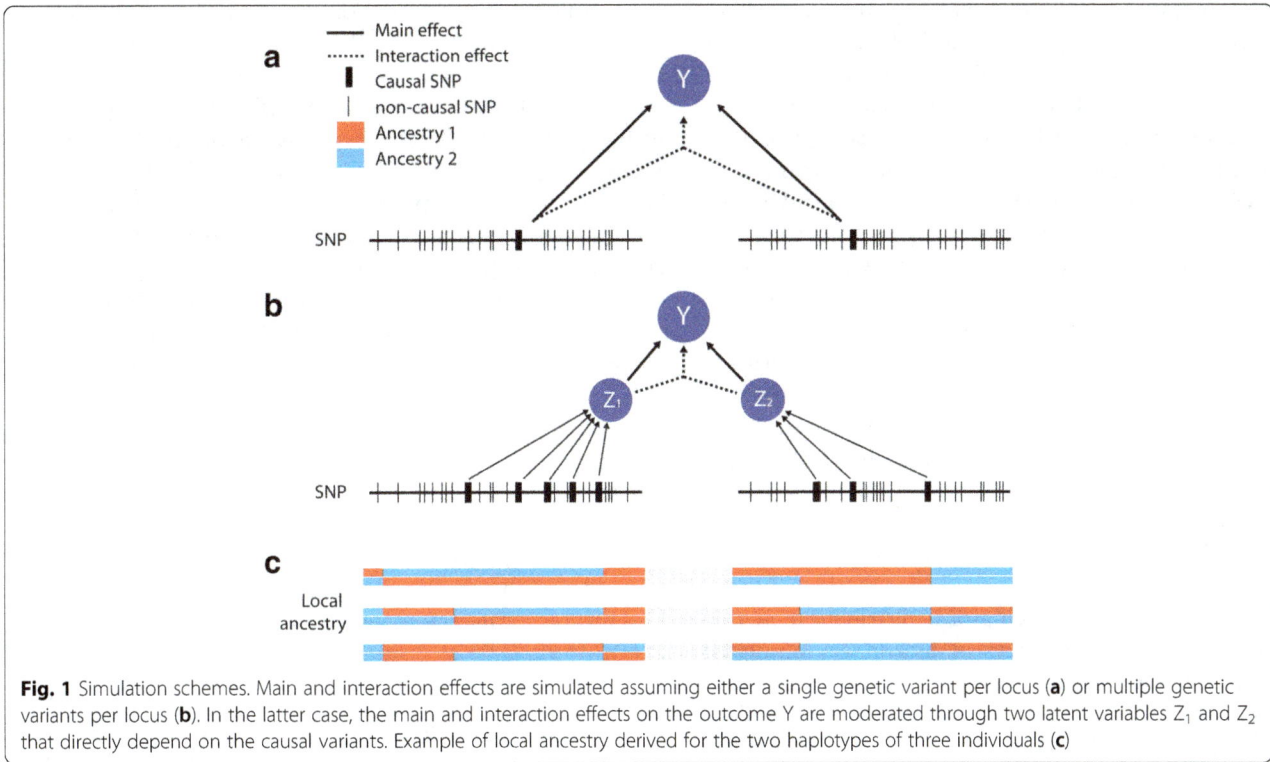

Fig. 1 Simulation schemes. Main and interaction effects are simulated assuming either a single genetic variant per locus (**a**) or multiple genetic variants per locus (**b**). In the latter case, the main and interaction effects on the outcome Y are moderated through two latent variables Z_1 and Z_2 that directly depend on the causal variants. Example of local ancestry derived for the two haplotypes of three individuals (**c**)

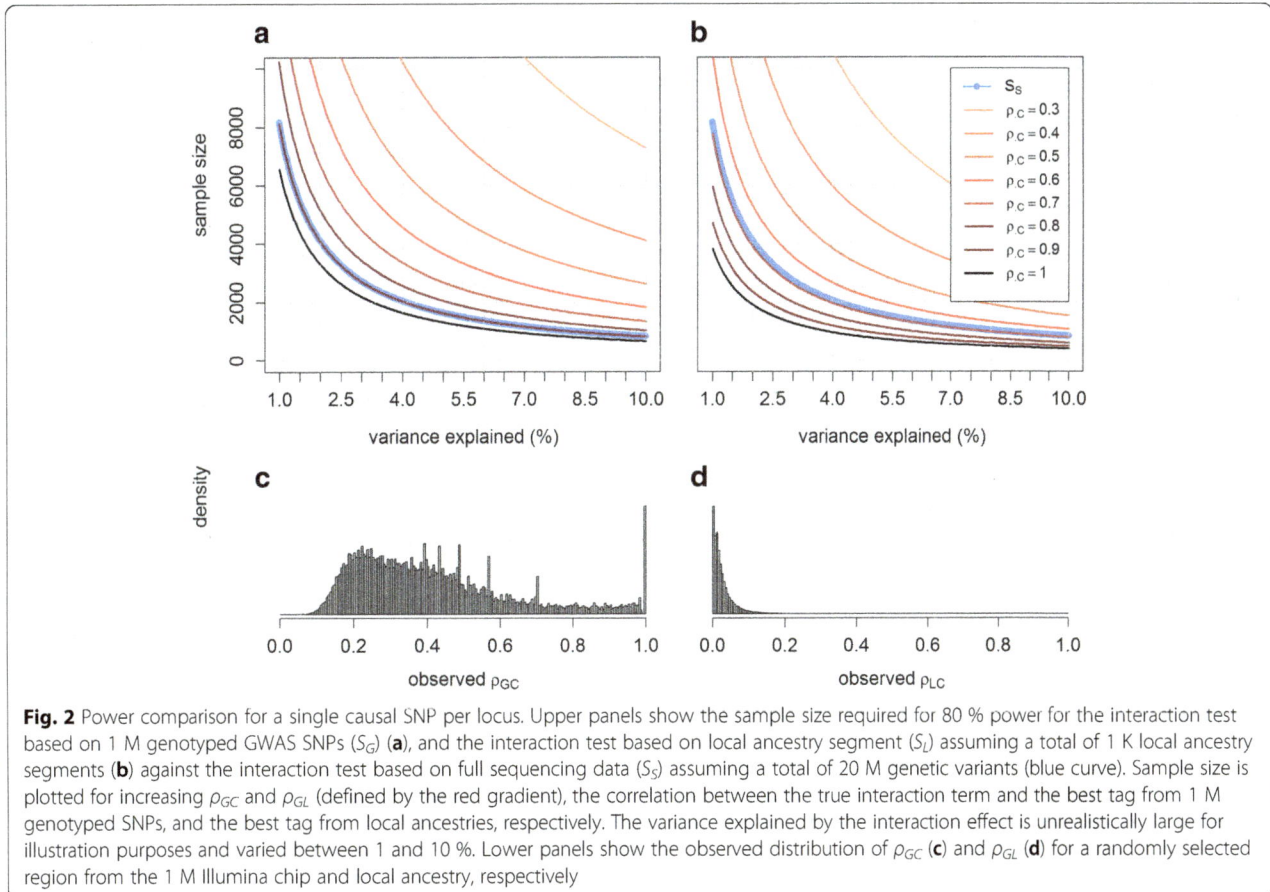

Fig. 2 Power comparison for a single causal SNP per locus. Upper panels show the sample size required for 80 % power for the interaction test based on 1 M genotyped GWAS SNPs (S_G) (**a**), and the interaction test based on local ancestry segment (S_L) assuming a total of 1 K local ancestry segments (**b**) against the interaction test based on full sequencing data (S_S) assuming a total of 20 M genetic variants (blue curve). Sample size is plotted for increasing ρ_{GC} and ρ_{GL} (defined by the red gradient), the correlation between the true interaction term and the best tag from 1 M genotyped SNPs, and the best tag from local ancestries, respectively. The variance explained by the interaction effect is unrealistically large for illustration purposes and varied between 1 and 10 %. Lower panels show the observed distribution of ρ_{GC} (**c**) and ρ_{GL} (**d**) for a randomly selected region from the 1 M Illumina chip and local ancestry, respectively

inferred ancestry ≥0.99), and therefore does not differ from true local ancestry. This will likely be the case for African Americans [15], but might be too optimistic for other populations such as Latino Americans (see below). Figure 2 shows that for GWAS-based test (S_G) to outperform the test based on the true causal variants (S_S), it requires ρ_{GC} to be above 0.9. For the local ancestry test, ρ_{LC} has to be above 0.7. Moreover, for ρ_{LC} above 0.8, the ancestry based interaction test would also outperform the GWAS-based interaction test even if the causal variants were genotyped. As expected, the maximum potential gain is achieved when the interaction term is perfectly tagged by either GWAS SNPs or local ancestry. We considered unrealistically large interaction effects in Fig. 2 for illustration purposes. When analyzing 20,000 samples, the smallest interaction effect (as measured by the proportion of variance explained) that can be detected with 80 % power is 0.8, 0.5 and 0.3 % for S_S, S_G and S_L, respectively.

We then estimated the empirical distribution of ρ_{GC} and ρ_{LC} using African-American individuals simulated using the 1,000 Genomes data (see Methods) [22]. From this simulation we randomly choose 20,000 independent SNPs, and built 10,000,000 hypothetical pairs of interacting SNPs. Bottom panels of Fig. 2 shows the distribution of these two correlation terms when using tagging SNPs from the 1 M Illumina chip and the simulated local ancestry. Despite the large potential increase in power shown in the upper panels of Fig. 2, improvement may actually exists only in very few real situations. For example we observed that the probability of ρ_{GC} to be above the 0.9 threshold is 0.05. For ρ_{LC} the "increased power" threshold of 0.7 is achieved only once in 10^7 times. Hence, even if interaction effects are extremely common in the architecture of complex trait, there is a low probability for the local ancestry–based test to do better than other approaches in the presence of a single causal variant per locus when assuming equal sample size for both tests.

Multiple causal SNPs per segment

Accounting for increasing evidence of multiple causal variants per locus [19–21], we then considered scenarios where gene-gene interaction effects involved multiple genetic variants per locus. For example when multiple SNPs contribute to gene transcript abundance and the interaction is taking place between the gene products. Such interaction would be challenging to identify using SNP data due to the vast search area among all possible combinations of SNPs. On the other hand, local ancestry offers a more appropriate and natural way to test for such models as it captures a form of an individual's genetic background at each locus (i.e. genetic variants share the same local ancestry at a given locus in the genome).

To evaluate this assumption we defined a simulation model where multiple SNPs at two independent loci contribute to two latent variables Z_1 and Z_2 that have an interaction effect on the outcome (Fig. 1b). The power of the pairwise SNPs test depends on the best tagging SNPs for Z_1 at locus 1 and for Z_2 at locus 2. This would be either the strongest causal variants for Z_1 or Z_2, or the best tag of these causals. The power the local ancestry-based test to detect this interaction depends on all parameters influencing $\rho_{L_i Z_i}$, the correlation between Z_i and L_i, the latent variable and the local ancestry at locus i, respectively. This includes the number of causal SNPs for Z_i, and the distribution of β_i, the effects of the causal SNPs of Z_i. Assuming the causal SNPs are a random sample of the variants in the segment, power is also bounded by the average difference in minor allele frequency between the two founder populations. Figure 3 presents the empirical distribution of this correlation in a simple scenario, when Z_i depends on 1 to 50 SNPs. Overall, $\rho_{L_i Z_i}$, increases with the number of SNPs involved and with increased homogeneity of genetic effect. For example if the $\beta_i = (\beta_{i1}, \dots \beta_{iK})$ are distributed around the null, the expected value of $\rho_{L_i Z_i}$ is null and a local ancestry-based test would have no power. Conversely, if the coded alleles from the causal variants tend to increase the outcome value (while the reference allele has no contribution), $\rho_{L_i Z_i}$ can be substantial (e.g. >0.2, Fig. 3b).

We performed a simulation study to compare the performance of the pairwise SNP-based approach (S_G), when using both genotyped and imputed common SNPs (MAF > 1 %), and the local ancestry approach (S_L) while increasing K the number of causal SNPs per locus. For simplicity we assumed the number of causal variants was the same in the two interacting regions, and only considered common variants (minor allele frequency, MAF > 0.10). We explored scenarios where the causal SNPs were either slightly differentiated or highly differentiated between the two ancestral populations. When assuming equal sample size, S_L is underpowered as compared to S_G, despite a dramatic increase in the total number of tests performed (Fig. 4). Hence when GWAS data exists, deriving local ancestry segments would be of limited interest for gene-gene interaction testing unless the number of causal variants is large (e.g. >10). However, when considering de novo genotyping with a fixed budget, an increase in sample size can be achieved when measuring local ancestry only, S_L can be more powerful than S_G. In particular, assuming a 6 fold decrease in cost, S_L outperform S_G if either the differentiation is moderate or the number of causals is large (>5), or if the causal SNPs are highly differentiated (e.g. correlation between local ancestry and the causal >0.5, Fig. 4c).

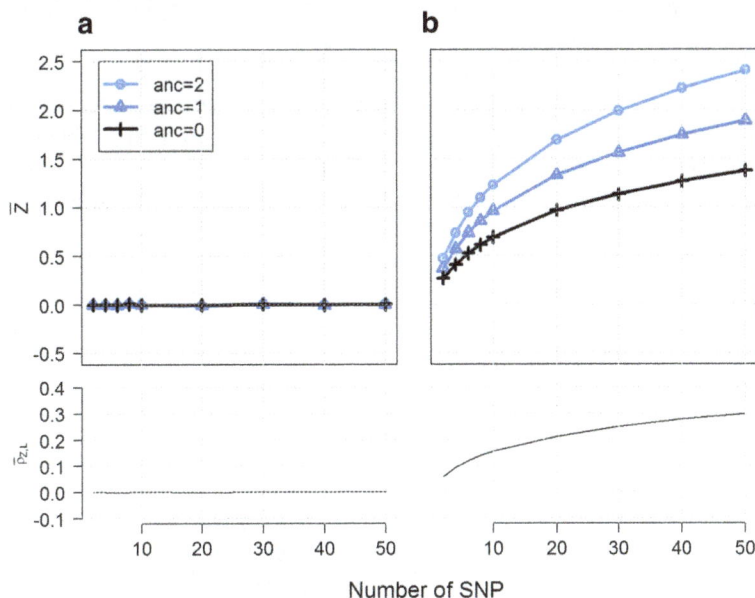

Fig. 3 Tagging interaction effects in a multiple causal model. A latent variable Z is generated as a function of an increasing number of SNPs at a single locus, explaining altogether 50 % of its variance. The average value of Z across 20,000 replicates of 10,000 admixed samples is plotted for each three local ancestry classes. The effect of the SNPs is drawn from a normal (**a**) and left-truncated normal (**b**) distribution with a mean of 0 (upper panel). When the SNP effects are null on average, the average values of Z do not differ by local ancestry and ρ_{ZL}, the correlation between Z and local ancestry, is also null on average. Conversely, when the average effect of the SNPs is not null, ρ_{ZL} increases with an increasing number of causal variants (lower panel)

We assumed in these simulations that true local ancestry is available. To evaluate the impact of additional noise introduced by ancestry inference, we analyzed the same data but using ancestry inferred using LAMP-LD. As shown in Additional file 1: Figure S1, using the inferred ancestry has only minor impact on power. This is partly expected thanks to the high accuracy of inference in African Americans [15]. The accuracy of SNP imputation could also impact the power of the SNP-based test, however the quality of imputation depends on more parameters, varying across allele frequencies and the chips used for genotyping [23]; it would therefore be

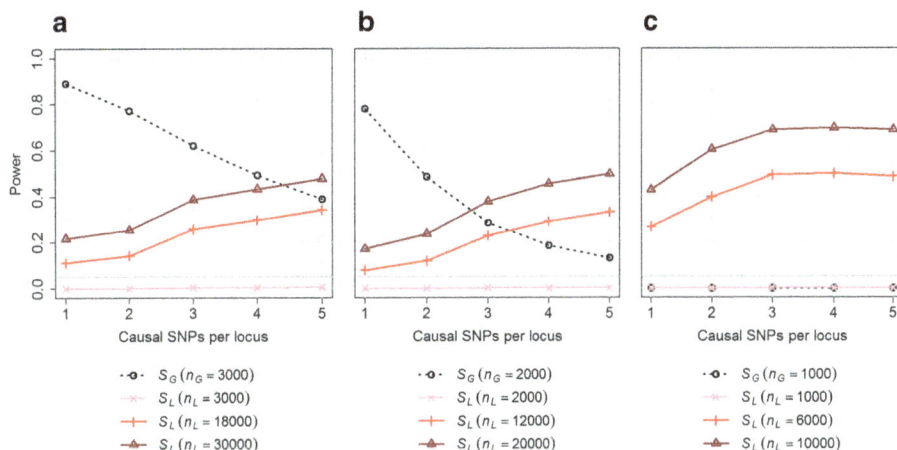

Fig. 4 Power comparison for multiple causal SNPs per locus. Power across 25,000 replicates using a Bonferroni correction resulting in p-value thresholds of 1×10^{-7} and 1×10^{-15} for the local ancestry-based interaction test (S_L) and the SNP-based interaction test (S_G), respectively. One to five common causals SNPs were selected per interacting locus while assuming either low (**a**), moderate (**b**) or high (**c**) differentiation of those SNPs between the two admixed populations. We considered three case scenarios for the additional increase in sample size that would be achieve when using local ancestry derived from AIMs, no increase (pink), a lower bound of six fold increase (light red) and an upper bound of 10 fold increase. We varied the baseline sample size (for S_G) across scenarios to emphasize the differences between the tests

more difficult to evaluate thoroughly. Instead we applied the SNP-based test using the genotyped SNPs only found on the Illumina Human1M-Duo BeadChip (Additional file 1: Figure S1). We observed a substantial decrease in power, highlighting both the importance of using imputed variants and the need for high quality imputation for the SNP-based test, which might be a concern for rare causal variants.

Finally, we evaluated how the relative power of the S_L and S_G tests is impacted when applying a two-steps procedure where SNPs and ancestry segments are first pre-selected for interaction testing based on their marginal effects [24]. For simplicity we assume that the vast majority of SNPs are not involved in interaction, so that for a p-value threshold t at step 1, the total number of interaction tests at step 2 can be approximated by $\binom{n \times t}{2}$, where n is the number of predictor (either SNP or local ancestry segment). Additional file 1: Figure S2 shows the results from this strategy when applied to a case similar to Fig. 4, but adding a main effect to each causal SNPs of the same magnitude as the interaction effect, and using $t = 0.01$ at step 1. In this specific scenario, the 2-step approach mostly benefits to the ancestry-based test, which outperform the SNP-based test in many more scenarios, including cases where sample size was the same for the 2 analyses.

Conclusions

We explored the performance of a local ancestry-based interaction test to capture non-linear effects from two independent loci. The strategy is similar to a standard SNP-based pairwise interaction screening but uses local ancestry segments instead of SNPs. One major underlying motivation for such an approach is that the total number of tests to be conducted is dramatically lower than for a standard pairwise SNP interaction test, reducing both the computational burden and the correction for multiple testing. We demonstrate that such a test would indeed capture interaction effects between two loci as long as the individual effects of the causal variants at each locus do not cancel each other. For existing datasets that only contain local ancestry data derived from AIMs and for de novo genotyping studies looking for the optimal cost/power ratio, our approach (S_L) can be highly relevant as it can outperform the pairwise SNP screening from standard GWAS data (S_G). Conversely, when GWAS genotyping data does exists, in most scenarios we explored, S_G outperforms S_L when the number of causal variants at the locus was small. As the number of causals grows beyond 10, the power of S_L increases but does not substantially exceed S_G unless the differentiation of the causal SNPs between the two populations

is very high. Interestingly, as the differentiation increases so does the relative power of S_G, which explains the underperformances of S_L. We found that, as differentiation increases, many genetic variants become good tags for local ancestry, and so S_G benefits from the increase in differentiation as well. Overall, the relative performances of our approach depends on the balance between the gain in power achieved thanks to the decreasing number of test and the decrease in power due to low correlation between local ancestry and the causal variant(s) at the interacting loci.

Furthermore, we used the whole genome sequence-based test (S_S) as a reference to compare the relative performance of the two alternative approaches. While such a test might have higher power than S_L and S_G (Fig. 2), testing all possible pairs of SNPs would requires extremely intensive computational power in practice, and the implementation of such tests, which have been rarely explored to our knowledge, would require substantial software development and hardware structures (e.g. graphics processing units [25]). This confirms that, as of today, GWAS-based pairwise interaction tests remain a relevant approach for identifying interactions as compared to whole genome sequence-based approaches.

Regarding power comparison between S_L and S_G, the power of the local-ancestry based interaction test was derived based on 1,000 local ancestry regions. However, the number of segments depends on the number of ancestral populations and the number of generations since admixture, and will therefore differ across admixed populations. Increase in the total number of segments can impact the correlation between local ancestry and the causal variants within these segments, as well as the total number of tests that have to be performed for an interaction screening. Using the inferred local ancestry had very limited impact on power in our simulation as the accuracy of inference is very high in African Americans. However, for other populations, the impact might be substantial. For example Brown et al. reported squared correlation between true and inferred local ancestry of 0.63, and 0.81 for Mexican and Puerto Rican population when using LANC-CSV, which had similar results to other methods [15]. While further analysis might explore such situations, we believe the results described in this study would remain valid. Finally, additional work might also include extensive explorations of scenarios where interaction factors (either single SNPs or single local ancestry segments) are selected based on their marginal association with the phenotype of interest. When applied in our simulation framework we observed a strong improvement of the ancestry-based test over the SNP-based test, however this needs to be confirmed across a broader range of scenarios.

Overall, while our approach shows some limitations when genome-wide genotypic data are available and when the number of causal variants per region is small and contains mostly undifferentiated variants, we highlight that genome-wide local-ancestry based interaction screening remains relevant. First, because some datasets only generated local ancestry data through AIMs, and do not have GWAS data. Second, considering budget constraint for de novo genotyping, and assuming a 6 fold decrease in cost for genotyping AIMs as compared to a standard GWAS, substantial additional gain in power can be achieved through local-ancestry based tests.

Methods

Genetic model

We considered a genetic model similar to the one described in Chatterjee et al. [26], which can be easily adapted to the local ancestry context. It consists in two independent sets of adjacent SNPs from two loci on different genomic regions, which represent in this study two local ancestry segments. Several SNPs within segment i have an indirect association with the outcome of interest through a latent variable Z_i, an unmeasured quantitative biological phenotype partially governed by SNPs within the locus. Interaction effects between the genetic variants on the outcome is introduced through an interaction term between the Z_1 and Z_2 variables (i.e. the cumulative effect of the genetic variants within a locus depends on the cumulative effect of the variants in a distant locus). More specifically the outcome Y is defined as follows:

$$Y = \theta_1 Z_1 + \theta_2 Z_2 + \theta_{12} Z_1 Z_2 + \varepsilon \qquad (1)$$

where Z_1 and Z_2 are the two latent variables that each depend on K SNPs and θ_1, θ_2, θ_{12} respectively represent the main effect of Z_1, Z_2, and the interaction of Z_1 and Z_2 on Y; ε is the residual noise and is normally distributed with mean 0. The latent variables Z_i are defined as follows:

$$Z_i = \sum_{k=1}^{K} G_{ik} \times \beta_{ik} \qquad (2)$$

where G_{ik} and β_{ik} is the standardized genotype of SNP k in locus i and its main effect respectively. The SNP effects $\beta_i = (\beta_{i1}, \dots \beta_{iK})$ were randomly drawn from either normal or left-truncated normal depending on the scenario explored. For simplicity, as the main effect of the latent variable has no impact on the interaction test [27], we set θ_1 and θ_2 to 0, and set θ_{12} to 1. Except when specified otherwise we scaled the variance of ε so that the proportion of the variance of Y explained by the

interaction equals 1 % (to reflect values observed in GWAS for common complex traits).

A single major causal SNP per segment

We first assumed the genetic effect of a segment is driven by a single causal variant, while the effect of other potential SNPs is null or negligible. This is equivalent to assuming $K = 1$ in equation (2). The effect of the SNPs on Y can then be re-written:

$$Y = \beta_{G_1} G_1 + \beta_{G_2} G_2 + \beta_{intG} G_1 G_2 + \varepsilon \qquad (3)$$

Assume G_1 and G_2 are tagged by L_1 and L_2 respectively, where L_i is the local ancestry measured at the segment harboring SNP G_i. Note that G_1 and G_2 do not necessarily need to be typed to correctly identify local ancestry (local ancestry spans many MB's in recently admixed populations and can be reliably identified using a small set of variants). For simplicity, we considered only the case of a two-way admixed population, so that local ancestry would be typically coded as an ordinal variable with value corresponding to the number of chromosomes harboring a particular ancestry. Hence, for African-American, L equals 0, 1 or 2. When the population under study is an admixture of more than two ancestries, testing for interaction would be more complex because of additional combinations of ancestries (e.g. for 3 ancestries A, B and C, an individual can have any of the six following ancestries at a given segment: AA, AB, AC, BB, BC and CC). To our knowledge there is no established standard to handle such situations, the simplest solution consists of testing one ancestry versus the rest [17].

We compared the relative performances of the standard test of interaction of β_{intG} (equation (3)) versus the test of β_{intL} (equation (4)), the interaction effect observed between L_1 and L_2 on Y, which can be obtained from the model:

$$Y = \beta_{L_1} L_1 + \beta_{L_2} L_2 + \beta_{intL} L_1 L_2 + \varepsilon' \qquad (4)$$

Note that both equation (3) and (4) are standard 1 ° of freedom tests of interaction effect estimates obtained from multiple linear regression (see our R script example in Additional file 2). The only difference being that the later test uses local ancestry instead of genotyped SNPs. When the causal SNPs are available (e.g. from whole genome sequence data) and have been standardized, the Wald test of β_{intG} is defined as $S_S = \left(\hat{\beta}_{intG} / \hat{\sigma}_{\beta_{intG}} \right)^2$. Similarly the test of β_{intL} is defined as $S_L = \left(\hat{\beta}_{intL} / \hat{\sigma}_{\beta_{intL}} \right)^2$. Under the null hypothesis of no association both S_S and S_L follow a *chi-square* distribution with one degree of freedom.

Let ρ denote the correlation between two variables. The two scores can be written as:

$$S_S = N * \rho^2(Y, G_1 \times G_2) \qquad (5)$$

$$S_L = N * \rho^2(Y, G_1 \times G_2) * \rho_{LC}^2 \qquad (6)$$

where $\rho_{LC}^2 = \rho^2(G_1 \times G_2, L_1 \times L_2)$ is the squared-correlation between the true interaction $G_1 \times G_2$ term and the local ancestry interaction term $L_1 \times L_2$.

In another scenario, one would test for interaction effects between pairs of SNPs from a standard GWAS chip, which implies that the tested SNPs, say G_1^* and G_2^* are tagging the two causals, but (likely) at a higher level than the local ancestry. We denote this test S_G:

$$S_G = N * \rho^2(Y, G_1 \times G_2) * \rho_{GC}^2 \qquad (7)$$

where $\rho_{GC}^2 = \rho^2(G_1 \times G_2, G_1^* \times G_2^*)$ is the squared-correlation between the true interaction $G_1 \times G_2$ term and the GWAS interaction term $G_1^* \times G_2^*$. The power of the three tests can be derived as:

$$\text{Power}_S = 1 - F\left(\chi_{1,1-\alpha,0}^2 | 1, S.\right) \qquad (8)$$

where $F(\chi^2 | d, S)$ is the cumulative probability function of the non-central chi-square distribution with d degrees of freedom and non-centrality parameter S; $\chi_{d,p,0}^2$ is the inverse of F under the null, i.e. the quantiles of the non-central chi-square distribution, and α is the type I error rate. The relative performances of the three strategies can then be evaluated by comparing the sample size N needed to identify the interaction at the 5 % significance level (α) after accounting for n_S, n_L, and n_G the number of tests that has to be performed respectively in the whole genome setting, the local ancestry setting and the GWAS setting, respectively. Therefore, the alpha levels for tests S_S S_L and S_G were set at α/n_S, α/n_L, and α/n_G, respectively.

For the test S_S, we assumed the true causal variants are available as part of a whole genome sequence data of $n_S = 20M$ SNPs, so that the total number of pairwise test equals 2×10^{-14}. We assumed a total of $n_L = 1,000$ local ancestry segments for the test S_L, and $n_G = 1M$ SNPs for test S_G. We considered values of ρ_{LC} and ρ_{GC} in the range [0.3; 1], so that the minimum squared-correlation was 0.09. A correlation of 1 corresponding to the highest potential increase in power that can be achieve, since it will be equivalent to testing the interaction with the true causals while dramatically reducing the multiple testing corrections.

Multiple causal SNPs per segment
We used models from equation (1) and (2) and generated outcome data across 1,000 replicates while using simulated genetic and local ancestry data (*see further*

section on simulation). For ease of computation, we considered the number of causal SNPs per segment to be equal between the two interacting loci. The β_i were randomly drawn from a left-truncated normal (cut at 0) distribution with mean 0 and variance 1. The causal alleles were chosen to be minor in one of the populations so that all effects go in the same direction in one population. We considered scenarios including 1 to 5 causal SNPs per locus, and selected the variants so that correlation between local ancestry and each SNP was either >0.1, >0.2 or >0.5, assuming low, moderate or high differentiation at the causal SNPs, respectively.

We perform both the local ancestry interaction and the pairwise SNP interaction test in a standard linear regression. However, instead of testing all combination of SNP, we first find the best combined tag in each locus (i.e. the single SNP j that maximizes $\sum_K$ $\left(\beta_{ik}\rho_{jk}\right)$ across the K causal variants at locus i, where ρ_{jk} is the correlation between the tag SNP G_j^* and the causal variant G_k) and then test only the product of that genotype with the best tag genotype at the second locus. Power was defined as the number of replicates for which the pairwise SNP interaction is significant at $p = 0.05 / \binom{n_G}{2}$, where n_G the number of SNPs tested equals 1 M when including genotyped SNPs only and equals 10 M when including common imputed SNPs from 1000 Genomes. For the local ancestry-based test, the threshold was $p = 0.05 / \binom{n_L}{2} = 1 \times 10^{-7}$, corresponding to the test of $n_L = 1,000$ local ancestry segments.

Simulation of admixed populations from the 1000 genome project
Similar to previous work [15], we simulated admixed chromosomes of African-Americans as a random walk over 1000 Genomes haplotypes [22]. CEU and FIN populations were used to represent European haplotypes and the YRI population represented the African haplotypes. We assumed that between any two base pairs there was a 10^{-8} chance of recombination. At a recombination event, the next haplotype was selected with a 20 % chance of being European and an 80 % chance of being African to reflect the estimated admixture proportions in the literature [12, 28]. Haplotypes were sampled with replacement. We simulated 20,000 haplotypes each for chromosomes 19 and 20 in this manner and added them together to form 10,000 unphased genotypes and true local ancestries. Local ancestry was then inferred

using LAMP-LD [13] with default settings and with the GBR and TSI populations representing Europeans and the LWK representing Africans. LAMP-LD was run only using variants found on the Human1M-Duo BeadChip.

Availability of supporting data
All supporting data are included as additional files.

Abbreviations
SNP: Single nucleotide polymorphisms; GWAS: Genome-wide association study; LD: Linkage disequilibrium; AIM: Ancestry informative marker; MAF: Minor allele frequency.

Competing interests
The authors declare that they have no competing interests.

Authors' contributions
HA and BP conceived the study. HA, BP, and AG participated in its design and coordination. HA, AG and RB conducted all analyses. HA, BP, AG and RB were involved in the drafting of the manuscript. All authors read and approved the final manuscript.

Acknowledgements
We acknowledge funding and support from the National Institute of Health, R03HG006720 (HA), T32HG002536 (RB), R01GM053275 (BP,RB), and R01HG006399 (BP).

Author details
[1]Department of Epidemiology, Harvard School of Public Health, Boston, MA, USA. [2]Bioinformatics Interdepartmental Program, University of California Los Angeles, Los Angeles, CA, USA. [3]Department of Pathology and Laboratory Medicine, University of California Los Angeles, Los Angeles, CA, USA. [4]Department of Human Genetics, University of California Los Angeles, Los Angeles, CA, USA.

References
1. Aschard H, Lutz S, Maus B, Duell EJ, Fingerlin TE, Chatterjee N, et al. Challenges and opportunities in genome-wide environmental interaction (GWEI) studies. Hum Genet. 2012;131(10):1591–613.
2. Cordell HJ. Detecting gene-gene interactions that underlie human diseases. Nat Rev Genet. 2009;10(6):392–404.
3. Mackay TF. Epistasis and quantitative traits: using model organisms to study gene-gene interactions. Nat Rev Genet. 2014;15(1):22–33.
4. Aldrich MC, Kumar R, Colangelo LA, Williams LK, Sen S, Kritchevsky SB, et al. Genetic ancestry-smoking interactions and lung function in African Americans: a cohort study. PLoS One. 2012;7(6):e39541.
5. Powell R, Davidson D, Divers J, Manichaikul A, Carr JJ, Detrano R, et al. Genetic ancestry and the relationship of cigarette smoking to lung function and per cent emphysema in four race/ethnic groups: a cross-sectional study. Thorax. 2013;68(7):634–42.
6. Pfaff CL, Parra EJ, Bonilla C, Hiester K, McKeigue PM, Kamboh MI, et al. Population structure in admixed populations: effect of admixture dynamics on the pattern of linkage disequilibrium. Am J Hum Genet. 2001;68(1):198–207.
7. Fejerman L, Chen GK, Eng C, Huntsman S, Hu D, Williams A, et al. Admixture mapping identifies a locus on 6q25 associated with breast cancer risk in US Latinas. Hum Mol Genet. 2012;21(8):1907–17.
8. Cheng CY, Kao WH, Patterson N, Tandon A, Haiman CA, Harris TB, et al. Admixture mapping of 15,280 African Americans identifies obesity susceptibility loci on chromosomes 5 and X. PLoS Genet. 2009;5(5):e1000490.
9. Nalls MA, Wilson JG, Patterson NJ, Tandon A, Zmuda JM, Huntsman S, et al. Admixture mapping of white cell count: genetic locus responsible for lower white blood cell count in the Health ABC and Jackson Heart studies. Am J Hum Genet. 2008;82(1):81–7.
10. Reich D, Patterson N, Ramesh V, De Jager PL, McDonald GJ, Tandon A, et al. Admixture mapping of an allele affecting interleukin 6 soluble receptor and interleukin 6 levels. Am J Hum Genet. 2007;80(4):716–26.
11. Zhu X, Luke A, Cooper RS, Quertermous T, Hanis C, Mosley T, et al. Admixture mapping for hypertension loci with genome-scan markers. Nat Genet. 2005;37(2):177–81.
12. Price AL, Tandon A, Patterson N, Barnes KC, Rafaels N, Ruczinski I, et al. Sensitive detection of chromosomal segments of distinct ancestry in admixed populations. PLoS Genet. 2009;5(6):e1000519.
13. Baran Y, Pasaniuc B, Sankararaman S, Torgerson DG, Gignoux C, Eng C, et al. Fast and accurate inference of local ancestry in Latino populations. Bioinformatics. 2012;28(10):1359–67.
14. Yang JJ, Li J, Buu A, Williams LK. Efficient inference of local ancestry. Bioinformatics. 2013;29(21):2750–6.
15. Brown R, Pasaniuc B. Enhanced methods for local ancestry assignment in sequenced admixed individuals. PLoS Comput Biol. 2014;10(4):e1003555.
16. Zhang J, Stram DO. The role of local ancestry adjustment in association studies using admixed populations. Genet Epidemiol. 2014;38(6):502–15.
17. Seldin MF, Pasaniuc B, Price AL. New approaches to disease mapping in admixed populations. Nat Rev Genet. 2011;12(8):523–8.
18. Zaitlen N, Pasaniuc B, Sankararaman S, Bhatia G, Zhang J, Gusev A, et al. Leveraging population admixture to characterize the heritability of complex traits. Nat Genet. 2014;46(12):1356–62.
19. Meyer KB, O'Reilly M, Michailidou K, Carlebur S, Edwards SL, French JD, et al. Fine-scale mapping of the FGFR2 breast cancer risk locus: putative functional variants differentially bind FOXA1 and E2F1. Am J Hum Genet. 2013;93(6):1046–60.
20. Trynka G, Hunt KA, Bockett NA, Romanos J, Mistry V, Szperl A, et al. Dense genotyping identifies and localizes multiple common and rare variant association signals in celiac disease. Nat Genet. 2011;43(12):1193–201.
21. Gusev A, Bhatia G, Zaitlen N, Vilhjalmsson BJ, Diogo D, Stahl EA, et al. Quantifying missing heritability at known GWAS loci. PLoS Genet. 2013;9(12):e1003993.
22. Genomes Project C, Abecasis GR, Auton A, Brooks LD, DePristo MA, Durbin RM, et al. An integrated map of genetic variation from 1,092 human genomes. Nature. 2012;491(7422):56–65.
23. International HapMap C, Altshuler DM, Gibbs RA, Peltonen L, Altshuler DM, Gibbs RA, et al. Integrating common and rare genetic variation in diverse human populations. Nature. 2010;467(7311):52–8.
24. Kooperberg C, Leblanc M. Increasing the power of identifying gene x gene interactions in genome-wide association studies. Genet Epidemiol. 2008;32(3):255–63.
25. Hemani G, Theocharidis A, Wei W, Haley C. EpiGPU: exhaustive pairwise epistasis scans parallelized on consumer level graphics cards. Bioinformatics. 2011;27(11):1462–5.
26. Chatterjee N, Kalaylioglu Z, Moslehi R, Peters U, Wacholder S. Powerful multilocus tests of genetic association in the presence of gene-gene and gene-environment interactions. Am J Hum Genet. 2006;79(6):1002–16.
27. Aschard H. A Perspective on Interaction Tests in Genetic Association Studies. bioRxiv 2015. doi:10.1101/019661".
28. Pasaniuc B, Zaitlen N, Lettre G, Chen GK, Tandon A, Kao WH, et al. Enhanced statistical tests for GWAS in admixed populations: assessment using African Americans from CARe and a Breast Cancer Consortium. PLoS Genet. 2011;7(4):e1001371.

9

Interaction of smoking and obesity susceptibility loci on adolescent BMI: The National Longitudinal Study of Adolescent to Adult Health

Kristin L. Young[1,2,7*], Misa Graff[1,2], Kari E. North[1,3], Andrea S. Richardson[2,6], Karen L. Mohlke[3,4], Leslie A. Lange[3,4], Ethan M. Lange[3,4], Kathleen M. Harris[2,5] and Penny Gordon-Larsen[2,6]

Abstract

Background: Adolescence is a sensitive period for weight gain and risky health behaviors, such as smoking. Genome-wide association studies (GWAS) have identified loci contributing to adult body mass index (BMI). Evidence suggests that many of these loci have a larger influence on adolescent BMI. However, few studies have examined interactions between smoking and obesity susceptibility loci on BMI. This study investigates the interaction of current smoking and established BMI SNPs on adolescent BMI. Using data from the National Longitudinal Study of Adolescent to Adult Health, a nationally-representative, prospective cohort of the US school-based population in grades 7 to 12 (12–20 years of age) in 1994–95 who have been followed into adulthood (Wave II 1996; ages 12–21, Wave III; ages 18–27), we assessed (in 2014) interactions of 40 BMI-related SNPs and smoking status with percent of the CDC/NCHS 2000 median BMI (%MBMI) in European Americans ($n = 5075$), African Americans ($n = 1744$) and Hispanic Americans ($n = 1294$).

Results: Two SNPs showed nominal significance for interaction ($p < 0.05$) between smoking and genotype with %MBMI in European Americans (EA) (rs2112347 (POC5): $\beta = 1.98$ (0.06, 3.90), $p = 0.04$ and near rs571312 (MC4R): β 2.15 (−0.03, 4.33) $p = 0.05$); and one SNP showed a significant interaction effect after stringent correction for multiple testing in Hispanic Americans (HA) (rs1514175 (TNNI3K): β 8.46 (4.32, 12.60), $p = 5.9E-05$). Stratifying by sex, these interactions suggest a stronger effect in female smokers.

Conclusions: Our study highlights potentially important sex differences in obesity risk by smoking status in adolescents, with those who may be most likely to initiate smoking (i.e., adolescent females), being at greatest risk for exacerbating genetic obesity susceptibility.

Keywords: Adolescence, Obesity, Smoking, Gene-environment interaction

Background

Adolescence is a sensitive period for weight gain and health risk behaviors, such as smoking [1, 2]. Obese smokers suffer 2.8–3.7 times greater mortality than those who are not obese and do not smoke [3]. In the US, nearly 90 % of adult daily smokers begin smoking in their teens [4], and 400,000 adolescents become daily smokers every year [5]. Many adolescents, particularly females, use smoking as an appetite control strategy [6, 7]. Females with greater body dissatisfaction are more likely to smoke [8], and obesity increases the likelihood of being highly addicted to nicotine during adolescence [9]. The effects of smoking differ by gender, in that smoking has a reported antiestrogenic effect in females, which may influence fat deposition [10, 11]. Adolescent smoking also varies by ethnicity, with Hispanic teens that have expressed concern about their weight being more likely to

* Correspondence: kristin.young@unc.edu
[1]Department of Epidemiology, Gillings School of Global Public Health, University of North Carolina, Chapel Hill, NC, USA
[2]Carolina Population Center, Gillings School of Global Public Health, University of North Carolina, Chapel Hill, NC, USA
Full list of author information is available at the end of the article

smoke than non-Hispanic teens [12]. While it has been demonstrated that weight is generally lower among adult smokers (ages 25–44 years), and higher among former adult smokers, this trend has not been observed in some younger smokers (ages 16–24 years) [13]. In addition, weight control effects of smoking may dissipate over time, as long-term smokers (20+ years) are heavier than never or former smokers, and heavy smokers are more likely to be obese than both other smokers and non-smokers [14, 15].

Genome-wide association studies (GWAS) have identified single nucleotide polymorphisms (SNPs) contributing to variation in adult body mass index (BMI) [16–21], and evidence suggests these loci may have the greatest influence on adolescent BMI [22–28]. While many studies of obesity control for smoking status [29–32], few have examined the interaction between smoking and obesity susceptibility loci on BMI [33–36]. However, smoking has been implicated in appetite suppression through the *POMC* neural pathway [37], and loci in this pathway *(POMC* and *MC4R)* increase obesity risk [18, 38]. Our study examines the interaction between current smoking and 40 GWAS-identified and replicated SNPs associated with BMI in European descent adults [16, 18, 19, 21] on adolescent BMI in a multiethnic nationally-representative cohort.

Results

Sample size, gender, mean age, percent median BMI (%MBMI), smoking status and other descriptives are presented by ancestry in Table 1. In the full sample, 11 % of participants aged 12–21 were obese (BMI ≥ 95th percentile), while a further 17 % were overweight (BMI ≥ 85th percentile). African Americans (AA) had the highest percent obese (15.8 %), while Hispanic Americans (HA) had the highest percent overweight (21.9 %). Two-sample t-tests showed significantly higher BMI and %MBMI in female, but not male, smokers than their non-smoking counterparts (Additional file 1: Table S1).

In main effects analyses of SNPs on %MBMI among European Americans (EA), 33 of the established 39 BMI SNPs were directionally consistent with previous results [18], and 19 of those showed nominally significant association with %MBMI (Additional file 2: Table S2). In AA, 12 out of 17 generalizable SNPs had effects on %MBMI that were directionally consistent with the published literature, and 5 of these were nominally associated with %MBMI (Additional file 3: Table S3). In our HA sample, 22 out of 31 established BMI loci in HA were directionally consistent with effects reported for BMI in EA adults, and 3 of these were nominally associated with %MBMI (Additional file 4: Table S4). Interaction analyses were subsequently performed for these

33, 12 and 22 directionally consistent SNPs in EA, AA and HA, respectively.

Two SNPs showed nominal ($p < 0.05$) evidence for interaction with smoking on %MBMI in EA adolescents [rs2112347 *(POC5)*: β = 1.98 (0.06, 3.90), $p = 0.04$ and near rs571312 *(MC4R)*: β 2.15 (−0.03, 4.33) $p = 0.05$]. One SNP had a significant interaction effect after the most stringent multiple test correction for 67 SNPs tested across three ancestries (0.05/67 = 7.5E-04) in HA adolescents [rs1514175 *(TNNI3K)*: β = 8.46 (4.32, 12.60), $p = 5.9E–05$] (Additional file 2: Tables S2, Additional file 3: Tables S3 and Additional file 4: Tables S4). Fig. 1 illustrates results from stratified analyses of these SNPs on %MBMI by smoking status. In all cases, the estimated effect of the BMI-increasing allele was more pronounced in smokers (Fig. 1 and Table 2). None of these SNPs showed a main effect on smoking status (Additional file 2: Table S2, and Additional file 4: Table S4).

Examination of three-way interactions (SNP x smoking status × sex) for these three SNPs revealed only *MC4R* had a nominally significant interaction effect [β = 5.44 (1.11, 9.77), $p = 0.014$]. Given the available sample sizes, it is not unexpected that statistical evidence supporting a three-way interaction would be difficult to detect. When we investigated SNP × smoking status interaction for *MC4R* in EA stratified by sex, we found a nominally significant interaction only in EA females [β = 4.75 (1.73, 7.77), $p = 2.0E-03$; EA males β = 1.09 (−4.23, 2.05), $p = 0.50$]. In addition, when we stratified the effect of the obesity-risk genotype by sex and smoking status, we noted differential association with %MBMI (Table 2). None of the three loci that showed nominal significance for interaction were associated ($p < 0.05$) with %MBMI in female nonsmokers, while only *MC4R* was nominally significant in male nonsmokers. Both *TNNI3K* [β = 6.41 (0.92, 11.90), $p = 0.02$] and *POC5* [β = 2.76 (0.55, 4.97), $p = 0.01$] were nominally significant in HA and EA female smokers, respectively. *MC4R* was significant after correction for multiple testing in EA female smokers [β = 5.48 (3.06, 7.88), $p = 8.4E-06$] (Fig. 2).

Discussion

While previous research has shown that some smoking-associated loci influence BMI in smokers but not never smokers [39], and some established BMI loci are associated with smoking [40], few studies have examined the interaction between smoking and genetic risk for obesity on adolescent BMI. In this nationally representative study of adolescents, we identify two nominally significant obesity susceptibility variants in EA, rs2112347 *(POC5)* and rs571312 *(MC4R)*, and one Bonferroni corrected significant variant in HA, rs1514175 *(TNNI3K)*, which showed a comparatively stronger association in

Table 1 Sex, age, BMI, %MBMI and smoking status by ethnicity in the Add Health analytic sample

Characteristic	All (N = 8113)	European Americans (N = 2065)		African Americans (N = 1744)		Hispanic Americans (N = 1294)	
	Mean [95 % CI] /N (%)	Smokers (N = 2065)	Nonsmokers (N = 3010)	Smokers (N = 324)	Nonsmokers (N = 1420)	Smokers (N = 367)	Nonsmokers (N = 927)
Female	4286 (52.8)	1102 (53.4)	1569 (52.1)	149 (46.0)	811 (57.1)	183 (49.9)	472 (50.9)
Age in years	16.36 [16.32,16.40]	16.60 [16.53, 16.68]	16.08 [16.02, 16.15]	16.75 [16.54, 16.95]	16.34 [16.24, 16.43]	16.66 [16.48, 16.84]	16.53 [16.41, 16.64]
BMI	23.45 [23.34, 23.57]	23.18 [22.96, 23.40]	22.94 [22.78, 23.12]	24.97 [24.31, 25.63]	24.13 [23.83, 24.43]	23.65 [24.12, 25.27]	24.70 [23.32, 23.99]
%MBMI	112.42 [111.88, 112.96]	110.40 [109.36, 111.45]	110.76 [109.93, 111.59]	118.52 [115.36, 121.67]	115.93 [114.49, 117.36]	112.83 [114.64, 120.06]	117.35 [111.25, 114.41]
Self-reported BMI	79 (0.01)	24 (0.01)	30 (0.01)	6 (0.02)	12 (0.01)	2 (0.005)	5 (0.005)
% Obese	11 %	11 %	9 %	18 %	14 %	17 %	11 %
% Overweight	17 %	17 %	16 %	19 %	20 %	22 %	19 %
Region of US							
West	1546 (19.1)	247 (12.0)	533 (17.7)	38 (11.7)	208 (14.6)	146 (39.8)	347 (40.3)
Midwest	2286 (28.2)	824 (39.9)	1034 (34.4)	65 (20.1)	268 (18.9)	39 (10.6)	56 (6.0)
South	3234 (39.8)	686 (33.2)	987 (32.8)	200 (61.7)	866 (61.0)	108 (29.4)	387 (41.8)
Northeast	1047 (12.9)	308 (14.9)	456 (15.1)	21 (6.5)	78 (5.50)	74 (20.2)	110 (11.9)
African Americans							
Highly Educated				49 (15.1)	291 (20.5)		
Hispanic Americans							
Ancestry							
Puerto Rican						89 (24.3)	134 (14.5)
Cuban						37 (10.0)	156 (16.8)
Mexican						181 (49.3)	475 (51.3)
Central/South American						27 (7.4)	92 (9.9)
Other Hispanic						33 (9.0)	70 (7.5)
Immigrant status							
US Born						325 (88.6)	702 (75.7)
Non-US born						42 (11.4)	225 (24.3)

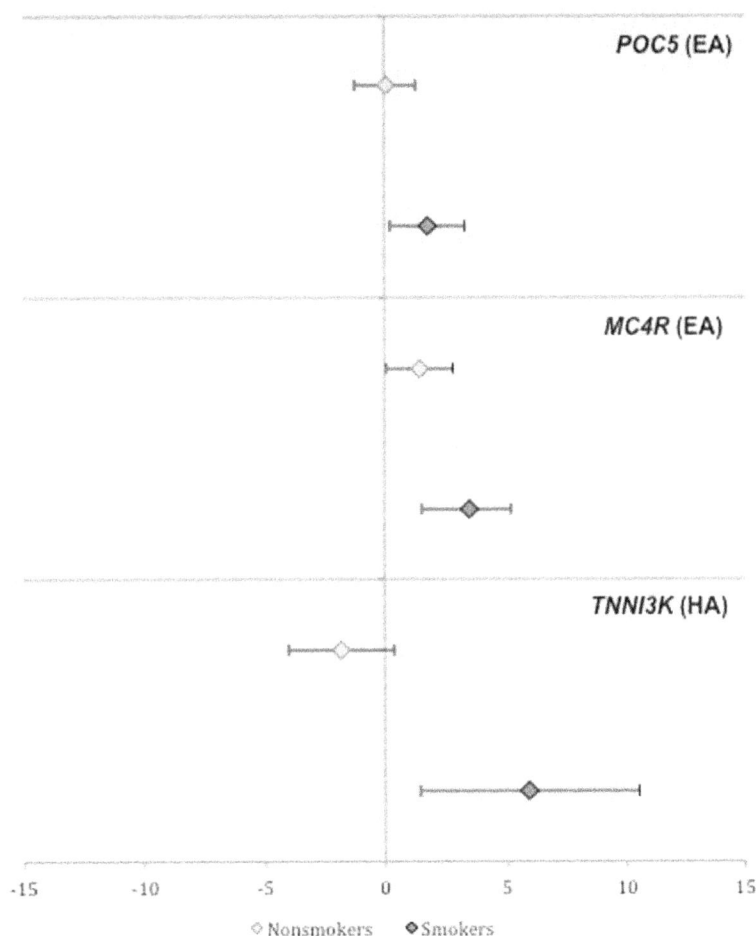

Fig. 1 Main effect of SNP on %MBMI, stratified by ethnicity and smoking status, for those SNPs which showed a nominally significant (p<0.05) interaction effect with smoking on %MBMI

smokers vs. nonsmokers. Sex-stratified analyses revealed that, in general, smoking had a greater estimated effect on %MBMI in adolescent females. In particular, EA female smokers who carry the *MC4R* obesity susceptibility allele had a %MBMI that was 5.48 % higher than nonsmokers that carry the allele ($p = 8.4E-06$).

Our results are consistent with previous literature, in that not all obesity susceptibility loci showed a greater estimated positive effect in smokers (data not shown). Among EA, 20 of 33 BMI loci (61 %) had a larger estimated effect in smokers versus nonsmokers, while 6 of 12 (50 %) and 12 of 22 BMI loci (55 %) had a larger estimated effect in smokers versus nonsmokers among AA and HA, respectively. In addition, the interaction effects we observed were generally more pronounced in women than in men. Previous analysis of 14 established BMI loci in EA and AA adults found no significant interaction ($p < 0.05$) between BMI

SNPs and smoking [33]. However, the authors noted a 3x increase in the estimated effect of the *FTO* (rs9939609) risk allele in EA female smokers, as well as a suggestive stronger estimated effect of the *TMEM18* risk allele in AA female former/never smokers. No differential effects were reported for men. In our analysis, EA female smokers had a 1.22x increase in the estimated effect of the *FTO* risk allele, while EA male smokers had a 1.17 increased estimated effect of the *FTO* risk allele, compared to nonsmokers. We did not examine the effect of *TMEM18* on BMI in AA, as that SNP did not generalize in the recent AA GWAS.

In our study, HA adolescent smokers carrying the obesity risk variant rs1514174 (near *TNNI3K*) were 8.46 %MBMI units larger than their non-smoking peers ($p = 5.9E-05$). The association of *TNNI3K* with obesity has been replicated in both in EA children

Table 2 Stratified analysis of nominally significant ($p < 0.05$) SNP-by-smoking interactions on %MBI in Add Health

	European American (EA) Nonsmokers		European American (EA) Smokers	
POC5 (rs2112347)	Beta [95 % CI]	*p*	Beta [95 % CI]	*p*
All	0.04 [−1.19, 1.27]	0.947	1.75 [0.22, 3.28]	**0.025**
Females	1.22 [−0.49, 2.93]	0.16	2.76 [0.55, 4.97]	**0.014**
Males	−1.48 [−3.26, 0.30]	0.102	1.03 [−1.01, 3.07]	0.324
MC4R (rs571312)	Beta [95 % CI]	*p*	Beta [95 % CI]	*p*
All	1.46 [0.09, 2.83]	**0.036**	3.50 [1.78, 5.22]	**7.54E−05**
Females	1.11 [−0.79, 3.01]	0.253	5.48 [3.07, 7.89]	**8.37E−06**
Males	2.05 [0.07, 4.03]	**0.042**	0.87 [−1.58, 3.32]	0.489
	Hispanic American (HA) Nonsmokers		Hispanic American (HA) Smokers	
TNNI3K (rs1514175)	Beta [95 % CI]	*p*	Beta [95 % CI]	*p*
All	−1.80 [−4.09, 0.49]	0.123	5.97 [2.36, 9.58]	**0.001**
Females	−2.00 [−5.21, 1.21]	0.223	6.41 [0.92, 11.90]	**0.022**
Males	−1.78 [−4.99, 1.43]	0.279	5.25 [0.39, 10.11]	**0.033**

Bold highlights nominally significant associations ($p \leq 0.05$). Mixed effects model, $BMI = \beta + \beta SNPxSMK + \beta SNP + \beta SMK + \beta age + \beta sex + f + s + \varepsilon$, Betas shown in table refer to $\beta SNPxSMK$. %MBMI = Percent of the CDC/NCHS 2000 median BMI

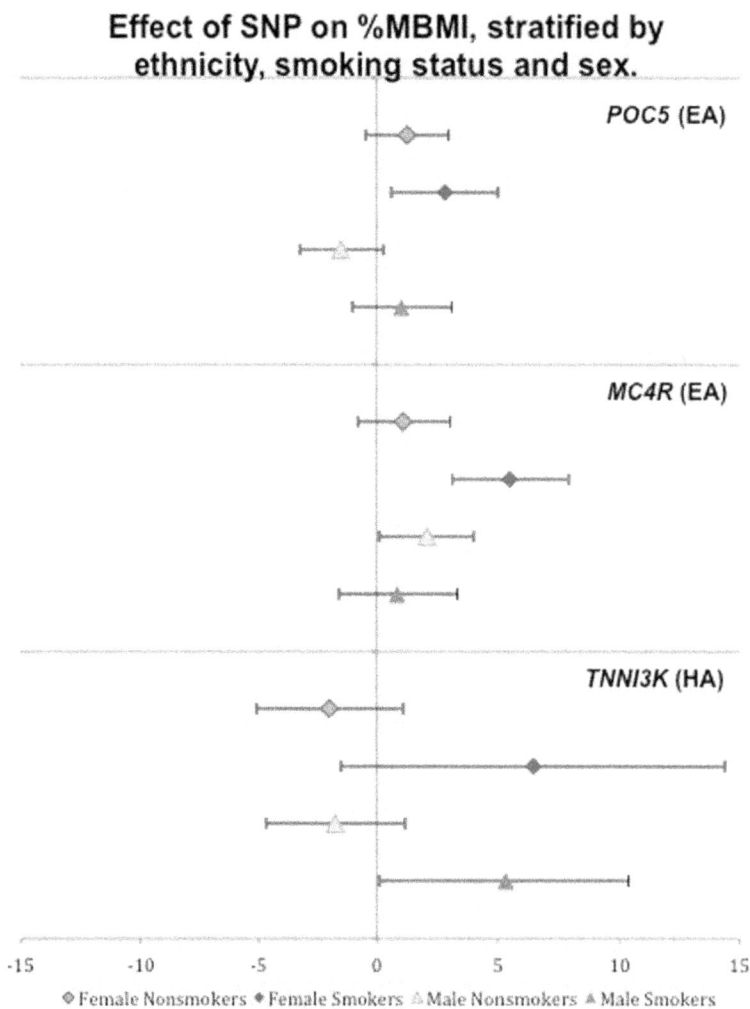

Fig. 2 Main effect of SNP on %MBMI, stratified by ethnicity, smoking status, and sex, for those SNPs which showed a nominally significant ($p<0.05$) interaction effect with smoking on %MBMI

[38, 41] and HA women [42]. *TNNI3K* has been associated with increased intake of fats and sugary foods in overweight or obese adults with Type 2 diabetes [43], and has been nominally associated ($p < 0.05$) with emotional and uncontrolled eating, suggesting a potential mechanism for influencing obesity [30]. In mouse models, *TNNI3K* expression has been linked to cardiac function and cardiac oxidative stress following myocardial infarction [44, 45]. Both smoking and obesity increase systemic oxidative stress [46] and risk of cardiovascular disease (CVD), and the influence of *TNNI3K* on cardiac function suggests a possible biological pathway for this interaction.

Two loci showed nominally significant effects for interaction in EA adolescents, *POC5* and *MC4R*. EA adolescents carrying the obesity risk variant rs2112347 (near *POC5*) were 1.98 %MBMI units larger than EA adolescent nonsmokers. Though the association between rs2112347 and BMI has been replicated [41, 47, 48], the biological mechanism through which rs2112347 influences obesity risk in not known [49]. This variant does lie, however, within 500 kb of *HMGCR*, a gene involved in lipid metabolism. Cigarette smoking increases dyslipidemia by inducing lipolysis in adipose tissue [50, 51], offering a promising avenue for future studies.

Finally, rs571312 near *MC4R* demonstrates the strongest influence on %MBMI in EA female smokers [$\beta = 5.48$ (3.06, 7.88), $p = 8.4E-06$], compared to EA female nonsmokers [β 1.11 (–0.79, 3.01), $p = 0.253$], EA male nonsmokers [β 2.05 (0.07, 4.03), $p = 0.042$], and EA male smokers [β 0.87 (–1.58, 3.32), $p = 0.489$] (Table 2, Fig. 2). Variants in *MC4R* are associated with monogenic obesity and show differential effects on BMI by sex and age, with a greater influence on adolescent females [22, 52]. *MC4R* is primarily expressed in the central nervous system [53], and plays a pivotal role in the leptin-melanocortin pathway regulating appetite, energy balance, and stress response [54]. Variants in and near *MC4R* have been linked to metabolic syndrome [55, 56], percent body fat [57, 58], eating behavior [59], higher fat intake [60], and lower energy expenditure [61, 62]. In animal and some human models, variants near *MC4R* have been shown to disproportionally affect adiposity in females [63–69]. While nicotine has been implicated in animal models as having a hypophagic effect on the leptin-melanocortin pathway influencing feeding behavior [37, 70], other research has shown a 2.9 fold increased risk of metabolic syndrome among smokers who carry a risk variant at a SNP (rs17782313) in high linkage disequilibrium (LD) with our *MC4R* SNP (rs571312, $R^2 = 0.955$) [69]. Rs17782313 has also been associated with a gender and temporal-specific effect on BMI, as well as smoking behavior [72]. Our results suggest *MC4R* obesity risk variants might mitigate the appetite suppressant effect of nicotine in adolescent female smokers.

Add Health represents a unique sample during a sensitive developmental period, when risky health behaviors are being established. Add Health is a nationally representative sample of US adolescents who are being followed into adulthood. As such, our results can be considered generalizable to American adolescents entering adulthood in the late 1990s-early 2000s, but likely are not generalizable to adolescents at other time periods or in other countries. While we are fortunate to have measured heights and weights for the majority of our sample, current smoking was self-reported, though the questions used to assess smoking status in Add Health have been validated among adolescents. Our study was also limited by the lack of established BMI loci in all ancestries, particularly HA. We also recognize that we were possibly underpowered to detect effects due to small sample size [73], and that our approach cannot account for SNPs with an interaction effect but no measurable marginal effect on %MBMI. Given our sample size ($N = 5075$) and other model parameters in EA, we have between 47 and 52 % power to detect nominally significant interaction effects as large as those seen for the variants near *POC5* (β 1.98) and *MC4R* (β 2.15). While our power is limited, pointing to the need to replicate our results in larger future studies, our results do suggest potential SNPs for further interrogation of the influence of smoking on BMI, particularly in adolescent females.

Conclusions

Our study highlights potentially important sex differences in obesity risk by smoking status in adolescents, with those who may be most likely to initiate smoking (i.e., adolescent females), being at greatest risk for poor health outcomes (exacerbating genetic obesity risk). Smoking influences central body fat distribution, and research suggests this effect could be particularly pronounced among women [74]. In addition, smokers have a greater risk of metabolic syndrome [71, 75] and dyslipidemia [76], as well as a much greater risk of mortality, particularly for CVD deaths among obese women under age 65 [3], highlighting the importance of targeting smoking early in adolescence to prevent poorer health in adulthood.

Methods
Study sample
The National Longitudinal Study of Adolescent to Adult Health (Add Health) is a nationally-representative, prospective cohort of adolescents from the US school-based population in grades 7 to 12 (12–20 years of age) in 1994–95 ($n = 20,745$) who have been followed into adulthood (Table 2). Add Health selected a systematic random sample of 80 high schools and 52 feeder middle

schools, representative of US schools with respect to region, urbanicity, school type and size, and student demographics. Written informed consent was obtained from participants or a parent/guardian if the participant was a minor at the time of recruitment. Respondents were followed through Wave II (1996, $n = 14,738$, age 13–21), Wave III (2001–2002, $n = 15,197$, age 18–26) and most recently Wave IV (2007–2008, $n = 15,701$, age 24–32), when respondents provided written informed consent for participation in genetic studies ($n = 12,234$). Add Health included a core sample plus subsamples of selected groups, including African American students with at least one parent with a college degree, collected under protocols approved by the Institutional Review Board at the University of North Carolina at Chapel Hill covering recruitment at all sites. The survey design and sampling frame have been described previously [77-79].

Race/ethnicity

Ancestry informative genetic markers were not available, so a self-reported race/ethnicity variable was constructed based on survey responses regarding ancestral background and family relationship status from both participants and their parents at Wave I. We used a three-category classification: non-Hispanic European American (EA), non-Hispanic African American (AA), and Hispanic American (HA). Within HA, we generated additional variables to account for subpopulation (Cuban, Puerto Rican, Central/South American, Mexican, or Other Hispanic), as well as foreign-born status (first generation immigrants versus those born in the US).

Sibling relatedness

Add Health oversampled related adolescents, resulting in 5524 related Wave I respondents living in 2639 households [80]. Familial relatedness was classified according to participant and parental self-report. Twin zygosity was confirmed by 11 molecular genetic markers [81].

Genetic characterization

The 40 SNPs genotyped in the current study were identified in published GWAS from the Genetic Investigation of Anthropometric Traits (GIANT) consortium for BMI in EA adults [16, 18, 19, 21]. Genotyping was performed using TaqMan assays and the ABI Prism 7900R Sequence Detection System (Applied Biosystems, Foster City, CA, USA). Primer sequences and TaqMan probes are available upon request. The genotype call rate ranged from 97.8 to 98.2 % and the discordance rate between blind duplicates was 0.3 %. SNPs that failed tests for Hardy-Weinberg Equilibrium (HWE) ($p < 0.001$) within race/ethnicity were excluded ($N = 1$, rs2922763) resulting in 39 SNPs for this analysis, as listed in Additional file 5: Table S5.

Criteria for generalizability

Across all groups, to the extent possible, generalizability was defined as similar direction of effect as reported in the literature and nominal statistical significance ($p < 0.05$) [24]. These criteria make generalization in the EA subpopulation straightforward, since these associations were defined in EA adults. A recent large AA GWAS [82], however, suggests that some SNPs fail to generalize, either due to limited power or because of linkage disequilibrium differences that fail to capture the signal of the functional variant. We thus excluded 15 SNPs in AA that have not shown evidence for generalization (i.e., SNP effect estimates were directionally inconsistent and evidence for association was $p > 0.20$ in the recent AA GWAS) [82]. Similar results were reported in a recent HA GWAS of postmenopausal women, where only 9 of 32 established BMI loci showed evidence for association. As this analysis was conducted in a limited sample, however, we chose to retain all directionally consistent loci in our HA analysis [42]. In addition, SNPs with insufficient cell size for analysis ($n < 10$ individuals per genotype) were excluded, leaving 33 SNPs in EA, 12 SNPs in AA, and 22 SNPs in HA for the interaction analyses (included SNPs highlighted in **bold** in Additional file 5: Table S5).

Analytic sample

At Wave IV, 59 % ($n = 12,234$) of Wave I ($n = 20,745$) respondents provided samples, with consent, from which DNA was extracted and genotyped ($n = 12,066$). To be eligible our study, individuals had to have at least 80 % of their 39 SNPs genotyped ($n = 11,448$) and be between the ages of 12 and 21 years at either Wave II or III ($n = 9129$). Among the 9129 eligible adolescents, we excluded: the monozygotic twin with fewer genotyped loci ($n = 139$), individuals of Native American ($n = 57$), Asian ($n = 436$) or unclassified ($n = 112$) race/ethnicity, pregnant ($n = 110$), disabled ($n = 47$), and those missing data for geographic region ($n = 67$), BMI ($n = 2$), or current smoking ($n = 46$). The analytic sample was selected from waves II or III to capture the age range of 12–21 years, and all covariates match the wave at which BMI was measured. Our final analytic sample ($n = 8113$) included 5075 EA, 1744 AA, and 1294 HA.

Body mass index (BMI)

Weight and height were measured during in-home surveys using standardized procedures. BMI (kg/m^2) was calculated using measured height and weight assessed at Waves II or III when participants were between the age of 12 and 21 years, with priority for younger age at measurement (Wave II: $n = 7681$), unless the respondent was not seen at Wave II and was still between the ages of 12–21 years at Wave III ($n = 432$). Self-reported heights and weights, which have been previously validated in Add Health, were substituted for those who refused measurement and/or weighed more than the scale capacity (Wave II $n = 55$; Wave III $n = 24$) [83]. Due to changes in weight and height with growth and development, BMI varies by age and sex, which necessitates using age- and sex-specific BMI Z-scores relative to a reference such as the US the CDC/NCHS 2000 growth curves [84]. However, these growth curves do not represent the tails of the distribution well, which is a particular issue in a cohort with considerable upward skew in distribution relative to the CDC/NCHS 2000 healthy reference. A strategy to deal with this is to use percent of the CDC/NCHS 2000 median [85], which also has the benefit of ease in interpretation relative to the Z-score. Accordingly, our outcome for all analyses was the percent of the CDC/NCHS 2000 median BMI (%MBMI).

Current smoking

Current smoking was based on self-report, which has been previously validated among adolescents [86], and was defined as smoking at least 1 day in the last 30 days. [2, 87, 88] Current smoking status was queried at Waves II ($N_{current\_smokers} = 2589$) and III ($N_{current\_smokers} = 2377$), to match the wave at which BMI was measured. To measure the effect of BMI-related SNPs on current smoking, we performed main effects logistic regression using smoking status as the outcome and SNP as predictor, stratified by ancestry (Additional file 2: Tables S2, Additional file 3: Tables S3 and Additional file 4: Tables S4).

Statistical analysis

In ancestry-stratified, multivariable interaction (SNPxsmoking) models with %MBMI as the outcome, we controlled for age, sex, geographic region, and self-reported heights and weights using Stata (v13.1, Stata Corp, College Station, Texas). In non-EA populations, we also controlled for oversampling of adolescents from highly-educated African American families ($n = 355$), and Hispanic subpopulation: Cuban ($n = 193$), Puerto Rican ($n = 223$), Central/South American ($n = 119$), Mexican ($n = 656$), and other Hispanic ($n = 102$), as well as an indicator for foreign-born status ($n = 267$). Sample design effects and familial relatedness were accounted for by including separate random effects for school and family. When a nominally significant interaction ($p < 0.05$) was detected, we ran additional interaction models (SNP x smoking status, stratified by sex; and SNP x smoking status x sex), and examined SNP effects in models stratified by smoking status and sex, to facilitate interpretation. To correct for multiple testing, we applied a Bonferroni correction equal to 0.05/number of SNPs tested in each group (0.05/33 = 0.0015 in EA, 0.05/22 = 0.0023 in HA, 0.05/12 = 0.0042 in AA).

Availability of data and materials

Add Health adheres to the NIH policy on data sharing, but due to the sensitive nature of Add Health data, access is limited and governed by the Add Health data management security plan to ensure respondent confidentiality. For this reason, the distribution of data is limited to a public-use dataset for a subset of respondents, and a restricted-use dataset distributed only to certified researchers committed to maintaining limited access. Add Health is currently in the process of submitting genetic data to dbGaP, which will be made available to researchers meeting both dbGaP and Add Health data use requirements. More information can be found here: http://www.cpc.unc.edu/projects/addhealth.

Additional files

Additional file 1: Table S1. Two-sample *t*-test of differences in BMI and %MBMI by smoking status, stratified by ancestry and sex. (DOCX 72 kb)

Additional file 2: Table S2. Results of SNPxSmoking on %MBMI (Interaction), SNP on %MBMI (Main effects), and SNP on smoking in European American adolescents in Add Health. (DOCX 41 kb)

Additional file 3: Table S3. Results of SNPxSmoking on %MBMI (Interaction), SNP on %MBMI (Main effects), and SNP on smoking in African American adolescents in Add Health. (DOCX 35 kb)

Additional file 4: Table S4. Results of SNPxSmoking on %MBMI (Interaction), SNP on %MBMI (Main effects), and SNP on smoking in Hispanic American adolescents in Add Health. (DOCX 40 kb)

Additional file 5: Table S5. Established BMI loci used in present analysis. (DOCX 31 kb)

Abbreviations

BMI: Body mass index; %MBMI: Percent median BMI; SNP: Single nucleotide polymorphism; EA: European American; AA: African American; HA: Hispanic American; CVD: Cardiovascular disease; GWAS: Genome Wide Assocation Study; HWE: Hardy Weinberg Equilibrium.

Competing interests

This work was funded by National Institutes of Health grant R01HD057194. There were no potential or real conflicts of financial or personal interest with the financial sponsors of the research project. Research sponsors had no role in study design; collection, analysis, or interpretation of data; writing the manuscript; or the decision to submit the manuscript for publication.

Authors' contributions

PGL, KMH, KEN, EML, and KLY contributed to study design; KLY, MG and ASR to data analysis, KLY, KEN, EML, LAL, KLM, and PGL contributed to data interpretation; and KLY, KEN, and PGL contributed to writing the manuscript. All authors provided critical evaluation of the manuscript, had full access to all data in the study, and take responsibility for data integrity and analysis accuracy. All authors read and approved the final manuscript.

Acknowledgement

We thank Amy Perou of the BioSpecimen Processing facility and Jason Luo of the Mammalian Genotyping Core at the University of North Carolina at Chapel Hill. This work was funded by National Institutes of Health grant R01HD057194. This research uses data from Add Health, a program project directed by Kathleen Mullan Harris and designed by J. Richard Udry, Peter S. Bearman, and Kathleen Mullan Harris at the University of North Carolina at Chapel Hill, and funded by grant P01-HD31921 from the Eunice Kennedy Shriver National Institute of Child Health and Human Development, with cooperative funding from 23 other federal agencies and foundations. Special acknowledgement is due to Ronald R. Rindfuss and Barbara Entwistle for assistance in the original design. Information on how to obtain Add Health data files is available on the Add Health website (http://www.cpc.unc.edu/addhealth). No direct support was received from grant P01-HD31921 for this analysis. We are grateful to the Carolina Population Center for general support. Preliminary analysis of this work has been presented at The Obesity Society Annual Meeting (2012).

Author details
[1]Department of Epidemiology, Gillings School of Global Public Health, University of North Carolina, Chapel Hill, NC, USA. [2]Carolina Population Center, Gillings School of Global Public Health, University of North Carolina, Chapel Hill, NC, USA. [3]Carolina Center for Genome Sciences, Gillings School of Global Public Health, University of North Carolina, Chapel Hill, NC, USA. [4]Department of Genetics, Gillings School of Global Public Health, University of North Carolina, Chapel Hill, NC, USA. [5]Department of Sociology, Gillings School of Global Public Health, University of North Carolina, Chapel Hill, NC, USA. [6]Department of Nutrition, Gillings School of Global Public Health, University of North Carolina, Chapel Hill, NC, USA. [7]137 East Franklin Street, Suite 306, Chapel Hill, NC 27514, USA.

References
1. Harris KM, Gordon-Larsen P, Chantala K, Udry JR. Longitudinal trends in race/ethnic disparities in leading health indicators from adolescence to young adulthood. Arch Pediatr Adolesc Med. 2006;160(1):74–81. doi:10.1001/archpedi.160.1.74.
2. Dierker L, Mermelstein R. Early emerging nicotine-dependence symptoms: A signal of propensity for chronic smoking behavior in adolescents. J Pediatr. 2010;156(5):818–22. doi:10.1016/j.jpeds.2009.11.044.
3. Freedman DM, Sigurdson AJ, Rajaraman P, Doody MM, Linet MS, Ron E. The mortality risk of smoking and obesity combined. Amer J Prev Med. 2006;31(5):355–62. doi:10.1016/j.amepre.2006.07.022.
4. U.S. Department of Health and Human Services. Preventing tobacco use among youth and young adults: A report from the Surgeon General. http://www.cdc.gov. 2012. Accessed August 12, 2013.
5. Administration Substance Abuse and Mental Health Services. Results from the 2011 National Survey on Drug Use and Health: Summary of National Findings. http://www.samhsa.gov/data/nsduh/2k11results/nsduhresults2011.htm. 2012. Accessed August 12, 2013.
6. Cawley J, Markowitz S, Tauras J. Lighting up and slimming down: The effects of body weight and cigarette prices on adolescent smoking initiation. J Health Econ. 2004;23(2):293–311.
7. Harakeh Z, Engels RCME, Monshouwer K, Hanssen PF. Adolescent's weight concerns and the onset of smoking. Subst Use Misuse. 2010;45(12):1847–60. doi:10.3109/10826081003682149.
8. Kaufman AR, Augustson EM. Predictors of regular cigarette smoking among adolescent females: Does body image matter? Nicotine Tob Res. 2008;10(8):1301–9. doi:10.1080/14622200802238985.
9. Hussaini AE, Nicholson LM, Shera D, Stettler N, Kinsman S. Adolescent obesity as a risk factor for high-level nicotine addiction in young women. J Adolesc Health. 2011;49(5):511–7. doi:10.1016/j.jadohealth.2011.04.001.
10. Pauly JR. Gender differences in tobacco smoking dynamics and the neuropharmacological actions of nicotine. Front Biosci. 2007;13:505–16.
11. Windham GC, Mitchell P, Anderson M, Lasley BL. Cigarette smoking and effects on hormone function in premenopausal women. Environ Health Perspect. 2005;113(10):1285–90.
12. Weiss JW, Merrill V, Gritz ER. Ethnic variation in the association between weight concern and adolescent smoking. Addict Behav. 2007;32(10):2311–6. doi:10.1016/j.addbeh.2007.01.020.
13. Mackay DF, Gray L, Pell JP. Impact of smoking and smoking cessation on overweight and obesity: Scotland-wide, cross-sectional study on 40,036 participants. BMC Public Health. 2013;13:348. doi:10.1186/1471-2458-13-348.
14. Chiolero A, Jacot-Sadowski I, Faeh D, Paccaud F, Cornuz J. Association of cigarettes smoked daily with obesity in a general adult population. Obesity. 2007;15(5):1311–8. doi:10.1038/oby.2007.153.
15. Clair C, Chiolero A, Faeh D, Cornuz J. Dose-dependent positive association between cigarette smoking, abdominal obesity and body fat: cross-sectional data from a population-based survey. BMC Public Health. 2011;11:23. doi:10.1186/1471-2458-11-23.
16. Willer CJ, Speliotes EK, Loos RJF, Li S, Lindgren CM, Heid IM, et al. Six new loci associated with body mass index highlight a neuronal influence on body weight regulation. Nat Genet. 2008;41(1):25–34. doi:10.1038/ng.287.
17. Heid IM, Jackson AU, Randall JC, Winkler TW, Qi L, Steinthorsdottir V, et al. Meta-analysis identifies 13 new loci associated with waist-hip ratio and reveals sexual dimorphism in the genetic basis of fat distribution. Nat Genet. 2010;42(11):949–60. doi:10.1038/ng.685.
18. Speliotes EK, Willer CJ, Berndt SI, Thorleifsson G, Jackson AU, Lango Allen H, et al. Association analyses of 249,796 individuals reveal 18 new loci associated with body mass index. Nat Genet. 2010;42(11):937–48. doi:10.1038/ng.686.
19. Heard-Costa NL, Zillikens MC, Monda KL, Johansson A, Harris TB, Fu M, et al. NRXN3 is a novel locus for waist circumference: A genome-wide association study from the CHARGE Consortium. PLoS Genet. 2009;5(6):e1000539.
20. Lindgren CM, Heid IM, Randall JC, Lamina C, Steinthorsdottir V, Qi L, et al. Genome-wide association scan meta-analysis identifies three loci influencing adiposity and fat distribution. PLoS Genet. 2009;5(6):e1000508. doi:10.1371/journal.pgen.1000508.
21. Thorleifsson G, Walters GB, Gudbjartsson DF, Steinthorsdottir V, Sulem P, Helgadottir A, et al. Genome-wide association yields new sequence variants at seven loci that associate with measures of obesity. Nat Genet. 2008;41(1):18–24.
22. Kvaløy K, Kulle B, Romundstad P, Holmen TL. Sex-specific effects of weight-affecting gene variants in a life course perspective–The HUNT Study, Norway. Int J Obes Relat Metab Disord. 2013;37(9):1221–9. doi:10.1038/ijo.2012.220.
23. Elks CE, Loos RJF, Hardy R, Wills AK, Wong A, Wareham NJ, et al. Adult obesity susceptibility variants are associated with greater childhood weight gain and a faster tempo of growth: the 1946 British Birth Cohort Study. Amer J Clin Nutr. 2012;95(5):1150–6. doi:10.3945/ajcn.111.027870.
24. Graff M, North KE, Mohlke KL, Lange LA, Luo J, Harris KM, et al. Estimation of genetic effects on BMI during adolescence in an ethnically diverse cohort: The National Longitudinal Study of Adolescent Health. Nutr Diab. 2012;2(9):e47–8. doi:10.1038/nutd.2012.20.
25. Graff M, Ngwa JS, Workalemahu T, Homuth G, Schipf S, Teumer A, et al. Genome-wide analysis of BMI in adolescents and young adults reveals additional insight into the effects of genetic loci over the life course. Hum Mol Genet. 2013;22(17):3597–607. doi:10.1093/hmg/ddt205.
26. Llewellyn CH, Trzaskowski M, Plomin R, Wardle J. Finding the missing heritability in pediatric obesity: the contribution of genome-wide complex trait analysis. Int J Obes Relat Metab Disord. 2013;37(11):1506–9. doi:10.1038/ijo.2013.30.
27. Zhao JH, Bradfield JP, Li MD, Wang K, Zhang HW, Kim CE, et al. The role of obesity-associated loci identified in genome-wide association studies in the determination of pediatric BMI. Obesity. 2009;17(12):2254–7. doi:10.1038/oby.2009.159.
28. Loos RJF, Lindgren CM, Li S, Wheeler E, Zhao JH, Prokopenko I, et al. Common variants near MC4R are associated with fat mass, weight and risk of obesity. Nat Genet. 2008;40(6):768–75. doi:10.1038/ng.140.
29. Bradshaw PT, Monda KL, Stevens J. Metabolic syndrome in healthy obese, overweight, and normal weight individuals: the Atherosclerosis Risk in Communities Study. Obesity. 2013;21(1):203–9. doi:10.1002/oby.20248.
30. Cornelis MC, Rimm EB, Curhan GC, Kraft P, Hunter DJ, Hu FB, et al. Obesity susceptibility loci and uncontrolled eating, emotional eating and cognitive restraint behaviors in men and women. Obesity. 2013:n/a–n/a. doi:10.1002/oby.20592.
31. Demerath EW, Lutsey PL, Monda KL, Linda Kao WH, Bressler J, Pankow JS, et al. Interaction of FTO and physical activity level on adiposity in African-American and European-American adults: the ARIC Study. Obesity. 2011;19(9):1866–72. doi:10.1038/oby.2011.131.
32. Dong C, Beecham A, Slifer S, Wang L, McClendon MS, Blanton SH, et al. Genome-wide linkage and peak-wide association study of obesity-related quantitative traits in Caribbean Hispanics. Hum Genet. 2011;129(2):209–19. doi:10.1007/s00439-010-0916-z.
33. Fesinmeyer MD, North KE, Lim U, Bůžková P, Crawford DC, Haessler J, et al. Effects of smoking on the genetic risk of obesity: The population architecture using genomics and epidemiology study. BMC Med Genet. 2013;14:6. doi:10.1186/1471-2350-14-6.

34. Velez Edwards DR, Naj AC, Monda K, North KE, Neuhouser M, Magvanjav O, et al. Gene-environment interactions and obesity traits among postmenopausal African-American and Hispanic women in the Women's Health Initiative SHARe Study. Hum Genet. 2012;132(3):323–36. doi:10.1007/s00439-012-1246-3.

35. Edwards TL, Velez Edwards DR, Villegas R, Cohen SS, Buchowski MS, Fowke JH, et al. *HTR1B, ADIPOR1, PPARGC1A*, and *CYP19A1* and obesity in a cohort of Caucasians and African Americans: an evaluation of gene-environment interactions and candidate genes. Am J Epidemiol. 2011;175(1):11–21. doi:10.1093/aje/kwr272.

36. Lee S, Kim CM, Kim HJ, Park HS. Interactive effects of main genotype, caloric intakes, and smoking status on risk of obesity. Asia Pac J Clin Nutr. 2011;20(4):563–71.

37. Mineur YS, Abizaid A, Rao Y, Salas R, DiLeone RJ, Gündisch D, et al. Nicotine decreases food intake through activation of *POMC* neurons. Science. 2011;332(6035):1330–2. doi:10.1126/science.1201889.

38. Bradfield JP, Taal HR, Timpson NJ, Scherag A, Lecoeur C, Warrington NM, et al. A genome-wide association meta-analysis identifies new childhood obesity loci. Nat Genet. 2012;44(5):526–31. doi:10.1038/ng.2247.

39. Freathy RM, Kazeem GR, Morris RW, Johnson PC, Paternoster L, Ebrahim S, et al. Genetic variation at *CHRNA5-CHRNA3-CHRNB4* interacts with smoking status to influence body mass index. Int J Epidemiol. 2011;40(6):1617–28. doi:10.1093/ije/dyr077.

40. Thorgeirsson TE, Stefansson K. Genetics of smoking behavior and its consequences: The role of nicotinic acetylcholine receptors. Biol Psychiatry. 2008;64(11):919–21. doi:10.1016/j.biopsych.2008.09.010.

41. Zhao J, Bradfield JP, Zhang H, Sleiman PM, Kim CE, Glessner JT, et al. Role of BMI-associated loci identified in GWAS meta-analyses in the context of common childhood obesity in European Americans. Obesity. 2011;19(12):2436–9. doi:10.1038/oby.2011.237.

42. Graff M, Fernández-Rhodes L, Liu S, Carlson C, Wassertheil-Smoller S, Neuhouser M, et al. Generalization of adiposity genetic loci to US Hispanic women. Nutr Diab. 2013;3:e85. doi:10.1038/nutd.2013.26.

43. McCaffery JM, Papandonatos GD, Peter I, Huggins GS, Raynor HA, Delahanty LM, et al. Obesity susceptibility loci and dietary intake in the Look AHEAD Trial. Amer J Clin Nutr. 2012;95(6):1477–86. doi:10.3945/ajcn.111.026955.

44. Lodder EM, Scicluna BP, Milano A, Sun AY, Tang H, Remme CA, et al. Dissection of a quantitative trait locus for PR interval duration identifies *TNNI3K* as a novel modulator of cardiac conduction. PLoS Genet. 2012;8:e1003113. doi:10.1371/journal.pgen.1003113.

45. Vagnozzi RJ, Gatto GJ, Kallander LS, Hoffman NE, Mallilankaraman K, Ballard VL, et al. Inhibition of the cardiomyocyte-specific kinase TNNI3K limits oxidative stress, injury, and adverse remodeling in the ischemic heart. Sci Transl Med. 2013;5(207):207ra141. doi:10.1126/scitranslmed.3006479.

46. Keaney JF, Larson MG, Vasan RS, Wilson PW, Lipinska I, Corey D, et al. Obesity and systemic oxidative stress clinical correlates of oxidative stress in the Framingham Study. Arterioscler Thromb Vasc Biol. 2003;23(3):434–9. doi:10.1161/01.ATV.0000058402.34138.11.

47. Juonala M, Juhola J, Magnussen CG, Würtz P, Viikari JS, Thomson R, et al. Childhood environmental and genetic predictors of adulthood obesity: The cardiovascular risk in young Finns study. J Clin Endocrinol Metab. 2011;96(9):E1542–9. doi:10.1210/jc.2011-1243.

48. Okada Y, Kubo M, Ohmiya H, Takahashi A, Kumasaka N, Hosono N, et al. Common variants at *CDKAL1* and *KLF9* are associated with body mass index in East Asian populations. Nat Genet. 2012;44(3):302–6. doi:10.1038/ng.1086.

49. Speakman JR. Functional analysis of seven genes linked to body mass index and adiposity by genome-wide association studies: A Review. Hum Hered. 2013;75(2–4):57–79. doi:10.1159/000353585.

50. Hellerstein MK, Benowitz NL, Neese RA. Effects of cigarette smoking and its cessation on lipid metabolism and energy expenditure in heavy smokers. J Clin Invest. 1994;93(1):265–72. doi:10.1172/JCI116955.

51. Audrain-McGovern J, Benowitz NL. Cigarette smoking, nicotine, and body weight. Clin Pharmacol Ther. 2011;90(1):164–8. doi:10.1038/clpt.2011.10562.

52. Grant SFA, Bradfield JP, Zhang H, Wang K, Kim CE, Annaiah K, et al. Investigation of the locus near *MC4R* with childhood obesity in Americans of European and African ancestry. Obesity. 2009;17(7):1461–5. doi:10.1038/oby.2009.53.

53. Garver WS, Newman SB, Gonzales-Pacheco DM, Castillo JJ, Jelinek D, Heidenreich RA, et al. The genetics of childhood obesity and interaction with dietary macronutrients. Genes Nutr. 2013;8(3):271–87. doi:10.1007/s12263-013-0339-5.

54. Horstmann A, Kovacs P, Kabisch S, Boettcher Y, Schloegl H, Tönjes A, et al. Common genetic variation near *MC4R* has a sex-specific impact on human brain structure and eating behavior. PLoS One. 2013;8(9):e74362. doi:10.1371/journal.pone.0074362.

55. Lubrano-Berthelier C, Cavazos M, Dubern B, Shapiro A, Stunff CL, Zhang S, et al. Molecular genetics of human obesity-associated *MC4R* mutations. Ann N Y Acad Sci. 2003;994:49–57. doi:10.1111/j.1749-6632.2003.tb03161.x.

56. Dušátková L, Zamrazilová H, Sedláčková B, Včelák J, Hlavatý P, Aldhoon Hainerová I, et al. Association of obesity susceptibility gene variants with metabolic syndrome and related traits in 1,443 Czech adolescents. Folia Biol (Praha). 2013;59(3):123–33.

57. den Hoed M, Luan J, Langenberg C, Cooper C, Sayer AA, Jameson K, et al. Evaluation of common genetic variants identified by GWAS for early onset and morbid obesity in population-based samples. Int J Obes Relat Metab Disord. 2013;37(2):191–6. doi:10.1038/ijo.2012.34.

58. Kilpeläinen TO, Zillikens MC, Stančáková A, Finucane FM, Ried JS, Langenberg C, et al. Genetic variation near *IRS1* associates with reduced adiposity and an impaired metabolic profile. Nat Genet. 2011;43(8):753–60. doi:10.1038/ng.866.

59. Branson R, Potoczna N, Kral JG, Lentes KU, Hoehe MR, Horber FF. Binge eating as a major phenotype of melanocortin 4 receptor gene mutations. N Engl J Med. 2003;348(12):1096–103. doi:10.1056/NEJMoa021971.

60. Qi L, Kraft P, Hunter D, Hu FB. The common obesity variant near *MC4R* gene is associated with higher intakes of total energy and dietary fat, weight change and diabetes risk in women. Hum Mol Genet. 2008;17(22):3502–8. doi:10.1093/hmg/ddn242.

61. Cole SA, Butte NF, Voruganti VS, Cai G, Haack K, Kent JW, et al. Evidence that multiple genetic variants of *MC4R* play a functional role in the regulation of energy expenditure and appetite in Hispanic children. Am J Clin Nutr. 2010;91(1):191–9. doi:10.3945/ajcn.2009.28514.

62. Cai G, Cole SA, Butte N, Bacino C, Diego V, Tan K, et al. A quantitative trait locus on chromosome 18q for physical activity and dietary intake in Hispanic children. Obesity. 2006;14(9):1596–604. doi:10.1038/oby.2006.184.

63. Hardy R, Wills AK, Wong A, Elks CE, Wareham NJ, Loos RJ, et al. Life course variations in the associations between *FTO* and *MC4R* gene variants and body size. Hum Mol Genet. 2010;19(3):545–52. doi:10.1093/hmg/ddp504.

64. Orkunoglu-Suer FE, Harmon BT, Gordish-Dressman H, Clarkson PM, Thompson PD, Angelopoulos TJ, et al. *MC4R* variant is associated with BMI but not response to resistance training in young females. Obesity. 2010;19(3):662–6. doi:10.1038/oby.2010.180.

65. Hinney A, Volckmar A-L, Knoll N. Melanocortin-4 receptor in energy homeostasis and obesity pathogenesis. Prog Mol Biol Transl Sci. 2013;114:147–91. doi:10.1016/B978-0-12-386933-3.00005-4.

66. Sina M, Hinney A, Ziegler A, Neupert T, Mayer H, Siegfried W, et al. Phenotypes in three pedigrees with autosomal dominant obesity caused by haploinsufficiency mutations in the melanocortin-4 receptor gene. Am J Hum Genet. 1999;65(6):1501–7. doi:10.1086/302660.

67. Dempfle A, Hinney A, Heinzel-Gutenbrunner M, Raab M, Geller F, Gudermann T, et al. Large quantitative effect of melanocortin-4 receptor gene mutations on body mass index. J Med Genet. 2004;41(10):795–800. doi:10.1136/jmg.2004.018614.

68. Dougkas A, Yaqoob P, Givens DI, Reynolds CK, Minihane AM. The impact of obesity-related SNP on appetite and energy intake. Br J Nutr. 2013;110(6):1151–6. doi:10.1017/S0007114513000147.

69. Vogel CIG, Boes T, Reinehr T, Roth CL, Scherag S, Scherag A, et al. Common variants near *MC4R*: exploring gender effects in overweight and obese children and adolescents participating in a lifestyle intervention. Obes Facts. 2011;4(1):67–75. doi:10.1159/000324557.

70. Martinez de Morentin PB, Whittle AJ, Fernø J, Nogueiras R, Diéguez C, Vidal-Puig A, et al. Nicotine induces negative energy balance through hypothalamic AMP-activated protein kinase. Diabetes. 2012;61(4):807–17. doi:10.2337/db11-1079.

71. Yang C-W, Li C-I, Liu C-S, Bau DT, Lin CH, Lin WY, et al. The joint effect of cigarette smoking and polymorphisms on *LRP5, LEPR*, near *MC4R* and *SH2B1* genes on metabolic syndrome susceptibility in Taiwan. Mol Biol Rep. 2012;40(1):525–33. doi:10.1007/s11033-012-2089-7.

72. Thorgeirsson TE, Gudbjartsson DF, Sulem P, Besenbacher S, Styrkarsdottir U, Thorleifsson G, et al. A common biological basis of obesity and nicotine addiction. Translational Psychiatry. 2013;3(10):e308–7. doi:10.1038/tp.2013.81.

73. Hunter D. Gene–environment interactions in human diseases. Nat Rev Genet. 2005;6(4):287–98. doi:10.1038/nrg1578.

74. Kwok S, Canoy D, Soran H, Ashton DW, Lowe GD, Wood D, et al. Body fat distribution in relation to smoking and exogenous hormones in British women. Clin Endocrinol (Oxf). 2012;77(6):828–33. doi:10.1111/j.1365-2265.2012.04331.x.

75. Sun K, Liu J, Ning G. Active smoking and risk of metabolic syndrome: A meta-analysis of prospective studies. PLoS One. 2012;7(10):e47791. doi:10.1371/journal.pone.0047791.

76. Ordovas JM, Robertson R, Cléirigh EN. Gene–gene and gene–environment interactions defining lipid-related traits. Curr Opin Lipidol. 2011;22(2):129–36. doi:10.1097/MOL.0b013e32834477a9.

77. Resnick MD, Bearman PS, Blum RW, Bauman KE, Harris KM, Jones J, et al. Protecting adolescents from harm. Findings from the National Longitudinal Study on Adolescent Health. JAMA. 1997;278(10):823–32.

78. Harris KM. An integrative approach to health. Demography. 2010;47(1):1–22.

79. Miller WC, Ford CA, Morris M, Handcock MS, Schmitz JL, Hobbs MM, et al. Prevalence of chlamydial and gonococcal infections among young adults in the United States. JAMA. 2004;291(18):2229–36. doi:10.1001/jama.291.18.2229.

80. Harris KM, Halpern CT, Haberstick BC, Smolen A. The National Longitudinal Study of Adolescent Health (Add Health): Sibling pairs data. Twin Res Hum Genet. 2013;16(1):391–8. doi:10.1017/thg.2012.137.

81. Team AHB. Biomarkers in Wave III of the Add Health Study. Carolina Population Center; 2005:1–57. Available at: hhttp://www.cpc.unc.edu/projects/addhealth/faqs/aboutdata/biomark.pdf/view. Accessed January 7, 2013.

82. Monda KL, Chen GK, Taylor KC, Palmer C, Edwards TL, Lange LA, et al. A meta-analysis identifies new loci associated with body mass index in individuals of African ancestry. Nat Genet. 2013;45(6):690–6. doi:10.1038/ng.2608.

83. Goodman E, Hinden BR, Khandelwal S. Accuracy of teen and parental reports of obesity and body mass index. Pediatrics. 2000;106(1 Pt 1):52–8.

84. Kuczmarski RJ, Ogden CL, Guo SS, Grummer-Strawn LM, Flegal KM, Mei Z, et al. 2000 CDC growth charts for the United States: Methods and development. Vital Health Stat. 2002;11(246):1–190.

85. Cole TJ, Faith MS, Pietrobelli A, Heo M. What is the best measure of adiposity change in growing children: BMI, BMI%, BMI z-score or BMI centile? Eur J Clin Nutr. 2005;59(3):419–25.

86. Patrick DL, Cheadle A, Thompson DC, Diehr P, Koepsell T, Kinne S. The validity of self-reported smoking: A review and meta-analysis. Am J Public Health. 1994;84(7):1086–93.

87. Centers for Disease Control and Prevention. Youth risk behavior surveillance—United States, 2011. Morb Mortal Wkly Rep Surveill Summ. 2012;61(4):1–162.

88. Rees DL, Sabia JJ. Body weight and smoking initiation: Evidence from Add Health. J Health Econ. 2010;29(5):774–7. doi:10.1016/j.jhealeco.2010.07.002.

10

Impact of QTL minor allele frequency on genomic evaluation using real genotype data and simulated phenotypes in Japanese Black cattle

Yoshinobu Uemoto[1*], Shinji Sasaki[1], Takatoshi Kojima[1], Yoshikazu Sugimoto[2] and Toshio Watanabe[1]

Abstract

Background: Genetic variance that is not captured by single nucleotide polymorphisms (SNPs) is due to imperfect linkage disequilibrium (LD) between SNPs and quantitative trait loci (QTLs), and the extent of LD between SNPs and QTLs depends on different minor allele frequencies (MAF) between them. To evaluate the impact of MAF of QTLs on genomic evaluation, we performed a simulation study using real cattle genotype data.

Methods: In total, 1368 Japanese Black cattle and 592,034 SNPs (Illumina BovineHD BeadChip) were used. We simulated phenotypes using real genotypes under different scenarios, varying the MAF categories, QTL heritability, number of QTLs, and distribution of QTL effect. After generating true breeding values and phenotypes, QTL heritability was estimated and the prediction accuracy of genomic estimated breeding value (GEBV) was assessed under different SNP densities, prediction models, and population size by a reference-test validation design.

Results: The extent of LD between SNPs and QTLs in this population was higher in the QTLs with high MAF than in those with low MAF. The effect of MAF of QTLs depended on the genetic architecture, evaluation strategy, and population size in genomic evaluation. In genetic architecture, genomic evaluation was affected by the MAF of QTLs combined with the QTL heritability and the distribution of QTL effect. The number of QTL was not affected on genomic evaluation if the number of QTL was more than 50. In the evaluation strategy, we showed that different SNP densities and prediction models affect the heritability estimation and genomic prediction and that this depends on the MAF of QTLs. In addition, accurate QTL heritability and GEBV were obtained using denser SNP information and the prediction model accounted for the SNPs with low and high MAFs. In population size, a large sample size is needed to increase the accuracy of GEBV.

Conclusion: The MAF of QTL had an impact on heritability estimation and prediction accuracy. Most genetic variance can be captured using denser SNPs and the prediction model accounted for MAF, but a large sample size is needed to increase the accuracy of GEBV under all QTL MAF categories.

Keywords: BovineHD, Genomic prediction, Heritability estimation, Japanese Black cattle, Minor allele frequency, Simulation study

* Correspondence: y0uemoto@nlbc.go.jp
[1]National Livestock Breeding Center, Nishigo, Fukushima 961-8511, Japan
Full list of author information is available at the end of the article

Background

The development of single nucleotide polymorphism (SNP) array technology has enhanced the genetic dissection of complex traits, and this SNP information can be directly utilized in cattle breeding programs using genomic selection [1, 2]. In addition, whole genome sequence (WGS) data are becoming increasingly available for cattle, and WGS data are expected to yield a better understanding of complex traits, which can capture all of the genetic variance and predict an accurate genomic estimated breeding value (GEBV), by accounting for all the variants including quantitative trait loci (QTLs) [3, 4].

A recent report showed that the SNPs significantly associated with a complex trait explain only a fraction of the phenotypic variance in human height, and this has been called the "missing heritability" problem [5]. It has been argued that missing heritability is due to imperfect linkage disequilibrium (LD) between SNPs and QTLs, and the extent of LD between SNPs and QTLs depends on differences in the minor allele frequency (MAF) between SNPs and QTLs [6]. SNPs with similar MAF can potentially have high LD, but SNPs with very different MAF cannot have high LD. In cattle populations, QTLs may have a lower MAF than SNPs on low-density SNP arrays, because these are designed to work in several different breeds. In this case, the genetic variation explained by SNPs will be lower than that due to low LD between SNPs and QTLs with low MAF. Meat from Japanese Black cattle is known to have the unique characteristic of a high degree of marbling; the cattle are genetically distant from other European breeds at the genome level [7]. The extent of LD between SNPs and QTLs in Japanese Black cattle may differ from that in other cattle breeds, and it is necessary to evaluate the impact of MAF of QTLs on the genomic evaluation in this target population.

Heritability estimation and GEBV prediction are measures of goodness-of-fit in reference populations and have predictive ability in test populations, respectively. The amount of genetic variance not captured by SNPs affects the maximum predictive ability [8]. On the other hand, increasing the goodness-of-fit will not necessarily increase the predictive ability, because of the model over-fitting problem [9]. The heritability estimation and prediction accuracy depend on several factors such as the genetic architecture of a trait (e.g., QTL heritability, number of QTLs, and distribution of QTL effect), the evaluation strategy (e.g., SNP marker density and prediction method), and population size [6, 9–12]. Therefore, it is important how heritability estimation and GEBV prediction depends on these factors in different MAF of QTLs.

The objective of this study was to evaluate the impact of MAF of QTLs on heritability estimation and accuracy of GEBV prediction, and how that depends on the genetic architecture (QTL heritability, number of QTLs, and distribution of QTL effect), the evaluation strategy (SNP density and prediction model), and population size. We performed a simulation analysis based on a reference-test validation design, which used real genotype data to account for the extent of LD in Japanese Black cattle.

Methods

Genotypes for this study were obtained from previously published data [13]. All animal experiments were performed according to the Guidelines for the Care and Use of Laboratory Animals of Shirakawa Institute of Animal Genetics, and this research was approved by Shirakawa Institute of Animal Genetics Committee on Animal Research (H21-2). We have obtained the written agreement from the cattle owners to use the samples.

Data

In this simulation analysis, real genotype data were used to account for the extent of LD in Japanese Black cattle. Complete descriptions of the experimental population and SNP information were reported previously by Uemoto et al. [13]. Briefly, a total of 1444 Japanese Black cattle, which were 653 steers from two slaughterhouses in Japan [14] and 791 cows from farms managed by a large cooperative farming company in Japan [15], were genotyped using the Illumina BovineHD BeadChip (HD) (Illumina, San Diego, CA, USA), and 593,696 SNPs on autosomal chromosomes assessed by the exclusion criteria of MAF < 0.01, call rate < 0.95, and Hardy–Weinberg equilibrium test < 0.001 were used in this study. To avoid having very close relatives in the data, the animals with large off-diagonal elements in the genomic relationship matrix (GRM) were excluded (a cut-off value of ± 0.4 for off-diagonal elements), and the SNPs were then reassessed by the same criteria. A total of 1368 animals and 592,034 SNPs were then used in the simulation study. These animals were low relatives with the progeny of 438 sires, and the mean, median, and maximum number of progenies per sire were 3.1, 2, and 24, respectively. The distribution of progenies per sire was shown in Additional file 1: Figure S1.

Simulation design

In this study, we simulated the true breeding value (TBV) and phenotypes under the different scenarios varying the following factors: different MAF categories, QTL heritability, number of QTLs, and distribution of QTL effect. After generating TBV and phenotypes, the QTL heritability was estimated and the prediction accuracy of GEBV was assessed under different conditions varying the following factors: different SNP densities, prediction models, and size of the reference-test populations by a reference-test validation design. The factors considered in the simulation study are summarized in

Table 1, and shown in detail below. The impact of the MAF of QTLs on genomic evaluation under different genetic architecture was evaluated in scenarios 1 and 2. In addition, the impact of the MAF of QTLs on genomic evaluation under different evaluation strategy and population size was evaluated in scenarios 3 and 4, respectively.

In this simulation, 36,478 and 6316 SNPs on the BovineSNP50v2 BeadChip (50 K) and the BovineLDv1.1 BeadChip (7 K) (Illumina, San Diego, CA, USA), respectively, were designated as SNP markers. The distribution density of MAF of SNPs on 7 K, 50 K, and HD is plotted in Fig. 1. The MAF distribution shows a low ratio of SNPs on 7 K and a high ratio of SNPs on 50 K and HD at low MAF. The remaining 555,556 SNPs that are present in the HD but not in the 50 K and 7 K were assumed as candidate QTLs. For SNP density, three types of SNPs were used in this simulation. First, SNPs on 7 K and 50 K were used, and this scenario involved imperfect LD between SNPs and QTLs (and named as the imperfect LD SNPs). Second, the HD genotype was imputed from SNPs on 50 K (50 K_to_HD) and 7 K (7K_to_HD) by the BEAGLE (v4.0) software [16]. We performed a 10-fold cross-validation to have imputed HD genotype in this population, and the detail of imputation was reported previously by Uemoto et al. [13]. The imputed SNPs were then reassessed by the same exclusion criteria as described above, and 585,015 and 588,547 SNPs were used in the 7K_to_HD and 50 K_to_HD, respectively. The detail of the imputation error ratio was shown by Uemoto et al. [13], and the average correlation between true and imputed genotypes were 0.98 in 50 K_to_HD and 0.93 in 7 K_to_HD. This scenario involved some SNPs being QTLs but with a low imputation error ratio (and named as the imputed SNPs). Third, all SNPs on the HD were used as SNPs, and this scenario assumed that WGS data were available and some SNPs were QTLs itself (and named as the perfect LD SNPs).

For candidate QTLs, three MAF categories were defined as follows: a low MAF group ($0.01 \leq \text{MAF} \leq 0.05$), a high MAF group ($0.05 < \text{MAF} \leq 0.5$), and an all MAF group ($0.01 \leq \text{MAF} \leq 0.5$). A total of 50, 100, 300, 500, 1000, and 2000 QTLs were randomly selected from candidate QTLs in each MAF group. Hill et al. [17] showed that the distribution of allele frequency affecting additive genetic variance is under the U-shaped distribution and $f(p) \propto \frac{1}{p(1-p)}$. For the all MAF group, the U-shaped distribution was assumed as the distribution of QTL allele frequency ($0.01 \leq p \leq 0.5$), and the ratio of the integrated values for low MAF, $\int_{0.01}^{0.05} f(p)dp$, and high MAF, $\int_{0.05}^{0.5} f(p)dp$, were 0.36 and 0.64, respectively. Therefore, QTLs with low and high MAFs in the all MAF group were randomly selected from the ratio 0.36:0.64, respectively.

We assumed the use of a polygenic model in the simulation, because this is a reasonable assumption for the majority of complex traits in cattle. The phenotype was simulated by summing all true QTL genotypic values and the residual effect, that is, $y_i = \sum_{j}^{m} x_{ij} b_j + e_i$, where m is the number of QTLs, x_{ij} is the genotype for the j-th QTL of the i-th animal (coded as 0, 1, or 2 for the homozygote, heterozygote, and the other homozygote, respectively), b_j is the allele substitution effect of the j-th QTL, and e_i is the residual effect generated from $N\left(0, \sigma_g^2(1/h^2-1)\right)$. $\sum_{j}^{m} x_{ij} b_j$ is TBV, σ_g^2 is the total genetic variance of TBV, and h^2 is the setting value of QTL heritability. Three setting values of QTL heritability ($h^2 = 0.20, 0.40,$ and 0.80) were used to generate phenotypes.

In this study, two different distributions of the QTL effect were assumed. The first model was a gamma distribution

Table 1 Factors for different scenarios in a simulation study

Factor	Scenario			
	1	2	3	4
MAF[a]	All, High, Low	All, High, Low	All, High, Low	All, High, Low
QTL heritability	0.2, 0.4, 0.8	0.4	0.4	0.4
Number of QTLs	500	50, 100, 300, 500, 1000, 2000	500	500
Distribution of QTL effect[b]	EquV	Gamma, EquV	EquV	EquV
SNP density[c]	50 K	50 K	7 K, 50 K, 7K_to_HD, 50 K_to_HD, HD	50 K
Prediction model[d]	Model (1) with G_Y	Model (1) with G_Y	Model (1) with G_V, G_Y, and G_S, Model (2)	Model (1) with G_Y
Size of reference set	1231	1231	1231	200, 400, 800, 1200
Size of test set	137	137	137	1168, 968, 568, 168

[a]MAF, Minor allele frequency; All, $0.01 \leq \text{MAF} \leq 0.5$; High, $0.05 < \text{MAF} \leq 0.5$; Low, $0.01 \leq \text{MAF} \leq 0.05$
[b]Gamma, Gamma distribution model; EquV, Equal variance model
[c]7K, 50 K and HD, Illumina infinium BovineLDv1.1, BovineSNP50v2, and BovineHD BeadChips, respectively; 7 K_to_HD and 50 K_to_HD, Imputations were performed from 7 K and 50 K to HD, respectively
[d]G_V, VanRaden's G matrix; G_Y, Yang's G matrix; G_S, Speed's G matrix

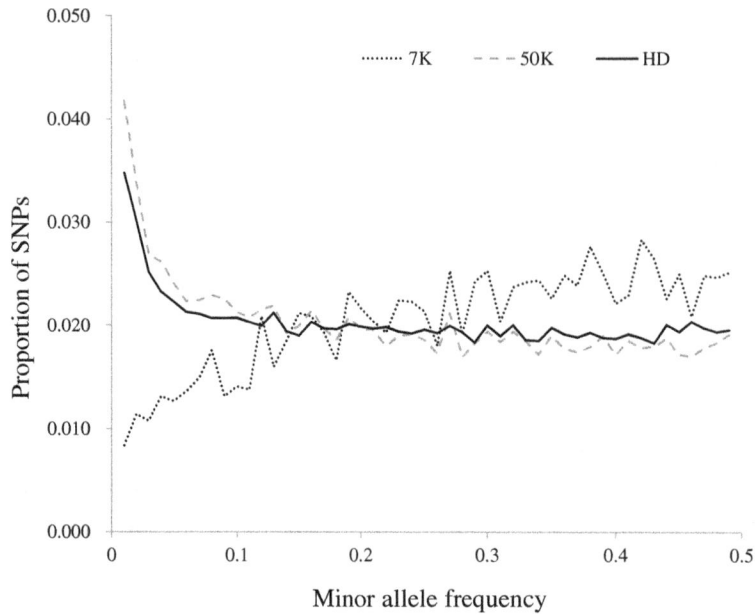

Fig. 1 Distribution of minor allele frequencies for SNPs under different SNP densities. The x-axis indicates the MAF of SNPs, and the y-axis represents the proportion of SNPs in each MAF category. 7 K, 50 K, and HD are SNP markers on Illumina infinium BovineLDv1.1, BovineSNP50v2, and BovineHD BeadChips, respectively

model in which the QTL effect was generated from a gamma distribution with a shape parameter of 0.4 and scale parameter of 1.66 [2]. The second model was an equal variance model in which the QTL effect was assumed as $b_j = \frac{1}{\sqrt{2p_j(1-p_j)}}$, where p_j is MAF of j-th QTL. In the equal variance model, the QTL effect was assumed in that all QTLs had contributed to QTL variance equally (Var(b_j) = 1 in this assumption) if linkage equilibrium was assumed among QTLs. The signs of QTL effects were randomly selected, and total QTL variance was adjusted to $100 \times h^2$ in both distribution models.

Statistical analysis

The generated data were analyzed by the genomic best linear unbiased prediction (GBLUP) method with the following model:

$$\mathbf{y} = \mathbf{1_n}\mu + \mathbf{Xu} + \mathbf{e} \qquad (1)$$

where $\mathbf{y}$ is the phenotypic values, $\mathbf{1_n}$ is a vector of n ones, μ is the mean, $\mathbf{X}$ is the design matrix for random effects, $\mathbf{u}$ is the additive genetic effect with $\mathbf{u} \sim N(0, \mathbf{G}\sigma_u^2)$, and $\mathbf{e}$ is the residual effect with $\mathbf{e} \sim N(0, \mathbf{I}\sigma_e^2)$. $\mathbf{G}$ is a GRM using all SNPs in each SNP density. σ_u^2 is the additive genetic variance, and σ_e^2 is residual variance. We also used the following model:

$$\mathbf{y} = \mathbf{1_n}\mu + \mathbf{Xu_L} + \mathbf{Xu_H} + \mathbf{e} \qquad (2)$$

where $\mathbf{u_L}$ is the additive genetic effect attributed to the low MAF SNPs with $\mathbf{u_L} \sim N\left(0, \mathbf{G_L}\sigma_{u_L}^2\right)$, and $\mathbf{u_H}$ is the

additive genetic effect attributed to the high MAF SNPs with $\mathbf{u_H} \sim N\left(0, \mathbf{G_H}\sigma_{u_H}^2\right)$. $\mathbf{G_L}$ is a GRM using SNPs with low MAF, and $\mathbf{G_H}$ is a GRM using SNPs with high MAF in each SNP density. $\sigma_{u_L}^2$ and $\sigma_{u_H}^2$ are the additive genetic variances attributed to the SNPs with low and high MAFs, respectively, and σ_e^2 is the residual variance. We defined three different GRMs as follows:

VanRaden's GRM ($\mathbf{G_V}$): The first GRM, $\mathbf{G_V}$, was proposed by VanRaden [18] and is calculated as follows:

$$\mathbf{G_V} = \frac{\mathbf{ZZ'}}{2\sum\limits_{j=1}^{m} p_j\left(1-p_j\right)}$$

where m is the number of SNPs, p_j is the frequency of the second allele of j-th SNP, and the elements of $\mathbf{Z}$ are calculated as follows:

$$z_{ij} = x_{ij} - 2p_j$$

where x_{ij} is the number of the second allele of the i-th individual at the j-th SNP.

Yang's GRM ($\mathbf{G_Y}$): The second GRM, $\mathbf{G_Y}$, was proposed by Yang et al. [6] and is computed as follows:

$$\mathbf{G_Y} = \frac{\bar{\mathbf{Z}}\bar{\mathbf{Z}}'}{m}$$

where $\bar{\mathbf{Z}}$ is the $\mathbf{Z}$ matrix but with each element scaled based on the allele frequency of each locus as follows:

$$\bar{z}_{ij} = \frac{z_{ij}}{\sqrt{2p_j\left(1-p_j\right)}}$$

Speed's GRM (G_S): The third GRM, G_S, was proposed by Speed et al. [19] and is calculated as follows:

$$G_S = \frac{WW'}{\sum_{j=1}^{m} k_j}$$

where k_j is the weighting factor of the j-th SNP accounted for LD and the elements of W are calculated as follows:

$$w_{ij} = \sqrt{k_j}\bar{z}_{ij}$$

Speed et al. [19] proposed a method for weighting markers to account for LD. Their method, linkage-disequilibrium adjusted kinships (LDAK), examines the local SNP correlation caused by LD and computes optimal SNP weights by solving a linear program. We calculated the weighting factor k_j and the LD-adjusted GRM (G_S) by the LDAK software with default parameters and LD decay function. When analyzing high density SNPs (i.e., imputed SNPs and perfect LD SNPs), the weighting factors were calculated twice as suggested.

After calculating these three GRMs, 0.00001 was added to diagonal elements of each GRM to avoid near singularity problems. We used the three GRMs in model (1) and G_Y in model (2). The QTL heritability h_1^2 and h_2^2 for model (1) and (2), respectively, are calculated as follows,

$$h_1^2 = \frac{\sigma_u^2}{\sigma_u^2 + \sigma_e^2}$$

$$h_2^2 = \frac{\sigma_{u_L}^2 + \sigma_{u_H}^2}{\sigma_{u_L}^2 + \sigma_{u_H}^2 + \sigma_e^2}$$

Validation test of heritability estimation and prediction accuracy

Under each scenario, we replicated a reference-test validation design 300 times. In each reference-test experiment, data were randomly split into two disjointed sets, that is, 137 animals (one-tenth of all animals) in the test population and the remaining 1231 animals in the reference population. In each replica, this approach was performed only one time. In addition, to evaluate the impact of MAF of QTLs under different population size, 200, 400, 800, and 1200 animals were randomly selected as the reference population, and the remaining 1168, 968, 568, and 168 animals were used as the test population, respectively. Phenotypes of animals in the test population were masked in each replicate, and we

estimated QTL heritability in the reference population and predicted the GEBV in the test population using the ASREML 3.0 program [20]. After predicting the GEBV, the prediction accuracy was assessed using Pearson's correlation between TBV and GEBV in each test population of the validation set. The mean and standard deviation (SD) of 300 replicates was then calculated.

Results

Extent of LD between SNPs and QTLs

Under all scenarios, three MAF categories were defined to evaluate the impact of MAF of QTLs. To evaluate the impact of MAF of QTLs on the extent of LD between SNPs and QTLs, the extent of LD between SNPs on 50 K and QTLs in each MAF category is shown in Fig. 2. The extent of LD between SNPs and QTLs was evaluated using the r^2 value, which is a measure of LD. The r^2 values between QTLs and both adjacent SNPs were calculated by PLINK software [21]. The maximum value of r^2 between two QTL-SNP intervals was chosen in each QTL, and the density distributions of r^2 for three MAF categories were then plotted. The parameters used were the same as those used in scenario 1. In this result, most QTLs with low MAF had a lower r^2 value than those with high MAF. The r^2 value of QTLs with all MAF was between that of QTLs with low and high MAFs. The mean values of r^2 for all, high, and low MAFs were 0.294, 0.360, and 0.184, respectively. This shows that the extent of LD between SNPs and QTLs is higher in the QTLs with high MAF than that in those with low MAF.

The genetic architecture

We evaluated the impact of MAF of QTLs on genomic evaluation under different QTL heritability in scenario 1, and the estimated QTL heritability and correlation between TBV and GEBV are shown in Fig. 3. The estimated QTL heritability was close to the setting value and a higher correlation was observed as the QTL heritability was increased in each MAF category. For the MAF of QTLs, the estimated QTL heritability and correlation between TBV and GEBV for QTLs with high MAF has the highest value, and the values of all MAF were between those of low and high MAFs in each setting value of QTL heritability. In addition, as the setting value was increased from 0.20 to 0.80, the differences in the results between high and low MAFs increased in QTL heritability (from 0.06 to 0.15, respectively) and correlation between TBV and GEBV (from 0.14 to 0.16, respectively).

We evaluated the impact of MAF of QTLs on genomic evaluation under different number of QTLs and distribution of the QTL effect in scenario 2, and the estimated QTL heritability and correlation between TBV and GEBV are shown in Fig. 4. For QTL number, the estimated QTL

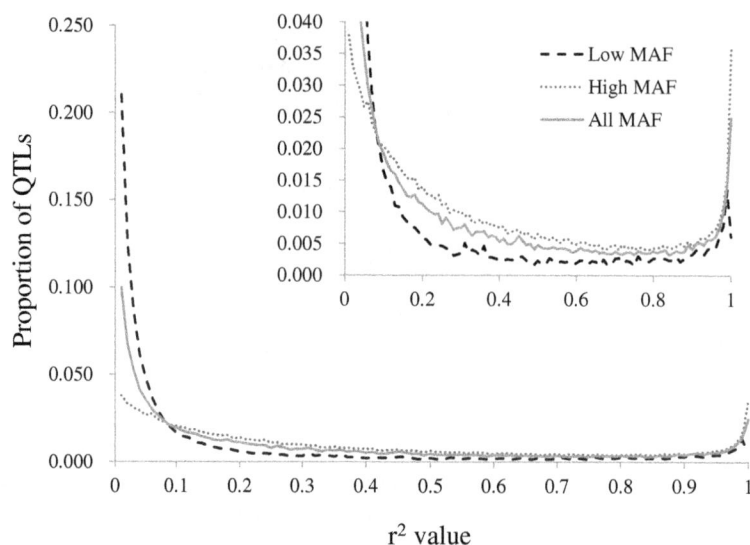

Fig. 2 Proportion of linkage disequilibrium value (r^2) between QTLs and adjacent SNPs. The plot on the right upper corner is the zoomed area of the bigger plot. The x-axis indicates the r^2 value between QTLs and SNPs, and the y-axis represents the proportion of QTLs in each minor allele frequency (MAF) category (All, Low, and High). The r^2 values between QTLs and both adjacent SNPs were calculated, and then the maximum value of r^2 between two QTL-SNP intervals was chosen to plot in each QTL. The parameters used were the same as those under scenario 1

heritability and correlation remained constant, regardless of the number of QTLs in each MAF category.

For the distribution of QTL effect, the results of the QTLs with high and low MAFs followed a similar trend between the two distribution models, whereas different results were observed between two distribution models in the QTLs with all MAFs. The results of high and all MAFs showed similar trends in the gamma distribution model, and the estimated QTL heritability and correlation between TBV and GEBV were about 0.39 and 0.50, respectively. On the other hand, the results of all MAFs were lower than those of high MAF in the equal variance model, and the values of estimated QTL heritability and correlation between TBV and GEBV were about 0.36 and 0.44 for all MAF and 0.39 and 0.50 for high MAF, respectively.

The evaluation strategy

We evaluated the impact of the MAF of QTLs on genomic evaluation under different evaluation strategy for SNP density and prediction model in scenario 3. Goodness-of-fit was measured by the Akaike information criterion (AIC) to compare the prediction models. The AIC is defined as $AIC = 2v - 2\ln(likelihood)$, where v is the number of variance components. This formula shows that the goodness of fit is high, if the AIC is low. The estimated QTL heritability, AIC, and correlation between TBV and GEBV are shown in Table 2, Table 3, and Table 4, respectively.

Differences in the SNP density have an impact on heritability estimation and GEBV prediction. For model (1) with G_Y, the results of 50 K were higher than those of

7 K in all MAF categories. For example, from the QTLs with all MAFs, the results of 50 K and 7 K were 0.36 and 0.30 for QTL heritability and 0.44 and 0.42 for correlation between TBV and GEBV, respectively. The results of imputed SNPs (i.e., 7 K_to_HD and 50 K_to_HD) were higher than those of 7 K and 50 K, and were very close to the results of perfect LD SNPs (i.e., HD) in all MAF categories. For example, from the QTLs with all MAFs, the results of both 50 K_to_HD and 7 K_to_HD were 0.37 for QTL heritability and 0.45 for correlation between TBV and GEBV, and the results of HD were 0.38 for QTL heritability and 0.45 for correlation between TBV and GEBV. These results indicate that heritability estimation and GEBV prediction depend on the SNP density. However, the different results among SNP densities in each MAF category depend on the prediction model.

For the prediction model, the result of model (1) with G_V was similar to that with G_Y in the QTL with high MAF, but the difference between the results obtained from G_V and G_Y increased in the QTL with low MAF. For example, the differences between G_V and G_Y in the AIC and correlation between TBV and GEBV with 50 K were 0 and 0.00 in the QTL with high MAF but 7 and 0.03 in the QTL with low MAF, respectively. The result of model (1) with G_S was similar to or better than that with G_Y in the QTL with all and low MAFs, but performed worse in the QTL with high MAF. In particular, the difference in the results between G_S and G_Y in the QTL with high MAF was increased at larger SNP density. For example, the difference between G_S and G_Y in AIC and correlation between TBV and GEBV were 1

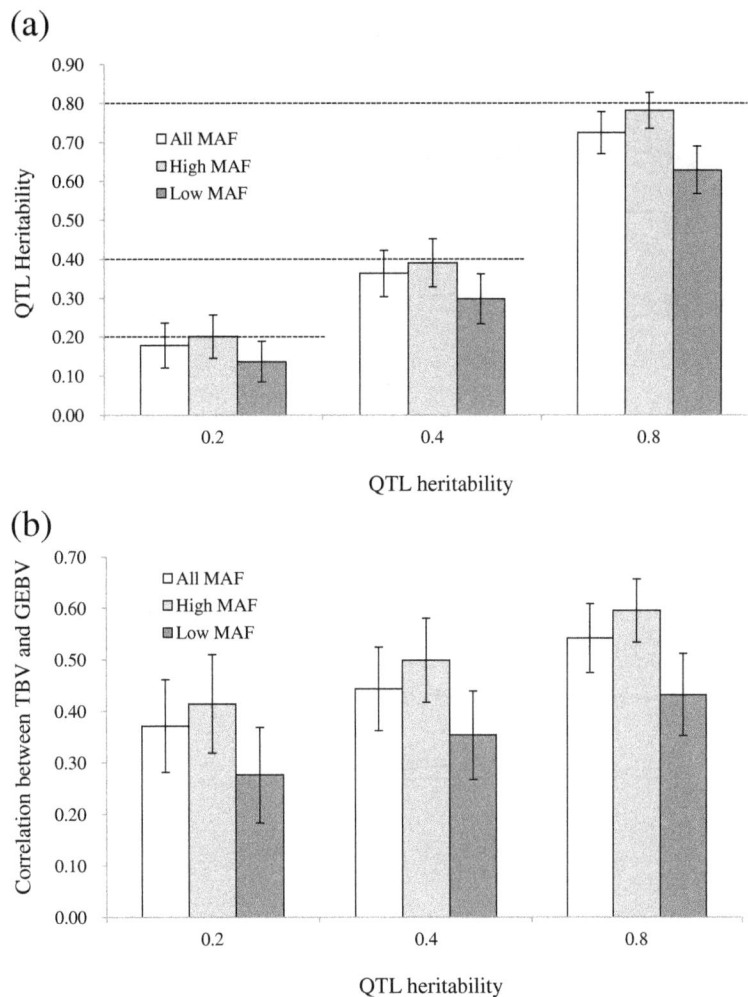

Fig. 3 Results obtained from scenario 1. Estimated QTL heritability and correlation between true breeding and genomic estimated breeding values are calculated. The x-axis indicates the true QTL heritability, and the y-axis represents mean values of 300 replicates for the estimated QTL heritability (**a**) and the correlation between true breeding value (TBV) and genomic estimated breeding value (GEBV) (**b**). The results of varying minor allele frequency (MAF) categories (All, Low, and High) and QTL heritabilities (0.20, 0.40, and 0.80) are shown. The whiskers represent the standard deviation of 300 replicates

and 0.00 in 7 K but 10 and 0.03 in HD. In addition, the results of G_S with HD in high MAF were 6147 in AIC and 0.48 in the correlation between TBV and GEBV, which represented the worst of all results by other models under the high MAF scenario. The results of model (2) were similar to or better than those of the other three models under all MAF categories. In particular, the results of model (2) with HD in low MAF, which were 6150 in AIC and 0.47 in correlation between TBV and GEBV, representing the best values in the low MAF results.

Population size

In this simulation, the impact of the MAF of QTLs on genomic evaluation under different population size was evaluated in scenario 4. The estimated QTL heritability and correlation between TBV and GEBV are shown in

Fig. 5. The results of heritability estimation and GEBV prediction followed a different trend. The mean values of estimated QTL heritability were close to the setting value (0.40) and were almost the same as those among different population sizes, but the SD of the estimated results decreased as the size of the population increased (e.g., from 0.47 to 0.07 in reference size from 200 to 1200, respectively, for all MAFs). The following trend of the results, the mean values of high MAF > all MAF > low MAF, was shown for QTL heritability, when the size of reference set was more than 800. These results indicated that the heritability estimates at lower population sizes are less precise than those at higher population sizes, even if the estimated value is close to the setting value. In addition, the impact of the MAF of QTLs was shown at larger population sizes.

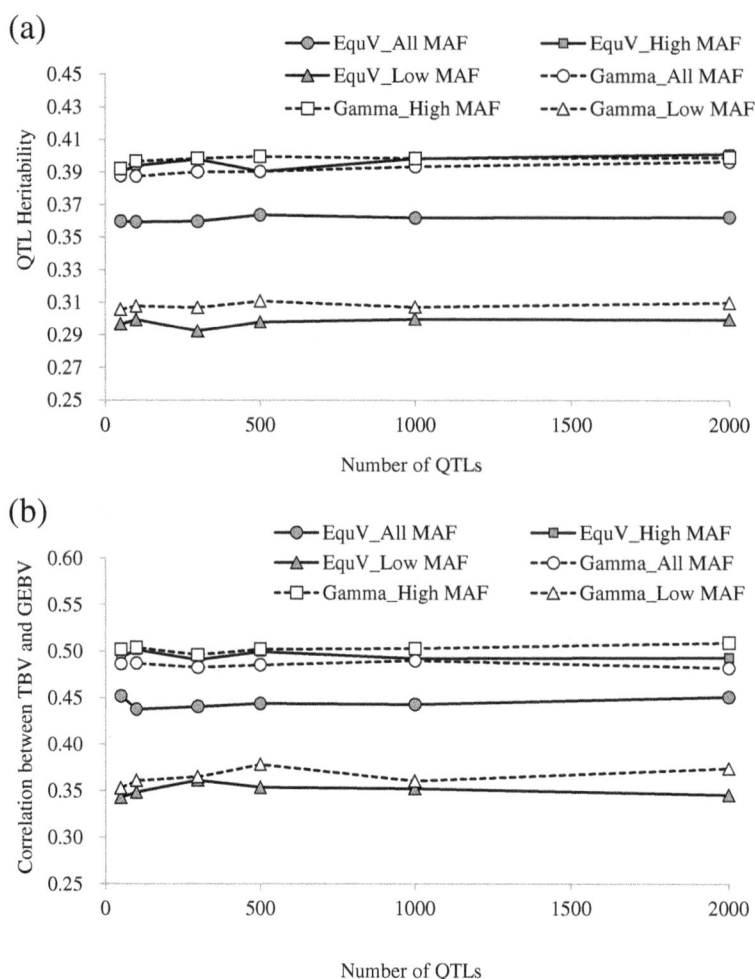

Fig. 4 Results obtained from scenario 2. Estimated QTL heritability and correlation between true breeding and genomic estimated breeding values are calculated. The x-axis indicates the number of QTLs, and the y-axis represents mean values of 300 replicates for the estimated QTL heritability (**a**) and the correlation between true breeding value (TBV) and genomic estimated breeding value (GEBV) (**b**). The results of varying minor allele frequency (MAF) categories (All, Low, and High), number of QTLs (50, 100, 300, 500, 1000, and 2000), and distribution of QTL allele substitution effect (Gamma, gamma distribution model; EquV, equal variance model) are shown

In the GEBV prediction, the correlations between TBV and GEBV were increased as the size of the reference increased (e.g., 0.11–0.41 at reference size 200–1200, respectively, for all MAFs). QTLs with high MAF had the highest value, and the values of all MAFs were between those with low and high MAFs in all reference sizes (e.g., 0.34, 0.41, and 0.50 for low, all, and high MAFs in reference size 1200, respectively). In addition, as the size of the reference increased from 200 to 1200, the difference between the high and low MAFs for the correlations between TBV and GEBV increased from 0.07 to 0.15, respectively.

Discussion
The genetic architecture
The differences in the QTL heritability and the distribution of QTL effect had an impact on heritability estimation and GEBV prediction under different MAF categories, but the

differences in the number of QTL did not have. The results of the correlation between TBV and GEBV for the number of QTL were the same as those described by Daetwyler et al. [10], because the accuracy of GBLUP is constant regardless of the number of QTLs. The trend of the results for QTL heritability was similar to that described by Yang et al. [6].

For the distribution of QTL effect, the genetic variance of the j-th QTL is theoretically calculated as $2p_j(1-p_j)\alpha_j^2$, where p_j is the allele frequency of QTLs and α_j is the QTL effect [22]. This formula shows that the QTL effect will increase as the allele frequency decreases, if the genetic variance is constant. Therefore, the QTLs with low MAF must have a higher QTL effect than those with high MAF to contribute to the total genetic variance. In a real data analysis, findings from a meta-analysis of human height showed that the QTLs with high MAF had

Table 2 Heritability estimation in scenario 3

SNP[b]	Prediction model[c]	All MAF[a] Mean	SD	High MAF[a] Mean	SD	Low MAF[a] Mean	SD
7 K	Model (1) with G_V	0.28	0.05	0.32	0.05	0.20	0.06
	Model (1) with G_Y	0.30	0.05	0.33	0.05	0.23	0.06
	Model (1) with G_S	0.30	0.05	0.33	0.05	0.24	0.06
	Model (2)	0.30	0.05	0.33	0.05	0.23	0.06
50 K	Model (1) with G_V	0.33	0.06	0.38	0.06	0.24	0.06
	Model (1) with G_Y	0.36	0.06	0.39	0.06	0.30	0.06
	Model (1) with G_S	0.38	0.06	0.40	0.06	0.34	0.07
	Model (2)	0.37	0.06	0.39	0.06	0.34	0.06
7K_to_HD	Model (1) with G_V	0.34	0.06	0.39	0.06	0.24	0.06
	Model (1) with G_Y	0.37	0.06	0.40	0.06	0.30	0.06
	Model (1) with G_S	0.41	0.07	0.41	0.07	0.39	0.07
	Model (2)	0.39	0.06	0.40	0.06	0.38	0.06
50K_to_HD	Model (1) with G_V	0.34	0.06	0.39	0.06	0.25	0.06
	Model (1) with G_Y	0.37	0.06	0.41	0.06	0.30	0.07
	Model (1) with G_S	0.41	0.07	0.42	0.07	0.40	0.07
	Model (2)	0.40	0.06	0.40	0.06	0.40	0.06
HD	Model (1) with G_V	0.35	0.06	0.39	0.06	0.25	0.06
	Model (1) with G_Y	0.38	0.06	0.41	0.06	0.31	0.07
	Model (1) with G_S	0.42	0.07	0.41	0.07	0.40	0.07
	Model (2)	0.40	0.06	0.40	0.06	0.41	0.06

[a]MAF, Minor allele frequency; All MAF, $0.01 \leq MAF \leq 0.5$; High MAF, $0.05 < MAF \leq 0.5$; Low MAF, $0.01 \leq MAF \leq 0.05$
[b]7K, 50 K and HD, Illumina infinium BovineLDv1.1, BovineSNP50v2, and BovineHD BeadChips, respectively; 7 K_to_HD and 50 K_to_HD, Imputations were performed from 7 K and 50 K to HD, respectively
[c]G_V, VanRaden's genome relationship matrix (GRM); G_Y, Yang's GRM; G_S, Speed's GRM

Table 3 Model fitness measured by Akaike information criterion (AIC) in scenario 3

SNP[b]	Prediction model[c]	All MAF[a] Mean	SD	High MAF[a] Mean	SD	Low MAF[a] Mean	SD
7 K	Model (1) with G_V	6164	63	6145	66	6191	61
	Model (1) with G_Y	6162	63	6145	66	6188	61
	Model (1) with G_S	6162	63	6146	66	6187	61
	Model (2)	6163	63	6147	66	6186	61
50 K	Model (1) with G_V	6159	63	6139	65	6188	62
	Model (1) with G_Y	6155	63	6139	65	6181	62
	Model (1) with G_S	6155	63	6142	65	6175	62
	Model (2)	6155	63	6140	65	6163	62
7K_to_HD	Model (1) with G_V	6158	63	6138	65	6189	62
	Model (1) with G_Y	6155	63	6138	65	6182	62
	Model (1) with G_S	6156	63	6147	65	6171	62
	Model (2)	6154	63	6139	65	6155	62
50K_to_HD	Model (1) with G_V	6157	63	6137	65	6188	62
	Model (1) with G_Y	6154	63	6137	65	6181	62
	Model (1) with G_S	6155	63	6146	65	6169	62
	Model (2)	6153	63	6138	65	6152	62
HD	Model (1) with G_V	6157	63	6136	65	6188	62
	Model (1) with G_Y	6154	63	6137	65	6180	62
	Model (1) with G_S	6155	63	6147	65	6168	62
	Model (2)	6152	63	6138	65	6150	62

[a]MAF, Minor allele frequency; All MAF, $0.01 \leq MAF \leq 0.5$; High MAF, $0.05 < MAF \leq 0.5$; Low MAF, $0.01 \leq MAF \leq 0.05$
[b]7K, 50 K and HD, Illumina infinium BovineLDv1.1, BovineSNP50v2, and BovineHD BeadChips, respectively; 7 K_to_HD and 50 K_to_HD, Imputations were performed from 7 K and 50 K to HD, respectively
[c]G_V, VanRaden's genome relationship matrix (GRM); G_Y, Yang's GRM; G_S, Speed's GRM

small phenotypic effects, whereas the QTLs with low MAF had large effects on this trait such as a function of the MAF [23]. Therefore, missing heritability has focused on the possible contribution of QTLs with low MAF, and the QTLs with low MAF could have an intermediate effect [24]. In this study, the factor for the distribution of QTL effect was evaluated to account for the low-MAF QTL with intermediate effect. In the gamma distribution model, the low-MAF QTL with intermediate effect cannot be defined in the QTL with all MAFs, because the QTL effect was randomly allocated to the MAF. On the other hand, this QTL can be defined in the equal variance model. As an example, the mean values for the QTL effect and QTL variance for the QTL with all MAFs as a function of MAF are shown in Additional file 1: Figure S2. Additional file 1: Figure S2 is drawn from a result of the randomly selected replica under scenario 2 with the parameters for the number of QTLs (500). In this result, no relationship between MAF, QTL effect, and QTL variance was observed in the gamma distribution model, whereas the QTL with low MAF had a higher QTL

effect and all QTLs had equal genetic variance in the equal variance model. Therefore, the results of the QTLs with all and high MAFs in Fig. 4 showed the same as that under the gamma distribution model, because of the low contribution of the QTLs with low MAF on the total genetic variance. This result was the similar trend as described by Wientjes et al. [25]. This result also shows that the equal variance model accounts for missing heritability in a simulation when the QTLs are composed of variants with low and high MAFs. If QTLs with a large effect do exist, they are at a low frequency and individually explain a small proportion of genetic variance [26]. Therefore, the equal variance model was used to evaluate the impact of MAF of QTL in all scenarios.

The evaluation strategy

In this study, three types of SNPs were assumed: the imperfect LD SNPs (7 K and 50 K), the imputed SNPs (7 K_to_HD and 50 K_to_HD), and the perfect LD SNPs

Table 4 Correlation between true breeding value and genomic breeding value in scenario 3

SNP[b]	Prediction model[c]	All MAF[a] Mean	SD	High MAF[a] Mean	SD	Low MAF[a] Mean	SD
7 K	Model (1) with G_V	0.41	0.08	0.48	0.08	0.30	0.09
	Model (1) with G_Y	0.42	0.08	0.48	0.08	0.32	0.09
	Model (1) with G_S	0.42	0.08	0.48	0.08	0.33	0.09
	Model (2)	0.42	0.08	0.48	0.08	0.33	0.09
50 K	Model (1) with G_V	0.43	0.08	0.50	0.08	0.32	0.09
	Model (1) with G_Y	0.44	0.08	0.50	0.08	0.35	0.09
	Model (1) with G_S	0.44	0.08	0.49	0.08	0.37	0.09
	Model (2)	0.44	0.08	0.50	0.08	0.41	0.09
7K_to_HD	Model (1) with G_V	0.44	0.08	0.50	0.08	0.32	0.09
	Model (1) with G_Y	0.45	0.08	0.50	0.08	0.35	0.09
	Model (1) with G_S	0.44	0.08	0.48	0.08	0.38	0.08
	Model (2)	0.45	0.08	0.50	0.08	0.44	0.08
50K_to_HD	Model (1) with G_V	0.44	0.08	0.51	0.08	0.32	0.09
	Model (1) with G_Y	0.45	0.08	0.51	0.08	0.36	0.09
	Model (1) with G_S	0.44	0.08	0.48	0.08	0.39	0.08
	Model (2)	0.46	0.08	0.51	0.08	0.46	0.08
HD	Model (1) with G_V	0.44	0.08	0.51	0.08	0.32	0.09
	Model (1) with G_Y	0.45	0.08	0.51	0.08	0.36	0.08
	Model (1) with G_S	0.44	0.08	0.48	0.08	0.39	0.08
	Model (2)	0.46	0.08	0.51	0.08	0.47	0.08

[a]MAF, Minor allele frequency; All MAF, $0.01 \leq MAF \leq 0.5$; High MAF, $0.05 < MAF \leq 0.5$; Low MAF, $0.01 \leq MAF \leq 0.05$
[b]7K, 50 K and HD, Illumina infinium BovineLDv1.1, BovineSNP50v2, and BovineHD BeadChips, respectively; 7 K_to_HD and 50 K_to_HD, Imputations were performed from 7 K and 50 K to HD, respectively
[c]G_V, VanRaden's genome relationship matrix (GRM); G_Y, Yang's GRM; G_S, Speed's GRM

(HD). Recently, WGS data are becoming increasingly available for use in cattle, and the 1000 bull genomes project provides annotated sequence variants and genotypes of key ancestor bulls [3]. One of the major advantages of WGS data is that they provide complete information on all the variants of an individual, which include many of the SNPs with low MAF that are not covered by the SNP array. Most of the low MAF variants are only accessible through WGS data, and this information could be important for genomic evaluation. WGS data can be obtained directly by next-generation sequencing techniques or indirectly by genotype imputation. When using imputed SNPs, the impact of imputation error on genomic evaluation must be investigated in genotype imputation. Therefore, the imputed SNPs (indirect information) and the perfect LD SNPs (direct information) were used to evaluate the effectiveness of using WGS data directly or indirectly. The results showed that differences in the SNP density have an impact on heritability estimation and GEBV prediction, especially in the low MAF

scenario. For the imperfect LD SNPs, the distribution of MAF for 7 K followed a different trend compared to HD. On the other hand, the distribution of MAF for 50 K had the different values but followed a similar trend compared to that for HD, especially at high MAF. Usually, all classes of MAFs are equally represented on a low density SNP array, while the low MAF class is overrepresented in the WGS data [27]. The difference in MAF distribution between QTL and SNPs indicates the difficulty of capturing genetic variance. Therefore, the results of 7 K were lower than those of 50 K. For the imputed SNPs and the complete LD SNPs, these results were higher than those with 7 K and 50 K in heritability estimation and GEBV prediction. The results of imputed SNPs were very close to those of the complete LD SNPs, even if the imputed SNPs were not in perfect LD with the QTL. The number of missing genotypes affects the accuracy, and the difference in imputation accuracy is larger at low MAFs [13]. However, our results showed that there was little difference in the results between 7K_to_HD and 50 K_to_HD under the low MAF scenario. A previous study reported that the accuracy of GEBV plateaus on increasing the number of SNPs [12]. On the other hand, GEBV prediction can achieve moderately high prediction accuracy under perfect LD between SNPs and QTLs in distantly related human data [28]. Therefore, using the SNPs related to QTLs directly or indirectly is effective for performing heritability estimation and GEBV prediction.

In this study, we showed that the differences of the result among SNP densities in each MAF category depend on the prediction model. For model (1) with G_V and G_Y, the difference of the results between G_V and G_Y was increased in the QTL with low MAF. Meuwissen et al. [29] suggested that weighted GRM by MAF would have a better result than unweighted GRM, when a high proportion of loci with low MAF are used. G_Y is corrected for variance of the allele frequency of each SNP, and gives weight to alleles with low MAF. On the other hand, G_V is corrected for the average frequency of heterozygotes, and gives less weight to alleles with low MAF. Therefore, the approach of G_Y was better than that of G_V, especially under the low MAF scenario. For the model (1) with G_S reflecting the degree of LD, the difference of the results between G_S and G_Y in the QTL with high MAF was increased at larger SNP densities, and the result using HD was the worse than that by other prediction models under the high MAF scenario. Lee et al. [30] reported that G_S generates biased heritability estimates through the use of denser SNPs, because of too much weight being attributed to the low MAF SNPs. This method accounts for the different extents of LD among SNPs, and weighted SNPs depend on the MAF distribution of SNPs. The distribution of MAF is different between the dense and sparse SNP data, because the

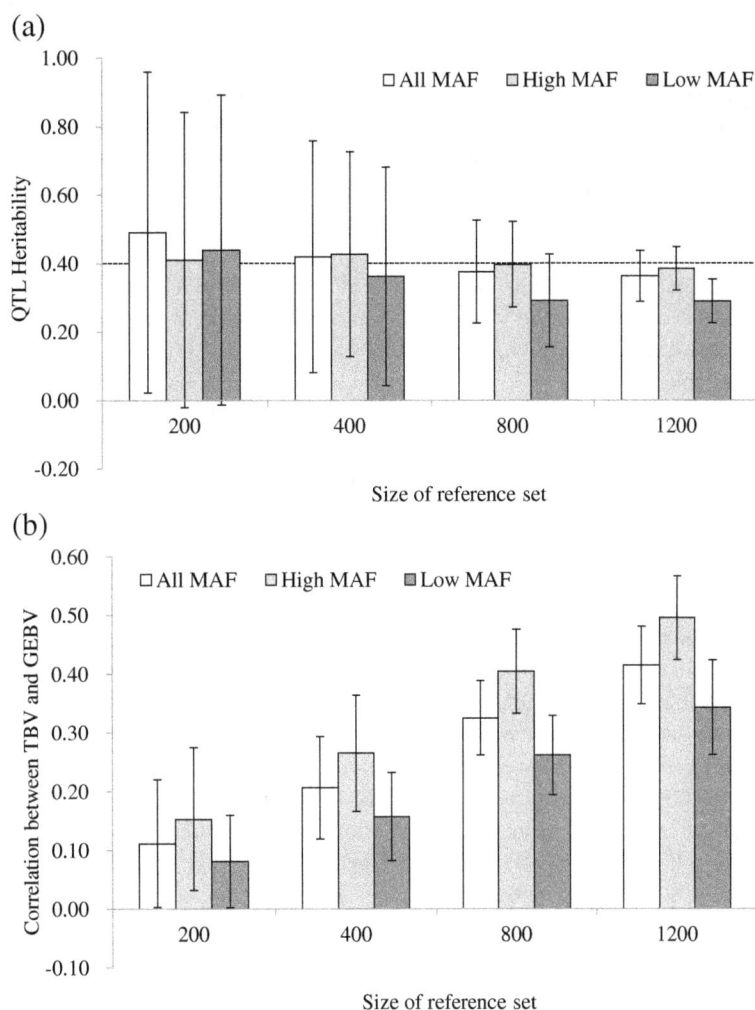

Fig. 5 Results obtained from scenario 3. Estimated QTL heritability and correlation between true breeding and genomic estimated breeding values are calculated. The x-axis indicates the size of the reference set, and the y-axis represents mean values of 300 replicates for the estimated QTL heritability (**a**) and the correlation between true breeding value (TBV) and genomic estimated breeding value (GEBV) (**b**). The results of varying minor allele frequency (MAF) categories (All, Low, and High) and size of the reference set (200, 400, 800, and 1200) are shown. The whiskers represent the standard deviation of 300 replicates

proportion of SNPs with low MAF increased as SNP density increased. The low MAF SNPs could be under low LD among SNPs and the proportion of weighted SNPs with low MAF will increase. In this simulation, the proportion of SNPs before and after weighting for G_S is different for sparse and dense SNP data, and the proportion of weighted SNPs with low MAF was higher than that with high MAF in imputed SNPs and perfect LD SNPs (Additional file 2: Table S1). Therefore, the result of the QTL with high MAF was the lowest, because many of the low MAF SNPs were weighted in G_S. To account for the degree of LD between SNPs at larger SNP densities, the haplotype model (such as Sun et al. [31]) could have significant average in low MAF QTL.

For model (2), the results were similar to or better than those from the other three models under all MAF categories, and represented the best value when a higher proportion of QTLs that had low MAF and SNPs on HD were used. This model means that QTL heritability is partitioned by the MAF of SNPs to provide insight into genetic architecture. Some studies have showed that different MAF categories could deliver estimates with little bias and high goodness-of-fit in human [30, 32] and chicken population [33, 34], because high LD is only possible between SNPs with similar MAFs. In addition, Lee et al. [30] also showed that the different MAF categories generate heritability estimates with higher goodness-of-fit in dense SNP data compared with sparse SNP data, especially when a higher proportion of QTLs have low MAF. In principle, fitting more MAF categories in a prediction model could more accurately represent the genetic architecture, but it brings the disadvantage of estimating more parameters. Therefore, Lee et al. [30] used five bins with MAF boundaries 0.1, 0.2, 0.3, 0.4, and 0.5. In this study, two MAF categories (high and low) were fitted in the model (2), because the QTL can be roughly classified into two types: an intermediate effect at low MAF and a small effect at high MAF under missing heritability. Therefore, more MAF categories were not fitted at high MAF. Our results using model (2) with HD were better than those based on five MAF categories in all MAF categories of this simulation (Additional file 2: Table S2). Therefore, model (2) is a robust method in all MAF categories and could capture more of the genetic variance under the low MAF scenario if the WGS data are available.

Population size

In this simulation, a high proportion of genetic variance was captured in the high MAF scenario. However, our results did not reflect high precision, and the accuracy of GEBV remained lower than that of the simulation study [2], even if the perfect LD SNP was used under the high MAF scenario. Under this simulation, the maximum correlation between TBV and GEBV for the QTLs with all, high, and low MAFs was 0.46, 0.51, and 0.47, respectively, for QTL heritability of 0.40. The main reason for low prediction accuracy could be the size of the reference population. The accuracy of GEBV depends on the size of the reference, and a large sample of animals is needed in the reference population if accurate GEBV prediction is desired [1]. Daetwyler et al. [35] also described the accuracy of GEBV deterministically as follows:

$$r = \sqrt{\frac{Nh^2}{Nh^2 + q}}$$

where N is the number of individuals in the reference population, h^2 is heritability, and q is the number of independent loci affecting a trait. This formula shows that the accuracy of GEBV depends on the size of the reference and the number of QTLs. Daetwyler et al. [35] also showed that GBLUP has a constant accuracy for a given N and h^2, regardless of q. Therefore, the accuracy of GEBV increases with a larger reference size.

In this study, the accuracy increased as the size of the reference increased in all MAF categories, and the accuracy did not plateau at the maximum size of the reference (Fig. 5). Under this simulation, there were insufficient DNA samples to evaluate the impact of a larger sample size. Therefore, further study is needed to evaluate the impact of population size on genomic prediction under different MAF scenarios.

Simulation based on real data

Populations containing related individuals (e.g., cattle) are expected to yield high LD between SNPs and QTLs than populations containing unrelated individuals (e.g., human), because of decreasing effective population size. The LD in cattle follows a similar pattern to that of humans at short distances, but is much larger than that of humans at long distances [1]. The extent of LD between SNPs and QTLs depends on a population structure, and the impact of MAF of QTL must be evaluated by the use of datasets accounting for the extent of LD in a target population. The effective population size of Japanese Black cattle has been reduced because of the intensive use of a few sires with high marbling [36], and the extent of LD could be higher than that in other cattle breeds. There has been little assessment of the MAF of QTLs that affect heritability estimation and GEBV prediction using real cattle data except for Wientjes et al. [25] in Dairy cattle population. Therefore, we used real genotype data in the simulation study, which accounted for the extent of LD in Japanese Black cattle. For the extent of LD in this target population, the extent of LD between SNPs and QTLs increased as the MAF of QTL increased in the scenario with 50 K. The main reason was that 87 % of SNPs in 50 K were high MAF (see Additional file 2: Table S1). On the other hand, only 13 % of SNPs were low MAF, and most of QTLs with low MAF were in low LD with SNPs in 50 K. This indicates that the concordance of MAF between QTL and SNP shows higher LD between two loci in this population.

Under this simulation, we showed that the results of these predicted models with different SNP densities depend on the genetic architecture of objective traits, especially the MAF of QTLs. Recently, some studies have investigated the proportion of genetic variance captured by SNPs and the prediction accuracy of GEBV in small Japanese Black cattle populations using the GBLUP method with G_V and 50 K [37–39]. In those studies, a higher proportion of genetic variance was captured for carcass traits, and these proportions were close to the genetic variance previously reported by estimations based on pedigree information. In our results with the scenario of all MAF category and setting QTL heritability (0.40), the genome heritability was 0.33 for the scenario with G_V and 50 K, whereas there was genome heritability of 0.40 for the scenario with model (2) and HD. In addition, previous studies show that the some proportion of the genetic variance are not captured by SNPs with high MAF in Holstein cattle population [40, 41] and in chicken population [33]. Therefore, the explanation of a high proportion may be that there are not only the QTLs with high MAF but also strong relationship structures in these populations, because our population was excluded very close relatives to obtain low relationship structure. The strong relationship structures also lead to capture more of genetic variance by SNPs with high MAF. To evaluate the genetic architecture of complex traits, we suggest that it is effective to compare among these prediction models with different SNP densities. However, it is not sufficient for the evaluation of predictive ability, because of the difficulty of obtaining TBV. It may not reflect the predictive ability, even if a higher proportion of genetic variance is captured. In this population, the extent of LD was still higher than that in other beef breeds [13], even if we excluded very close relatives to evaluate the minimum value of QTL heritability and correlation in Japanese Black cattle. These results show that a higher proportion of genetic variance was captured under the high MAF scenario. However, the accuracy of GEBV remained low, and the goodness-of-fit did not increase the prediction accuracy. Therefore, further study including additional animals in the reference population is needed to increase the prediction accuracy.

Conclusions

The current study evaluated the impact of MAF of QTL on genomic evaluation in a simulation study by assuming different MAFs of QTLs and several factors in Japanese Black cattle. The extent of LD between SNPs and QTLs was higher in the QTLs with high MAF than in those with low MAF. The MAF of QTLs had an impact on heritability estimation and prediction accuracy and that depended on the genetic architecture, evaluation strategy and population size. The genetic architecture results showed that genomic evaluation was affected by the MAF of QTLs combined with the QTL heritability and the distribution of QTL effect. The number of QTL was not affected on genomic evaluation if the number of QTL was more than 50. For the evaluation strategy, different SNP densities and prediction models affected the heritability estimation and genomic prediction, and these depended on the MAF of QTLs. The genetic dissection of complex traits would be possible by comparing the results of these predicted models with different SNP densities. In addition, accurate QTL heritability and GEBV were obtained by using denser SNP information and model (2) under all MAF categories. However, it may not reflect the predictive ability, and a larger sample size is needed to increase the accuracy of GEBV.

Availability of supporting data

The data sets supporting the results of this article are included within the article and its additional file.

Abbreviations
AIC: Akaike information criterion; GBLUP: Genomic best linear unbiased prediction; GEBV: Genomic estimated breeding value; GRM: Genomic relationship matrix; LD: Linkage disequilibrium; MAF: Minor allele frequency; QTL: Quantitative trait locus; SNP: Single nucleotide polymorphism; TBV: True breeding value; WGS: Whole genome sequence.

Competing interests
The authors declare that they have no competing interests.

Authors' contributions
YU participated in the design of the study, performed the statistical analysis, and drafted the manuscript. SS, YS and TW participated in the design of the study, provided the raw data, and contributed in writing and improving the manuscript. TK participated in the design of the study, and contributed in writing and improving the manuscript. All authors read and approved the final manuscript.

Acknowledgements
The authors thank the staff of the Shirakawa Institute of Animal Genetics for technical assistance. This work was supported by the Japan Racing and Livestock Promotion to YS.

Author details
[1]National Livestock Breeding Center, Nishigo, Fukushima 961-8511, Japan. [2]Shirakawa Institute of Animal Genetics, Japan Livestock Technology Association, Nishigo, Fukushima 961-8511, Japan.

References
1. Goddard ME, Hayes BJ. Mapping genes for complex traits in domestic animals and their use in breeding programmes. Nat Rev Genet. 2009;10:381–91.
2. Meuwissen THE, Hayes BJ, Goddard ME. Prediction of total genetic value using genome-wide dense marker maps. Genetics. 2001;157:1819–29.
3. Daetwyler HD, Capitan A, Pausch H, Stothard P, Van Binsbergen R, Brøndum RF, et al. Whole-genome sequencing of 234 bulls facilitates mapping of monogenic and complex traits in cattle. Nat Genet. 2014;46:858–67.
4. Meuwissen THE, Goddard M. Accurate prediction of genetic values for complex traits by whole-genome resequencing. Genetics. 2010;185:623–31.
5. Maher B. Personal genomes: the case of missing heritability. Nature. 2008;456:18–21.
6. Yang J, Benyamin B, McEvoy BP, Gordon S, Henders AK, Nyholt DR, et al. Common SNPs explain a large proportion of the heritability for human height. Nat Genet. 2010;42:565–9.
7. Nishimura S, Watanabe T, Ogino A, Shimizu K, Morita M, Sugimoto Y, et al. Application of highly differentiated SNPs between Japanese Black and Holstein to a breed assignment test between Japanese Black and F1 (Japanese Black x Holstein) and Holstein. Anim Sci J. 2013;84:1–7.
8. Dekkers JC. Prediction of response to marker-assisted and genomic selection using selection index theory. J Anim Breed Genet. 2007;124:331–41.
9. Makowsky R, Pajewski NM, Klimentidis YC, Vazquez AI, Duarte CW, Allison DB, et al. Beyond missing heritability: prediction of complex traits. PLoS Genet. 2011;7:e1002051.
10. Daetwyler HD, Pong-Wong R, Villanueva B, Woolliams JA. The impact of genetic architecture on genome-wide evaluation methods. Genetics. 2010;185:1021–31.
11. Goddard ME. Genomic selection: prediction of accuracy and maximization of long term response. Genetica. 2009;136:245–57.

12. Hayes BJ, Bowman PJ, Chamberlain AJ, Goddard ME. Invited review: genomic selection in dairy cattle: progress and challenge. J Dairy Sci. 2009;92:433–43.

13. Uemoto Y, Sasaki S, Sugimoto Y, Watanabe T. Accuracy of high-density genotype imputation in Japanese Black cattle. Anim Genet. 2015;46:388–94.

14. Nishimura S, Watanabe T, Mizoshita K, Tatsuda K, Fujita T, Watanabe N, et al. Genome-wide association study identified three major QTL for carcass weight including the PLAG1-CHCHD7 QTN for stature in Japanese Black cattle. BMC Genet. 2012;13:40.

15. Sasaki S, Ibi T, Watanabe T, Matsuhashi T, Ikeda S, Sugimoto Y. Variants in the 3'UTR of General Transcription Factor IIF, polypeptide 2 affect female calving efficiency in Japanese Black cattle. BMC Genet. 2013;14:41.

16. Browning BL, Browning SR. Improving the accuracy and efficiency of identity-by-descent detection in population data. Genetics. 2013;194:459–71.

17. Hill WG, Goddard ME, Visscher PM. Data and theory point to mainly additive genetic variance for complex traits. PLoS Genet. 2008;4:e1000008.

18. VanRaden PM. Efficient methods to compute genomic predictions. J Dairy Sci. 2008;91:4414–23.

19. Speed D, Hemani G, Johnson MR, Balding DJ. Improved heritability estimation from genome-wide SNPs. Am J Hum Genet. 2012;91:1011–21.

20. Gilmour AR, Gogel BJ, Cullis BR, Thompson R. ASRreml User Guide Release 3.0. 2009.

21. Purcell S, Neale B, Todd-Brown K, Thomas L, Ferreira MA, Bender D, et al. PLINK: A tool set for whole-genome association and population-based linkage analyses. Am J Hum Genet. 2007;81:559–75.

22. Falconer DS, Mackay TF. Introduction to quantitative genetics. 4th ed. Harlow. UK: Longman; 1996.

23. Lanktree MB, Guo Y, Murtaza M, Glessner JT, Bailey SD, Onland-Moret NC, et al. Meta-analysis of dense genecentric association studies reveals common and uncommon variants associated with height. Am J Hum Genet. 2011;88:6–18.

24. Manolio TA, Collins FS, Cox NJ, Goldstein DB, Hindorff LA, Hunter DJ, et al. Finding the missing heritability of complex diseases. Nature. 2009;461:747–53.

25. Wientjes YC, Calus MP, Goddard ME, Hayes BJ. Impact of QTL properties on the accuracy of multi-breed genomic prediction. Genet Sel Evol. 2015;47:42.

26. Hayes BJ, Pryce J, Chamberlain AJ, Bowman PJ, Goddard ME. Genetic architecture of complex traits and accuracy of genomic prediction: coat colour, milk-fat percentage, and type in Holstein cattle as contrasting model traits. PLoS Genet. 2010;6:e1001139.

27. Eynard SE, Windig JJ, Leroy G, van Binsbergen R, Calus MPL. The effect of rare alleles on estimated genomic relationships from whole genome sequence data. BMC Genet. 2015;16:24.

28. Berger S, Pérez-Rodríguez P, Veturi Y, Simianer H, los Campos G. Effectiveness of shrinkage and variable selection methods for the prediction of complex human traits using data from distantly related individuals. Ann Hum Genet. 2015;79:122–35.

29. Meuwissen THE, Luan T, Woolliams JA. The unified approach to the use of genomic and pedigree information in genomic evaluations revisited. J Anim Breed Genet. 2011;128:429–39.

30. Lee SH, Yang J, Chen GB, Ripke S, Stahl EA, Hultman CM, et al. Estimation of SNP heritability from dense genotype data. Am J Hum Genet. 2013;93:1151–5.

31. Sun X, Fernando RL, Garrick DJ, Dekkers J. Improved accuracy of genomic prediction for traits with rare QTL by fitting haplotypes. In: Proceedings of the 10th World Congress of Genetics Applied to Livestock Production; Vancouver. 2014. p. 209.

32. Lee SH, DeCandia TR, Ripke S, Yang J, the Schizophrenia Psychiatric Genome-Wide Association Study Consortium, The International Schizophrenia Consortium, et al. Estimating the proportion of variation in susceptibility to schizophrenia captured by common SNPs. Nat Genet. 2012;44:247–50.

33. Abdollahi-Arpanahi R, Pakdel A, Nejati-Javaremi A, Moradi-Shahrbabak M, Morota G, Valente BD, et al. Dissection of additive genetic variability for quantitative traits in chickens using SNP markers. J Anim Breed Genet. 2014;131:183–93.

34. Abdollahi-Arpanahi R, Nejati-Javaremi A, Pakdel A, Moradi-Shahrbabak M, Morota G, Valente BD, et al. Effect of allele frequencies, effect sizes and number of markers on prediction of quantitative traits in chickens. J Anim Breed Genet. 2014;131:123–33.

35. Daetwyler HD, Villanueva B, Woolliams JA. Accuracy of predicting the genetic risk of disease using a genome-wide approach. PLoS One. 2008;3:e3395.

36. Nomura T, Honda T, Mukai F. Inbreeding and effective population size of Japanese Black cattle. J Anim Sci. 2001;79:366–70.

37. Ogawa S, Matsuda H, Taniguchi Y, Watanabe T, Nishimura S, Sugimoto Y, et al. Effects of single nucleotide polymorphism marker density on degree of genetic variance explained and genomic evaluation for carcass traits in Japanese Black beef cattle. BMC Genet. 2014;15:15.

38. Onogi A, Ogino A, Komatsu T, Shoji N, Simizu K, Kurogi K, et al. Genomic prediction in Japanese Black cattle: application of a single-step approach to beef cattle. J Anim Sci. 2014;92:1931–8.

39. Watanabe T, Matsuda H, Arakawa A, Yamada T, Iwaisaki H, Nishimura S, et al. Estimation of variance components for carcass traits in Japanese Black cattle using 50 K SNP genotype data. Anim Sci J. 2014;85:1–7.

40. Haile-Mariam M, Nieuwhof GJ, Beard KT, Konstatinov KV, Hayes BJ. Comparison of heritabilities of dairy traits in Australian Holstein-Friesian cattle from genomic and pedigree data and implications for genomic evaluations. J Anim Breed Genet. 2013;130:20–31.

41. Loberg A, Dürr JW, Fikse WF, Jorjani H, Crooks L. Estimates of genetic variance and variance of predicted genetic merits using pedigree or genomic relationship matrices in six Brown Swiss cattle populations for different traits. J Anim Breed Genet. 2015. doi:10.1111/jbg.12142.

11

Genome-wide evolutionary and functional analysis of the Equine Repetitive Element 1: an insertion in the myostatin promoter affects gene expression

Marco Santagostino[1†], Lela Khoriauli[1†], Riccardo Gamba[1†], Margherita Bonuglia[2], Ori Klipstein[1], Francesca M. Piras[1], Francesco Vella[1], Alessandra Russo[2], Claudia Badiale[1], Alice Mazzagatti[1], Elena Raimondi[1], Solomon G. Nergadze[1*] and Elena Giulotto[1*]

Abstract

Background: In mammals, an important source of genomic variation is insertion polymorphism of retrotransposons. These may acquire a functional role when inserted inside genes or in their proximity. The aim of this work was to carry out a genome wide analysis of ERE1 retrotransposons in the horse and to analyze insertion polymorphism in relation to evolution and function. The effect of an ERE1 insertion in the promoter of the myostatin gene, which is involved in muscle development, was also investigated.

Results: In the horse population, the fraction of ERE1 polymorphic loci is related to the degree of similarity to their consensus sequence. Through the analysis of ERE1 conservation in seven equid species, we established that the level of identity to their consensus is indicative of evolutionary age of insertion. The position of ERE1s relative to genes suggests that some elements have acquired a functional role. Reporter gene assays showed that the ERE1 insertion within the horse myostatin promoter affects gene expression. The frequency of this variant promoter correlates with sport aptitude and racing performance.

Conclusions: Sequence conservation and insertion polymorphism of ERE1 elements are related to the time of their appearance in the horse lineage, therefore, ERE1s are a useful tool for evolutionary and population studies. Our results suggest that the ERE1 insertion at the myostatin locus has been unwittingly selected by breeders to obtain horses with specific racing abilities. Although a complex combination of environmental and genetic factors contributes to athletic performance, breeding schemes may take into account ERE1 insertion polymorphism at the myostatin promoter.

Keywords: Horse genome, SINEs, Equids, Myostatin gene expression

Background

A large fraction of the genome of mammals is occupied by interspersed repeats that were generated during evolution by the propagation of transposable elements [1–3]. Short INterspersed Elements (SINEs) are non-autonomous retrotransposons that make use of a transposition process in which an RNA intermediate is reverse transcribed and the resulting cDNA is inserted into a new genomic location [4, 5]. Sequence analysis of SINE elements suggested that most of them derive from ancestral tRNAs, but there are examples of 5S- or 7SL-like sequences [6]. These elements are characterized by two internal RNA-polymerase III promoters that make them transcriptionally independent, but their retrotranscription and integration processes are catalyzed by enzymes encoded by autonomous Long INterspesed Elements (LINEs) [4, 5]. The primate Alu family is an example of SINE; Alu repeats are the most abundant transposable elements in the human genome accounting

* Correspondence: solomon.nergadze@unipv.it; elena.giulotto@unipv.it
†Equal contributors
[1]Dipartimento di Biologia e Biotecnologie "Lazzaro Spallanzani", Università di Pavia, Via Ferrata 1, 27100 Pavia, Italy
Full list of author information is available at the end of the article

for more than one million copies [7–9]. The majority of human Alu elements are present in all individuals because they were inserted in the genome before the radiation of extant humans; however, some Alu elements, that were integrated recently in the human lineage, are characterized by insertion polymorphism [9–12]. In humans, an inverse correlation between the evolutionary age of Alu subfamilies and the percentage of polymorphic elements was demonstrated: 20–25 % of the elements belonging to the youngest subfamily (AluY) are polymorphic [13].

Because of their abundance and mechanism of origin, transposable elements were considered "junk DNA", albeit, in a number of examples it was shown that they can acquire a functional role, a process termed "exaptation" [14–17]; in particular, the insertion of transposable elements inside genes or in their proximity may alter gene structure or expression through gene interruption, introduction of promoter sequences or splice sites [18–20]. In some rare cases, transposons are implicated in genetic disease or cancer [21–23].

In the present paper, taking advantage of the published horse genome sequence [24], we carried out a genome wide analysis of the perissodactyl-specific SINE family of Equine Repetitive Elements (ERE) focusing our attention on insertion polymorphism in relation to sequence conservation. ERE retrotransposons derive from tRNA^{ser} and occupy about 4 % of the horse genome [25, 26]; to date, four main ERE subfamilies were identified: ERE1-4 [27, 28]. To our knowledge, before the present study, no data were available on the involvement of horse transposable elements in the modulation of gene expression. The description of a polymorphic ERE1 insertion in the promoter of the myostatin gene [29] prompted us to investigate the possible functional role of this insertion.

Myostatin or growth/differentiation factor 8, a member of the transforming growth factor-β family, is a repressor of muscle growth that regulates myoblast proliferation and differentiation. It has been shown previously that mutations in the myostatin gene can cause muscle hypertrophy in a range of mammals such as mice [30], cattle [31, 32] and sheep [33]. In 2004, Schuelke and collaborators reported the case of an extraordinarily muscular child whose mother appeared muscular, although not to the extent observed in her son, and was a professional athlete [34]. The authors discovered that the boy carried a single base substitution in both copies of the myostatin gene generating a premature termination codon while the mother was heterozygous for the mutation. Particularly relevant in this context is also the "bully" phenotype in whippet racing dogs, which depends on a frameshift mutation causing the production of a truncated protein. Individuals homozygous for the mutation show a double-muscle-phenotype, called

"bully", while heterozygotes display an intermediate phenotype. While heterozygous animals have significantly greater racing ability than wild-type and mutated homozygous dogs, the excessive muscle mass of homozygotes for the mutation is detrimental for performance [35].

In the horse, the myostatin gene, which comprises three exons and two introns, is located on chromosome 18; several sequence variants were identified in this gene and in its flanking regions [29, 36–41]; among these variants the SNP g.66493737C > T, which is contained within the first intron, was associated with regulation of gene expression in Thoroughbred race horses and proposed as the best predictor of optimum racing distance [29, 38, 42]. The same variant was also associated with high values of body weight/withers height ratio, which, in the horse, is considered a good indicator of skeletal muscle mass [43]. Four additional SNPs, located in the regions adjacent to the myostatin gene, have been identified on chromosome 18 and were associated to performance [43–45]. Finally, as mentioned above, the insertion of an ERE1 element within the promoter region of the myostatin gene was described in some Thoroughbreds [37]. Recently, the presence of this insertion has been associated with a different muscle fiber composition [40, 46]. In the present paper we tested whether this insertion affects gene expression, contributes to breed differentiation and is relevant for sport aptitude and racing performance.

Results and discussion
Insertion polymorphism of ERE loci in the reference genome

A large body of evidence suggests that the horse genome is in a state of rapid evolution [24, 47–50]. Therefore, we may expect that several transposon insertions may have occurred in the horse lineage in relatively recent evolutionary times.

A preliminary in silico analysis of the four ERE subfamilies (ERE1 to ERE4) was carried out. To this purpose, the consensus of each ERE subfamily [27, 28] was used as query for a BLAT search (BLAST-Like Alignment Tool) in the reference sequence of the horse [51, 52], which derives from the assembly of the genomic sequence of the Thoroughbred horse named Twilight [24]. From each ERE subfamily, the 200 loci with the highest identity to their consensus were analyzed in search of empty alleles (i.e., alleles in which the ERE element is not present, ERE–) that may be present in the reference genome, thus identifying heterozygous loci in the genome of Twilight. ERE– alleles were found for 3.5 % of the ERE1, 0.5 % of the ERE2 and none of the ERE3 and ERE4 loci. Since the frequency of insertion polymorphism of transposable elements is related to the age of their insertion in the host genome [11], these results strongly suggest that ERE1s are the elements that were inserted most recently in the horse genome. It must be underlined that, since the reference sequence derives from the genome of a single horse, the frequencies of polymorphic loci reported above are largely underestimated being based on the analysis of two alleles per locus.

We then focused on the youngest subfamily, the ERE1, and carried out an extensive genome wide search of these elements in the reference genome sequence (Broad/equCab2). A list of 45,713 ERE1 loci was obtained using the consensus sequence deposited at the RepBase database as query [53] for a BLAST search (Additional file 1: Table S1A). The sequences were then filtered to include only elements with sizes similar to the ERE1 consensus (225 bp ± 10 bp) and with minimum identity of 84 % to the consensus. This operation left 34,131 loci (Additional file 1: Table S1B). The ERE1 sequences located inside other repetitive elements were also excluded from the analysis to avoid false positive results; this operation left 27,396 loci (Additional file 1: Table S1C). In order to obtain a comprehensive view of polymorphic ERE1 loci in Twilight, we analyzed the horse trace database, which includes unassembled traces [54] (center_project number G836). The sequence of each one of the 27,396 ERE1 loci was used as query for a BLAST search. The results of this analysis showed that Twilight is heterozygous at 377 ERE1 loci, possessing an ERE1+ and an ERE1– allele. A complete list of these polymorphic loci is reported in Additional file 2: Table S2. It is important to point out that an undefined number of ERE1 insertions, that are present in the horse population, is not detectable in the reference genome because Twilight may carry two ERE1– empty alleles at such loci. A clear example of this situation is the insertion in the myostatin gene promoter described below.

Since the fixation of insertion elements in the genome of a phylogenetic lineage requires many generations, the presence of empty alleles suggests that the insertion event occurred in relatively recent evolutionary times. In addition, mutations tend to accumulate in the inserted element and therefore a high degree of sequence conservation is considered indicative of a young evolutionary age of insertion, as previously shown for primate and rodent interstitial telomeric sequences [55, 56] and for human transposable elements [57]. In light of these considerations, we can hypothesize that ERE1 elements with higher identities to the consensus may have greater probabilities of being polymorphic compared to less conserved elements. To test this hypothesis, we evaluated the frequency of polymorphic loci in eight classes of ERE1 elements, characterized by different degrees of identity to the consensus (Fig. 1 and 2; data file 1: Table S3). In the class including ERE1 loci with the highest identity to the consensus (98–100 %), the percentage of loci that are polymorphic in Twilight is surprisingly high (4.6 %); this fraction decreases with the decrease of identity to the consensus reaching values as low as 0.1 % (Fig. 1). The correlation between fraction of polymorphic loci and percentage of identity to the ERE1 consensus sequence is highly significant (Pearson's correlation $\rho = 0.93$, $p = 8.5 \times 10^{-4}$). These results suggest that sequence conservation and insertion polymorphism of ERE elements are both related to the time of their appearance in the horse lineage.

Insertion polymorphism in the horse population, evolutionary history and sequence conservation of ERE1 loci

To evaluate the frequency of insertion polymorphism in the horse population, we analyzed 80 ERE1 loci in 30 unrelated domestic horses of different origin (see Materials and Methods). The 80 loci were chosen randomly

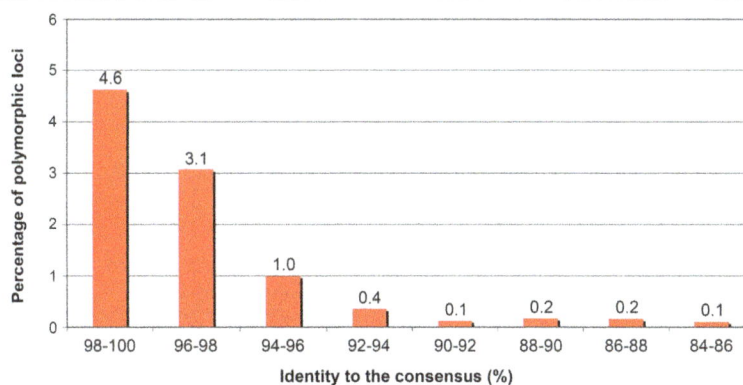

Fig. 1 Percentage of ERE1 polymorphic loci in the horse genome reference sequence. The ERE1 elements were grouped in eight classes according to their identity to the ERE1 consensus sequence published in Repbase. The percentage of polymorphic loci in each class is reported

from four classes (20 loci per class) with different degrees of identity to the ERE1 consensus sequence ($\geq$98, 95, 90 and 85 % identity). For each locus, a primer pair flanking the ERE1 element was designed (Additional file 2: Table S4) and the genomic DNA of the 30 horses was amplified by PCR. The analysis of these loci in the 30 horses is summarized in Fig. 2, where different colours indicate the genotypes of each individual: ERE1+/+, green; ERE1+/−, yellow; ERE1−/−, red. For 71 loci (Fig. 2) only individuals homozygous for the presence of the ERE1 element (ERE1+/+) were found, suggesting that either the insertion is fixed in the population or the frequency of ERE1- alleles is very low. The remaining 9 loci were characterized by insertion polymorphism (Fig. 2). At these 9 loci, the fraction of ERE1- alleles per locus is highly variable ranging from 1.7 (locus 51) to 97 % (locus 11). Although the number of loci analyzed in each class as well as the number of individuals are relatively small, the results are in agreement with the *in silico* results described above: polymorphic loci are more represented in the class with the highest similarity to the ERE1 consensus sequence (6 loci out 20) whereas no polymorphic loci were identified in the class with the lowest identity to the consensus. These results confirm the observation, reported in the previous paragraph, that elements with high similarity to the consensus sequence, have a greater probability of being polymorphic compared to less conserved elements. We previously observed a high frequency of insertion polymorphism in the horse, involving NUMT elements (NUclear sequences of MiTochondrial origin) [49]. Similarly to NUMT sequences, the fraction of ERE1 polymorphic loci described here is particularly high compared to that reported for SINE elements in the human genome [9], thus providing further evidence for the rapid evolution of the horse genome.

We also analyzed the 80 loci in 20 Przewalski's horses, in three individuals from *E. asinus* and in one individual each from *E. burchellii*, *E.grevyi*, *E. zebra hartmannae*, *E. kiang* and *E. hemionus onager*, respectively (Fig. 2); since the results of the three *E. asinus* individuals were identical, only one column is reported in Fig. 2. As shown in Fig. 3, from the evolutionary point of view, ERE1 loci can be classified in three groups: elements which are conserved in all species of the genus *Equus* (53 loci) and thus were inserted in a common ancestor of all extant equids, at least 3.8 Ma ago (Mya); elements which are conserved in all analyzed horses (*E. caballus* and *E. przewalskii*) but absent in the other *Equus* species (25 loci), thus inserted after the separation of the horse lineage, that is about 3.8 Mya [58, 59]; elements which are present in *E. caballus* only (two loci: 11 and 35 in Fig. 2) and therefore were probably inserted after the separation of the two horse species. To this regard, it must be

pointed out that, in the middle of the twentieth century, Przewalski's horses were close to extinction and the extant population derives from a very limited number of individuals [60]; therefore, the absence of an ERE1 element in Przewalski's horses may be related either to the date of its insertion or to genetic drift. Nine loci (number 1, 6, 9, 11, 13, 15, 28, 35, 51 in Fig. 2) are polymorphic in one or both horse species and absent in the other species, suggesting that these insertions occurred in a relatively recent evolutionary time, after the separation of the horse lineages, and are not yet fixed.

In conclusion, these results showed that the fraction of ERE1 insertions conserved in all *Equus* species increases with the decrease of their identity to the consensus (Fig. 3): only 3 out of the 20 horse ERE1 elements with 98–100 % identity were present in the other species while 13, 17 and 20 loci out of 20 were conserved in the classes with 95, 90 and 85 % identity, respectively (Fig. 3). On the contrary, the majority of ERE1s that are present in the horse lineage only (16/20) share a high identity to the consensus (98–100 %). The loci that were conserved in all *Equus* species were not polymorphic in the horse (Fig. 2) confirming that they were inserted earlier during evolution, in a common ancestor of the extant *Equus* lineages. Since only three individuals from *E. asinus* and one individual from *E. burchellii*, *E. grevyi*, *E. zebra hartmannae*, *E. kiang* and *E. hemionus onager* were analyzed, we cannot exclude that, at some ERE1 loci, insertion polymorphism may be present in one or more *Equus* species, however, the results confirm that the level of identity to the consensus not only is related to their polymorphism but is also indicative of their evolutionary age. Therefore, ERE1 insertion polymorphism can be used for evolutionary analyses and population studies.

Position of ERE1 loci relative to genes

Since transposable elements, when inserted within or near genes, may influence gene expression, we used an algorithm developed in our laboratory (see Material and Methods) to classify ERE1 elements according to their position relative to genes. The coordinates of the horse genes were obtained using the tool "UCSC Table Browser" [61, 62]. Horse genes are poorly mapped, therefore we included in the analysis the coordinates of putative horse genes listed in a table generated by UCSC, based on homology with human and bovine genes. The results (Fig. 4) showed that 45.4 % of ERE1 elements were located inside introns of validated or putative genes. The fraction of the human genome occupied by introns has been estimated to be between 26 and 38 % [2, 63–67]; since no data are available for the horse, we are unable to conclude whether the fraction of ERE1 elements contained within introns is simply due to random insertion. Given the high number of ERE1 elements within introns, it is possible that some have acquired a functional role by modifying the splicing pattern as documented for other SINEs [68–70]. The remaining ERE1s (54.6 %) were located at variable distances from genes. Our data suggest that there are no hotspots for ERE1 integration sites in the horse genome and that insertion events may have occurred at random. Counter-selection may be responsible for the lack of insertions within exons. Moreover, only 170 ERE1 insertions (0.5 %) were found at less than 1 kb from the 5' end of validated or putative genes suggesting that some of them may affect gene expression.

Sequence organization of the myostatin gene promoter and mechanism of ERE1 insertion

As mentioned above, a polymorphic ERE1 insertion was identified at the myostatin locus [29]. In Fig. 5, the wild type myostatin locus (Fig. 5a), the ERE1+ allele (Fig. 5e), and a model for the transposition mechanism (Fig. 5b–d) are shown. At the wild type myostatin locus, the regulatory elements, located upstream and in close proximity of the putative transcription start site (Fig. 5a), comprise: two TATA boxes (TATA box1 and 2, located 24 and 1 bp upstream the transcription start site, respectively) and one CAAT box (70 bp upstream the transcription start site). In addition, two E-boxes (E1 and E2), which are muscle gene control elements [71, 72], are located 49 and 16 bp upstream the transcription start site, respectively. Given their position relative to the putative transcription start site, the TATA Box 1 and the CAAT box are likely to constitute the core promoter directing transcription of the horse wild type myostatin gene.

Sequence comparison of the wild type and ERE1+ alleles suggested that this insertion may have occurred according to the previously proposed mechanism of SINE elements retrotransposition in the human genome leading to a direct duplication of the target site [16, 73, 74]. According to this model, during the first step of the process (Fig. 5a), the target site was cleaved inside the TATA box 1 (black arrowhead); the 3' end of the ERE1 RNA (light blue) annealed through microhomology to the single-stranded 5'-TTTTT-3' sequence generated after the nick in the TATA box 1 (Fig. 5b). The free 3'OH group created after the cleavage was then used to prime the reverse transcription of the ERE1 RNA and synthesize the first strand of the cDNA (dark blue, Fig. 5b). The second strand of the DNA was then cleaved one bp downstream the E-box E2 (black arrowhead, Fig. 5c), producing a 3' end that was used to prime the synthesis of the second strand of the ERE1 DNA (Fig. 5d). Through a gap filling reaction, the entire ERE1 sequence was integrated into the myostatin promoter with the formation of the Target Site Duplication. Fig. 5e shows the ERE1+ allele of the myostatin promoter

obtained as a result of the retrotransposition event. The inserted ERE1 (dark blue) is located 29 bp upstream the transcription start site. The size of the Target Site Duplication (14 bp) falls into the range described for SINE elements in the human genome [16, 73, 74]. The consequence of the ERE1 insertion was a modification of the core promoter with the formation of a variant TATA Box 1 and the displacement of the CAAT box. This rearrangement likely affects the strength of the core promoter.

Reporter gene assay of the two variants of the myostatin gene promoter

To test the hypothesis that the ERE1 insertion alters the expression of the myostatin gene, we performed a reporter gene assay using a plasmid containing the enhanced Green Fluorescent Protein (eGFP) gene and the puromycin resistance gene. The two variants of the myostatin promoter (ERE1+ and ERE1-) were cloned from the genomic DNA of a heterozygous Thoroughbred horse and inserted into the plasmid cloning site upstream of the eGFP reporter gene. The ERE1- variant plasmid contained a 2042 bp genomic fragment comprising 31 bp from the myostatin UTR; the ERE1+ plasmid contained an insert differing from the previous one only for the ERE1 insertion.

To test whether the ERE1 insertion can affect promoter strength the two plasmids were transfected in human HeLa cells and in a horse fibroblast cell line that we immortalized using the procedure described in Vidale et al. [75]. Since transfection efficiency in horse fibroblasts is extremely low (3–5 %), transient short term transfections could not be performed. Long-term selection with puromycin had to be carried out in order to isolate stably transfected cell populations. The expression of eGFP was evaluated by fluorescence microscopy, western blotting and quantitative real-time PCR (Fig. 6). Both in human and in horse cells, the ERE1 insertion caused a reduction of eGFP fluorescence signals to almost undetectable levels (Fig. 6a). The effect of the insertion on promoter strength was also demonstrated by immunoblotting of protein extracts with an anti-eGFP antibody (Fig. 6b): while a strong band could be detected in protein extracts from cells transfected with the plasmid containing the ERE1- promoter, only a very faint band could be observed in extracts from cells transfected with the ERE1+ plasmid. We then carried out a quantitative real-time PCR reaction using eGFP specific primers (Additional file 2: Table S4B) to amplify reverse transcribed mRNA from the transfected cell lines (Fig. 6c): in human cells transfected with the ERE1+ plasmid the expression level of the reporter gene showed a 6.4-fold reduction compared with that observed in cells transfected with the vector carrying the ERE1-

promoter; similarly, a 4.9-fold reduction was observed in horse fibroblasts. These results demonstrate that the ERE1 insertion affects the ability of the myostatin gene promoter to drive transcription of a reporter gene and strongly suggest that the myostatin gene may be under-expressed in horses containing this variant promoter sequence.

ERE1 insertion polymorphism at the myostatin locus: sport aptitude and racing performance

Given the role of myostatin in the regulation of muscle development and considering the relevance of muscular mass in athletic performance, we wondered whether the genotype of horses relative to the ERE1 insertion may influence their sport aptitude and racing abilities.

Using primers flanking the myostatin gene promoter (Additional file 2: Table S4B), we set up a PCR assay to identify the two alleles: the ERE1 containing allele, ERE1+, produces a 441 bp band, while the allele lacking the insertion, ERE1-, produces a 214 bp band. We then ana-lyzed the frequency of the two alleles, in 5 horse breeds (Quarter Horse, Andalusian, Lipizzaner, Norwegian Fjord and Icelandic Pony) and in Przewalski's horse. As shown in Table 1A, in Quarter horses, although the number of individuals analyzed is limited (20), the frequency of the ERE1+ allele seems particularly high (57 %). In the Andalusian breed, the ERE1+ allele was observed only in 3 heterozygous individuals, while in the other breeds and in Przwelaski's horse the ERE1+ variant was not present. Since the ERE1 insertion was present only in horse populations in which Thoroughbred blood is known to have been introduced (Quarters, Andalu-sians, Show Jumpers), it is likely that it appeared recently in the horse lineage and probably occurred in a Thor-oughbred ancestor, as previously suggested [46].

Although the number of individuals tested for each breed is relatively small (19–23 animals per breed), the striking frequency variation of the two alleles suggests that the two variants may have been under selection during the establishment and improvement of some breeds in relation to specific aptitude and performance traits. In particular, the high frequency of ERE1+ alleles in Quarter horses suggests that this variant may favor the ability of sprinting short distances. To this regard, it is im-portant to point out that the name of this breed came from its excellence in races of a quarter mile or less.

Therefore, to test the hypothesis that the ERE1 inser-tion at the myostatin locus may affect the aptitude for specific sport abilities, we initially analyzed the fre-quency of the two allelic variants in 30 horses competing in show-jumping at various levels, in 90 horses regis-tered in the Italian Trotter studbook, bred for harness racing, and in 75 horses registered in the Italian Thor-oughbred studbook mainly bred for flat racing (Table 1B).

Although Italian Trotters derive from English Thorough-bred stallions crossed with mares of different origins, and Thoroughbreds have been introduced in several bloodlines of Show Jumpers, the allelic frequencies in the three groups were strikingly different (Table 1B): the ERE1+ allele was completely absent in the Trotters and, in the Show Jumpers, only one individual was heterozy-gous for the variant; on the contrary, among the flat ra-cing horses, the percentage of ERE1+ alleles was 43. These observations suggest that the ERE1+ allele may have been selected in the Thoroughbreds and in the Quarter Horses together with flat racing aptitude.

To test whether the ERE1+ variant may influence ra-cing performance in the Thoroughbreds, we selected a group of 117 elite horses classified in the top three places in at least one high level race in Italy in the period ranging from 2005 to 2011. In this selected group, the ERE1+ allele was significantly more frequent compared to the general Thoroughbred population ($p = 9.31 \times 10^{-6}$, Table 1 B). To test whether the ERE1 insertion influ-ences performance relatively to race distance, the elite horses were grouped according to Best Race Distance, defined as the distance of the highest grade race won. When multiple races of the same grade were won, the distance of the race with the most valuable prize was considered. The results of this analysis are shown in Fig. 7: in short distance races (1000 and 1200 m), the majority of winning horses (18 out of 30) were homozy-gous for the ERE1+ allele and no homozygous individ-uals for the ERE1- allele were found; in the long distance races (>2000 m), only heterozygotes and ERE1- homozy-gotes were observed and, in medium distance races (1400–2000 m), all the three genotypes were represented although the ERE1+ homozygotes were relatively more frequent in the groups winning up to 1600 m races com-pared to horses winning 1700–2000 m races. When the genotypic frequencies in horses winning short distance (1000–1200 m), medium distance (1400–2000 m) and long distance (>2000 m) races were compared, the dif-ferences were highly significant ($p = 1.94 \times 10^{-6}$).

Since the ERE1+ variant is associated with better per-formance in short distance races, it may have been artifi-cially selected through breeding, consequently, its frequency increased in the Thoroughbred population, al-though it was not fixed. The empty allele might also have been subjected to artificial selection. Thorough-breds are also used for long distance races, in which in-dividuals homozygous for the ERE1- alleles have the best performance while heterozygous animals seem to be advantaged in average distance races. It should be pointed out that among the Italian Trotters, a breed de-rived from English Thoroughbreds, no ERE1+ allele was identified. This is probably due to the fact that Italian Trotters are bred for harness racing at a trot gait in relatively long distance races and this artificial selection led to the loss of the ERE1+ allele. Finally, although Quarter Horses derive from the crossing of Thoroughbreds with horses from other breeds, the frequency of the ERE1+ allele was even higher than in the Thoroughbreds themselves (Table 1); this observation can be related to the fact that these horses have been selected for their sprinting ability in flat races of a quarter mile or less.

As mentioned in the introduction, the g.66493737C > T SNP in the first intron of the myostatin gene was shown to be predictive of athletic performance [29, 37]: C/C horses are suited for short-distance, C/T for middle-distance and T/T for long-distance races.

Comparing the ERE1 and the g.66493737C > T genotypes (Fig. 7), we observed that in 112 out of 117 horses the two genotypes were concordant, with the C SNP allele associated with ERE1+ and the T SNP allele associated with the ERE1- promoter. These results show that the two polymorphic loci are tightly linked, as expected by their close proximity in the genome (1605 bp). Although the ERE1 insertion was previously described [37], its influence on myostatin gene expression was not investigated. In the present work, we demonstrate that the ERE1 insertion affects gene expression supporting the hypothesis that this is the genotype that drove selection [46]. In particular, we showed that the ERE1 insertion causes a 5–6 fold decrease in the

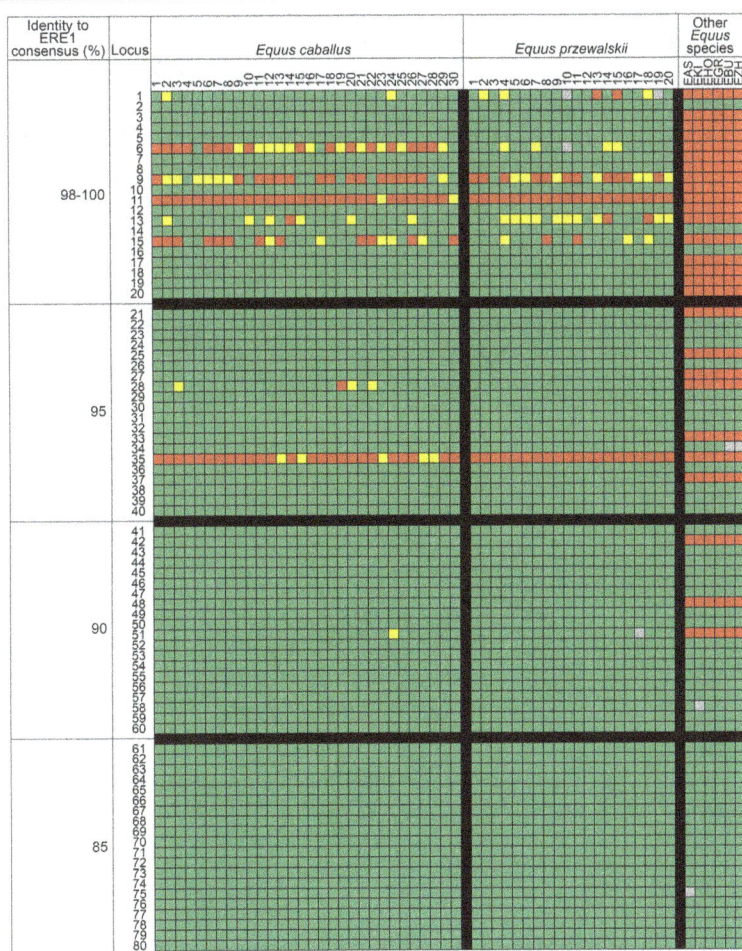

Fig. 2 Insertion polymorphism of 80 ERE1 loci in equids. The insertion polymorphism of 80 random ERE1 loci with different percentage of identity to the ERE1 consensus were analysed: 20 loci with 98–100 %, 20 loci with 95 %, 20 loci with 90 % and 20 loci with 85 % identity. The analysis was carried out in 30 individuals from *E. caballus*, 20 individuals from *E. przewalskii*, three individuals from *E. asinus*, EAS, and one individual from each one of the following species: *E. kiang*, EKI; *E. hemionus onager*, EHO; *E. grevyi*, EGR; *E. burchellii*, EBU; *E. zebra hartmannae*, EZH. The position of each locus in the horse genome is reported in the left column. Each column reports data from the animal indicated on top. Each table cell shows the genotype of an individual at a specific locus. Genotypes are indicated using different colours: green, homozygous for the ERE1+ allele; red, homozygous for the ERE1- allele; yellow, heterozygous; grey, no data

transcription of the reporter gene (Fig. 6), providing the first example of a SINE element influencing gene expression in the horse genome.

Although the g.66493737C > T SNP showed an association with racing performance [29], this sequence variation does not provide an immediate functional explanation of this trait. On the contrary, our experimental data strongly suggest a direct influence of the ERE1 insertion on myostatin expression. Since the g.66493737C > T SNP is located only 1605 bp away from the ERE1 insertion site in the promoter, the ERE1 insertion, rather than the g.66493737C > T SNP (located in the first intron), may functionally influence racing performance, the two polymorphisms being in linkage disequilibrium ($r^2 = 0.73$) as previously observed [29, 46]. In other words, the results presented here on myostatin expression provide a physiological interpretation of the correlation between ERE1 insertion and racing performance; moreover, the previously described correlation among the g.66493737C > T SNP, muscle mass [43] and muscle fiber composition [46] can also be reinterpreted on the basis of the linkage disequilibrium between the two polymorphic loci.

Conclusions

In the work presented here we provide a catalogue of the most abundant SINE retrotransposons, ERE1, in the horse genome. Through the analysis of sequence conservation, insertion polymorphism and presence in other equids, we provide an evolutionary dating of ERE1

elements appearance in the *Equus* lineage. Therefore, similarly to other mammalian SINE elements, ERE1 insertion polymorphism can be used for evolutionary analyses and population studies.

The analysis of ERE1s position relative to genes suggests that some may have acquired a functional role by modifying the splicing pattern, when interrupting an intron, or by altering gene expression, when inserted inside regulatory regions. To this regard, we studied the effect of an ERE1 insertion in the promoter of the myostatin gene showing that it causes a reduction of promoter strength in a reporter gene assay. Therefore, we suggest that this ERE1 insertion may decrease the levels of myostatin thus modifying muscle development.

The ERE1 insertion at the myostatin locus is polymorphic in the horse population and seems to be related to specific racing aptitude, the ERE1+ allele being particularly common in breeds characterized by sprinting ability, such as the Quarter Horse, and absent in other breeds, such as the Italian Trotter, which are used for long distance racing. In a sample of Thoroughbred elite horses, classified in the top three places in at least one high level race in Italy, we observed a statistically significant correlation between the ERE1+ variant and good performance in short distance races; on the other hand, the empty allele was more frequent in Thoroughbreds winning long distance races. We propose that the two variants have been unwittingly selected by breeders in order to obtain horses with specific racing abilities.

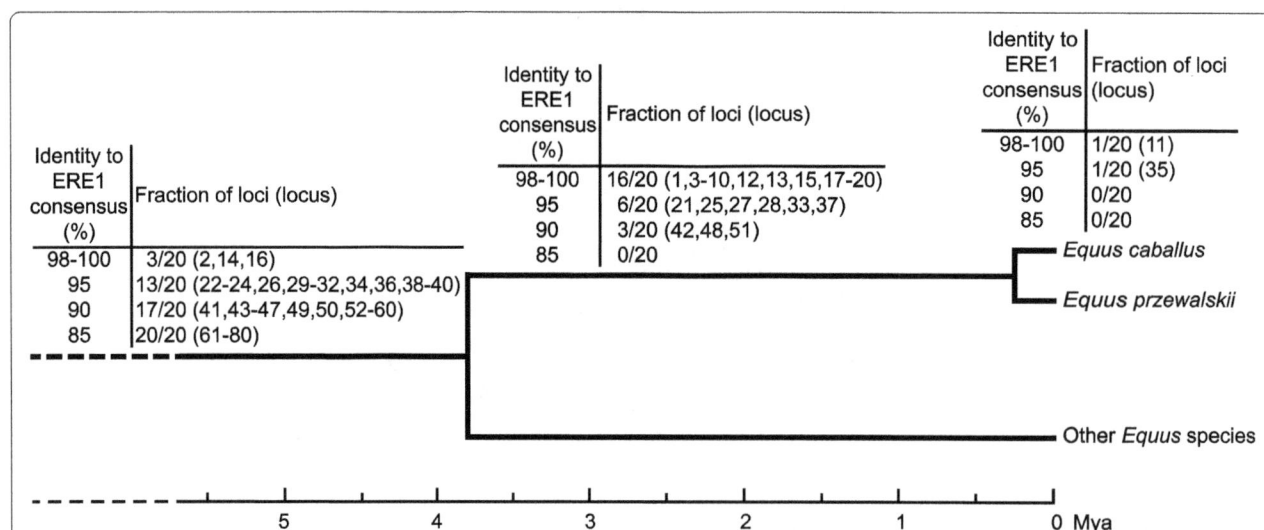

Fig. 3 Phylogenetic tree of equids. The time of insertion of each one of the 80 ERE1s is marked on the phylogenetic tree (adapted from [58, 59]). ERE1 loci are classified according to the percentage of identity to the consensus sequence, the fraction of inserted loci in each class of identity is shown. Each ERE1 is indicated by a unique locus number (see Fig. 2 and Additional data file 1: Table S3A). The lineage "Other *Equus* species" comprises the following non-horse species: *E. asinus, E. kiang, E. hemionus onager, E. burchellii, E. grevyi, E. zebra hartmannae*

Table 1 ERE1+ and ERE1- genotyping at the myostatin locus

		Number of individuals	Number of alleles (%)		Homozygous individuals (%)		Heterozygous individuals (%)
			ERE1+	ERE1-	ERE1+/+	ERE1−/−	ERE1+/−
A	Quarter Horse	20	23 (57.5)	17 (42.5)	9 (45)	6 (30)	5 (25)
	Andalusian	20	3 (7.5)	37 (92.5)	0	17 (85)	3 (15)
	Lipizzaner	23	0	46 (100)	0	23 (100)	0
	Norwegian Fjord	20	0	40 (100)	0	20 (100)	0
	Icelandic Pony	19	0	38 (100)	0	19 (100)	0
	Przewalski's Horse	20	0	40 (100)	0	20 (100)	0
B	Show Jumpers	30	1 (1.7)	59 (98.3)	0	29 (96.7)	1 (3.3)
	Italian Trotters	90	0	180 (100)	0	90 (100)	0
	Unselected Italian Thoroughbreds	75	65 (43.3)	85 (56.7)	18 (24.0)	28 (37.3)	29 (38.7)
	Elite Italian Thoroughbreds	117	135 (57.7)	99 (42.3)	33 (28.2)	15 (12.8)	69 (59.0)

(A) Analysis of individuals from five breeds of the domestic horse and from Przewalski's horse. (B) Analysis of individuals bred for different sport aptitude.

Our results indicate that, although racing performance is certainly influenced by environmental factors, like training and nutrition, and by several genetic factors, breeding schemes may also take into account the differential effect of these two ERE1 allelic variants.

Methods
Ethics statement
Horse blood and hair samples were collected in the stables where the animals were kept, during veterinary practices carried out for routine clinical analysis, animal care or registration requirements. Since blood samples were not collected for experimental purposes, according to the Italian law (Decreto Legislativo 4/03/2014 n.26), the procedures do not require approval by an ethical committee. Written consent from the owners was not required because the identity of horses and owners cannot be established from the data presented in this work.

DNA samples from endangered Equus species were shipped to Italy from the San Diego zoo together with the appropriate international CITES permit. Horse fibroblast cell lines were established from skin samples taken from animals not specifically sacrificed for this study; the animals were being processed as part of the normal work of the abattoirs.

Preliminary *in silico* analysis of the polymorphism of the four ERE subfamilies
The consensus sequences of the ERE subfamilies ERE1 (accession number: D26566) [53], ERE2 [76], ERE3 [77], ERE4 [78] were downloaded from Repbase [27, 28] and used as queries for a BLAT search against the horse genome reference sequence (September 2007 Broad/equCab2.0 assembly) [51, 52]. For each ERE subfamily the 200 loci with the highest identity to their consensus sequence were identified. Their sequence was used as query for a BLAST search against the horse Trace

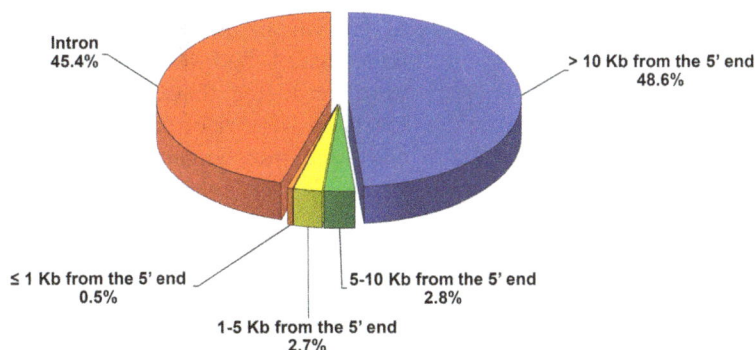

Fig. 4 Distribution of ERE1 elements relative to genes. The percentage of ERE1 loci located in introns of validated or putative genes (red) and in non-genic regions is indicated. ERE1 elements located in non-genic regions are classified according to their distance from the 5' end of the nearest gene (>10 Kb, 5–10 Kb, 1–5 Kb, ≤ 1 Kb)

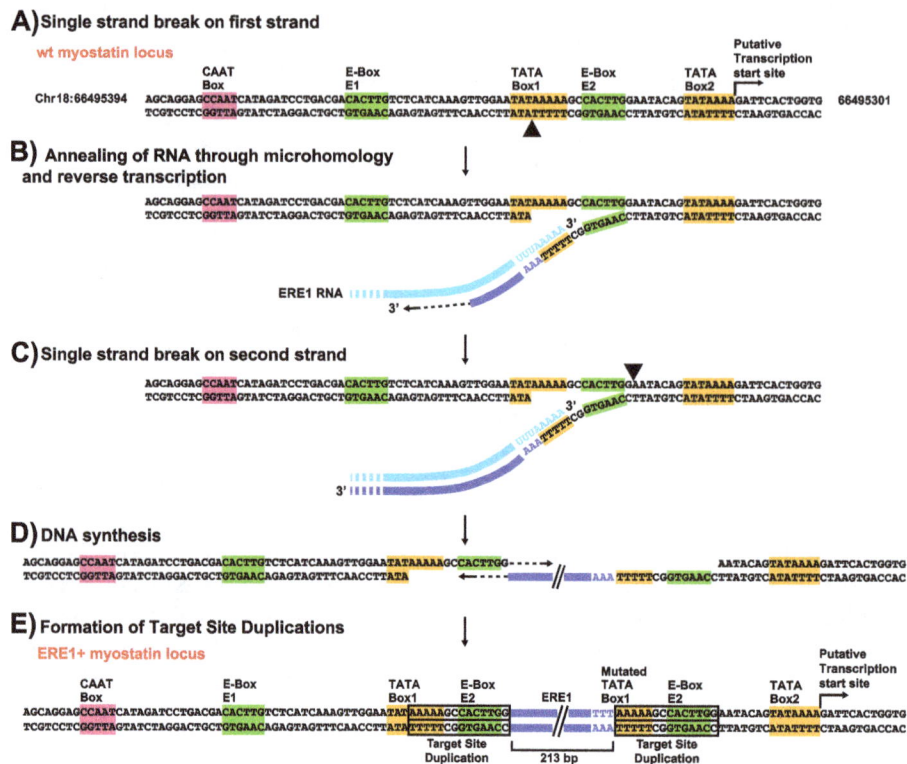

Fig. 5 Model for ERE1 integration via retrotransposition at the myostatin promoter. **a** Sequence of the empty wild type myostatin locus, the coordinates of the locus in the horse genome reference sequence (equCab2.0) are indicated on both sides. The promoter elements located upstream the transcription start site are shown: CAAT-box (pink), E-boxes (green), TATA boxes (yellow). The black arrowhead points to the position of the single strand break. **b** Annealing through microhomology (TTTTT/AAAAA) of the ERE1 RNA (light blue) to the single stranded DNA end originated after the nick and reverse transcription. The cDNA produced by retrotranscription is shown in dark blue. **c** Cleavage of the second DNA strand (black arrowhead). **d** Synthesis of the second strand of ERE1 cDNA and gap filling. **e** Sequence of the ERE1+ myostatin locus: the ERE1 element is integrated (dark blue) and the target site is duplicated (boxed nucleotides)

database [54], which is a collection of short sequences (<1 Kb) generated during large-scale sequencing projects. From the Trace database we selected the dataset Equus caballus-WGS, which contains reads that were not included in the final assembly of the horse genome reference sequence. We then used the sequences flanking each ERE insertion as query to search for traces corresponding to the same loci but lacking the ERE insertion (empty alleles).

Search of ERE1 loci characterized by insertion polymorphism in the horse genome reference sequence

Our preliminary search, based on the analysis of 200 loci from each ERE subfamily, showed that ERE1s have the highest proportion of empty alleles. We then focused further analyses on this subfamily.

In order to obtain a comprehensive catalog of ERE1 polymorphic loci in the horse genome reference sequence, we developed a pipeline using the C# programming language (Microsoft Visual Studio 2008) and Microsoft SQL Server 2008 as the database management system. The ERE1 consensus sequence downloaded from

RepBase (accession number D26566) [53] was used as query for a BLAST search against the horse genome reference sequence (September 2007 Broad/equCab2.0 assembly) [79]. The BLAST search was performed using "megablast" as optimization algorithm and standard search parameters. Results were downloaded as hit table. Only the loci with identity to the consensus greater than 84 % were considered. To exclude loci that were subject to deletions or insertions, only the hits with length similar to that of the ERE1 consensus sequence (225 ± 10 bp) were considered. Since the coordinates of the hits inside the table were referred to contig sequences, they were converted into genomic coordinates using the conversion table "seq_contig.md" at [80]. ERE1s located inside unplaced regions were discarded. Since our method is based on similarity, ERE1s inserted inside other transposons could give rise to false positive hits because several uninterrupted transposons are scattered through the genome. Therefore, before starting the search for polymorphic loci we identified and discarded ERE1 elements inserted inside other transposons. To this purpose, we downloaded the list of the horse transposable elements

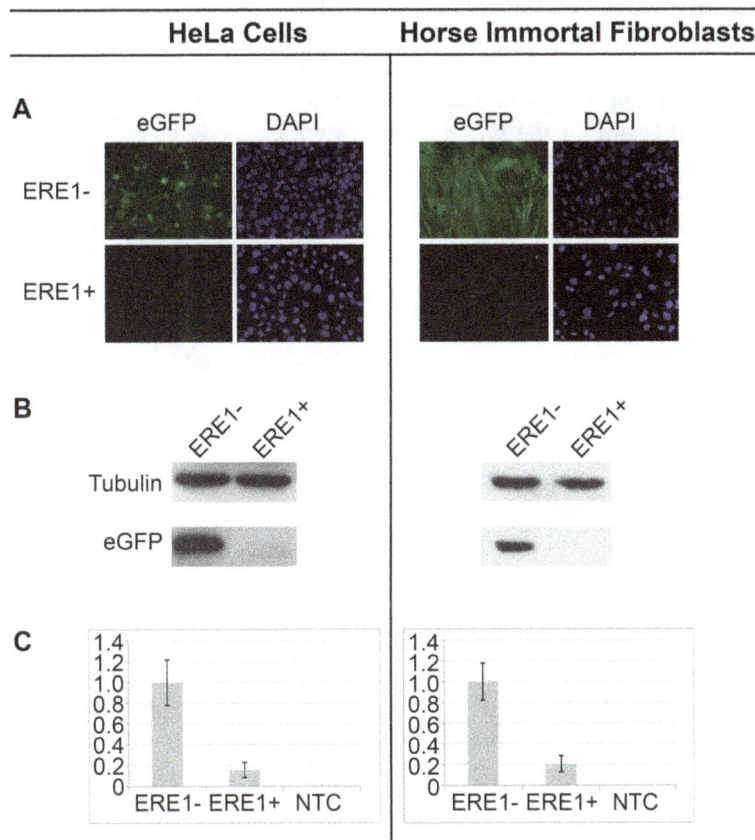

Fig. 6 Reporter gene assay. The reporter gene assay (*eGFP* expression) was carried out in human HeLa cells (left) and in horse immortalized fibroblasts (right). **a** Fluorescence microscopy images of cells transfected with the two constructs containing the *eGFP* gene under the control of the ERE1- (top) and ERE1+ (bottom) promoter. DAPI-staineing is shown on the right of each image. **b** Western blots with anti-tubulin (loading control, top) and anti-eGFP (bottom) antibodies using protein extracts from cells transfected with the plasmid containing the ERE1- (left) and the ERE1+ (right) promoter. **c** Quantification of *eGFP* expression by quantitative RT-PCR. The expression levels of the eGFP transcript are indicated in arbitrary units. eGFP levels in cells transfected with the ERE1- plasmid were used as reference and set to 1.0. NTCs, no-template controls

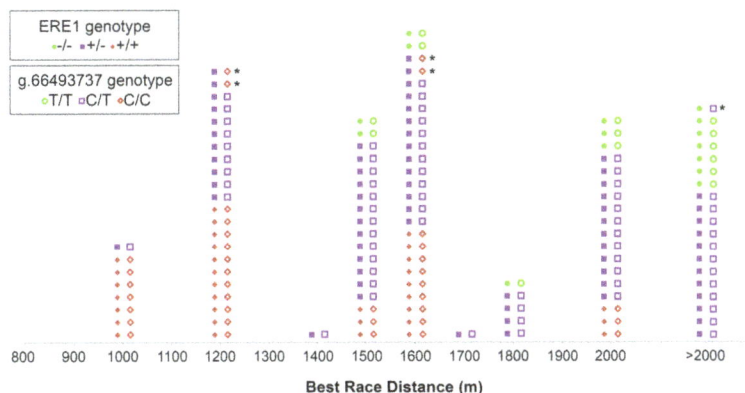

Fig. 7 Genotyping of elite Thoroughbred horses. Elite Thoroughbreds (*n* = 117) were classified according to the distance of the highest grade race won (Best Race Distance), expressed in meters. For each individual, the myostatin promoter (full symbols) and the SNP g.66493737C > T (empty symbols) genotypes are shown. The majority of the ERE1−/−, ERE1+/− and ERE1+/+ individuals have a T/T, C/T and C/C SNP genotype, respectively; the five individulas with a different combination of genotypes at the two loci are marked with an asterisk

from the site UCSC Genome Bioinformatics using the tool "Table Browser" [61, 62]. The list of transposons is found in the data table called "rmsk" (Group "Variation and Repeats", Track "RepeatMasker") that was generated using the software RepeatMasker [81] during the horse genome sequencing project [24]. The coordinates of each ERE1 were compared with those of the boundaries of other transposable elements. If an ERE1 interrupted a repetitive element the locus was discarded.

To identify empty alleles, for each locus we downloaded a 2.2 Kb sequence from UCSC Genome Browser [24, 82, 83] containing the transposon (about 225 bp), 1 Kb from the 5' flanking region and 1 Kb from the 3' flanking region. These sequences were then used as queries for a BLAST search [54] against the horse "Traces – WGS sequence" database. The BLAST search was performed using "megablast" as optimization algorithm and standard search parameters. If the hit contained a 225 ± 10 bp gap and was at least 98 % identical to the sequences flanking the transposon, it was considered an ERE1- locus. Only traces from the reference genome of Twilight were considered identifying them as belonging to "center_project number" G836. The specificity of each trace sequence was manually checked using BLAT [51, 52] and MultAlin [84, 85]. In order to focus on the loci inserted in single copy sequences, the ERE1 loci that were found at multiple positions during the BLAT search, and were probably located inside segmental duplications, were discarded. The complete list of single copy polymorphic ERE1 loci and the accession codes of the traces (Trace id) corresponding to the empty alleles is reported in Additional file 2: Table S2.

In silico localization of ERE1 elements relative to genes

The position of ERE1 elements relative to horse genes was defined using the genomic coordinates of known horse validated and putative genes. Horse validated genes and their coordinates are listed in the data table "refGene" (assembly "Sep. 2007") downloaded from the site UCSC Genome Bioinformatics using the tool "Table Browser" [61, 62]. The "refGene" table contains, among other information, the name of each gene, the coordinates of the transcription start and stop sites, the coordinates of the boundaries of each exon. Since the number of known horse genes is relatively small, we also included in the search the genomic coordinates of putative genes defined by sequence homology with those from human and bovine as listed in the data table "Other RefSeq (xenoRefGene)". The data table (xenoRefGene) was downloaded from using the tool "Table Browser" [61, 62] and was used to define the coordinates of the beginning and end of putative genes in horse that are orthologous to those from human and bovine. This track was prepared by the UCSC genome browser group as

described in the information page (https://genome.ucsc.edu/cgi-bin/hgTrackUi?hgsid=442242277_zw0eu9-Hy93E8wlE62c8BxvE3BJox&c=chr11&g=xenoRefGene): as stated in the information page, this track shows known protein-coding and non-protein-coding genes for organisms other than horse. The RNAs were aligned against the horse genome using blat. This track was produced at UCSC from RNA sequence data generated by scientists worldwide and curated by the NCBI RefSeq project.

Genomic DNA samples

Genomic DNA was extracted from blood or hair samples, or from cultured primary fibroblasts using standard protocols. The 30 *E. caballus* samples shown in Fig. 2 derive from: peripheral blood of 22 show jumping horses which, according to their pedigree chart, do not share common ancestors up to the third generation (they were also used for the analysis of the myostatin gene polymorphism shown in Table 1, see below); fibroblast cell lines established from the skin of 8 slaughtered animals which were shown to be unrelated by microsatellite analysis as described in [86]. The *E. asinus* samples derive from fibroblast cell lines established from the skin of 3 slaughtered animals. The *E. grevyi* sample derives from a fibroblast cell line purchased from Coriell Repositories and *E. burchellii* fibroblasts were a kind gift from Mariano Rocchi (University of Bari, Italy) [50, 87]. *E. zebra hartmannae*, *E. kiang* and *E. hemionus onager* fibroblasts were provided by Oliver Ryder (Genetics Division of San Diego Zoo, San Diego, California, USA) [48]. DNA samples from Quarter Horses, Andalusian, Norwegian Fjord, Icelandic Ponies (Table 1) and *E. przewalskii* (Fig. 2 and Table 1) were provided by Cecilia Penedo (UC Davis, California, USA). Lipizzaner DNA samples (Table 1) were described in [88]. The 30 Show Jumpers in Table 1, which comprise the 22 *E. caballus* individuals of Fig. 2, were animals kept in Italian sport riding stables and competing at the National and International level; they derived from different stud farms in Italy, France, Germany, Holland, Belgium and were chosen by the owners for their show jumping aptitude. Genomic DNA from Italian Trotters and Italian Thoroughbreds was extracted from blood spotted on FTA® filter papers (Whatman Bioscience, Cambridge, UK). All samples came from horses belonging to the Italian Stud Book of MiPAAF (Ministero Delle Politiche Agricole Alimentari e Forestali). The performance information were provided by ANAC (Associazione Nazionale Allevatori Cavalli Purosangue).

PCR and SNP analysis

Eighty ERE1 insertions with different degrees of identity relative to the consensus sequence were randomly selected from the list of 27,396 loci obtained by *in silico* analysis. The coordinates of the 80 loci are reported in Additional file 2: Table S4A together with the sequence of the primers deduced from the sequences flanking the transposon (Additional file 2: Table S4A). Twenty ng of genomic DNA were used as template for PCR experiments performed in a 10 μl-final volume with 8 pmoles of each primer, 0.2 mM dNTPs, 1× Green Buffer (Promega) and 0.4 units of GoTaq DNA polymerase (Promega). After a denaturation step at 95 °C for 2 min, the following amplification cycle was performed 3 times: 95 °C for 50 s, appropriate annealing temperature (Additional file 2: Table S4A) for 45 s, 72 °C for 1 min. The first 3 cycles were followed by 27 cycles: 95 ° C for 30 s, appropriate annealing temperature for 35 s, 72 °C for 1 min. Final extension was carried out at 72 °C for 5 min. PCR products were checked by electrophoresis in 1 % agarose gel.

To analyze the ERE1 insertion polymorphism at the myostatin promoter, we amplified genomic DNAs using primers from the sequences flanking the insertion site (MyostProm-F0 and MyostProm-R, Additional file 2: Table S4B). The expected length of the PCR products from the ERE1+ and the ERE1- alleles were 441 and 214 bp, respectively. The reactions were carried out as described above.

The Analysis of SNP g.66493737C > T was performed using the "Custom TaqMan SNP Assay" (Applied Biosystems) on a 7500 Fast Real Time PCR Instrument.

Preparation of plasmids for reporter gene assay

In order to clone the entire promoter and the transcription start site of the myostatin gene we PCR-amplified the locus chr18:66495283–66497324 (equCab2.0) from the genomic DNA of a horse heterozygous for the ERE1 insertion.

PCR reaction was performed using the primers MyostProm-F and MyostProm-R (Additional file 2: Table S4B), which contain *Hind*III and *Bam*HI restriction sites, respectively. After a denaturation step at 95 °C for 2 min, the following amplification cycle was repeated for 30 times: 94 °C for 40 s, 65 °C for 40 s and 72 °C for 4 min. The final extension was carried out at 72 °C for 10 min. The reaction products corresponding to the ERE1- and the ERE1+ allele (2058 and 2285 bp, respectively) were separated by electrophoresis on 1 % agarose gel and purified using the Wizard SV Gel and PCR Clean-Up System (Promega). The two alleles differed only for the presence of the ERE1 element and the target site duplication (see Fig. 3).

The purified PCR products were digested with *Hind*III and *Bam*HI and then cloned, upstream of the enhanced Green Fluorescent Protein (*eGFP*) cDNA, into an expression vector that was previously constructed in our laboratory [89]. Our vector contains the puromycin and ampicillin resistance genes. All constructs were checked by Sanger sequencing.

Cell culture and transfection

Horse Immortal Fibroblasts [75] and HeLa (human cervical carcinoma) cells were cultured in high-glucose D-MEM supplemented with 10 % fetal calf serum (Euroclone), 2 % non-essential amino acids, 2 mM L-glutamine and 1× penicillin-streptomycin (Sigma). For primary fibroblast cell lines, the culture medium was supplemented with 20 % fetal calf serum. Cells were routinely cultured at 37 °C in 5 % CO_2.

Plasmid DNA for promoter reporter assays was prepared using QIAGEN Plasmid Midi kit. Transfections were carried out using the Lipofectamine 2000 reagent (Invitrogen) according to the manufacturer's protocol.

Twenty-four hours post-transfection, cells were selected adding 300 ng/ml (horse immortal fibroblasts) or 1 μy/ml (HeLa cells) puromycin to the medium. Cells were cultured with selective medium until the emergence of drug-resistant colonies, that is 3 weeks for horse fibroblasts and 2 weeks for HeLa cells. Pools of about 50 colonies were obtained and grown as stably transfected cell populations.

Western Blot experiments

Protein extracts were prepared from samples three million cells as follows: the cells were washed twice with ice cold 1xPBS, resuspended in lysis buffer (50 mM Tris–HCl pH 6.8, 86 mM β-mercaptoethanol, 2 % SDS) and boiled for 10 min. Proteins were separated by 10 % SDS-PAGE and transferred to nitrocellulose membranes (Amersham Protran Premium 0.45 μm NC) through wet transfer. Membranes were incubated with a rat monoclonal antibody against eGFP (Chromotek, code 3H9), diluted 1:1000, and with a mouse monoclonal antibody against tubulin (NeoMarkers, Ab-4, code MS-719-P1ABX), diluted 1:3000. Secondary antibodies, conjugated to horseradish peroxidase, were a chicken anti-rat IgG-HRP (Santa Cruz Biotechnology, code sc-2956), diluted 1:5000, and an ImmunoPure goat anti-mouse monoclonal (H + L) (Pierce, code 31430), diluted 1:10,000. Detection was performed using Immun-Star WesternC Kit (Bio-Rad) according to the manufacturer's protocol. Pre-incubation of the membranes and dilutions of the antibodies were performed in 1xPBS containing 0,05 % Tween20 and 7.5 % skim milk.

eGFP fluorescence analysis

Cells for eGFP fluorescence analysis were grown on coverslips (24 × 24 mm), washed with cold 1xPBS and fixed in 2 % paraformaldehyde in PBS for 10 min. Fixed cells were then stained with DAPI (4,6-

diamidino-2-phenylindole) and observed with a ZEISS Axioplan fluorescence microscope at 63× magnification. Pictures were captured using a CoolSNAP CCD camera (RS Photometrics) and processed using the software IPLab 3.5.5 (Scanalytics inc).

RNA preparation and quantitative RT-PCR

Total RNA from transfected HeLa and horse fibroblast cells was extracted using TRizol Reagent (Invitrogen) according to the manufacturer's protocol. The extracted RNA was purified using the RNA Clean & Concentrator-25 kit (Zymo Research) and treated three times with RQ1 RNase-free DNase (Promega).

For quantitative RT-PCR experiments we reverse transcribed 2.5 μg of total RNA using oligo-d(T)$_{17}$ primers and Revert Aid Premium First Strand cDNA synthesis kit (Fermentas) according to the manufacturer's protocol.

The cDNA was PCR amplified using GoTaq qPCR Master Mix (Promega) containing the appropriate oligonucleotides (Additional file 2: Table S4B). Oligonucleotides eGFP-F and eGFP-R were used to detect the eGFP transcript. *GAPDH* (glyceraldehyde 3-phosphate dehydrogenase, primer pair GAPDH-F and GAPDH-R) or *PRKCI* (protein kinase C iota, primer pair humcavPRKC-RealT-F and cavPRKC-RealT-R) were used as control genes for quantitative RT-PCRs carried out with the cDNA from HeLa cells or horse immortal fibroblasts, respectively. Each sample was prepared in triplicate. Negative controls (No template controls, NTCs) were included in the experiments. Reactions were carried out using an Opticon 2 System instrument (MJ Research). Cycling parameters comprised an initial denaturation at 95 °C for 2 min followed by 50 cycles at 95 °C for 15 s, 62 °C for 30 s and 72 °C for 30 s coupled to fluorescence detection. Experiments were repeated twice for each transfected cell line. Data were analyzed with the Opticon Monitor 3 software. Levels of expression were calculated using the standard ΔΔCq method [90], the level of expression in cells transfected with the plasmid containing the wild type allele was used as reference.

Statistical analysis

The correlation between the percentage of identity of the ERE1 loci and the natural logarithm of the frequency of polymorphic loci in each class was tested calculating Pearson's product moment correlation coefficient.

The significance of the difference of the allelic frequencies at the myostatin promoter in the populations of Elite and Unselected Thoroughbreds was tested using a Chi-Square test goodness of fit. The allelic frequencies in the 75 Unselected Thoroughbreds were adopted as expected values.

The significance of the correlation between the Best Race Distance and the genotype of the 117 Elite Thoroughbreds for the ERE1 insertion at the myostatin promoter was tested using a Chi-Square test for independence.

All statistical analyses were performed using R [91].

Availability of supporting data

The data sets supporting the results of this article are included within the article and its additional files.

Additional files

Additional file 1: Table S1.A lists the 45,713 loci identified using the ERE1 consensus sequence deposited at the RepBase database as query for a BLAST search against the horse genome reference sequence. **Table S1B** reports the 34,131 ERE1 loci with sizes similar to the ERE1 consensus (225 ± 10 bp) and with minimum identity to the consensus of 84 %. **Table S1.C** lists the 27,396 ERE1 loci that are not located inside other repetitive elements. (XLSX 3293 kb)

Additional file 2: Table S2. lists the ERE1 polymorphic loci identified in the horse reference genome sequence. **Table S3.** reports the frequency of ERE1 polymorphic loci in eight classes of ERE1 elements grouped according to consensus identity. The values reported in this table were used to draw Fig. 1. **Table S4A** lists the genomic position of the 80 ERE1 loci analysed in Fig. 2 and the sequence of the primers used for each locus. **Table S4B** lists the primers used to clone the myostatin promoter region and those used to perform quantitative RT-PCR experiments for reporter gene assay. (PDF 184 kb)

Abbreviations

BLAST: Basic Local Alignment Search Tool; BLAT: BLAST-Like Alignment Tool; DAPI: 4′,6-diamidino-2-phenylindole; EAS: Equus asinus; EBU: Equus burchellii; eGFP: Enhanced Green Fluorescent Protein; EGR: Equus grevyi; EZH: Equus zebra hartmannae; EHO: Equus hemionus onager; EKI: Equus kiang; ERE: Equine Repetitive Element; GAPDH: Glyceraldehyde 3-phosphate dehydrogenase; LINE: Long INterspersed Element; Mya: Million years ago; NTC: No Template Control; NUMT: NUclear sequences of MITochondrial origin; PCR: Polymerase Chain Reaction; PRKCI: Protein kinase C iota; RT-PCR: Real Time Polymerase Chain Reaction; SINE: Short INterspersed Element; SNP: Single Nucleotide Polymorphism..

Competing interests

The authors declare that there are no competing interests.

Authors' contributions

MS carried out sequence analysis and alignment, performed some PCR experiments, contributed to data analysis, manuscript drafting and figure preparation. LK carried out gene reporter assays, including immunoassays and quantitative RT-PCR, and participated to figure preparation. RG carried out some PCR experiments, participated to gene reporter assays, performed the statistical analysis and participated to figure preparation. MB and AR carried out PCR and SNP analysis in racing Thoroughbreds and in Italian trotters. OK contributed to bioinformatic analysis. FMP and CB established fibroblast cell lines from horses and donkeys. FV and AM carried out some PCR experiments and sequence analysis. ER contributed to conception and design and to manuscript preparation. SGN carried out and supervised molecular and bioinformatics studies, contributed to the design of the study and to data analysis and interpretation. EG conceived the study, designed and coordinated the experiments, prepared the manuscript. All authors have read and approved the manuscript.

Acknowledgements

We thank Antonella Lisa (Istituto di Genetica Molecolare, CNR, Pavia) for helpful suggestions on the statistical analysis and Oliver Ryder (San Diego Zoo, Institute for Conservation Research, San Diego, CA, USA) for providing cell lines from *E. zebra hartmannae, E. kiang* and *E. hemionus onager.* We are

grateful to Cecilia M. Penedo for providing DNA samples from horse breeds (Veterinary Genetics Laboratory, University of California, Davis, Old Davis Road, Davis, 95616 California, USA), Cesare Rognoni, DVM, (Equicenter Monteleone, Inverno e Monteleone, Pavia, Italy) and Simone Vignati, DVM, for their precious collaboration.
This work was supported by grants from the Ministero dell'Istruzione dell'Università e della Ricerca (MIUR-PRIN) and from the Consiglio Nazionale delle Ricerche (CNR-Progetto Bandiera Epigenomica).

Author details
[1]Dipartimento di Biologia e Biotecnologie "Lazzaro Spallanzani", Università di Pavia, Via Ferrata 1, 27100 Pavia, Italy. [2]Laboratorio di Genetica Forense Veterinaria, UNIRELAB srl, Via A. Gramsci 70, 20019 Settimo Milanese (MI), Italy.

References
1. Smit AF. Interspersed repeats and other mementos of transposable elements in mammalian genomes. Curr Opin Genet Dev. 1999;9:657–63.
2. Lander ES, Linton LM, Birren B, Nusbaum C, Zody MC, Baldwin J, et al. Initial sequencing and analysis of the human genome. Nature. 2001;409:860–921.
3. Deininger PL, Moran JV, Batzer MA, Kazazian HH. Mobile elements and mammalian genome evolution. Curr Opin Genet Dev. 2003;13:651–8.
4. Kramerov DA, Vassetzky NS. Short retroposons in eukaryotic genomes. Int Rev Cytol. 2005;247:165–221.
5. Jurka J, Kapitonov VV, Kohany O, Jurka MV. Repetitive sequences in complex genomes: structure and evolution. Annu Rev Genomics Hum Genet. 2007;8:241–59.
6. Luchetti A, Mantovani B. Conserved domains and SINE diversity during animal evolution. Genomics. 2013;102:296–300.
7. Schmid CW, Jelinek WR. The Alu family of dispersed repetitive sequences. Science. 1982;216:1065–70.
8. Mighell AJ, Markham AF, Robinson PA. Alu sequences. FEBS Lett. 1997;417:1–5.
9. Batzer MA, Deininger PL. Alu repeats and human genomic diversity. Nat Rev Genet. 2002;3:370–9.
10. Roy-Engel AM, Carroll ML, El-Sawy M, Salem A-H, Garber RK, Nguyen SV, et al. Non-traditional Alu evolution and primate genomic diversity. J Mol Biol. 2002;316:1033–40.
11. Salem A-H, Kilroy GE, Watkins WS, Jorde LB, Batzer MA. Recently integrated Alu elements and human genomic diversity. Mol Biol Evol. 2003;20:1349–61.
12. Wang J, Song L, Gonder MK, Azrak S, Ray DA, Batzer MA, et al. Whole genome computational comparative genomics: A fruitful approach for ascertaining Alu insertion polymorphisms. Gene. 2006;365:11–20.
13. Carroll ML, Roy-Engel AM, Nguyen SV, Salem AH, Vogel E, Vincent B, et al. Large-scale analysis of the Alu Ya5 and Yb8 subfamilies and their contribution to human genomic diversity. J Mol Biol. 2001;311:17–40.
14. Bejerano G, Pheasant M, Makunin I, Stephen S, Kent WJ, Mattick JS, et al. Ultraconserved elements in the human genome. Science. 2004;304:1321–5.
15. Santangelo AM, de Souza FSJ, Franchini LF, Bumaschny VF, Low MJ, Rubinstein M. Ancient exaptation of a CORE-SINE retroposon into a highly conserved mammalian neuronal enhancer of the proopiomelanocortin gene. PLoS Genet. 2007;3:1813–26.
16. Cordaux R, Batzer MA. The impact of retrotransposons on human genome evolution. Nat Rev Genet. 2009;10:691–703.
17. Okada N, Sasaki T, Shimogori T, Nishihara H. Emergence of mammals by emergency: exaptation. Genes Cells. 2010;15:801–12.
18. Speek M. Antisense promoter of human L1 retrotransposon drives transcription of adjacent cellular genes. Mol Cell Biol. 2001;21:1973–85.
19. Wheelan SJ, Aizawa Y, Han JS, Boeke JD. Gene-breaking: a new paradigm for human retrotransposon-mediated gene evolution. Genome Res. 2005;15:1073–8.
20. Mätlik K, Redik K, Speek M. L1 antisense promoter drives tissue-specific transcription of human genes. J Biomed Biotechnol. 2006;2006:71753.
21. Druker R, Whitelaw E. Retrotransposon-derived elements in the mammalian genome: a potential source of disease. J Inherit Metab Dis. 2004;27:319–30.
22. Bejerano G, Lowe CB, Ahituv N, King B, Siepel A, Salama SR, et al. A distal enhancer and an ultraconserved exon are derived from a novel retroposon. Nature. 2006;441:87–90.
23. Lee E, Iskow R, Yang L, Gokcumen O, Haseley P, Luquette LJ, et al. Landscape of somatic retrotransposition in human cancers. Science. 2012;337:967–71.
24. Wade CM, Giulotto E, Sigurdsson S, Zoli M, Gnerre S, Imsland F, et al. Genome sequence, comparative analysis, and population genetics of the domestic horse. Science. 2009;326:865–7.
25. Sakagami M, Ohshima K, Mukoyama H, Yasue H, Okada N. A novel tRNA species as an origin of short interspersed repetitive elements (SINEs). Equine SINEs may have originated from tRNA(Ser). J Mol Biol. 1994;239:731–5.
26. Gallagher PC, Lear TL, Coogle LD, Bailey E. Two SINE families associated with equine microsatellite loci. Mamm Genome. 1999;10:140–4.
27. RepBase. http://www.girinst.org/repbase/ .
28. Jurka J, Kapitonov VV, Pavlicek A, Klonowski P, Kohany O, Walichiewicz J. Repbase Update, a database of eukaryotic repetitive elements. Cytogenet Genome Res. 2005;110:462–7.
29. Hill EW, McGivney BA, Gu J, Whiston R, MacHugh DE. A genome-wide SNP-association study confirms a sequence variant (g.66493737C > T) in the equine myostatin (MSTN) gene as the most powerful predictor of optimum racing distance for Thoroughbred racehorses. BMC Genomics. 2010;11:552.
30. Szabó G, Dallmann G, Müller G, Patthy L, Soller M, Varga L. A deletion in the myostatin gene causes the compact (Cmpt) hypermuscular mutation in mice. Mamm Genome. 1998;9:671–2.
31. Grobet L, Martin LJ, Poncelet D, Pirottin D, Brouwers B, Riquet J, et al. A deletion in the bovine myostatin gene causes the double-muscled phenotype in cattle. Nat Genet. 1997;17:71–4.
32. McPherron AC, Lee SJ. Double muscling in cattle due to mutations in the myostatin gene. Proc Natl Acad Sci U S A. 1997;94:12457–61.
33. Clop A, Marcq F, Takeda H, Pirottin D, Tordoir X, Bibé B, et al. A mutation creating a potential illegitimate microRNA target site in the myostatin gene affects muscularity in sheep. Nat Genet. 2006;38:813–8.
34. Schuelke M, Wagner KR, Stolz LE, Hübner C, Riebel T, Kömen W, et al. Myostatin mutation associated with gross muscle hypertrophy in a child. N Engl J Med. 2004;350:2682–8.
35. Mosher DS, Quignon P, Bustamante CD, Sutter NB, Mellersh CS, Parker HG, et al. A mutation in the myostatin gene increases muscle mass and enhances racing performance in heterozygote dogs. PLoS Genet. 2007;3, e79.
36. Dall'Olio S, Fontanesi L, Nanni Costa L, Tassinari M, Minieri L, Falaschini A. Analysis of horse myostatin gene and identification of single nucleotide polymorphisms in breeds of different morphological types. J Biomed Biotechnol. 2010;2010:542945.
37. Hill EW, Gu J, Eivers SS, Fonseca RG, McGivney BA, Govindarajan P, et al. A Sequence Polymorphism in MSTN Predicts Sprinting Ability and Racing Stamina in Thoroughbred Horses. PLoS ONE. 2010;5, e8645.
38. Tozaki T, Miyake T, Kakoi H, Gawahara H, Sugita S, Hasegawa T, et al. A genome-wide association study for racing performances in Thoroughbreds clarifies a candidate region near the MSTN gene. Anim Genet. 2010;41 Suppl 2:28–35.
39. Baron EE, Lopes MS, Mendonça D, da Câmara MA. SNP identification and polymorphism analysis in exon 2 of the horse myostatin gene. Anim Genet. 2012;43:229–32.
40. Petersen JL, Mickelson JR, Rendahl AK, Valberg SJ, Andersson LS, Axelsson J, et al. Genome-wide analysis reveals selection for important traits in domestic horse breeds. PLoS Genet. 2013;9, e1003211.
41. Li R, Liu D-H, Cao C-N, Wang S-Q, Dang R-H, Lan X-Y, et al. Single nucleotide polymorphisms of myostatin gene in Chinese domestic horses. Gene. 2014;538:150–4.
42. McGivney BA, Browne JA, Fonseca RG, Katz LM, Machugh DE, Whiston R, et al. MSTN genotypes in Thoroughbred horses influence skeletal muscle gene expression and racetrack performance. Anim Genet. 2012;43:810–2.
43. Tozaki T, Sato F, Hill EW, Miyake T, Endo Y, Kakoi H, et al. Sequence variants at the myostatin gene locus influence the body composition of Thoroughbred horses. J Vet Med Sci. 2011;73:1617–24.
44. Binns MM, Boehler DA, Lambert DH. Identification of the myostatin locus (MSTN) as having a major effect on optimum racing distance in the Thoroughbred horse in the USA. Anim Genet. 2010;41 Suppl 2:154–8.
45. Tozaki T, Hill EW, Hirota K, Kakoi H, Gawahara H, Miyake T, et al. A cohort study of racing performance in Japanese Thoroughbred racehorses using genome information on ECA18. Anim Genet. 2012;43:42–52.

46. Petersen JL, Valberg SJ, Mickelson JR, McCue ME. Haplotype diversity in the equine myostatin gene with focus on variants associated with race distance propensity and muscle fiber type proportions. Anim Genet. 2014;45:827–35.

47. Trifonov VA, Stanyon R, Nesterenko AI, Fu B, Perelman PL, O'Brien PCM, et al. Multidirectional cross-species painting illuminates the history of karyotypic evolution in Perissodactyla. Chromosome Res. 2008;16:89–107.

48. Piras FM, Nergadze SG, Poletto V, Cerutti F, Ryder OA, Leeb T, et al. Phylogeny of horse chromosome 5q in the genus Equus and centromere repositioning. Cytogenet Genome Res. 2009;126:165–72.

49. Nergadze SG, Lupotto M, Pellanda P, Santagostino M, Vitelli V, Giulotto E. Mitochondrial DNA insertions in the nuclear horse genome. Anim Genet. 2010;41 Suppl 2:176–85.

50. Piras FM, Nergadze SG, Magnani E, Bertoni L, Attolini C, Khoriauli L, et al. Uncoupling of satellite DNA and centromeric function in the genus equus. PLoS Genet. 2010;6, e1000845.

51. BLAT. http://genome.ucsc.edu/cgi-bin/hgBlat .

52. Kent WJ. BLAT–the BLAST-like alignment tool. Genome Res. 2002;12:656–64.

53. Jurka J. ERE1. http://www.girinst.org/protected/repbase_extract.php?access=ERE1 .

54. BLAST Trace database. https://blast.ncbi.nlm.nih.gov/Blast.cgi?PROGRAM=blastn&BLAST_SPEC=TraceArchive&PAGE_TYPE=BlastSearch&PROG_DEFAULTS=on .

55. Nergadze SG, Rocchi M, Azzalin CM, Mondello C, Giulotto E. Insertion of telomeric repeats at intrachromosomal break sites during primate evolution. Genome Res. 2004;14:1704–10.

56. Nergadze SG, Santagostino MA, Salzano A, Mondello C, Giulotto E. Contribution of telomerase RNA retrotranscription to DNA double-strand break repair during mammalian genome evolution. Genome Biol. 2007;8:R260.

57. Giordano J, Ge Y, Gelfand Y, Abrusán G, Benson G, Warburton PE. Evolutionary history of mammalian transposons determined by genome-wide defragmentation. PLoS Comput Biol. 2007;3, e137.

58. Steiner CC, Ryder OA. Molecular phylogeny and evolution of the Perissodactyla. Zool J Linn Soc. 2011;163:1289–303.

59. Trifonov VA, Musilova P, Kulemsina AI. Chromosome evolution in Perissodactyla. Cytogenet Genome Res. 2012;137:208–17.

60. Wakefield S, Knowles J, Zimmermann W, van Dierendonck M. Chapter 7: status and action plan for the Przewalski's horse (equus ferus przewalskii). In: Moehlman PD, editor. Equids: zebras, asses and horses: status survey and conservation action plan. Gland: IUCN; 2002. p. 82–92.

61. Table Browser. https://genome.ucsc.edu/cgi-bin/hgTables .

62. Karolchik D, Hinrichs AS, Furey TS, Roskin KM, Sugnet CW, Haussler D, et al. The UCSC Table Browser data retrieval tool. Nucleic Acids Res. 2004;32:D493–6.

63. Sakharkar MK, Chow VTK, Kangueane P. Distributions of exons and introns in the human genome. In Silico Biol. 2004;4:387–93.

64. Fedorova L, Fedorov A. Puzzles of the human genome: Why Do We need Our introns? Curr Genomics. 2005;6:589–95.

65. Gregory TR. Synergy between sequence and size in large-scale genomics. Nat Rev Genet. 2005;6:699–708.

66. Patrushev LI, Minkevich IG. The problem of the eukaryotic genome size. Biochem Mosc. 2008;73:1519–52.

67. Shepard S, McCreary M, Fedorov A. The peculiarities of large intron splicing in animals. PLoS ONE. 2009;4, e7853.

68. Krull M, Brosius J, Schmitz J. Alu-SINE exonization: en route to protein-coding function. Mol Biol Evol. 2005;22:1702–11.

69. Keren H, Lev-Maor G, Ast G. Alternative splicing and evolution: diversification, exon definition and function. Nat Rev Genet. 2010;11:345–55.

70. Ponicsan SL, Kugel JF, Goodrich JA. Genomic gems: SINE RNAs regulate mRNA production. Curr Opin Genet Dev. 2010;20:149–55.

71. Apone S, Hauschka SD. Muscle gene E-box control elements. Evidence for quantitatively different transcriptional activities and the binding of distinct regulatory factors. J Biol Chem. 1995;270:21420–7.

72. Spiller MP, Kambadur R, Jeanplong F, Thomas M, Martyn JK, Bass JJ, et al. The myostatin gene is a downstream target gene of basic helix-loop-helix transcription factor MyoD. Mol Cell Biol. 2002;22:7066–82.

73. Jurka J. Sequence patterns indicate an enzymatic involvement in integration of mammalian retroposons. Proc Natl Acad Sci U S A. 1997;94:1872–7.

74. Szak S, Pickeral O, Makalowski W, Boguski M, Landsman D, Boeke J. Molecular archeology of L1 insertions in the human genome. Genome Biol. 2002;3:research0052.

75. Vidale P, Magnani E, Nergadze SG, Santagostino M, Cristofari G, Smirnova A, et al. The catalytic and the RNA subunits of human telomerase are required to immortalize equid primary fibroblasts. Chromosoma. 2012;121:475–88.

76. Smit AF. ERE2. http://www.girinst.org/protected/repbase_extract.php?access=ERE2 .

77. Jurka J. ERE3. http://www.girinst.org/protected/repbase_extract.php?access=ERE3 .

78. Wade CM. ERE4. http://www.girinst.org/protected/repbase_extract.php?access=ERE4.

79. Equus caballus (horse) Nucleotide BLAST. http://blast.ncbi.nlm.nih.gov/Blast.cgi?PAGE_TYPE=BlastSearch&PROG_DEF=blastn&BLAST_PROG_DEF=megaBlast&BLAST_SPEC=OGP__9796__11760 .

80. seq_contig.md. ftp://ftp.ncbi.nih.gov/genomes/Equus_caballus/mapview/seq_contig.md.gz.

81. Smit AFA, Hubley R, Green P. RepeatMasker. http://www.repeatmasker.org/.

82. UCSC Genome Browser ftp. ftp://hgdownload.cse.ucsc.edu/goldenPath/equCab2/chromosomes/.

83. Karolchik D, Barber GP, Casper J, Clawson H, Cline MS, Diekhans M, et al. The UCSC Genome Browser database: 2014 update. Nucleic Acids Res. 2014;42:D764–70.

84. MultAlin. http://multalin.toulouse.inra.fr/multalin/.

85. Corpet F. Multiple sequence alignment with hierarchical clustering. Nucleic Acids Res. 1988;16:10881–90.

86. Purgato S, Belloni E, Piras FM, Zoli M, Badiale C, Cerutti F, et al. Centromere sliding on a mammalian chromosome. Chromosoma. 2015;124:277–87.

87. Carbone L, Nergadze SG, Magnani E, Misceo D, Francesca Cardone M, Roberto R, et al. Evolutionary movement of centromeres in horse, donkey, and zebra. Genomics. 2006;87:777–82.

88. Anglana M, Bertoni L, Giulotto E. Cloning of a polymorphic sequence from the nontranscribed spacer of horse rDNA. Mamm Genome Off J Int Mamm Genome Soc. 1996;7:539–41.

89. Nergadze SG, Farnung BO, Wischnewski H, Khoriauli L, Vitelli V, Chawla R, et al. CpG-island promoters drive transcription of human telomeres. RNA. 2009;15:2186–94.

90. Livak KJ, Schmittgen TD. Analysis of relative gene expression data using real-time quantitative PCR and the 2(–Delta Delta C(T)) Method. Methods. 2001;25:402–8.

91. R Development Core Team. R. A language and environment for statistical computing. Vienna: R Foundation for Statistical Computing; 2008. http://www.R-project.org.

12

Genome-wide association and genomic prediction of breeding values for fatty acid composition in subcutaneous adipose and *longissimus lumborum* muscle of beef cattle

Liuhong Chen[1,2], Chinyere Ekine-Dzivenu[1], Michael Vinsky[2], John Basarab[3], Jennifer Aalhus[2], Mike E. R. Dugan[2], Carolyn Fitzsimmons[1,2], Paul Stothard[1] and Changxi Li[1,2*]

Abstract

Background: Identification of genetic variants that are associated with fatty acid composition in beef will enhance our understanding of host genetic influence on the trait and also allow for more effective improvement of beef fatty acid profiles through genomic selection and marker-assisted diet management. In this study, 81 and 83 fatty acid traits were measured in subcutaneous adipose (SQ) and *longissimus lumborum* muscle (LL), respectively, from 1366 purebred and crossbred beef steers and heifers that were genotyped on the Illumina BovineSNP50 Beadchip. The objective was to conduct genome-wide association studies (GWAS) for the fatty acid traits and to evaluate the accuracy of genomic prediction for fatty acid composition using genomic best linear unbiased prediction (GBLUP) and Bayesian methods.

Results: In total, 302 and 360 significant SNPs spanning all autosomal chromosomes were identified to be associated with fatty acid composition in SQ and LL tissues, respectively. Proportions of total genetic variance explained by individual significant SNPs ranged from 0.03 to 11.06 % in SQ, and from 0.005 to 24.28 % in the LL muscle. Markers with relatively large effects were located near fatty acid synthase (*FASN*), stearoyl-CoA desaturase (*SCD*), and thyroid hormone responsive (*THRSP*) genes. For the majority of the fatty acid traits studied, the accuracy of genomic prediction was relatively low (<0.40). Relatively high accuracies (> = 0.50) were achieved for 10:0, 12:0, 14:0, 15:0, 16:0, 9c-14:1, 12c-16:1, 13c-18:1, and health index (HI) in LL, and for 12:0, 14:0, 15:0, 10 t,12c-18:2, and 11 t,13c + 11c,13 t-18:2 in SQ. The Bayesian method performed similarly as GBLUP for most of the traits but substantially better for traits that were affected by SNPs of large effects as identified by GWAS.

Conclusions: Fatty acid composition in beef is influenced by a few host genes with major effects and many genes of smaller effects. With the current training population size and marker density, genomic prediction has the potential to predict the breeding values of fatty acid composition in beef cattle at a moderate to relatively high accuracy for fatty acids that have moderate to high heritability.

Keywords: Fatty acid composition, Beef cattle, Genome-wide association study, Genomic prediction, Single nucleotide polymorphism

* Correspondence: changxi.li@agr.gc.ca
[1]Department of Agricultural, Food and Nutritional Science, University of Alberta, Edmonton, AB T6G 2P5, Canada
[2]Lacombe Research Centre, Agriculture and Agri-Food Canada, 6000 C&E Trail, Lacombe, AB T4L 1 W1, Canada
Full list of author information is available at the end of the article

Background

Dietary fats influence risks for developing cardiovascular disease, obesity and various forms of cancer, and have led to recommendations to limit consumption of some foods including beef [1]. Recommendations to limit beef consumption are mainly related to its relatively high content of saturated fatty acids (SFAs) as SFA consumption is believed to have negative effects on human health [2, 3]. Beef, however, is also a natural source of polyunsaturated fatty acid (PUFA) biohydrogenation intermediates (BHI) including vaccenic acid (11 t-18:1) and conjugated linoleic acids (CLAs), which have a number of purported health benefits [4–6]. In addition, beef is rich in monounsaturated fatty acids (MUFAs), in particular oleic acid 9c-18:1, which is the main fatty acid found in healthy Mediterranean diets, and may also contribute positively to beef flavour and tenderness [7]. Considerable efforts have, therefore, gone into improving beef fatty acid profiles in beef to meet the consumers' growing demand for more nutritious, healthier and more palatable meat. Diet is known to have a major influence on beef fatty acid composition [8], but the use of genomic technologies to improve beef fatty acid profiles have not been thoroughly investigated [9].

The fatty acid composition of beef is a complex trait with heritability estimates ranging from near 0 to 0.73, depending on populations and the types of fatty acid [10–18]. To further elucidate the genetic control of host animals on fatty acid composition, chromosomal regions or quantitative trait loci (QTL) and candidate genes that are associated with fatty acid composition in beef cattle have been identified on multiple chromosomes based on low density DNA markers [19–21], and based on candidate gene DNA marker association analyses [22–36]. Genome-wide association studies (GWAS) using a relatively high density of single nucleotide polymorphism (SNP) markers (e.g. Illumina BovineSNP50 Beadchip) in recent years have assisted in the search for DNA markers associated with the fatty acid composition of beef, but studies are limited to a small number of fatty acids in certain beef breeds [7, 37, 38]. Saatchi *et al.* [16] analyzed 49 fatty acid traits in steaks of Angus beef cattle and reported results of GWAS and genomic prediction of direct genomic breeding values of the fatty acid traits. Onogi et al. [39] also reported genomic prediction for 8 fatty acid traits in Japanese Black cattle. Many fatty acids with potential health value (i.e. PUFA-BHI) were, however, not reported in those studies. In this study, we comprehensively analyzed fatty acid profiles and report GWAS and genomic prediction of breeding values for 81 and 83 individual and grouped fatty acids in subcutaneous adipose (SQ) and *longissimus lumborum* muscle (LL), respectively, in Canadian beef cattle populations.

Results and discussion

Descriptive statistics and genomic heritability estimates

Summary statistics and genomic heritability estimates for the 81 fatty acid traits of SQ and 83 fatty acid traits of LL are presented in Table 1. In general, the estimates of heritability for the same fatty acids are comparable in both the adipose and muscle tissues, with a correlation coefficient of 0.61. Relatively higher (>0.40) heritability estimates were found for 10:0, 12:0, 18:0, ai15:0, 9c-14:1, 9c-16:1, 13c-18:1, 18:3n-3, 18:2n-6, n-3, n-6, sumtrans 18:1, total PUFA, P/S, and P/(S + B) in the SQ tissue, and for 12:0, 14:0, 16:0, 9c-14:1, 9c-16:1, 12c-16:1, 9c-18:1, 13c-18:1, SFA, SFA + BFA, MUFA, n-6/n-3, and health index (HI) in the LL muscle, which suggests greater direct host genetic effects on these traits in the corresponding tissues. Very low (<0.05) or zero heritability were observed for 22:0, 7c-17:1, 12 t-18:1, 15 t-18:1, 6 t,8 t:18:2, 7 t,9 t-18:2, 9 t,11 t-18:2, 10 t,12 t-18:2, 12 t,14 t-18:2, and n-6/n-3 in the SQ tissue, and for 7c-17:1, 15 t-18:1, 6 t,8 t-18:2, 7 t,9 t-18:2, 7 t,9c-18:2, 8 t,10 t-18:2, 12 t,14 t-18:2 in the LL muscle, which indicates weak host direct genetic control on these traits. In general, the heritability estimates for the fatty acids in this study are in line with those reported in other studies [10, 12, 15]. Therefore, the genomic estimated additive genetic variance and heritability for the fatty acid traits were further used in the calculation of total genetic variance explained by significant markers identified in GWAS and in the derivation of realised accuracy of genomic prediction in this study.

Genome-wide association study

In total, 302 and 360 significant SNPs spanning all autosomal chromosomes were identified to be associated with one or more fatty acid traits in the SQ and LL tissues, respectively, at the genome-wise empirical significance threshold at α = 0.05. Significant SNPs and their distributions over the genome varied for different fatty acid traits. Manhattan plots of posterior probability of inclusion (PPI) were provided in Additional file 1 for all fatty acid traits in the two tissues. Proportions of genotypic variance explained by individual significant SNPs ranged from 0.03 to 11.06 % in SQ, and from 0.005 to 24.28 % in LL. Among these, 28 and 41 SNPs individually explained greater than 1 % of total genetic variance for at least one fatty acid trait in the SQ and LL tissues, respectively. Figures 1 and 2 showed these SNPs and their associated traits in SQ and LL, respectively. Of these SNPs, SNP *rs41921177* at the location of BTA19:51326750 had the largest effects on multiple fatty acid traits in both tissues, followed by SNP *rs42714483* at BTA29:18090509 and SNP *rs42090719* at BTA26:20903573. Details of all significant SNPs including SNP name, chromosome position, allele substitution effect, percentage of total genetic variance explained, and PPI were also provided in

Table 1 Summary statistics of mean, standard deviation (SD), additive genetic variance (σ_a^2), and heritability estimates ($h^2 \pm SE$)

Trait[a]	Subcutaneous adipose			Longissimus lumborum		
	Mean (SD)	$\sigma_a^2 \times 10^4$	$h^2 \pm SE$	Mean (SD)	$\sigma_a^2 \times 10^4$	$h^2 \pm SE$
10:0	0.051 (0.014)	0.66	0.43 ± 0.06	0.056 (0.012)	0.60	0.37 ± 0.09
12:0	0.071 (0.017)	1.60	0.53 ± 0.08	0.072 (0.015)	1.30	0.57 ± 0.07
13:0	0.028 (0.010)	0.28	0.26 ± 0.07	0.027 (0.009)	0.14	0.16 ± 0.06
14:0	3.204 (0.599)	1605.35	0.35 ± 0.11	2.804 (0.486)	1292.53	0.53 ± 0.10
15:0	0.642 (0.167)	90.38	0.22 ± 0.10	0.502 (0.111)	38.96	0.23 ± 0.10
16:0	25.092 (2.594)	12965.10	0.21 ± 0.06	24.607 (2.056)	11278.40	0.42 ± 0.08
17:0	1.709 (0.445)	575.54	0.32 ± 0.11	1.548 (0.329)	326.58	0.31 ± 0.12
18:0	10.545 (1.956)	14701.00	0.41 ± 0.07	12.406 (1.417)	7779.92	0.38 ± 0.09
19:0	0.108 (0.032)	0.90	0.06 ± 0.03	0.090 (0.029)	0.82	0.08 ± 0.04
20:0	0.082 (0.019)	0.75	0.20 ± 0.06	0.089 (0.016)	0.55	0.18 ± 0.06
22:0	0.033 (0.009)	0.000036	0.00 ± 0.03	0.069 (0.021)	0.73	0.09 ± 0.05
24:0	0.035 (0.015)	0.60	0.22 ± 0.07	0.151 (0.070)	7.65	0.21 ± 0.07
SFA	41.598 (3.469)	36697.00	0.29 ± 0.07	42.421 (2.695)	25765.90	0.43 ± 0.08
iso14:0	0.031 (0.012)	0.14	0.12 ± 0.05	0.027 (0.008)	0.050	0.10 ± 0.05
iso15:0	0.109 (0.026)	1.11	0.22 ± 0.06	0.082 (0.015)	0.52	0.25 ± 0.07
ai15:0	0.180 (0.048)	11.26	0.52 ± 0.08	0.140 (0.028)	3.07	0.30 ± 0.10
iso16:0	0.177 (0.041)	5.97	0.34 ± 0.08	0.140 (0.027)	1.98	0.22 ± 0.08
iso17:0	0.382 (0.062)	12.74	0.31 ± 0.08	0.345 (0.062)	5.48	0.11 ± 0.05
ai17:0	0.672 (0.095)	22.50	0.22 ± 0.07	0.489 (0.076)	15.93	0.19 ± 0.08
iso18:0	0.163 (0.037)	3.25	0.23 ± 0.07	0.133 (0.028)	1.95	0.21 ± 0.08
BFA	1.714 (0.261)	225.65	0.29 ± 0.08	1.356 (0.203)	94.08	0.16 ± 0.07
SFA + BFA	43.312 (3.561)	38438.00	0.28 ± 0.07	43.777 (2.681)	25797.30	0.43 ± 0.08
9c-14:1	1.046 (0.390)	542.74	0.43 ± 0.08	0.640 (0.184)	185.50	0.59 ± 0.07
9c-15:1	0.034 (0.012)	0.38	0.23 ± 0.08	0.026 (0.009)	0.073	0.08 ± 0.04
7c-16:1	0.140 (0.025)	1.90	0.34 ± 0.07	0.136 (0.019)	1.20	0.30 ± 0.09
9c-16:1	4.247 (1.096)	4450.52	0.42 ± 0.07	3.408 (0.564)	1730.54	0.64 ± 0.07
11 t-16:1	0.047 (0.012)	0.21	0.12 ± 0.05	0.042 (0.012)	0.12	0.06 ± 0.04
12c-16:1	0.239 (0.090)	23.77	0.36 ± 0.07	0.168 (0.044)	9.70	0.53 ± 0.07
7c-17:1	0.023 (0.009)	0	0	0.025 (0.013)	0.040	0.02 ± 0.03
9c-17:1	1.377 (0.340)	267.08	0.19 ± 0.09	1.191 (0.298)	130.35	0.15 ± 0.07
9c-18:1	37.917 (4.343)	30916.80	0.13 ± 0.05	36.679 (2.997)	35344.80	0.47 ± 0.07
11c-18:1	1.960 (1.736)	1591.98	0.05 ± 0.04	1.836 (0.244)	152.64	0.30 ± 0.10
12c-18:1	0.260 (0.079)	6.88	0.09 ± 0.04	0.227 (0.071)	6.70	0.13 ± 0.06
13c-18:1	0.487 (0.159)	83.84	0.44 ± 0.07	0.396 (0.089)	40.17	0.57 ± 0.07
14c-18:1	0.053 (0.011)	0.17	0.12 ± 0.05	0.048 (0.009)	0.13	0.19 ± 0.06
15c-18:1	0.248 (0.060)	13.68	0.38 ± 0.07	0.204 (0.044)	5.62	0.28 ± 0.07
6 t + 8 t-18:1	0.275 (0.121)	38.18	0.34 ± 0.06	0.193 (0.085)	14.30	0.21 ± 0.07
9 t-18:1	0.291 (0.094)	18.74	0.27 ± 0.06	0.232 (0.067)	9.04	0.22 ± 0.07
10 t-18:1	2.908 (1.685)	10759.90	0.38 ± 0.11	2.028 (1.119)	4230.39	0.36 ± 0.10
11 t-18:1	0.546 (0.234)	119.36	0.17 ± 0.07	0.441 (0.164)	70.08	0.26 ± 0.08
12 t-18:1	0.184 (0.172)	0.00012	0	0.137 (0.029)	1.06	0.13 ± 0.05
15 t-18:1	0.169 (0.179)	1.45	0.00 ± 0.03	0.130 (0.079)	2.25	0.04 ± 0.05
16 t-18:1	0.113 (0.037)	2.51	0.11 ± 0.05	0.092 (0.026)	1.78	0.19 ± 0.08
sumtrans18:1	4.486 (1.687)	10976.60	0.42 ± 0.09	3.252 (1.131)	4357.97	0.37 ± 0.09

Table 1 Summary statistics of mean, standard deviation (SD), additive genetic variance (σ_a^2), and heritability estimates (h^2 ± SE) *(Continued)*

9c-20:1	0.107 (0.019)	0.82	0.22 ± 0.08	0.090 (0.013)	0.20	0.08 ± 0.04
11c-20:1	0.270 (0.075)	15.80	0.37 ± 0.07	0.198 (0.035)	5.20	0.38 ± 0.08
MUFA	52.941 (3.583)	37044.00	0.26 ± 0.06	48.565 (2.691)	26915.70	0.47 ± 0.06
9c,13 t + 8 t,12c-18:2	0.242 (0.045)	6.07	0.30 ± 0.09	0.165 (0.029)	1.94	0.24 ± 0.08
9c,15c-18:2	0.184 (0.044)	5.29	0.27 ± 0.07	0.178 (0.036)	3.28	0.25 ± 0.08
8 t,13c-18:2	0.165 (0.043)	4.38	0.19 ± 0.08	0.121 (0.025)	0.63	0.10 ± 0.05
11 t,15c-18:2	0.162 (0.100)	32.52	0.32 ± 0.08	0.122 (0.070)	17.25	0.37 ± 0.08
9c,11 t + 9 t,11c-18:2	0.471 (0.358)	33.83	0.25 ± 0.06	0.257 (0.062)	9.26	0.16 ± 0.06
6 t,8 t-18:2	0.0024 (0.003)	0	0	0.0019 (0.004)	0.0002	0.00 ± 0.03
7 t,9c-18:2	0.108 (0.083)	11.38	0.25 ± 0.09	0.060 (0.060)	1.22	0.03 ± 0.03
12 t,14 t-18:2	0.0088 (0.011)	0.041	0.02 ± 0.01	0.0061 (0.008)	0.023	0.03 ± 0.02
11 t,13 t-18:2	0.0083 (0.006)	0.017	0.10 ± 0.04	0.0060 (0.002)	0.0056	0.15 ± 0.05
10 t,12 t-18:2	0.010 (0.005)	0	0	0.0061 (0.002)	0.0049	0.12 ± 0.05
9 t,11 t-18:2	0.014 (0.010)	0	0	0.0092 (0.003)	0.0083	0.08 ± 0.04
8 t,10 t-18:2	0.0026 (0.002)	0.0020	0.08 ± 0.04	0.0017 (0.002)	0.00084	0.04 ± 0.04
7 t,9 t-18:2	0.0069 (0.005)	0.0020	0.02 ± 0.02	0.0041 (0.003)	0.0010	0.01 ± 0.02
12 t,14c + 12c,14 t −18:2	0.012 (0.010)	0.042	0.07 ± 0.03	0.0068 (0.003)	0.018	0.21 ± 0.07
11 t,13c + 11c,13 t −18:2	0.024 (0.024)	0.15	0.15 ± 0.05	0.012 (0.005)	0.031	0.14 ± 0.06
10 t,12c-18:2	0.025 (0.012)	0.32	0.24 ± 0.07	0.018 (0.010)	0.065	0.07 ± 0.04
8 t,10c-18:2	0.012 (0.010)	0.022	0.11 ± 0.05	0.0079 (0.003)	0.018	0.21 ± 0.07
Total CLA	0.704 (0.493)	77.85	0.30 ± 0.06	0.395 (0.080)	11.21	0.15 ± 0.05
18:2n-6	1.876 (0.587)	1032.59	0.53 ± 0.08	4.387 (1.612)	5072.90	0.39 ± 0.08
18:3n-3	0.211 (0.055)	11.32	0.43 ± 0.07	0.297 (0.082)	8.60	0.20 ± 0.06
18:3n-6	0.0023 (0.006)	0.027	0.06 ± 0.05	0.043 (0.016)	0.51	0.16 ± 0.07
20:2n-6	0.038 (0.013)	0.22	0.19 ± 0.05	0.068 (0.022)	0.41	0.11 ± 0.05
20:3n-6	0.059 (0.017)	0.32	0.08 ± 0.04	0.292 (0.098)	23.30	0.32 ± 0.08
20:3n-9	0.017 (0.016)	0.12	0.06 ± 0.04	0.066 (0.025)	1.28	0.20 ± 0.06
20:4n-6	0.040 (0.012)	0.22	0.15 ± 0.05	1.000 (0.412)	277.29	0.25 ± 0.07
20:5n-3	*ND*	*NA*	*NA*	0.029 (0.009)	0.049	0.05 ± 0.04
22:4n-6	0.031 (0.011)	0.19	0.13 ± 0.05	0.136 (0.045)	4.67	0.20 ± 0.08
22:5n-3	0.017 (0.010)	0.05	0.11 ± 0.05	0.332 (0.126)	28.06	0.20 ± 0.07
22:6n-3	*ND*	*NA*	*NA*	0.046 (0.023)	0.78	0.16 ± 0.05
PUFA	2.290 (0.627)	1252.63	0.51 ± 0.08	6.695 (2.231)	8788.56	0.31 ± 0.08
n-3	0.228 (0.054)	10.59	0.40 ± 0.07	0.704 (0.208)	58.08	0.17 ± 0.06
n-6	2.046 (0.602)	1058.05	0.51 ± 0.08	5.926 (2.107)	8016.81	0.34 ± 0.08
n-6/n-3	9.263 (5.078)	2605.22	0.01 ± 0.02	8.628 (2.526)	9313.42	0.42 ± 0.09
P/S	0.056 (0.016)	0.77	0.50 ± 0.07	0.160 (0.058)	6.60	0.30 ± 0.09
P/(S + B)	0.053 (0.015)	0.69	0.50 ± 0.08	0.155 (0.056)	6.03	0.30 ± 0.09
HI	1.488 (0.265)	236.45	0.32 ± 0.08	1.566 (0.232)	201.90	0.47 ± 0.08

[a]The concentrations of fatty acids were expressed as a percentage of fatty acid methyl esters (FAME) quantified. c = cis, t = trans. SFA = 10:0 + 12:0 + 13:0 + 14:0 + 15:0 + 16:0 + 17:0 + 18:0 + 19:0 + 20:0 + 22:0 + 24:0; BFA = iso14:0 + iso15:0 + ai15:0 + iso16:0 + iso17:0 + ai17:0 + iso18:0; SFA + BFA: sum of SFA and BFA; sumtrans18:1 = 6 t/8 t-18:1 + 9 t-18:1 + 10 t-18:1 + 11 t-18:1 + 12 t-18:1 + 15 t-18:1 + 16 t-18:1; MUFA = 9c-14:1 + 9c-15:1 + 7c-16:1 + 9c-16:1 + 11 t-16:1 + 12c-16:1 + 7c-17:1 + 9c-17:1 + 9c-18:1 + 11c-18:1 + 12c-18:1 + 13c-18:1 + 14c-18:1 + 15c-18:1 + 9c-20:1 + 11c-20 + 6 t/8 t-18:1 + 9 t-18:1 + 10 t-18:1 + 11 t-18:1 + 12 t-18:1 + 15 t-18:1 + 16 t-18:1; Total CLA = 9c,11 t + 9 t,11c-18:2 + 6 t,8 t-18:2 + 7 t,9c-18:2 + 12 t,(14 t-18:2 + 11 t,13 t)-18:2 + 10 t,12 t-18:2 + 9 t,11 t-18:2 + 8 t,10 t-18:2 + 7 t,9 t-18:2 + (12 t,14c + 12c,14 t) -18:2 + (11 t,13c + 11c,13 t)-18:2 + 10 t,12c-18:2 + 8 t,10c-18:2; PUFA = 18:2n-6 + 18:3n-6 + 18:3n-3 + 20:2n-6 + 20:3n-9 + 20:3n-6 + 20:4n-6 + 22:4n-6 + 22:5n-3+ 22:6n-3; n-3 = 18:3n-3 + 20:5n-3 + 22:5n-3 + 22:6n-3; n-6 = 18:2n-6 + 18:3n-6 + 20:2n-6 + 20:3n-6 + 20:4n-6 + 22:4n-6; n-6/n-3: ratio between n-6 and n-3; P/S = PUFA/SFA; P/(S + B) = PUFA/(SFA + BFA); HI = (MUFA + PUFA) / (4 × 14:0 + 16:0)
ND not detected, *NA* not applicable

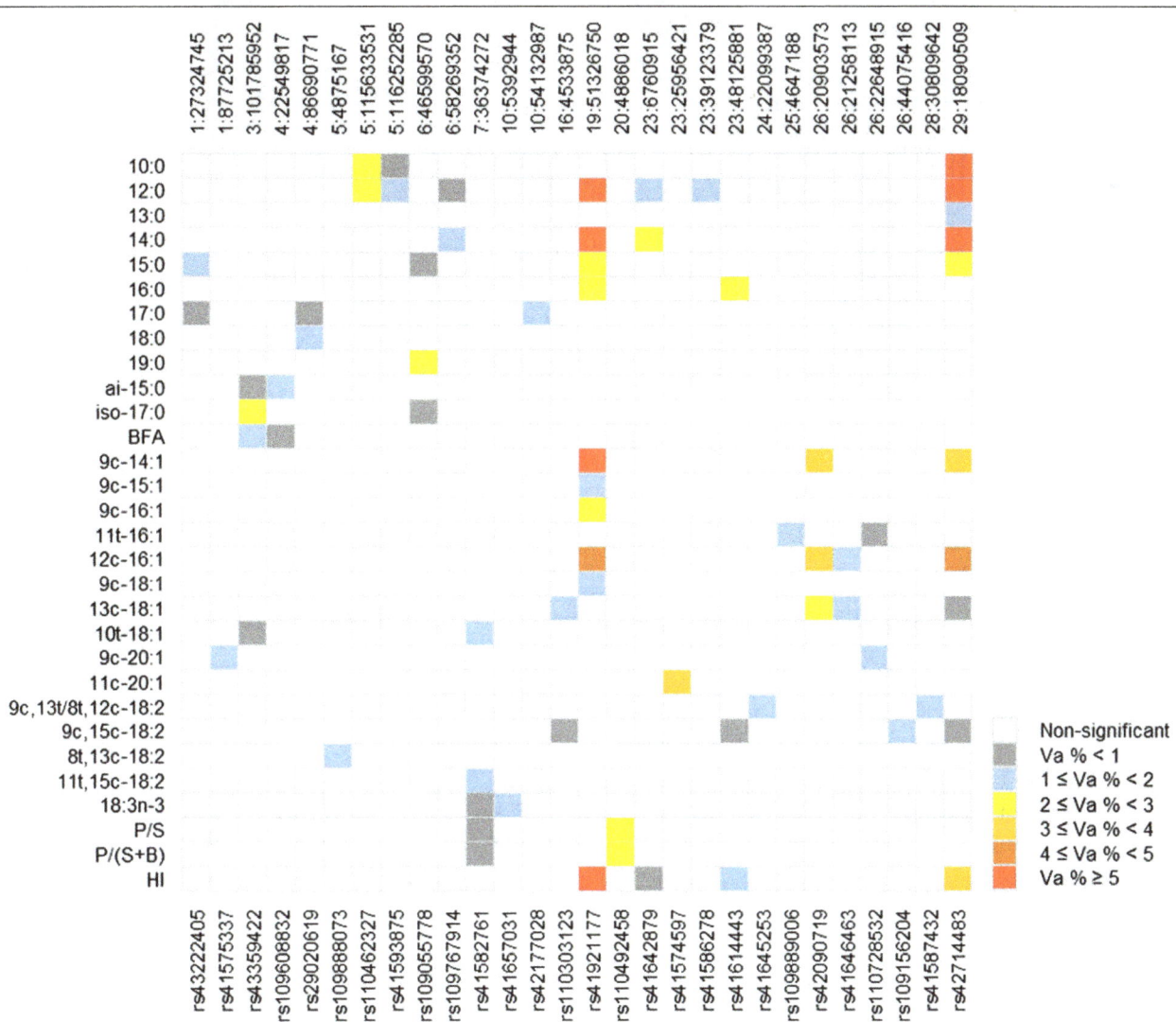

Fig. 1 Summary of fatty acid trait associations across genomic regions (SNPs) and percentage of genetic variance explained by significant SNPs in the subcutaneous adipose tissue (SQ). Each row represents a trait and each column represents a SNP. Only traits with at least one significant SNP explaining greater than 1 % of genetic variance were listed, and only SNPs that explain greater than 1 % of genetic variance for at least one trait were shown. The top of the figure shows the chromosome and position and the bottom shows the name of the SNP. Va %: percentage of genetic variance

additional files (Additional file 2 for SQ and Additional file 3 for LL). Candidate genes within 1 mega base pair (Mb) region centering the significant SNPs were provided separately (Additional file 4).

SNP *rs41921177* was significantly associated with 19 individual and grouped fatty acids in the LL muscle including SFAs 10:0, 12:0, 13:0, 14:0, 15:0, 16:0, 18:0, branched fatty acids (BFAs) ai 15:0 and iso 18:0, MUFAs 9c-14:1, 9c-15:1, 9c-16:1, 12c-16:1, 9c-18:1, 11c-20:1, grouped fatty acids total SFA, SFA + BFA, total MUFA and HI, with genetic variance explained from 1.37 % (18:0) to 24.28 % (14:0). The same SNP also showed significant associations with 11 of the above fatty acids in SQ including saturated fatty acids 12:0, 14:0, 15:0, 16:0, SFA, monounsaturated

fatty acids 9c-14:1, 9c-15:1, 9c-16:1, 9c-18:1, 12c-16:1 and HI, explaining 0.33 % (SFA) to 11:06 % (14:0) of the genetic variance (also see Additional file 2). This chromosomal region was previously identified to be associated with 14:0, 16:0, 16:1 and 18:1 in adipose and muscle tissues of a Jersey and Limousin crossbred beef cattle [21], with 9c-18:1, and 14:0, 14:1, 16:0, 16:1 in intramuscular fat of Japanese Black cattle [7, 37], with 14:0, 16:0, 9c-14:1 and 9c-18:1 in adipose tissue of an Australian multi-breed beef population [38], with 14:0, 14:1, 16:0, 16:1, 9c-18:1, MUFA, SFA, and Atherogenic index (AI, the inverse of HI) in muscle of American Angus beef cattle [16]. The association of this chromosomal region with the fatty acid traits was therefore confirmed

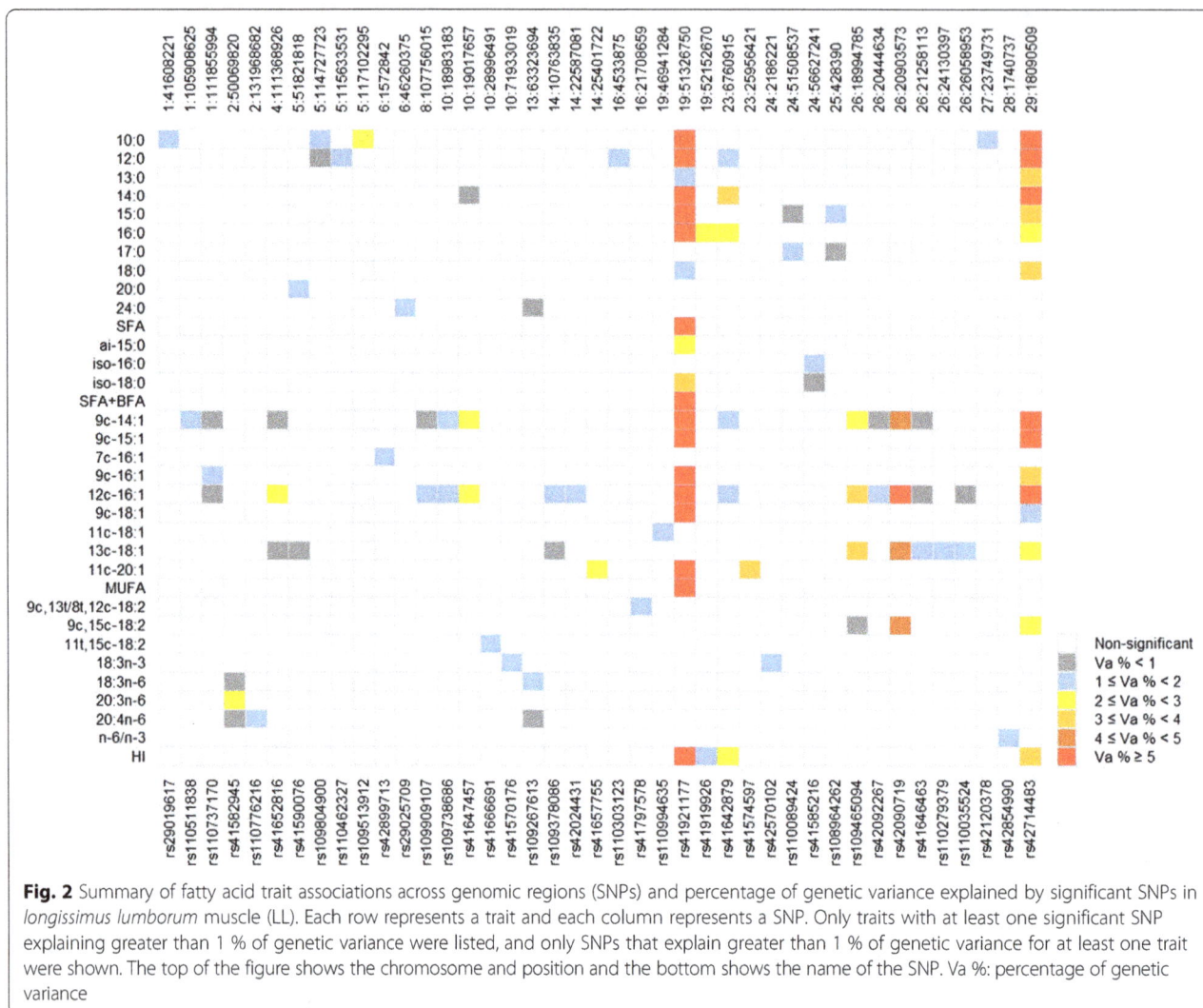

Fig. 2 Summary of fatty acid trait associations across genomic regions (SNPs) and percentage of genetic variance explained by significant SNPs in *longissimus lumborum* muscle (LL). Each row represents a trait and each column represents a SNP. Only traits with at least one significant SNP explaining greater than 1 % of genetic variance were listed, and only SNPs that explain greater than 1 % of genetic variance for at least one trait were shown. The top of the figure shows the chromosome and position and the bottom shows the name of the SNP. Va %: percentage of genetic variance

in both the SQ and LL tissues of a Canadian beef population of diverse breed compositions, indicating a strong host genetic effect on the fatty acid composition in beef tissues. Multiple genes are within 1 Mb region centering the SNP (see Additional file 4), with *FASN* being a strong candidate gene due to its function in fatty acid synthesis [40, 41]. Different SNPs of the *FASN* gene have also been reported to be associated with concentrations of saturated and monounsaturated fatty acids in various beef and dairy cattle populations [7, 16, 18, 22, 37, 38, 40–45].

SNP *rs42714483* showed significant associations with concentrations of 15 fatty acids in the LL tissue and 10 fatty acids in SQ including 10:0, 12:0, 13:0, 14:0, 15:0, 9c-14:1, 12c-16:1, 13c-18:1, 9c,15c-18:2, and HI in both the tissues, and 16:0, 18:0, 9c-15:1, 9c-16:1, and 9c-18:1 in the LL tissue. Saatchi *et al.* [16] also identified the same chromosomal region associated with fatty acids 14:0, 9c-14:1, 16:0, 16:1, 18:0, 9c-18:1, and AI, and Kelly *et al.* [38] found SNPs in the same chromosomal region that

were associated with fatty acids 14:0, 9c-14:1 in subcutaneous adipose tissue of an Australian multi-breed beef population [38]. These results strongly support that the chromosome region on BTA 29 harbors host genes that influence fatty acid composition of beef tissues. In this study, the SNP at BTA29:18090509 is a missense mutation (*T/C*) of the thyroid hormone responsive gene (*THRSP*), causing amino acid change from isoleucine to valine (I16V). Recently *THRSP* has been considered as a candidate gene for fatty acid composition in beef [16, 46]. Substitution of allele *T* with *C* of this missense mutation was associated with decrease of 10:0, 12:0, 13:0, 14:0, 15:0, 16:0, 9c-14:1, 9c-15:1, 9c-16:1, 12c-16:1, 13c-18:1, 9c,15c-18:2, and increase of 18:0, 9c-18:1, and HI (see Additional files 2 and 3). The direction of the allele substitution effect on different fatty acid traits also coincided with that of SNP *rs41921177*, which is close to *FASN* gene, suggesting possible co-ordinations between *THRSP* and *FASN* genes in fatty acid synthesis.

SNP rs42090719 at BTA26: 20903573 was found to be significantly associated with 9c-14:1, 12c-16:1, 13c-18:1 in both the LL and SQ tissues, 9c,15c-18:2 and CLA isomers 11 t,13c + 11c,13 t-18:2 (also see Additional file 3) in the LL tissue. In addition, SNP rs41646463 at BTA26:21258113 also showed significant associations with 13c-18:1 in both the LL and SQ tissues. In the nearby chromosomal region of BTA26:18994785, SNP rs109465094 was significantly associated with 9c-14:1, 12c-16:1, 13c-18:1 and 9c,15c-18:2 in LL. The chromosomal regions on BTA 26 were previously found associated with a variety of fatty acids in muscle of American Angus and in adipose tissue in an Australian multi-breed beef population, and stearoyl-CoA desaturase gene (SCD) was suggested as a candidate gene [16, 38]. The two SNPs, rs42090719 and rs41646463, are within 250 kilo base pairs (Kb) of SCD. The other SNP rs109465094 is more than 2 Mb distant from SCD, indicating a possible alternative candidate gene or its association could merely be due to LD with SCD. Linkage disequilibrium between SNPs around the SCD gene were analysed and visualised using the Haploview software [47] and results are shown in Additional file 5. Indeed, the three significant SNPs are in moderate to high LD with SNPs in a LD block containing the SCD gene. The SCD gene is involved in the synthesis of particular MUFA and CLA isomers, in creating a double bond at the Δ^9 position of fatty-acyl CoA [48, 49]. SCD has been reported to be associated with both meat and milk fatty acid composition in cattle [7, 16, 18, 32, 34, 37, 38, 42, 50–56]. The present study showed that SNPs close to the SCD gene were associated with many MUFAs and several CLA isomers but none of the SFAs, which supports the proposed role of SCD in fatty acid composition in beef. However, in this study none of the SNPs around SCD were associated with oleic acid, 9c-18:1, the most abundant MUFA in beef. This could be partly due to lack of SNPs in the current panel that are in a high LD with SCD to capture all its effects. Interestingly, several other studies also showed no associations between SCD and oleic acid in various beef and dairy cattle populations, using different SNP panels or SCD gene SNP [7, 16, 38, 50, 51, 57], although two other studies have reported significant associations between SCD SNP variants and oleic acid concentrations in Japanese Black cattle [18, 32]. The role of SCD on the concentration of oleic acid in beef is worthy of further investigation.

Other SNPs on BTA 1, 3, 4, 5, 6, 7, 10, 16, 20, 23, 24, 25, 28 and on 1, 2, 4, 5, 6, 8, 10, 13, 14, 16, 23, 24, 25, 27, 28 were found significantly associated with one or more fatty acid concentrations in the SQ and LL, respectively, but with relatively smaller effects (Figs. 1 and 2). The SNP rs41642879 at BTA23:6760915 was associated with 12:0, 14:0, and HI of both the LL and SQ, and with 16:0, 9c-14:1 in LL tissue. There are several genes within

the 1 Mb window centering the SNP. One possible candidate gene is the glutamate-cysteine ligase catalytic subunit gene (GCLC), which is involved in the synthesis of glutathione (GSH) [58]. Glutathione has an antioxidant function by oxidizing itself into Glutathione disulfide (GSSG), which in turn is reduced to GSH at the expense of nicotinamide adenine dinucleotide phosphate (NADPH) oxidase [58]. The latter is essential for fatty acid synthesis [40]. The SNP rs41574597 at BTA23:25956421 was found to be associated with 11c-20:1 in both tissues. Several genes belonging to the butyrophilin family are located nearby. Butyrophilin is the major protein associated with milk fat droplets and has been reported to be related to milk quality in cattle [59]. However, it was suggested that butyrophilin is specific to mammary tissue [60] hence its role in meat fatty acid production remains unclear.

Fewer SNPs were identified for PUFAs (90 in SQ and 87 in LL) in comparison to the number of significant SNPs for SFAs (121 in SQ and 117 in LL) and MUFAs (174 in SQ and 120 in LL). One SNP rs41582945 at BTA2:50069820 explained 2.45 % of genetic variance for dihomo-gamma-linolenic acid (Dihomo-GLA, 20:3n-6) in LL. However, no known genes exist in the 1 Mb region of this SNP. Several SNPs were also found to be associated with the intermediate product of Dihomo-GLA, arachidomic acid in LL (20:4n-6). The most significant SNP rs110776216 on BTA2:131968682 explained 1.91 % of total genetic variance of 20:4n-6 and was located within the endothelin converting enzyme 1 gene (ECE1) which encodes the enzyme that converts big endothelin-1 to endothelin-1. Endothelin-1 was previously found to stimulate arachidonic acid release in human pericardial smooth muscle cells [61, 62]. In this study, no significant SNPs were found to be associated with iso14:0, 7c-17:1, 15 t-18:1, 9c-20:1, 6 t,8 t-18:2, 7 t,9c-18:2, 9 t,11 t-18:2, 8 t,10 t-18:2, 7 t,9 t-18:2, 20:5n3 in LL and 22:0, 7c-17:1, 11c-18:1, 12 t-18:1, 15 t-18:1, 6 t,8 t-18:2, 10 t,12 t-18:2, 9 t,11 t-18:2, 8 t,10 t-18:2, 7 t,9 t-18:2, 18:3n6, 20:3n9, and n-6/n-3 in SQ. These fatty acid traits had very low or near zero heritability estimates (Table 1), therefore their concentrations were less likely influenced by host direct genetic effects.

Genomic prediction

Realized accuracies of genomic prediction measured as the Pearson's correlation coefficients between genomic estimated breeding values (GEBV) and adjusted phenotypic values of fatty acid traits divided by square root of heritability are presented in Table 2. Accuracies of breeding values estimated from the pedigree-based BLUP method (PBLUP) are also presented in Table 2 as comparisons. The realized accuracy of genomic prediction ranged from –0.05 for 15 t-18:1 to 0.73 for 14:0 in the LL

Table 2 Realised accuracy (±SE) of breeding value prediction for fatty acid traits in the subcutaneous adipose and *longissimus lumborum* musle

Trait[a]	Subcutaneous adipose			Longissimus lumborum		
	PBLUP	GBLUP	BayesCπ	PBLUP	GBLUP	BayesCπ
10:0	0.32 ± 0.05	0.37 ± 0.05	0.45 ± 0.05	0.27 ± 0.05	0.37 ± 0.07	0.53 ± 0.06
12:0	0.26 ± 0.04	0.42 ± 0.03	0.58 ± 0.02	0.23 ± 0.03	0.31 ± 0.06	0.53 ± 0.04
13:0	0.06 ± 0.06	0.35 ± 0.04	0.34 ± 0.04	−0.05 ± 0.06	0.31 ± 0.07	0.32 ± 0.07
14:0	0.34 ± 0.04	0.39 ± 0.05	0.61 ± 0.04	0.27 ± 0.04	0.45 ± 0.06	0.73 ± 0.04
15:0	0.38 ± 0.06	0.55 ± 0.06	0.62 ± 0.06	0.26 ± 0.06	0.57 ± 0.08	0.69 ± 0.08
16:0	0.34 ± 0.05	0.28 ± 0.05	0.31 ± 0.05	0.23 ± 0.03	0.36 ± 0.05	0.50 ± 0.05
17:0	0.30 ± 0.05	0.33 ± 0.05	0.34 ± 0.05	0.29 ± 0.05	0.40 ± 0.04	0.46 ± 0.04
18:0	0.15 ± 0.05	0.29 ± 0.06	0.29 ± 0.07	0.24 ± 0.05	0.35 ± 0.05	0.38 ± 0.04
19:0	0.33 ± 0.10	0.40 ± 0.09	0.40 ± 0.10	0.18 ± 0.11	0.18 ± 0.13	0.18 ± 0.13
20:0	0.05 ± 0.06	0.22 ± 0.08	0.21 ± 0.08	0.25 ± 0.05	0.24 ± 0.05	0.24 ± 0.05
22:0	0.24 ± 0.07	0.36 ± 0.05	0.38 ± 0.05	0.17 ± 0.06	0.44 ± 0.07	0.43 ± 0.08
24:0	-	-	-	0.23 ± 0.06	0.27 ± 0.06	0.28 ± 0.06
SFA	0.29 ± 0.06	0.28 ± 0.07	0.29 ± 0.07	0.14 ± 0.04	0.37 ± 0.06	0.43 ± 0.06
iso14:0	0.16 ± 0.10	0.25 ± 0.10	0.24 ± 0.09	0.30 ± 0.05	0.18 ± 0.08	0.18 ± 0.08
iso15:0	0.20 ± 0.05	0.24 ± 0.07	0.25 ± 0.07	0.17 ± 0.03	0.20 ± 0.07	0.18 ± 0.07
ai15:0	0.21 ± 0.03	0.30 ± 0.04	0.31 ± 0.04	0.18 ± 0.04	0.37 ± 0.05	0.39 ± 0.05
iso16:0	0.20 ± 0.07	0.27 ± 0.06	0.28 ± 0.05	0.15 ± 0.04	0.28 ± 0.06	0.29 ± 0.05
iso17:0	0.18 ± 0.06	0.27 ± 0.05	0.27 ± 0.05	0.25 ± 0.07	0.29 ± 0.06	0.29 ± 0.06
ai17:0	0.14 ± 0.08	0.33 ± 0.08	0.31 ± 0.08	0.15 ± 0.04	0.34 ± 0.04	0.34 ± 0.05
iso18:0	0.12 ± 0.05	0.22 ± 0.05	0.22 ± 0.05	0.21 ± 0.05	0.33 ± 0.06	0.37 ± 0.06
BFA	0.20 ± 0.06	0.31 ± 0.05	0.31 ± 0.05	0.17 ± 0.05	0.32 ± 0.04	0.32 ± 0.04
SFA + BFA	0.30 ± 0.06	0.29 ± 0.07	0.31 ± 0.07	0.14 ± 0.04	0.36 ± 0.06	0.42 ± 0.06
9c-14:1	0.19 ± 0.06	0.31 ± 0.04	0.43 ± 0.04	0.27 ± 0.06	0.34 ± 0.04	0.55 ± 0.03
9c-15:1	0.16 ± 0.06	0.23 ± 0.07	0.23 ± 0.08	0.24 ± 0.11	0.36 ± 0.10	0.37 ± 0.10
7c-16:1	0.19 ± 0.05	0.18 ± 0.07	0.20 ± 0.07	0.16 ± 0.07	0.22 ± 0.05	0.24 ± 0.05
9c-16:1	0.13 ± 0.04	0.25 ± 0.06	0.26 ± 0.07	0.29 ± 0.03	0.37 ± 0.03	0.49 ± 0.02
11 t-16:1	0.14 ± 0.08	0.33 ± 0.05	0.33 ± 0.05	0.08 ± 0.08	0.12 ± 0.09	0.13 ± 0.10
12c-16:1	0.22 ± 0.05	0.34 ± 0.03	0.43 ± 0.03	0.25 ± 0.05	0.32 ± 0.04	0.55 ± 0.02
7c-17:1	-	-	-	−0.02 ± 0.15	0.33 ± 0.24	0.32 ± 0.21
9c-17:1	0.34 ± 0.07	0.48 ± 0.06	0.48 ± 0.05	0.37 ± 0.08	0.40 ± 0.08	0.40 ± 0.08
9c-18:1	0.35 ± 0.10	0.30 ± 0.10	0.29 ± 0.10	0.14 ± 0.03	0.27 ± 0.06	0.37 ± 0.05
11c-18:1	0.57 ± 0.18	0.57 ± 0.18	0.55 ± 0.18	0.29 ± 0.02	0.44 ± 0.07	0.45 ± 0.07
12c-18:1	0.00 ± 0.09	0.11 ± 0.10	0.11 ± 0.09	0.11 ± 0.10	0.18 ± 0.10	0.19 ± 0.10
13c-18:1	0.19 ± 0.05	0.36 ± 0.04	0.41 ± 0.04	0.21 ± 0.03	0.36 ± 0.05	0.51 ± 0.04
14c-18:1	0.10 ± 0.09	0.28 ± 0.11	0.28 ± 0.11	0.15 ± 0.09	0.17 ± 0.07	0.17 ± 0.07
15c-18:1	0.20 ± 0.04	0.30 ± 0.04	0.31 ± 0.04	0.19 ± 0.05	0.26 ± 0.05	0.25 ± 0.06
6 t + 8 t-18:1	0.30 ± 0.06	0.35 ± 0.04	0.34 ± 0.04	0.20 ± 0.07	0.29 ± 0.04	0.30 ± 0.04
9 t-18:1	0.30 ± 0.09	0.33 ± 0.06	0.33 ± 0.06	0.26 ± 0.08	0.31 ± 0.04	0.32 ± 0.04
10 t-18:1	0.28 ± 0.03	0.42 ± 0.03	0.43 ± 0.03	0.26 ± 0.03	0.41 ± 0.03	0.42 ± 0.03
11 t-18:1	0.35 ± 0.08	0.43 ± 0.09	0.42 ± 0.09	0.29 ± 0.04	0.38 ± 0.04	0.35 ± 0.05
12 t-18:1	-	-	-	0.05 ± 0.07	0.27 ± 0.10	0.26 ± 0.10
15 t-18:1	-	-	-	−0.26 ± 0.15	−0.05 ± 0.16	−0.01 ± 0.16
16 t-18:1	0.41 ± 0.07	0.65 ± 0.13	0.64 ± 0.13	0.33 ± 0.05	0.47 ± 0.07	0.48 ± 0.07

Genome-wide association and genomic prediction of breeding values for fatty acid composition...

135

Table 2 Realised accuracy (±SE) of breeding value prediction for fatty acid traits in the subcutaneous adipose and *longissimus lumborum* musle *(Continued)*

sumtrans18:1	0.22 ± 0.03	0.35 ± 0.04	0.35 ± 0.03	0.25 ± 0.02	0.36 ± 0.03	0.37 ± 0.03
9c-20:1	0.33 ± 0.06	0.45 ± 0.07	0.45 ± 0.07	−0.03 ± 0.10	0.02 ± 0.09	0.00 ± 0.09
11c-20:1	0.20 ± 0.05	0.27 ± 0.04	0.30 ± 0.04	0.35 ± 0.06	0.37 ± 0.05	0.47 ± 0.04
MUFA	0.29 ± 0.07	0.29 ± 0.08	0.30 ± 0.08	0.11 ± 0.03	0.25 ± 0.07	0.31 ± 0.06
9c,13 t + 8 t,12c-18:2	0.22 ± 0.05	0.34 ± 0.07	0.33 ± 0.07	0.14 ± 0.06	0.36 ± 0.08	0.36 ± 0.08
9c,15c-18:2	0.18 ± 0.07	0.33 ± 0.05	0.35 ± 0.05	0.22 ± 0.04	0.38 ± 0.06	0.45 ± 0.06
8 t,13c-18:2	0.23 ± 0.05	0.38 ± 0.06	0.37 ± 0.06	0.02 ± 0.10	0.30 ± 0.04	0.31 ± 0.04
11 t,15c-18:2	0.25 ± 0.05	0.31 ± 0.06	0.32 ± 0.06	0.25 ± 0.03	0.34 ± 0.04	0.36 ± 0.04
9c,11 t + 9 t,11c-18:2	0.05 ± 0.08	0.25 ± 0.07	0.24 ± 0.07	0.26 ± 0.05	0.26 ± 0.07	0.25 ± 0.06
7 t,9c-18:2	0.58 ± 0.14	0.63 ± 0.12	0.64 ± 0.13	0.43 ± 0.22	0.31 ± 0.25	0.28 ± 0.26
12,14 t-18:2	0.06 ± 0.28	0.44 ± 0.33	0.41 ± 0.34	0.25 ± 0.19	0.50 ± 0.23	0.48 ± 0.22
11 t,13 t-18:2	0.48 ± 0.16	0.47 ± 0.11	0.46 ± 0.12	0.11 ± 0.10	0.06 ± 0.09	0.06 ± 0.09
10 t,12 t-18:2	-	-	-	0.09 ± 0.08	0.08 ± 0.11	0.08 ± 0.11
9 t,11 t-18:2	-	-	-	0.11 ± 0.10	0.01 ± 0.08	0.02 ± 0.08
8 t,10 t-18:2	0.35 ± 0.14	0.28 ± 0.14	0.28 ± 0.13	0.24 ± 0.13	−0.01 ± 0.10	0.05 ± 0.11
7 t,9 t-18:2	0.41 ± 0.44	0.10 ± 0.47	0.04 ± 0.48	-	-	-
12,14c + 12c,14 t −18:2	−0.01 ± 0.26	0.32 ± 0.25	0.34 ± 0.24	0.06 ± 0.09	0.04 ± 0.07	0.04 ± 0.07
11 t,13c + 11c,13 t −18:2	0.54 ± 0.06	0.56 ± 0.08	0.54 ± 0.08	0.16 ± 0.10	0.29 ± 0.06	0.30 ± 0.06
10 t,12c-18:2	0.44 ± 0.09	0.52 ± 0.07	0.52 ± 0.08	0.04 ± 0.12	0.21 ± 0.14	0.22 ± 0.14
8 t,10c-18:2	0.33 ± 0.17	0.32 ± 0.14	0.31 ± 0.14	0.27 ± 0.08	0.15 ± 0.07	0.13 ± 0.07
Total CLA	0.25 ± 0.09	0.37 ± 0.09	0.36 ± 0.10	0.12 ± 0.06	0.06 ± 0.06	0.06 ± 0.06
18:2n-6	0.21 ± 0.02	0.35 ± 0.04	0.36 ± 0.04	0.15 ± 0.05	0.29 ± 0.04	0.32 ± 0.04
18:3n-3	0.27 ± 0.04	0.38 ± 0.04	0.38 ± 0.04	0.28 ± 0.06	0.33 ± 0.04	0.34 ± 0.04
18:3n-6	0.10 ± 0.11	0.09 ± 0.09	0.11 ± 0.09	0.20 ± 0.05	0.36 ± 0.05	0.41 ± 0.04
20:2n-6	0.20 ± 0.07	0.06 ± 0.05	0.06 ± 0.06	0.07 ± 0.11	0.10 ± 0.07	0.11 ± 0.07
20:3n-6	0.10 ± 0.11	0.24 ± 0.11	0.22 ± 0.11	0.14 ± 0.03	0.32 ± 0.04	0.34 ± 0.05
20:3n-9	0.20 ± 0.16	−0.03 ± 0.11	−0.05 ± 0.13	0.12 ± 0.04	0.33 ± 0.07	0.33 ± 0.07
20:4n-6	−0.10 ± 0.04	0.10 ± 0.09	0.09 ± 0.08	0.09 ± 0.03	0.29 ± 0.06	0.31 ± 0.05
20:5n-3	-	-	-	0.32 ± 0.06	0.10 ± 0.11	0.09 ± 0.11
22:4n-6	0.10 ± 0.04	0.19 ± 0.08	0.19 ± 0.08	0.15 ± 0.06	0.38 ± 0.08	0.40 ± 0.08
22:5n-3	0.04 ± 0.10	0.14 ± 0.08	0.14 ± 0.08	0.15 ± 0.04	0.39 ± 0.07	0.39 ± 0.07
22:6n-3	-	-	-	0.00 ± 0.05	0.10 ± 0.08	0.10 ± 0.07
PUFA	0.21 ± 0.02	0.35 ± 0.04	0.36 ± 0.04	0.14 ± 0.04	0.30 ± 0.04	0.33 ± 0.04
n-3	0.27 ± 0.04	0.37 ± 0.05	0.38 ± 0.04	0.22 ± 0.04	0.40 ± 0.05	0.40 ± 0.05
n-6	0.20 ± 0.02	0.35 ± 0.04	0.36 ± 0.04	0.14 ± 0.04	0.29 ± 0.04	0.32 ± 0.04
n-6/n-3	-	-	-	0.11 ± 0.04	0.16 ± 0.04	0.14 ± 0.04
P/S	0.19 ± 0.03	0.33 ± 0.04	0.34 ± 0.04	0.14 ± 0.03	0.38 ± 0.04	0.39 ± 0.03
P/(S + B)	0.19 ± 0.03	0.33 ± 0.04	0.34 ± 0.04	0.14 ± 0.09	0.38 ± 0.04	0.39 ± 0.04
HI	0.31 ± 0.05	0.30 ± 0.05	0.38 ± 0.05	0.22 ± 0.05	0.41 ± 0.06	0.59 ± 0.06

[a]The concentrations of fatty acids were expressed as a percentage of fatty acid methyl esters (FAME) quantified. c = cis, t = trans. SFA = 10:0 + 12:0 + 13:0 + 14:0 + 15:0 + 16:0 + 17:0 + 18:0 + 19:0 + 20:0 + 22:0 + 24:0; BFA = iso14:0 + iso15:0 + ai15:0 + iso16:0 + iso17:0 + ai17:0 + iso18:0; SFA + BFA: sum of SFA and BFA; sumtrans18:1 = 6 t/8 t-18:1 + 9 t-18:1 + 10 t-18:1 + 11 t-18:1 + 12 t-18:1 + 15 t-18:1 + 16 t-18:1; MUFA = 9c-14:1 + 9c-15:1 + 7c-16:1 + 9c-16:1 + 11 t-16:1 + 12c-16:1 + 7c-17:1 + 9c-17:1 + 9c-18:1 + 11c-18:1 + 12c-18:1 + 13c-18:1 + 14c-18:1 + 15c-18:1 + 9c-20:1 + 11c-20 + 6 t/8 t-18:1 + 9 t-18:1 + 10 t-18:1 + 11 t-18:1 + 12 t-18:1 + 15 t-18:1 + 16 t-18:1; Total CLA = 9c,11 t + 9 t,11c-18:2 + 6 t,8 t-18:2 + 7 t,9c-18:2 + 12 t,(14 t-18:2 + 11 t,13 t)-18:2 + 10 t,12 t-18:2 + 9 t,11 t-18:2 + 8 t,10 t-18:2 + 7 t,9 t-18:2 + (12 t,14c + 12c,14 t) -18:2 + (11 t,13c + 11c,13 t)-18:2 + 10 t,12c-18:2 + 8 t,10c-18:2; PUFA = 18:2n-6 + 18:3n-6 + 18:3n-3 + 20:2n-6 + 20:3n-6 + 20:4n-6 + 22:5n-3+ 22:6n-3; n-3 = 18:3n-3 + 20:5n-3 + 22:5n-3 + 22:6n-3; n-6 = 18:2n-6 + 18:3n-6 + 20:2n-6 + 20:3n-6 + 20:4n-6 + 22:4n-6; n-6/n-3: ratio between n-6 and n-3; P/S = PUFA/SFA; P/(S + B) = PUFA/(SFA + BFA); HI = (MUFA + PUFA) / (4 × 14:0 + 16:0). '-' = not calculated due to a zero heritability

muscle, and varied from −0.05 for 20:3n-9 to 0.65 for 16 t-18:1 in the SQ tissue. Averaged across all traits, accuracies from PBLUP, GBLUP, and the Bayesian method were 0.23, 0.32, and 0.35, respectively, in SQ, and 0.17, 0.39, and 0.46, respectively, in LL. These results suggested the effectiveness of genomic prediction using either GBLUP or the Bayesian method. However, the incompleteness of the pedigree (only one generation) may largely contribute to the low accuracy for the PBLUP method. It should be noted that the realized accuracy could be overestimated when heritability is underestimated as pointed out by Lourenco *et al.* [63]. Accuracies that were substantially overestimated tended to have relatively large SE (>0.10) as shown in Table 2. Additionally, Pearson's correlation coefficient between estimated breeding values and adjusted phenotypes, and regression coefficient by regressing adjusted phenotypes on estimated breeding values were also calculated and provided in Additional file 6. The correlation coefficients averaged 0.11, 0.15, and 0.15 for PBLUP, GBLUP, and the Bayesian method, respectively, in SQ, and averaged 0.08, 0.14, and 0.16, respectively in LL. The average regression coefficients in SQ were 1.02, 0.77, and 0.92, and were 1.04, 0.90, and 0.83 in LL for PBLUP, GBLUP, and the Bayesian method, respectively. The regression coefficient is expected to be 1 if the estimated breeding values were unbiased predictions of the true breeding values. Nevertheless, for most of the fatty acid traits, the accuracy of genomic prediction were relatively low (<0.40), which was expected given the low heritability estimates and the small sample size used in this study [64]. Relatively higher accuracy $(r_{(GEBV,y)}/h \geq 0.50$ with SE < 0.10) were achieved for 10:0 (0.53), 12:0 (0.53), 14:0 (0.73), 15:0 (0.69), 16:0 (0.50), 9c-14:1 (0.55), 12c-16:1 (0.55), 13c-18:1 (0.51), and HI (0.59) in LL, and for 12:0 (0.58), 14:0 (0.61), 15:0 (0.62), 10 t,12c-18:2 (0.52), and 11 t,13c + 11c,13 t-18:2 (0.56) in SQ. The relatively higher accuracy for certain saturated and monounsaturated fatty acids, and HI, and relatively lower accuracy for CLAs and other PUFAs in muscle were compatible with the magnitude of their estimated heritability (Table 1). The correlations between heritability estimates and realised accuracy of genomic prediction in LL were 0.61 and 0.39 for Bayesian and GBLUP methods, respectively. However, in SQ such correlations were only 0.10 for GBLUP and 0.23 for the Bayesian method, which is likely due to many overestimations of realised accuracy for traits with low and inaccurate heritability estimates. Genomic prediction from the Bayesian method performed similarly as GBLUP for most of the traits, but substantially better for several traits in LL muscle such as 10:0 (0.37 for GBLUP vs 0.53 for BayesCπ), 12:0 (0.31 vs 0.53), 14:0 (0.45 vs 0.73), 15:0 (0.57 vs 0.69), 16:0 (0.36 vs 0.50), 9c-14:1 (0.34 vs 0.55), 9c-16:1 (0.37 vs 0.49), 12c-16:1 (0.32 vs 0.55), 9c-18:1 (0.27 vs 0.37), 13c-18:1 (0.36

vs 0.51), and HI (0.41 vs 0.59), and for traits in SQ including 12:0 (0.42 vs 0.58), 14:0 (0.39 vs 0.61), and 9c-14:1 (0.31 vs 0.43). These traits have been shown to have SNPs with larger effects from GWAS results (Figs. 1 and 2). The Bayesian method adopted in this study allows a fraction of SNPs to take relatively large effects, which may better characterize the genetic architecture of traits that have QTL of larger effects than the GBLUP method [65], which assumes all SNPs have the same genetic variance.

Fatty acid composition is a complex trait and it is difficult and expensive to measure, making it a good candidate trait for genomic selection. To date, genomic prediction for fatty acid composition in beef cattle has only been reported by Saatchi *et al.* [16] for 24 individual and grouped/ratio of fatty acids in steaks of American Angus beef cattle, and by Onogi *et al.* [39] for 8 fatty acid traits in *musculus trapezius* of Japanese Black cattle. Relatively higher prediction accuracies were found for 14:0 (0.57), 16:0 (0.53), total long chain saturated fatty acids (0.57), total medium chain saturated fatty acids (0.57), 9c-18:1 (0.35), 12c-18:1 (0.35), total MUFA (0.38), (14:0 + 16:0)/all (0.55), and AI (0.56) in Saatchi's study, compared to other fatty acid traits. In this study, relatively higher accuracies were also obtained for SFAs 12:0, 14:0, and 15:0 in both the LL and SQ tissues, and for 10:0, 16:0, 9c-14:1, 12c-16:1, 13c-18:1, and HI in LL (Table 2), suggesting strong host genetic controls on synthesis of these SFAs and MUFAs. Saatchi *et al.* [16] reported genomic prediction accuracies for 12 PUFAs and all were very low (<0.30). In this study, we analyzed 32 and 34 PUFAs and PUFA-BHI (including CLAs and 11 t-18:1) in the adipose and muscle tissues, respectively, and found moderate accuracies (between 0.30 and 0.45) for 11 t-18:1, 9c,13 t + 8 t,12c-18:2, 9c,15c-18:2, 8 t,13c-18:2, 11 t,15c-18:2, 18:2n-6, 18:3n-3, n-3, n-6, and total PUFA in both the adipose and muscle tissues, moderate accuracies for 20:3n-6, 20:3n-9, 22:4n-6, 22:6n-3 in the muscle, and relatively high accuracies (>0.50) for 12 t,14c + 12c,14 t-18:2, and 11 t,13c + 11c,13 t-18:2 in the adipose tissue, suggesting considerable host genetic influence on these fatty acids. Different beef cattle populations, environments where the animals were raised, sample sizes and statistical models may also contribute to the differences of genomic prediction accuracy observed between different studies. Although most dietary PUFAs are biohydrogenated by rumen bacteria [66], a portion of PUFAs and PUFA-BHI may escape and deposit into body fat of beef. In addition, some PUFAs can be endogenously synthesized, for example CLAs can be synthesized from one of the PUFA-BHI, vaccenic acid (11 t-18:1) by the host [67]. Therefore, contents of both PUFAs and PUFA-BHI are potentially influenced by host genetics and thus predictable by genomic prediction. Onogi *et al.*

[39] also reported a relatively high accuracy (0.56) for PUFA C18:2 in Japanese Black cattle. Although it would be worthwhile to further verify the genomic prediction accuracy in other beef cattle populations, the moderate to relatively high genomic prediction accuracies achieved in this study for the HI, several individual SFAs, MUFAs, PUFAs and PUFA-BHI suggest that genomic selection is a promising tool for genetic improvement of fatty acid profiles in beef cattle to produce healthier meat. Therefore, as consumers' demand for healthier meat continues to grow, beef producers may get more premiums by producing meat with enhanced fatty acid profiles, which can be achieved by incorporating fatty acid composition traits into a multi-trait selection index for selection and/or by genetic based diet management.

Conclusions

Fatty acid composition in beef tissues is a polygenic trait that is controlled by a few major host genes and many genes of small effects. Several genes, including *FASN*, *SCD*, and *THRSP*, are major candidate genes for variations of fatty acid contents in beef cattle. Accuracy of genomic prediction was low for most of the fatty acid traits investigated. Moderate accuracy was obtained for SFAs 10:0, 12:0, 13:0, 14:0, 15:0, 16:0, MUFAs 9c-14:1, 12c-16:1, 13c-18:1, and HI in LL, and for SFAs 12:0, 14:0, 15:0, and CLA isomers 10 t,12c-18:2, and 11 t,13c + 11c,13 t-18:2 in SQ. The Bayesian method performed similarly as GBLUP for most of the traits, but substantially better for fatty acid traits that are influenced by QTL of larger effects. The moderate genomic prediction accuracy achieved in this study for HI in LL and several individual fatty acids in LL and SQ tissue suggest that it is possible to genetically improve fatty acid profiles in beef cattle to produce healthier meat through genomic selection. Further investigations on the identification of causal mutations for variations of fatty acid contents in beef tissues and on improvement of genomic prediction accuracy are required.

Methods

Animal populations, tissue collection and fatty acid analyses

A total of 1366 steers and heifers born between 2008 and 2011 were used in this study. The animals were from four different herds including three commercial herds and one experimental herd located in Alberta of Canada. All dietary treatments and experimental procedures were approved by the AAFC Lacombe Research Centre Animal Care Committee and animals were cared for as outlined under the guidelines established by the Canadian Council on Animal Care [68]. Breed compositions of the 1366 steers and heifers were represented by purebred Angus (ANAN, $n = 6$), Hereford-Angus crossbreds (HEAN, $n =$

120), Charolais-Red Angus crossbreds (CHAR, $n = 93$), crossbreds produced by mating Hereford-Angus to Gelbvieh-Angus crossbreds (HEANGV, $n = 209$), and calves produced from crosses between a composite terminal bull strain which was derived from Hereford, Black Angus, Red Angus, and Limousin, and crossbred cows with a mixed background of Angus, Red Angus, Hereford, Simmental, Charolais, Limousin and Gelbvieh (TXX, $n = 938$). A more detailed description of breeding and management of the herds have been described previously [69–71]. A written consent from the owner of the commercial herds was obtained for the use of cattle data in this study. After weaning, animals were raised under one of four production systems: (1) calf-fed, growth implant; (2) calf-fed; no growth implant; (3) yearling-fed, growth implant; (4) yearling-fed, no growth implant [70, 72]. All animals were fed high concentration diets for finishing and were targeted to be slaughtered at a constant back fat thickness of 9 to 10 mm measured between the 12th and 13th ribs.

After slaughter, the *longissimus luborum* muscle (LL) of each animal was taken from the left striploin at 48 h post-mortem, vacuum packed and then chilled at 2 °C. The striploin samples were then transported by a refrigerated truck to a meat lab of the AAFC Lacombe Research Centre where a sub-sample of approximately 10 grams of LL muscle and 5 grams of subcutaneous adipose (SQ) tissue from the side of the striploin of each animal was taken, vacuum packed and frozen at –80 °C for subsequent fatty acid analyses. The two tissues have distinct metabolism roles involving fat usage: muscle is mainly for energy expenditure to produce force and motion while adipose including intramuscular fat within muscle is the main tissue for fat storage [73]. The two tissues were selected mainly because they are major parts of carcass that are consumed as beef products by humans. Fatty acid analyses of LL and SQ tissues were based on the protocols described previously with some modifications [10]. Briefly, lipid was extracted from the LL muscle tissue using Folch's method [74] as outlined by Cruz-Hernandez et al. [75] and from the SQ tissue based on the procedures described in [75] and [76]. Fatty acid methyl esters (FAME) were then derivatized using sodium methoxide from the lipid extracts for quantification of fatty acid composition. Gas chromatography (GC) and silver-ion high performance liquid chromatography (Ag + HPLC) analyses were conducted to separate and quantify individual fatty acids as outlined in [77] using a two-step GC procedure and in [75] using Ag + HPLC. Individual fatty acids were expressed as a percentage of the total FAME. Concentrations of groups of fatty acids, including total saturated fatty acids (SFA), branched fatty acids (BFA), sum of SFA and BFA (SFA + BFA), mono-unsaturated fatty acids (MUFA), poly-unsaturated fatty acids (PUFA), sum of trans 18:1 fatty acids (sumtrans18:1), total conjugated linoleic acid (Total CLA), n-3, and n-6, were measured by summing up the percentages of individual fatty acids within the fatty acid group. Ratios between PUFA and SFA (P/S), PUFA and sum of SFA and BFA (P/(S + B)), and between n-6 and n-3 (n-6/n-3) were also calculated. A health index (HI), proposed in [35], was computed as HI = (MUFA + PUFA) / (4 × 14:0 + 16:0). A total of 83 individual and grouped/ratio fatty acid traits in the LL muscle and 81 in the SQ tissue were quantified. Two fatty acids, 20:5n3 and 22:6n3 were not detected in SQ in this study due to their extremely low concentrations in the tissue.

Single nucleotide polymorphism genotyping

All animals were genotyped on the Illumina BovineSNP50 Beadchip comprised of 54,609 SNP markers. Markers with minor allele frequency less than 0.05, missing rate greater than 0.20, extremely deviated from Hardy-Weinberg equilibrium test ($P < 10^{-6}$), or in high correlation with another SNP ($r \geq 0.95$) were removed. After filtering, 35,446 SNPs were kept for analyses. Sporadically missing genotypes represented 0.14 % of the total genotypes and were imputed via Beagle 3.3.2 [78].

Genome-wide association study

Phenotypic values were adjusted for fixed effects and random contemporary group effects using a linear mixed model which included fixed effects of breed type, gender, production system, linear covariates of animal's age at slaughter, days between slaughter and fatty acid extraction, and metabolic energy of diet, and random effects of contemporary groups defined as combinations of feedlot location and year, additive genetic effects and residual errors. Fatty acid traits in LL muscle were also adjusted for intramuscular fat content by including the marbling score as an additional fixed linear covariate. A genomic relationship matrix for additive genetic effects was constructed from SNP marker genotypes using the first method of VanRaden [79]. Variance components and heritability were estimated using the above model and average-information REML algorithm implemented via ASReml v3.0 software package [80].

The adjusted phenotypes were subsequently analysed using the BayesCπ method [81] for genome-wide association studies. The model can be described as follows:

$$y_i = \mu + \sum_{j=1}^{M} x_{ij} a_j + e_i,$$

where y_i is the adjusted phenotypic value of the ith animal, μ is the general mean, x_{ij} is the jth SNP genotype of animal i and was coded as 0, 1 or 2 depending on copies of an arbitrarily specified allele, M is the total number of SNP

markers, a_j is the allele substitution effect of SNP j, and e_i is the random residual effect.

A mixture distribution was assumed for a_j so that $(a_j | \pi, \sigma_a^2) \sim (1 - \pi) N(0, \sigma_a^2) + \pi \delta_0(a_j)$, where $N(0, \sigma_a^2)$ is a normal distribution with mean 0 and variance σ_a^2, and $\delta_0(a_j)$ denotes a distribution concentrated at zero, and $(1 - \pi)$ and π are the weights for the two distributions. A latent indicator variable γ_j was introduced for each SNP so that when $\gamma_j = 1$, $a_j \sim N(0, \sigma_a^2)$, and when $\gamma_j = 0$, $a_j = 0$. Prior distribution for γ_j follows a Bernoulli distribution with probability $(1 - \pi)$, and the joint prior density for γ is $f(\gamma | \pi) = \prod_j \pi^{(1-\gamma_j)} (1-\pi)^{\gamma_j}$. Residual error e_i was assumed from a normal distribution $N(0, \sigma_e^2)$. The prior distribution for σ_a^2 (or σ_e^2) is a scaled inverse Chi-square distribution with degree of freedom v_a (or v_e) and a scale parameter S_a^2 (or S_e^2). The hyper-parameter v_a (or v_e) was arbitrarily set to 4 (or 10), and S_a^2 (or S_e^2) was set to $\sigma_u 2(v_a -2)/[v_a(1-\pi)\sum 2p_j(1-p_j)]$ (or $\hat{\sigma}_0^2(v_e-2)/v_e$), where p_j is allele frequency for marker j, $\hat{\sigma}_u^2$ and $\hat{\sigma}_0^2$ were total additive genetic and residual variances obtained from the analyses described previously. A Gibbs sampling algorithm was used for generating samples for unknown parameters from their joint posterior distribution. The computer program was self-written in C language using the computing algorithm as described by Chen et al. [82]. The Gibbs chain length was 45,000 with the first 5000 discarded as burn-in. Posterior inclusion probability for each SNP was estimated as sample mean of the latent indicator variable for that SNP, and was used as a signal of association. To declare the significance of a SNP effect, empirical genome-wise significance threshold at α = 0.05 was determined by 1000 permutation analyses according to the procedure of Churchill and Doerge [83]. Briefly, the adjusted phenotype values of each fatty acid were randomly shuffled and assigned back to the animals for the BayesCπ analyses while the genotype data remained intact. The process was repeated 1000 times and the largest SNP posterior inclusion probability from each permutation analysis was kept and ordered in ascending order, and the 950th value was defined as the genome-wise significance threshold. Candidate genes in the window of 1 Mb centering the significant SNPs were obtained by querying the Ensemble gene database using SNP locations from the bovine UMD3.1 genome assembly via the SNP annotation tool in the NGS-SNP suite [84].

Genomic prediction

Genomic best linear unbiased prediction (GBLUP) and BayesCπ methods were used for genomic prediction. A ten-fold cross validation was used to evaluate the accuracy of genomic prediction. The data was first split into 10 approximately equal-sized groups according to sires of the animals so that no sire families overlapped between

Genome-wide association and genomic prediction of breeding values for fatty acid composition...

139

any two groups. For each breed type, the number of animals in each cross-validation group was kept approximately the same so that each breed in the validation group was also represented in the training population. For each cross validation, nine groups were used for training and the remaining one was used as the validation population. For GBLUP, animals in the validation population were assumed with no phenotypic values, and the animals in the training and validation populations were then combined to estimate the breeding values for animals in the validation population using a linear animal model, which can be written as:

$$\mathbf{y}^* = \mathbf{1}\mu + \mathbf{Za} + \mathbf{e},$$

Where y^* is the vector of adjusted fatty acid phenotypic values from animals in the training population, μ is the overall mean, $\mathbf{a}$ is the vector of breeding values for all animals, $\mathbf{e}$ is the vector of random residuals and $\mathbf{Z}$ is the incidence matrix relating $\mathbf{a}$ to $\mathbf{y}^*$. The additive genomic relationship matrix for all animals was derived from the SNP markers using the first method of VanRaden [79], and ASReml 3.0 [80] was used to estimate the breeding values. For the BayesCπ method, SNP effects were estimated based on the training population using the statistical model as described in the GWAS analyses. The GEBV for animal i in the validation population was predicted by summing up SNP effects over all loci as follows: $\text{GEBV}_i = \sum_{j=1}^{M} x_{ij} a_j$, where a_j is the estimated effect for SNP j. For comparisons, a pedigree based BLUP method (PBLUP) was also used to estimate the breeding values, assuming no phenotypic values for validation animals. However, only one generation of the pedigree was available for construction of the additive genetic relationship matrix. Realized accuracy of genomic prediction was measured as the correlation between estimated breeding values and the adjusted phenotype in the validation groups divided by square root of heritability and was averaged across the ten cross-validations.

Availability of supporting data

The data sets supporting the results of this article are included within the article and its additional files. The original data sets used in this study are available upon request as part of the data is not public.

Additional files

Additional file 1: Manhattan plot of posterior probability of inclusion of single nucleotide polymorphisms (SNP) for 81 fatty acid composition traits in subcutaneous adipose (SQ) and 83 traits in *longissimus lumborum* muscle (LL). Dashed line indicates the significance threshold at genome-wise empirical threshold at α = 0.05 determined from a 1000 permutation analysis. (PDF 13567 kb)

Additional file 2: List of significant single nucleotide polymorphisms (SNP) for 81 fatty acid composition traits in subcutaneous adipose (SQ). Results include trait name, SNP name, chromosome, position on the UMD3.1 genome assembly, alleles, allele substitution effect, percentage of total genetic variance explained, and posterior probability of inclusion of SNP. (CSV 24 kb)

Additional file 3: List of significant single nucleotide polymorphisms (SNP) for 81 fatty acid composition traits in *longissimus lumborum* muscle (LL). Results include trait name, SNP name, chromosome, position on the UMD3.1 genome assembly, alleles, allele substitution effect, percentage of total genetic variance explained, and posterior probability of inclusion of SNP. (CSV 27 kb)

Additional file 4: Genes within 1 Mb region centering significant single nucleotide polymorphisms (SNP). Candidate genes were obtained from the Ensemble database by querying the SNP locations against the bovine UMD3.1 genome assembly using the SNP annotation tool in the NGS-SNP suite. (CSV 1037 kb)

Additional file 5: Linkage disequilibrium (LD) beween single nucleotide polymorphisms (SNP) around the SCD gene. Linkage disequilibrium was measured as D prime. The red triangle indicates the location of the SCD gene. An LD block was shown surrounding the SCD gene. Names of the SNPs significantly associated with fatty acid composition were highlighted in green. (PDF 971 kb)

Additional file 6: Pearson's correlation coefficients between estimated breeding values and adjusted phenotypes and regression coefficients of adjusted phenotypes on estimated breeding values. (DOCX 35 kb)

Abbreviations

AAFC: Agriculture and Agri-food Canada; Ag + HPLC: silver-ion high performance liquid chromatography; AI: atherogenic index; BFA: branched fatty acid; BHI: biohydrogenation intermediates; BTA: *Bos taurus* autosome; CLA: conjugated linoleic acid; FAME: fatty acid methyl esters; GBLUP: genomic best linear unbiased prediction; GC: gas chromatography; GEBV: genomic estimated breeding value; GWAS: genome-wide association study; HI: health index; LD: linkage disequilibrium; LL: *Longissimus lumborum* muscle; MME: mixed model equations; MUFA: monounsaturated fatty acid; PBLUP: pedigree-based best linear unbiased prediction; PPI: posterior probability of inclusion; PUFA: polyunsaturated fatty acid; QTL: quantitative trait loci; SFA: saturated fatty acid; SNP: single nucleotide polymorphism; SQ: subcutaneous adipose.

Competing interests

The authors declare that they have no competing interests.

Authors' contributions

CL conceived, designed and oversaw the study. LC performed statistical analyses. LC and CL wrote the paper. CED participated in data analyses. MV carried out quantification of fatty acids. JB, JA, MD, CF participated in the design of the study and contributed to acquisition of data. PS performed SNP and gene annotation. All authors read, commented and approved the final manuscript.

Acknowledgements

The authors thank Ivy Larsen for assistance on beef tissue collection, Shurong Xiong for fatty acid analyses, Ereddad Kharraz and Jonathan Curtis for assistance on fatty acid analyses. This work was supported by the Alberta Livestock and Meat Agency (ALMA) (2010R038R). Support (in kind) from Beefbooster Inc., Alberta, Canada to the project was also highly appreciated. This research has been enabled by the use of computing resources provided by WestGrid and Compute/Calcul Canada.

Author details

[1]Department of Agricultural, Food and Nutritional Science, University of Alberta, Edmonton, AB T6G 2P5, Canada. [2]Lacombe Research Centre, Agriculture and Agri-Food Canada, 6000 C&E Trail, Lacombe, AB T4L 1 W1, Canada. [3]Lacombe Research Centre, Alberta Agriculture and Forestry, 6000 C & E Trail, Lacombe, AB T4L 1 W1, Canada.

References

1. Pan A, Sun Q, Bernstein AM, Schulze MB, Manson JE, Stampfer MJ, et al. Red meat consumption and mortality results from 2 prospective cohort studies. Arch Intern Med. 2012;172(7):555–63.

2. Gormley TR, Downey G, O'Beirne D. Food, health and the consumer. London: Elsevier Applied Science Publishers Ltd.; 1987.

3. Raes K, de Smet S, Demeyer D. Effect of double-muscling in Belgian Blue young bulls on the intramuscular fatty acid composition with emphasis on conjugated linoleic acid and polyunsaturated fatty acids. Anim Sci. 2001;73:253–60.

4. Bassett CMC, Edel AL, Patenaude AF, McCullough RS, Blackwood DP, Chouinard PY, et al. Dietary vaccenic acid has antiatherogenic effects in LDLr(−/−) mice. J Nutr. 2010;140(1):18–24.

5. Corl BA, Barbano DM, Bauman DE, Ip C. Cis-9, trans-11 CLA derived endogenously from trans-11 18:1 reduces cancer risk in rats. J Nutr. 2003;133(9):2893–900.

6. Wijendran V, Hayes KC. Dietary n-6 and n-3 fatty acid balance and cardiovascular health. Annu Rev Nutr. 2004;24:597–615.

7. Uemoto Y, Abe T, Tameoka N, Hasebe H, Inoue K, Nakajima H, et al. Whole-genome association study for fatty acid composition of oleic acid in Japanese Black cattle. Anim Genet. 2011;42(2):141–8.

8. De Smet S, Raes K, Demeyer D. Meat fatty acid composition as affected by fatness and genetic factors: a review. Anim Res. 2004;53(2):81–98.

9. Vahmani P, Mapiye C, Prieto N, Rolland DC, McAllister TA, Aalhus JL, et al. The scope for manipulating the polyunsaturated fatty acid content of beef: a review. J Anim Sci Biotechnol. 2015;6(1):1–13.

10. Ekine-Dzivenu C, Chen L, Vinsky M, Aldai N, Dugan M, McAllister T, et al. Estimates of genetic parameters for fatty acids in brisket adipose tissue of Canadian commercial crossbred beef steers. Meat Sci. 2014;96(4):1517–26.

11. Inoue K, Kobayashi M, Shoji N, Kato K. Genetic parameters for fatty acid composition and feed efficiency traits in Japanese Black cattle. Animal. 2011;5(07):987–94.

12. Kelly M, Tume R, Newman S, Thompson J. Genetic variation in fatty acid composition of subcutaneous fat in cattle. Anim Prod Sci. 2013;53(2):129–33.

13. Malau-Aduli A, Edriss M, Siebert B, Bottema C, Pitchford W. Breed differences and genetic parameters for melting point, marbling score and fatty acid composition of lot-fed cattle. J Anim Physiol Anim Nutr. 2000;83(2):95–105.

14. Nogi T, Honda T, Mukai F, Okagaki T, Oyama K. Heritabilities and genetic correlations of fatty acid compositions in longissimus muscle lipid with carcass traits in Japanese Black cattle. J Anim Sci. 2011;89(3):615–21.

15. Pitchford W, Deland M, Siebert B, Malau-Aduliand A, Bottema C. Genetic variation in fatness and fatty acid composition of crossbred cattle. J Anim Sci. 2002;80(11):2825–32.

16. Saatchi M, Garrick DJ, Tait RG, Mayes MS, Drewnoski M, Schoonmaker J, et al. Genome-wide association and prediction of direct genomic breeding values for composition of fatty acids in Angus beef cattle. BMC Genomics. 2013;14(1):730.

17. Tait Jr RG, Zhang S, Knight T, Bormann JM, Strohbehn DR, Beitz DC, et al. Heritability estimates for fatty acid concentration in Angus beef. Anim Industry Rep. 2007;653(1):17.

18. Yokota S, Sugita H, Ardiyanti A, Shoji N, Nakajima H, Hosono M, et al. Contributions of FASN and SCD gene polymorphisms on fatty acid composition in muscle from Japanese Black cattle. Anim Genet. 2012;43(6):790–2.

19. Alexander L, MacNeil M, Geary T, Snelling W, Rule D, Scanga J. Quantitative trait loci with additive effects on palatability and fatty acid composition of meat in a Wagyu–Limousin F2 population. Anim Genet. 2007;38(5):506–13.

20. Gutierrez-Gil B, Wiener P, Richardson R, Wood J, Williams J. Identification of QTL with effects on fatty acid composition of meat in a Charolais × Holstein cross population. Meat Sci. 2010;85(4):721–9.

21. Morris C, Bottema C, Cullen N, Hickey S, Esmailizadeh A, Siebert B, et al. Quantitative trait loci for organ weights and adipose fat composition in Jersey and Limousin back-cross cattle finished on pasture or feedlot. Anim Genet. 2010;41(6):589–96.

22. Abe T, Saburi J, Hasebe H, Nakagawa T, Misumi S, Nade T, et al. Novel mutations of the FASN gene and their effect on fatty acid composition in Japanese Black beef. Biochem Genet. 2009;47(5–6):397–411.

23. Bartoň L, Kott T, Bureš D, Řehák D, Zahradkova R, Kottova B. The polymorphisms of stearoyl-CoA desaturase (SCD1) and sterol regulatory element binding protein-1 (SREBP-1) genes and their association with the fatty acid profile of muscle and subcutaneous fat in Fleckvieh bulls. Meat Sci. 2010;85(1):15–20.

24. Han C, Vinsky M, Aldai N, Dugan MER, McAllister TA, Li C. Association analyses of DNA polymorphisms in bovine SREBP-1, LXR alpha, FADS1 genes with fatty acid composition in Canadian commercial crossbred beef steers. Meat Sci. 2013;93(3):429–36.

25. Hoashi S, Ashida N, Ohsaki H, Utsugi T, Sasazaki S, Taniguchi M, et al. Genotype of bovine sterol regulatory element binding protein-1 (SREBP-1) is associated with fatty acid composition in Japanese Black cattle. Mamm Genome. 2007;18(12):880–6.

26. Hoashi S, Hinenoya T, Tanaka A, Ohsaki H, Sasazaki S, Taniguchi M, et al. Association between fatty acid compositions and genotypes of FABP4 and LXR-alpha in Japanese Black cattle. BMC Genet. 2008;9(1):84.

27. Oh D, La B, Lee Y, Byun Y, Lee J, Yeo G, et al. Identification of novel single nucleotide polymorphisms (SNPs) of the lipoprotein lipase (LPL) gene associated with fatty acid composition in Korean cattle. Mol Biol Rep. 2013;40(4):3155–63.

28. Oh D, Lee Y, Lee C, Chung E, Yeo J. Association of bovine fatty acid composition with missense nucleotide polymorphism in exon7 of peroxisome proliferator-activated receptor gamma gene. Anim Genet. 2012;43(4):474.

29. Oh DY, Lee YS, La BM, Lee JY, Park YS, Lee JH, et al. Identification of exonic nucleotide variants of the thyroid hormone responsive protein gene associated with carcass traits and fatty acid composition in Korean cattle. Asian Austral J Anim. 2014;27(10):1373–80.

30. Oh DY, Lee YS, La BM, Yeo JS. Identification of the SNP (single nucleotide polymorphism) for fatty acid composition associated with beef flavor-related FABP4 (fatty acid binding protein 4) in Korean cattle. Asian Austral J Anim. 2012;25(7):913–20.

31. Oh DY, Lee YS, Yeo JS. Identification of the SNP (single necleotide polymorphism) of the stearoyl-CoA desaturase (SCD) associated with unsaturated fatty acid in Hanwoo (Korean cattle). Asian Austral J Anim. 2011;24(6):757–65.

32. Ohsaki H, Tanaka A, Hoashi S, Sasazaki S, Oyama K, Taniguchi M, et al. Effect of SCD and SREBP genotypes on fatty acid composition in adipose tissue of Japanese Black cattle herds. Anim Sci J. 2009;80(3):225–32.

33. Orru L, Cifuni G, Piasentier E, Corazzin M, Bovolenta S, Moioli B. Association analyses of single nucleotide polymorphisms in the LEP and SCD1 genes on the fatty acid profile of muscle fat in Simmental bulls. Meat Sci. 2011;87(4):344–8.

34. Taniguchi M, Utsugi T, Oyama K, Mannen H, Kobayashi M, Tanabe Y, et al. Genotype of stearoyl-CoA desaturase is associated with fatty acid composition in Japanese Black cattle. Mamm Genome. 2004;15(2):142–8.

35. Zhang S, Knight TJ, Reecy JM, Wheeler TL, Shackelford SD, Cundiff LV, et al. Associations of polymorphisms in the promoter I of bovine acetyl-CoA carboxylase-alpha gene with beef fatty acid composition. Anim Genet. 2010;41(4):417–20.

36. Sevane N, Armstrong E, Cortés O, Wiener P, Wong RP, Dunner S, et al. Association of bovine meat quality traits with genes included in the PPARG and PPARGC1A networks. Meat Sci. 2013;94(3):328–35.

37. Ishii A, Yamaji K, Uemoto Y, Sasago N, Kobayashi E, Kobayashi N, et al. Genome-wide association study for fatty acid composition in Japanese Black cattle. Anim Sci J. 2013;84(10):675–82.

38. Kelly MJ, Tume RK, Fortes M, Thompson JM. Whole-genome association study of fatty acid composition in a diverse range of beef cattle breeds. J Anim Sci. 2014;92(5):1895–901.

39. Onogi A, Ogino A, Komatsu T, Shoji N, Shimizu K, Kurogi K, et al. Whole-genome prediction of fatty acid composition in meat of Japanese Black cattle. Anim Genet. 2015;46(5):557-559.

40. Menendez JA, Lupu R. Fatty acid synthase and the lipogenic phenotype in cancer pathogenesis. Nat Rev Cancer. 2007;7(10):763–77.

41. Morris CA, Cullen NG, Glass BC, Hyndman DL, Manley TR, Hickey SM, et al. Fatty acid synthase effects on bovine adipose fat and milk fat. Mamm Genome. 2007;18(1):64–74.

42. Matsuhashi T, Maruyama S, Uemoto Y, Kobayashi N, Mannen H, Abe T, et al. Effects of bovine fatty acid synthase, stearoyl-coenzyme a desaturase, sterol regulatory element-binding protein 1, and growth hormone gene polymorphisms on fatty acid composition and carcass traits in Japanese Black cattle. J Anim Sci. 2011;89(1):12–22.

43. Oh D, Lee Y, La B, Yeo J, Chung E, Kim Y, et al. Fatty acid composition of beef is associated with exonic nucleotide variants of the gene encoding FASN. Mol Biol Rep. 2012;39(4):4083–90.

44. Roy L. Association of polymorphisms in the bovine FASN gene with milk-fat content (vol 37, pg 215, 2006). Anim Genet. 2009;40(1):126.

45. Schennink A, Bovenhuis H, Leon-Kloosterziel KM, van Arendonk JAM, Visker MHPW. Effect of polymorphisms in the FASN, OLR1, PPARGC1A, PRL and STAT5A genes on bovine milk-fat composition. Anim Genet. 2009;40(6):909–16.

46. La B, Oh D, Lee Y, Shin S, Lee C, Chung E, et al. Association of bovine fatty acid composition with novel missense nucleotide polymorphism in the thyroid hormone-responsive (THRSP) gene. Anim Genet. 2013;44(1):118.

47. Barrett JC, Fry B, Maller J, Daly MJ. Haploview: Analysis and visualization of ld and haplotype maps. Bioinformatics. 2005;21(2):263–5.

48. Kim Y-C, Ntambi JM. Regulation of stearoyl-CoA desaturase genes: role in cellular metabolism and preadipocyte differentiation. Biochem Biophys Res Commun. 1999;266(1):1–4.

49. Bauman D, Baumgard L, Corl B, Griinari dJ. Biosynthesis of conjugated linoleic acid in ruminants. Proc Am Soc Anim Sci. 1999;1999:1–14.

50. Li C, Aldai N, Vinsky M, Dugan MER, McAllister TA. Association analyses of single nucleotide polymorphisms in bovine stearoyl-CoA desaturase and fatty acid synthase genes with fatty acid composition in commercial cross-bred beef steers. Anim Genet. 2012;43(1):93–7.

51. Maharani D, Jung Y, Jung WY, Jo C, Ryoo SH, Lee SH, et al. Association of five candidate genes with fatty acid composition in Korean cattle. Mol Biol Rep. 2012;39(5):6113–21.

52. Mele M, Conte G, Castiglioni B, Chessa S, Macciotta NPP, Serra A, et al. Stearoyl-coenzyme a desaturase gene polymorphism and milk fatty acid composition in Italian Holsteins. J Dairy Sci. 2007;90(9):4458–65.

53. Milanesi E, Nicoloso L, Crepaldi P. Stearoyl CoA desaturase (SCD) gene polymorphisms in Italian cattle breeds. J Anim Breed Genet. 2008;125(1):63–7.

54. Moioli B, Contarini G, Avalli A, Catillo G, Orru L, De Matteis G, et al. Short communication: effect of stearoyl-coenzyme a desaturase polymorphism on fatty acid composition of milk. J Dairy Sci. 2007;90(7):3553–8.

55. Narukami T, Sasazaki S, Oyama K, Nogi T, Taniguchi M, Mannen H. Effect of DNA polymorphisms related to fatty acid composition in adipose tissue of Holstein cattle. Anim Sci J. 2011;82(3):406–11.

56. Rincon G, Islas-Trejo A, Castillo AR, Bauman DE, German BJ, Medrano JF. Polymorphisms in genes in the SREBP1 signalling pathway and SCD are associated with milk fatty acid composition in Holstein cattle. J Dairy Res. 2012;79(1):66–75.

57. Bouwman AC, Bovenhuis H, Visker MH, van Arendonk JA. Genome-wide association of milk fatty acids in Dutch dairy cattle. BMC Genet. 2011;12(1):43.

58. Lu SC. Glutathione synthesis. Bba-Gen Subjects. 2013;1830(5):3143–53.

59. Bhattacharya TK, Misra SS, Sheikh FD, Sukla S, Kumar P, Sharma A. Effect of butyrophilin gene polymorphism on milk quality traits in crossbred cattle. Asian Austral J Anim. 2006;19(7):922–6.

60. Franke WW, Heid HW, Grund C, Winter S, Freudenstein C, Schmid E, et al. Antibodies to the major insoluble milk fat globule membrane-associated protein: Specific location in apical regions of lactating epithelial cells. J Cell Biol. 1981;89(3):485–94.

61. Simonson M, Dunn M. The molecular mechanisms of cardiovascular and renal regulation by endothelin peptides. J Lab Clin Med. 1992;119(6):622–39.

62. Wu-Wong JR, Dayton BD, Opgenorth TJ. Endothelin-1-evoked arachidonic acid release: a Ca (2+)-dependent pathway. Am J Physiol Cell Physiol. 1996;271(3):C869–77.

63. Lourenco D, Tsuruta S, Fragomeni B, Masuda Y, Aguilar I, Legarra A, et al. Genetic evaluation using single-step genomic best linear unbiased predictor in American Angus. J Anim Sci. 2015;93(6):2653–62.

64. Goddard M. Genomic selection: Prediction of accuracy and maximisation of long term response. Genetica. 2009;136(2):245–57.

65. Hayes BJ, Pryce J, Chamberlain AJ, Bowman PJ, Goddard ME. Genetic architecture of complex traits and accuracy of genomic prediction: coat colour, milk-fat percentage, and type in Holstein cattle as contrasting model traits. PLoS Genet. 2010;6(9):e1001139.

66. Harfoot C, Hazlewood G: Lipid metabolism in the rumen. In: The rumen microbial ecosystem. London: Blackie Academic & Professional; 1997: 382–426.

67. Grinari J, Bauman DE. Biosynthesis of conjugated linoleic acid and its incorporation into meat and milk in ruminants. Adv Conjugated Linoleic Acid Res. 1999;1:180–200.

68. Canadian Council on Animal Care. Guide to the care and use of experimental animals. Olfert EB, Cross BM and McWilliam AA, eds. Vol. 1, 2nd ed. Ottawa, ON: CCAC; 1993.

69. Basarab J, Colazo M, Ambrose D, Novak S, McCartney D, Baron V. Residual feed intake adjusted for backfat thickness and feeding frequency is independent of fertility in beef heifers. Can J Anim Sci. 2011;91(4):573–84.

70. Basarab J, McCartney D, Okine E, Baron V. Relationships between progeny residual feed intake and dam productivity traits. Can J Anim Sci. 2007;87(4):489–502.

71. López-Campos Ó, Aalhus JL, Okine EK, Baron VS, Basarab JA. Effects of calf-and yearling-fed beef production systems and growth promotants on production and profitability. Can J Anim Sci. 2013;93(1):171–84.

72. Akanno E, Plastow G, Li C, Miller S, Basarab J. Accuracy of molecular breeding values for production and efficiency traits of Canadian crossbred beef cattle using a cross-validation approach. In: 10th World Congress on Genetics Applied to Livestock Production: August 17-22 2014; Vancouver, BC, Canada: Asas; 2014.

73. Lee H-J, Park H-S, Kim W, Yoon D, Seo S. Comparison of metabolic network between muscle and intramuscular adipose tissues in Hanwoo beef cattle using a systems biology approach. Int J Genomics. 2014;2014.

74. Folch J, Lees M, Stanley GHS. A simple method for the isolation and purification of total lipids from animal tissues. J Biol Chem. 1957;226(1):497–509.

75. Cruz-Hernandez C, Deng ZY, Zhou JQ, Hill AR, Yurawecz MP, Delmonte P, et al. Methods for analysis of conjugated linoleic acids and trans-18: 1 isomers in dairy fats by using a combination of gas chromatography, silver-ion thin-layer chromatography/gas chromatography, and silver-ion liquid chromatography. J AOAC Int. 2004;87(2):545–62.

76. Dugan MER, Kramer JKG, Robertson WM, Meadus WJ, Aldai N, Rolland DC. Comparing subcutaneous adipose tissue in beef and muskox with emphasis on trans 18:1 and conjugated linoleic acids. Lipids. 2007;42(6):509–18.

77. Kramer JKG, Hernandez M, Cruz-Hernandez C, Kraft J, Dugan MER. Combining results of two GC separations partly achieves determination of all cis and trans 16:1, 18:1, 18:2 and 18:3 except CLA isomers of milk fat as demonstrated using Ag-Ion SPE fractionation. Lipids. 2008;43(3):259–73.

78. Browning SR, Browning BL. Rapid and accurate haplotype phasing and missing-data inference for whole-genome association studies by use of localized haplotype clustering. Am J Hum Genet. 2007;81(5):1084–97.

79. VanRaden PM. Efficient methods to compute genomic predictions. J Dairy Sci. 2008;91(11):4414–23.

80. Gilmour AR, Gogel BJ, Cullis BR, Thompson R. ASREML user guide release 3.0. In. Hemel Hempstead, HP1 1ES, UK: VSN International Ltd.; 2009. www.vsni.co.uk.

81. Habier D, Fernando RL, Kizilkaya K, Garrick DJ. Extension of the Bayesian alphabet for genomic selection. BMC Bioinformatics. 2011;12.

82. Chen L, Li C, Schenkel F. An alternative computing strategy for genomic prediction using a Bayesian mixture model. Can J Anim Sci. 2015;95(1):1–11.

83. Churchill GA, Doerge RW. Empirical threshold values for quantitative trait mapping. Genetics. 1994;138(3):963–71.

84. Grant JR, Arantes AS, Liao X, Stothard P. In-depth annotation of SNPs arising from resequencing projects using NGS-SNP. Bioinformatics. 2011;27(16):2300–1.

A forest-based feature screening approach for large-scale genome data with complex structures

Gang Wang, Guifang Fu* and Christopher Corcoran

Abstract

Background: Genome-wide association studies (GWAS) interrogate large-scale whole genome to characterize the complex genetic architecture for biomedical traits. When the number of SNPs dramatically increases to half million but the sample size is still limited to thousands, the traditional *p*-value based statistical approaches suffer from unprecedented limitations. Feature screening has proved to be an effective and powerful approach to handle ultrahigh dimensional data statistically, yet it has not received much attention in GWAS. Feature screening reduces the feature space from millions to hundreds by removing non-informative noise. However, the univariate measures used to rank features are mainly based on individual effect without considering the mutual interactions with other features. In this article, we explore the performance of a random forest (RF) based feature screening procedure to emphasize the SNPs that have complex effects for a continuous phenotype.

Results: Both simulation and real data analysis are conducted to examine the power of the forest-based feature screening. We compare it with five other popular feature screening approaches via simulation and conclude that RF can serve as a decent feature screening tool to accommodate complex genetic effects such as nonlinear, interactive, correlative, and joint effects. Unlike the traditional *p*-value based Manhattan plot, we use the Permutation Variable Importance Measure (PVIM) to display the relative significance and believe that it will provide as much useful information as the traditional plot.

Conclusion: Most complex traits are found to be regulated by epistatic and polygenic variants. The forest-based feature screening is proven to be an efficient, easily implemented, and accurate approach to cope whole genome data with complex structures. Our explorations should add to a growing body of enlargement of feature screening better serving the demands of contemporary genome data.

Keywords: Feature screening, GWAS, Epistasis, Random forest, Large-scale modeling

Background

High-throughput genotyping techniques and large data repository capability give genome-wide association studies (GWAS) great power to unravel the genetic etiology of complex traits. With the number of Single Nucleotide Polymorphisms (SNPs) per DNA array growing from 10,000 to 1 million [1], ultra-high dimensionality is one of the grand challenges in GWAS. The prevailing strategies of GWAS focus on single-locus model [2, 3]. However, most complex traits are regulated by polygenetic variants, which decreases the power of most popular traditional *p*-value based approaches [4–7].

Epistasis [2, 8, 9], defined as the interactive effects of two or more genetic variants (i.e. the effect of one genetic variant is suppressed or enhanced by other genetic variants), has received growing attention in GWAS due to increasing evidence of its important role in the development of complex diseases [7, 10–12]. Epistasis will likely bring key breakthroughs for detecting more susceptible loci for various real life scenarios and for explaining larger heritability of traits [13–16]. Many approaches have already

*Correspondence: guifang.fu@usu.edu
Department of Mathematics and Statistics, Utah State University, 3900 Old Main, 84322 Logan, UT, US

been developed for detecting epistasis [17–20]. Despite the fact that these approaches work nicely for detecting epistasis with a moderate number of SNPs ($n > p$), they quickly lose power and suffer from computational burden when the dimension is ultrahigh ($n >> p$) [12].

There exists a big gap between current statistical modeling of big data and the real demand of contemporary entire genome data. Fan et al. elaborately introduced the unusually big challenges in computational cost, statistical estimation accuracy, and algorithm stability caused by ultrahigh dimensional data [21–23]. The population covariance matrix may become ill conditioned as dimension grows as multicollinearity grows with dimensionality. As a result, the number and extent of spurious correlations between a feature and response increase rapidly with increasing dimension because unimportant features are often highly correlated with a truly important one. What increases the difficulty is that multiple genetic variants affect the phenotype in an interactive or correlative manner but each have a weak marginal signal. Additionally, without any priori information, modeling and searching all possible pairwise and higher order interactions is intractable when the number of features is very large. For example, there will be around 8 million pairs involved when simply considering 2-way interactions for only 4000 SNPs [24].

Feature Screening brings about a revolutionary time in statistics due to its advantages in handling ultrahigh dimensional data. It also fills the gap between traditional statistical approaches and demands of contemporary genomics [25]. The sparsity principle (only a small number of SNPs associate with the phenotype) of the whole genome data matches well with the goal of the feature screening. It has been confirmed that the computational speed and estimation accuracy are both improved after dimension is reduced from ultrahigh to moderate size [26]. The computational burden reduces dramatically, from a huge scale (say $\exp\{O(n^h)\}$) to $o(n)$. Most important of all, aforementioned traditional statistical approaches regain their power and feasibility after feature screening removes the majority of confounding noises. Fan and Lv proposed sure independence screening (SIS) and iterated sure independence screening (ISIS) [26] to overcome the challenges of ultra-high dimension. SIS is shown to have the sure screening property (all truly important predictors can be selected with the probability tending to one as the sample size asymptotically diverges to ∞ [26, 27]) for the case of $n >> p$. Fan and Song developed SIS for generalized linear models [28]. Li et al. proposed distance correlation learning (DC-SIS) without assuming linear relation or restricting data type [27, 29]. Liu et al. proposed conditional correlation sure independence screening (CC-SIS) to adjust the confounding effect of a covariate [30].

Although the advantages of the feature screening have been sufficiently shown, almost all current feature screening approaches assign univariate rankings to consider the individual effect of each feature and hence neglect features that have weak marginal but strong joint or interactive effects. In addition, most existing feature screening approaches are not well-designed for examining two, three, or higher-order interactive structures and nonlinear structures. As an alternative direction, Random Forest (RF) overcomes the aforementioned drawbacks of feature screening. RF uncovers interactive effects even if the relevant features only have weak marginal signals [31]. Each hierarchical decision tree within the RF explicitly represents the attribute interaction of features through the branches of the tree. As a result, as more and higher order interactive SNPs are added to the model, the superiority of RF increases. In particular, RF was claimed to outperform Fisher's exact test when interactive effects exist [32]. RF can be flexibly modeled to both continuous and categorical phenotype and nonlinear structures without assuming any model structure or interaction forms.

The aim of this article is to assess the performance of a forest-based feature screening approach for large-scale whole genome data with complex genetic structures such as epistatic, polygenic, correlative, and nonlinear effects. The key problem that we emphasize is to select a manageable number of important candidates from an ultrahigh dimension of SNP pool, while keeping the case of strong marginal signal, the case of of weak marginal but strong interactive or correlative SNPs, and keeping both linear and nonlinear structures. Unlike the traditional p-value based Manhattan plot, we view the significance of SNPs using permutation variable importance measure (PVIM). The PVIM based Manhattan plot can provide as much helpful information as the traditional p-value based Manhattan plot, additionally it considers the individual effect of each SNP as well as accounting for the mutual joint effects of all other SNPs in a multivariate sense. In current literature, a few studies have already assessed the performance of RF for detecting epistasis [32–35], but they all focused on binary/case-control phenotype. Additionally, current literature simply consideres two-way interaction simulations and it is not clear whether or not RF can perform well for more complex interactions. Instead, we explored the performance of RF for quantitative/continuous traits and additionally increased the complexity level by considering nonlinearity, correlation, and more difficult interaction simultaneously.

Results and discussion
Power simulation
To illustrate the power of RF as a feature screening tool for detecting correlative, nonlinear, and interactive effects, we designed four different simulation settings to control

linear vs nonlinear, constant vs functional, and additive vs interactive features. We compare RF with five popular feature screening tools, SIS [26], ISIS [26], CC-SIS [30], ICC-SIS [30], and DC-SIS [27]. In order to make the comparisons fair, we keep some of their original simulation settings the same, as well as design other settings different to accommodate the emphasis of this study.

The sample size n is set to be 200. Let $X = (x_1, \ldots, x_p)^T \sim N(\mathbf{0}, \Sigma)$ be the feature matrix with dimension $p = 1000$. By controlling the component $\sigma_{ij} = \rho^{|i-j|}, i, j = 1, \ldots, p$ of covariance matrix Σ, the correlations among features are introduced. All the values of βs are zero, except the truly causative features. Among the 1000 features, we set the first five to be truly associated with phenotype and all others be noise by letting

$$Y = \beta_1 x_1 + \beta_2 x_2 + \beta_3 x_3 + \beta_4 x_4 x_5 + \epsilon, \qquad (1)$$

for the linear and moderate interactive setting, and

$$Y = \beta_1 x_1^2 + \beta_2 x_2 x_3 + \beta_3 x_4 x_5 + \epsilon. \qquad (2)$$

for the nonlinear and strong interactive setting. The noise ϵ is randomly generated from white noise $N(0, 1)$.

Simulation 1

For Sim 1, we consider three linear and one interactive terms with constant parameters. i.e. Y is generated based on Eq. (1), $\rho = 0.4$, and βs are set to be $\beta = (0.5, 0.8, 1, 2)$.

Simulation 2

For Sim 2, we consider one nonlinear and two interactive terms with constant parameters. i.e. Y is generated based on Eq. (2), $\rho = 0.4$, and βs are set to be $\beta = (2, 3, 4)$.

Simulation 3

For Sim 3, we consider three linear and one interactive terms with functional parameters. i.e. Y is generated based on Eq. (1), $\rho = 0.4$, and βs are generated by $\beta_1 = 2 + (u+1)^3$, $\beta_2 = \frac{2u^2+3}{2}$, $\beta_3 = e^{\frac{4u}{u+4}}$, and $\beta_4 = \cos\left(\frac{8u^2}{2}\right) + 2$. In order to introduce the correlation between each feature and a covariate u, we generate $(u^*, X) \sim N(\mathbf{0}, \Sigma^*)$, here Σ^* is $(p+1) \times (p+1)$ dimension using similar AR(1) structure as above Σ. Then we generate u by $u = \Phi(u^*)$, here $\Phi(.)$ is the cumulative distribution function (cdf) of the standard normal distribution. By the theoretical properties of cdf, u follows a uniform distribution $U(0, 1)$ and is correlated with X. The functional parameter $\beta(u)$ is useful to explain personalized covariate effects that vary

for different individuals due to different genetic information and other factors [30].

Simulation 4

For Sim 4, we consider one nonlinear and two interactive terms with functional parameters. i.e. Y is generated based on Eq. (2), $\rho = 0.4$, and βs are generated by $\beta_1 = 2 + \cos\left(\frac{\pi(6u-5)}{3}\right)$, $\beta_2 = (4 - 4u)e^{\frac{3u^2}{3u^2+1}}$, and $\beta_3 = u + 2$. u and X are generated using the same rule as Sim 3. This setting has the hardest conditions that hinder most approaches from detecting the truly causative features.

The comparisons were assessed based on 100 simulation replications. Three traditional criteria that frequently appeared in feature screening literature [27], R, p, and M, are used to compare the performances of six approaches.

- R_j, $j = 1, \ldots, 5$, is defined as the average rank of each causative feature x_j for 100 replications. Since the most important feature is ranked as top one, smaller R for causative features means better performance.
- $M = \max R_j$, $j = 1, \ldots, 5$, is defined as the minimum size of the candidate containing all five causative features. Therefore, M close to five means good performance. Like other feature screening studies, we also compared the 5, 25, 50, 75, and 95 % quantiles of M for the 100 replications. These quantiles display how effective each approach is during selection process.
- d is defined as the pre-specified number of candidates that will be chosen as important. In real life data, we do not know the minimum size containing all causative features. Liu et al. [30] suggested to use the multiplier of the integer part of $d = \left[n^{4/5} / \log\left(n^{4/5}\right)\right]$. i.e. for $n = 200$, d is suggested to be 16, 32, and 48, and so on. We use the same values to make the comparisons fair.
- p_j, $j = 1, \ldots, 5$, is defined as the percentage of each x_j being successfully selected within size d among 100 replications. The larger p_j, the more accurate (higher individual power).
- p_a is defined as the percentage of all five causative features being successfully selected within size d among 100 replications. The larger p_a, the more accurate (higher overall power).

The comparative results of the constant parameters for Sim 1 and Sim 2 are summarized in Tables 1, 2 and 3. Table 1 reports the average rank of all five causative features. For Sim 1, the first three features have linear marginal effects but x_4 and x_5 have interactive effects. The marginal effect of x_1 is designed to be smaller than

Table 1 The average rank of each causative feature, R_j, for Simulation 1 & 2

METHOD	Sim1					Sim2				
	R1	R2	R3	R4	R5	R1	R2	R3	R4	R5
SIS	12.21	1.56	1.51	143.14	322.16	359.17	360.41	398.89	340.45	428.30
ISIS	39.29	1.56	1.51	250.98	412.43	432.97	456.97	481.98	426.94	502.13
CC-SIS	12.81	1.59	1.48	60.31	179.77	168.57	242.27	242.85	258.39	369.68
ICC-SIS	43.75	1.59	1.48	129.80	259.34	237.70	362.12	382.58	368.86	400.27
DC-SIS	5.95	1.59	1.48	7.93	19.58	3.51	21.07	32.86	7.44	14.79
RF	8.63	1.91	1.67	3.72	4.06	2.80	8.59	10.70	4.66	7.85

that of x_2 or x_3 by setting $\beta_1 = 0.5, \beta_2 = 0.8$, and $\beta_3 = 1$. For the simplest scenario (strong linear marginal effects of x_2 and x_3), all six approaches achieve remarkable results with the average ranks R_2 and R_3 all less than 2. It means that all six feature screening approaches successfully locate these two causative features as the top two. For the weak linear marginal effect of x_1, it seems that the iterative approaches perform worse than their corresponding original approaches, say ISIS 39.29 versus SIS 12.21 and ICC-SIS 43.75 versus CC-SIS 12.81. In the reports of Fan et al. and Liu et al., the iterative procedure greatly improved the results compared to that of previous iterative procedures under all their reported scenarios [26, 30]. Therefore, we still agree with the advantages of iterative approaches, but maintain that our new findings can help readers gain insight about the pitfalls and benefits of each approach. The six approaches behave dramatically different for the interactive terms x_4 and x_5. Both R_4 and R_5 obtained from the first four approaches are very large, which means that they rank hundreds of other candidates before these two causative features. Compared to the 412.43 of ISIS and 179.77 of CC-SIS, RF achieves a rank as small as 4.06. Observing the last row of Table 1, we conclude that RF detects all five causative features using the smallest number of candidates (less than 9 in average). One more thing worth mentioning is that RF ranks the features with strong interactive but weak marginal effects (3.72 for x_4 and 4.06 for x_5) more important than features with weak marginal effects (8.65 for x_1). The overall importance rank of RF combines all related effects rather than simply considering marginal importance.

For Sim 2, x_1 has a nonlinear effect and all other four features have interactive effects. This setting is much more difficult than Sim 1. As a result, all five ranks achieved by the first four approaches dramatically increased from decades in Sim 1 to hundreds in Sim 2. RF consistently performs best for this harder condition by locating all five causative features with complex structures within 11 candidates on average. Compared the results of Sim 1 and Sim 2 in Table 1, all six approaches get worse in harder conditions, but the differences of RF is negligible, with 8.63 versus 10.70. It indicates that RF is more robust than the other five approaches under harder conditions.

Table 2 reports five quantiles of M, the minimum size of candidates containing all the five truly causative features, among 100 simulation replicates. The first four approaches have a 95 % quantile as large as 958 for Sim 1 and 986 for Sim 2, meaning the detection of interactive terms fails. Among the 100 simulation replicates, the five quantiles of RF are relatively unchanged. To be more specific, 50 % of the replicates locate all five truly causative features using 5 candidates (a perfect match), 75 % of the replicates locate all five truly causative features by 8 candidates, and 95 % of the replicates locate truth by 17 candidates. Comparing the span from 5–95 % of these six approaches, we conclude that RF is very effective and accurate in locating important causative features.

Table 3 reports the powers achieved by three different pre-specified sizes $d = 16, 32$ and 48. For a small size $d = 16$, RF already achieves a power as large as 93 %, while the first four approaches only a power of 15 %. When d triples, the power of DC-SIS increases from 77–94 % but

Table 2 The quantiles of M, for Simulation 1 & 2

METHOD	Sim1					Sim2				
	5 %	25 %	50 %	75 %	95 %	5 %	25 %	50 %	75 %	95 %
SIS	15.60	72.25	339.50	646.00	887.75	257.25	681.00	817.50	888.00	970.20
ISIS	14.65	331.75	597.50	756.25	958.00	555.85	766.75	875.00	954.75	986.15
CC-SIS	7.90	34.75	107.00	288.50	703.80	131.85	357.25	605.50	812.75	957.30
ICC-SIS	7.90	150.50	357.50	530.25	838.60	387.35	614.50	784.00	865.75	951.25
DC-SIS	5.00	6.00	8.00	16.25	55.20	7.00	16.50	31.00	66.50	152.60
RF	5.00	5.00	5.00	8.00	17.05	5.00	7.75	11.00	22.00	67.15

Table 3 The overall and individual power, p_a and p_j, for Simulation 1 & 2

d	METHOD	Sim1						Sim2					
		p_1	p_2	p_3	p_4	p_5	p_a	p_1	p_2	p_3	p_4	p_5	p_a
	SIS	0.97	0.97	0.97	0.48	0.09	0.08	0.09	0.04	0.01	0.01	0.01	0.00
	ISIS	0.95	0.95	0.95	0.41	0.11	0.06	0.09	0.02	0.01	0.03	0.01	0.00
	CC-SIS	0.95	0.95	0.95	0.61	0.17	0.15	0.34	0.13	0.09	0.07	0.01	0.00
16	ICC-SIS	0.91	0.91	0.91	0.52	0.16	0.11	0.32	0.09	0.09	0.04	0.01	0.00
	DC-SIS	0.99	0.99	0.99	0.92	0.79	0.77	0.95	0.67	0.56	0.81	0.72	0.30
	RF	0.93	0.93	0.93	0.93	0.93	0.93	0.99	0.88	0.84	0.95	0.89	0.67
	SIS	0.97	0.97	0.97	0.55	0.18	0.14	0.09	0.05	0.04	0.03	0.02	0.00
	ISIS	0.95	0.95	0.95	0.42	0.12	0.07	0.09	0.02	0.01	0.03	0.01	0.00
	CC-SIS	0.95	0.95	0.95	0.67	0.29	0.22	0.34	0.17	0.15	0.08	0.04	0.00
32	ICC-SIS	0.91	0.91	0.91	0.55	0.22	0.14	0.32	0.09	0.10	0.05	0.01	0.00
	DC-SIS	0.99	0.99	0.99	0.94	0.90	0.86	0.95	0.78	0.67	0.91	0.84	0.48
	RF	0.93	0.93	0.93	0.93	0.93	0.93	0.99	0.94	0.92	0.97	0.95	0.82
	SIS	0.97	0.97	0.97	0.57	0.20	0.16	0.09	0.05	0.04	0.03	0.02	0.00
	ISIS	0.95	0.95	0.95	0.42	0.13	0.07	0.09	0.02	0.02	0.03	0.01	0.00
	CC-SIS	0.95	0.95	0.95	0.74	0.35	0.30	0.34	0.19	0.17	0.08	0.05	0.00
48	ICC-SIS	0.91	0.91	0.91	0.60	0.24	0.16	0.32	0.10	0.11	0.07	0.03	0.00
	DC-SIS	0.99	0.99	0.99	0.97	0.96	0.94	0.95	0.85	0.75	0.94	0.92	0.64
	RF	0.93	0.93	0.93	0.93	0.93	0.93	0.99	0.96	0.95	0.99	0.95	0.88

the power of RF keeps all the same as 93 %. Additionally, the five individual powers of RF do not differ much like other approaches. These findings confirm that RF detects all true causative features with high efficiency and high accuracy for complex structures.

The comparative results of the functional parameters for Sim 3 and Sim 4 are summarized in Tables 4, 5 and 6. Closely inspecting the results of Tables 4, 5 and 6, we find that the superiorities of RF over all other five approaches are similar as summarized in Tables 1, 2 and 3. For Sim 3, the first three features have linear marginal effects but x_4 and x_5 have interactive effect. The parameter βs are designed to be nonlinear and complex functions of a covariate u. For Sim 4, x_1 is in nonlinear form, and the

interactions are very strong because x_2 interacts with x_3 and x_4 interacts with x_5. The βs are designed to be more complex functions of u. The six approaches all do well for x_1 through x_3 under Sim 3, but RF beats all other five approaches under the remaining scenarios (see Tables 4, 5 and 6). DC-SIS has performed as better as RF in the first two simulations but lost its power for Sim 3 and Sim 4.

Summarized from Tables 1, 2, 3, 4, 5 and 6, we conclude that RF performs uniformly best among the six feature screening approaches. In particular, RF stands out under harder conditions. We know that Sim 2 and Sim 4 have more harsh conditions than that of Sim 1 and Sim 3. However, if comparing the left panel and right panel of these tables, we notice that while the majority of approaches get

Table 4 The average rank of each causative feature, R_j, for Simulation 3 & 4

METHOD	Sim3					Sim4				
	R_1	R_2	R_3	R_4	R_5	R_1	R_2	R_3	R_4	R_5
SIS	1.00	2.00	3.00	160.15	379.58	262.87	369.88	392.07	363.33	494.10
ISIS	1.00	2.00	3.00	353.95	518.05	311.39	416.71	485.63	428.34	461.79
CC-SIS	1.00	2.00	3.00	140.46	376.82	26.58	155.47	199.26	269.65	409.99
ICC-SIS	1.00	2.00	3.00	285.10	429.93	44.73	305.71	316.91	344.90	417.34
DC-SIS	1.00	2.01	2.99	111.32	228.75	1.35	16.93	27.88	30.67	57.52
RF	1.00	2.01	3.14	59.87	107.06	1.25	6.58	13.66	14.98	26.18

Table 5 The quantiles of M, for Simulation 3 & 4

METHOD	Sim3					Sim4				
	5 %	25 %	50 %	75 %	95 %	5 %	25 %	50 %	75 %	95 %
SIS	24.95	176.00	380.50	711.25	960.75	384.20	599.75	787.00	926.75	992.05
ISIS	225.30	425.00	624.00	827.00	956.05	361.70	688.25	796.50	917.00	983.05
CC-SIS	31.90	165.50	393.00	662.75	883.60	43.70	330.00	623.00	811.50	959.55
ICC-SIS	95.45	321.00	538.00	754.50	936.90	209.75	479.00	721.00	867.25	961.35
DC-SIS	15.00	57.50	205.50	445.25	692.85	10.00	29.00	61.50	115.00	228.30
RF	7.00	14.50	65.00	189.25	603.05	6.00	12.00	19.50	42.75	149.75

caught by the traps of complexity, RF obtains either similar or even better results.

Mice HDL GWAS project

Epidemiological studies have consistently shown that the level of plasma high density lipoprotein (HDL) cholesterol is negatively correlated with the risks of coronary artery disease and gallstones [36–38]. Therefore, there has been considerable interest in understanding genetic mechanisms contributing to variations in HDL levels. Zhang et al. published an open resource outbred mouse database with 288 Naval Medical Research Institute (NMRI) mice and 44,428 unique SNP genotypes (available at http://cgd.jax.org/datasets/datasets.shtml) [39]. A total of 581,672 high density SNP were initially genotyped by the Novartis Genomics Factory using the Mouse Diversity Genotyping Array [40]. Quality control was made and only polymorphic SNPs with minor allele frequency greater than 2 %, Hardy-Weinberg equilibrium $\chi^2 < 20$, and missing values less than 40 % were retained [41]. Moreover, identical SNPs within a 2 Mb interval were collapsed. This left 44,428 unique SNP genotypes for final analysis.

We implemented RF as the feature screening tool to this data to compare our findings with the highly validated discoveries in current literature. Figure 1 depicts the PVIM for each SNP as a function of the SNP location (in Mb) for 19 chromosomes. The two dramatic peaks detected by RF are located at *Chr1* at *Mb173* and *Chr5* at *Mb125*, which

Table 6 The overall and individual power, p_a and p_j, for Simulation 3 & 4

d	METHOD	Sim3						Sim4					
		p_1	p_2	p_3	p_4	p_5	p_a	p_1	p_2	p_3	p_4	p_5	p_a
	SIS	1.00	1.00	1.00	0.41	0.03	0.02	0.22	0.08	0.02	0.03	0.03	0.00
	ISIS	1.00	1.00	1.00	0.31	0.01	0.00	0.16	0.06	0.03	0.03	0.02	0.00
	CC-SIS	1.00	1.00	1.00	0.37	0.04	0.03	0.86	0.30	0.22	0.06	0.07	0.00
16	ICC-SIS	1.00	1.00	1.00	0.27	0.02	0.02	0.83	0.24	0.16	0.06	0.05	0.00
	DC-SIS	1.00	1.00	1.00	0.42	0.09	0.09	1.00	0.73	0.66	0.58	0.35	0.11
	RF	1.00	1.00	1.00	0.60	0.37	0.32	1.00	0.94	0.83	0.79	0.69	0.49
	SIS	1.00	1.00	1.00	0.48	0.08	0.06	0.22	0.08	0.04	0.04	0.03	0.00
	ISIS	1.00	1.00	1.00	0.32	0.02	0.00	0.16	0.06	0.03	0.03	0.02	0.00
	CC-SIS	1.00	1.00	1.00	0.50	0.08	0.06	0.86	0.40	0.31	0.14	0.11	0.02
32	ICC-SIS	1.00	1.00	1.00	0.30	0.05	0.02	0.83	0.27	0.18	0.12	0.08	0.00
	DC-SIS	1.00	1.00	1.00	0.54	0.22	0.16	1.00	0.87	0.73	0.76	0.55	0.28
	RF	1.00	1.00	1.00	0.66	0.47	0.37	1.00	0.97	0.93	0.88	0.80	0.66
	SIS	1.00	1.00	1.00	0.54	0.12	0.08	0.22	0.08	0.04	0.04	0.04	0.00
	ISIS	1.00	1.00	1.00	0.32	0.03	0.00	0.16	0.06	0.04	0.03	0.02	0.00
	CC-SIS	1.00	1.00	1.00	0.54	0.13	0.08	0.86	0.46	0.33	0.20	0.12	0.04
48	ICC-SIS	1.00	1.00	1.00	0.33	0.08	0.02	0.83	0.28	0.19	0.14	0.10	0.01
	DC-SIS	1.00	1.00	1.00	0.60	0.28	0.22	1.00	0.91	0.81	0.81	0.65	0.40
	RF	1.00	1.00	1.00	0.71	0.58	0.41	1.00	0.99	0.94	0.94	0.88	0.77

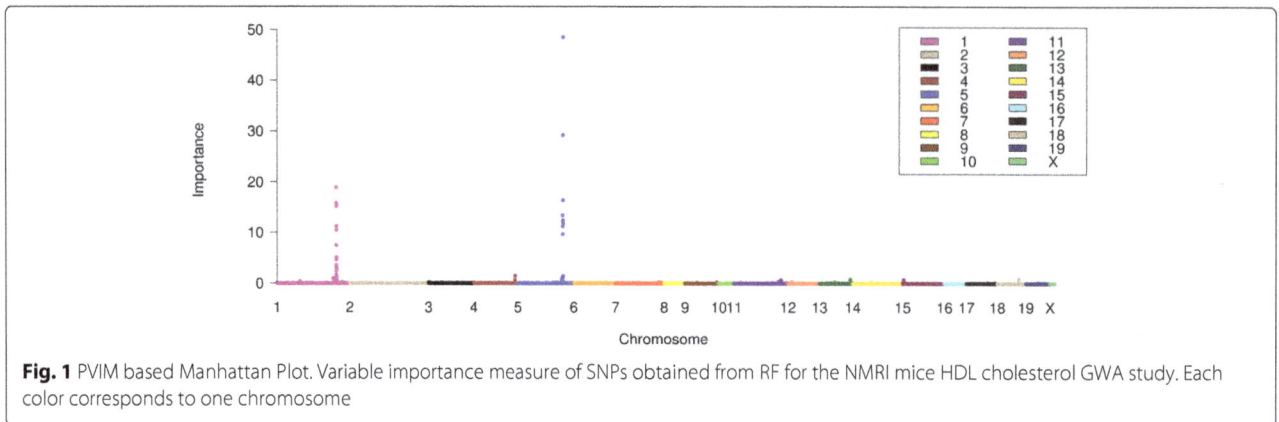

Fig. 1 PVIM based Manhattan Plot. Variable importance measure of SNPs obtained from RF for the NMRI mice HDL cholesterol GWA study. Each color corresponds to one chromosome

are exactly the same as other reports for the same data, but with a couple of advantages. First, type I error is not a problem here. In traditional p-value based Manhattan plots, there exist lots of signals surrounding the peaks and these signals can be so dense and strong (slightly above the threshold line) that it is hard to determine them as type I error or not. However, we notice that the signals in Fig 1 are polar opposites, with only two peaks standing out and all other SNPs shrinking towards zero. With such a clear trend, no one will doubt whether all SNPs other than the two peaks are type I error or truly causative genetic variants. Second, we achieve the same results more directly. Zhang et al. identified three loci as significant, with two loci on Chromosome 1 (Chr 1) and a single locus on Chromosome 5 (Chr 5) (see Fig. 3 of [39]). However, after an extensive comparisons of three analysis, linear trend test, two way ANOVA, and EMMA, they claimed that the significant findings in Mb182 of Chr1 were spurious [39]. Third, we achieve the same results with much less computational speed and burden. Zhang et al. made multiple correction by using a simulation approach [42] as well as the permutation approach [43], both of which are very time consuming by generating thousands of replication samples.

There is one difference in findings worth mentioning here. Zhang et al. had the highest peak achieved at Chr 1 and the second highest peak at Chr 5. We found the opposite. The p-values obtained from single-locus models (linear trend test, two way ANOVA, and EMMA) all found that the peak at Chr 1 has smaller p-values and hence is more significant than that of Chr 5. However, single-locus models only rank features by their marginal effects without considering interactive, correlative, and polygenic effects. On the contrary, RF gives a rank based on the overall importance, considering the individual effect of each SNP as well as accounting for the mutual joint effects of all other SNPs in a multivariate sense. Confirmed from Tables 1, 2, 3, 4, 5 and 6 of the simulation results, we think that RF ranks the peak of Chr 5 the highest because it is

more important in terms of its overall effects (marginal, interactive, correlative, and polygenic effects) for the phenotype.

The two dramatic peaks detected by RF are also highlighted by a *Nature Reviews Genetics* report [44]. Chr5 locus at Mb125, the highest peak in Fig. 1, is located in the same locus as QTL *Hdlq1* found by Su et al. and Korstanje et al. [45, 46]. In addition, they conclude that *Scarb1*, the well known gene involved in HDL metabolism, is the causal gene underlying *Hdlq1* by haplotype analysis, gene sequencing, expression studies, and a spontaneous mutation [47, 48]. Chr1 locus at Mb173, the second highest peak in Fig. 1, is the major determinant of HDL, which has been detected as QTL *Hdlq15* in inbred mouse strains multiple times. Numerous mouse crosses have linked HDL to this region, and *Apoa2* has been identified as the gene underlying this QTL [37, 38, 45].

The Manhattan plot using $-\log_{10}(p)$ as the rule to test significance of each SNP has been widely used in almost all current GWAS literature [16, 44, 49–52]. Instead, we make Manhattan plot from PVIM as an alternative rule to judge significance. A possible argument may come from the threshold or cutoff level used to determine the significance. If using p-value, the traditional determination is to judge if $-\log_{10}(p)$ passes the threshold of $-log_{10}(0.05/p)$. However, the threshold is quite controversial in RF area. There is no a clear solution for it yet. Chen et al. combined the PVIM with permutation to compute the p-values so that the threshold can be available [13]. However, they did not support it using solid theoretical derivations and simulation verifications.

Although the threshold of PVIM of RF is not feasible, it does not affect us to use PVIM based Manhattan plot to draw importance conclusions given the following concerns. 1) The threshold determination is not the key interest of the feature screening approach. Like aforementioned five popular feature screening approaches, a prespecified number of candidates is picked and there is no requirement of close parameter estimating or significance determining in feature screening. 2) Jiang et al. compared RF with the p-values got from B statistic and reported an extremely strong consistency between the p-value and the importance measure. They claimed that larger importance corresponds to smaller p-value of B statistic [11, 33]. It indicated that the importance of RF can give an alternative significance measure of association between SNPs and phenotype. 3) Lunetta et al. found that RF outperforms Fisher's Exact test when interactive effects exist, in terms of power and type I error [32]. It again illustrated the comparable performance of PVIM with a p-value approach. 4) The threshold of p-value approach is obtained by multiple correction, which may not be reliable for a ultra-high dimensional number of SNPs. For example, Bonferroni correction was claimed to be too conservative for large number of tests. The PVIM avoids the multiple correction issue. 5) After having a closer investigation on the Fig 1, we notice that the difference between significance vs non-signifiance is very obvious. Therefore, it is not necessary to use thresholds to determine significance versus non-significance. The two polarized separate is not an accidental because RF tends to have small type I error without losing power.

Conclusion

In this article, we investigated the performance of a forest-based feature screening approach for detecting epistatic, correlative, and polygenic effects for large-scale genome data. Besides the difficulties caused by high dimension, the challenges of epistasis are tripled when hundreds of thousands of SNPs are genotyped. The most popular single-locus models are lack of power, mainly because they ignore the complex mutual effects among SNPs. Extensive studies have already been performed to handle epistasis, such as Brute-force search, exhaustive search, greedy search, MDR, CPM, and so on. However they mainly target for manageable number of features and will lose power for ultrahigh dimension of features. Marchini et al. proposed to exhaustively search all possible 2-way interactive combinations [2]. We agree that this exhaustive search is able to detect all important 2-way interactions. However, it cannot track higher order interactions or more complex structures. Additionally, the search load will be astronomical if the dimension is ultrahigh.

Due to its high efficiency, easy implementation, and great accuracy, feature screening has received much attention for reducing the number of features from huge to moderate through importance rankings [26]. However, majority current feature screening approaches rank the features by univariate measure and neglect the features with weak marginal but complex overall effects. By controlling the difficulty levels through four different monte carlo simulation studies, we compared RF with five other popular feature screening approaches. To make the comparisons consistent, we used the same criteria, same simulation design, and same simulated data for all six approaches. We conclude that the forest-based feature screening performs nicely when nonlinear, interactive, correlative, and other complex associations of response and features exist. In addition, we noticed that the advantages of RF are more manifested when the data conditions are more harsh. We also examined a real mice HDL whole genome data and further confirmed the advantages of RF compared to other current studies for the same data. The human data can be easily extended.

Methods

The purpose of feature screening is to recognize a small set of features that are truly associated with response from a big pool with ultrahigh dimension. By individually defining a surrogate measure for underlying association between response and each feature, feature screening ranks features from the most important to the least important.

Sure independence screening (SIS)

SIS ranks features based on componentwise regression or correlation learning. Each feature is used independently to decide how useful it is for predicting the response variable. Let $w = (w_1, \ldots, w_p)^T = X^T y$ be a vector that is obtained by component wise regression, where X is the standardized feature matrix. Then, w is the measure of marginal correlations of features with the response. The features are sorted based on the componentwise magnitude of the absolute value of w in a decreasing order [26].

Iterative sure independence screening (ISIS)

Fan and Lv pointed out the drawbacks of the SIS: an important feature marginally uncorrelated but jointly correlated with the response can not be picked by SIS. The spurious features not directly associate with the response but in high correlation with a causative feature will likely be selected by SIS [26]. The iterative SIS (ISIS) was proposed to address these drawbacks. The idea of ISIS is to iterate the SIS procedure conditional on previously selected features. To be more specific, first select a small subset k_1 of features, then regress the response over these features. Treat the residuals as the new response and apply the same method to the remaining $p - k_1$ features to pick another small subset k_2 of features. Keep on the iteration until the union of all steps achieve the prespecified size [26].

Conditional correlation sure independence screening (CC-SIS)

Consider how the case effect of response on a feature is related with a covariate, i.e. the parameter β can be a function of certain important covariate u. Now the conditional correlation between the response and each feature is defined as

$$\rho(x_j, y|u) = \frac{cov(x_j, y|u)}{\sqrt{cov(x_j, x_j|u)\, cov(y, y|u)}}, \; j = 1, \ldots, p.$$

Define the marginal measure as $w = (w_1, \ldots, w_p)^T = E\left\{\rho^2(x_j, y|u)\right\}$ and rank the importance of features based on the estimated value of w in a decreasing order [30].

Iterative conditional correlation sure independence screening (ICC-SIS)

Since CC-SIS is based on the top of SIS, it also exists similar drawbacks of the SIS. In order to select the marginally uncorrelated but jointly correlated features and also reduce the effect of collinearity, ICC-SIS was proposed. The idea of ICC-SIS is exactly same as ISIS, but performs CC-SIS during each iteration of residual fitting [30].

Distance correlation sure independence screening (DC-SIS)

The dependence strength between two random vectors can be measured by the distance correlation (Dcorr) [29]. Szekely et al. showed that the Dcorr of two random vectors equals zero if and only if these two random vectors are independent. The distance covariance is defined as

$$dcov^2(y, x_j) = \int ||\phi_{y,x_j}(t, s) - \phi_y(t)\phi_{x_j}(s)||^2 \, w(t, s) dt ds,$$

where $\phi_y(t)$ and $\phi_{x_j}(s)$ are the respective characteristic functions of y and x_j, and $\phi_{y,x_j}(t, s)$ is the joint characteristic function of (y, x_j), and

$$w(t, s) = \left\{c_1^2 \, ||t||^2 \, ||s||^2\right\}^{-1},$$

with $c_1 = \pi$, and $||\cdot||$ stands for the Euclidean norm. Then the Dcorr is defined as

$$dcorr(y, x_j) = \frac{dcov(y, x_j)}{\sqrt{dcov(y, y)\, dcov(x_j, x_j)}}.$$

DC-SIS approach does not assume any parametric model structure and works well for both linear and non-linear associations. In addition, it works well for both categorical and continuous data without making assumptions about the data type.

Random forest (RF)

RF has been widely used for modeling complex joint and interactive associations between response and multiple features [12, 32, 33, 53]. In particular, many nice properties of RF make it an extremely attractive tool for genome studies: the data structure of response and features can be a mixture of categorical and continuous variables; it can nonparametrically incorporate complex nonlinear associations between feature and response; it can implicitly incorporate joint and unknown complex interactions

among a large number of features (higher orders or any structure); it is able to handle big data with a large number of features but limited sample size; it can implicitly accommodate highly correlated features; it is less prone to over-fitting; it has good predictive performance even when the majority of features are noise; it is invariant to monotone transformations of the features; it is robust to changes in its tuning parameters; it performs internal estimation of error, so does not need to assess classification performance by cross-validation, and hence greatly reduces computational time [13, 32, 53, 54].

Using an ensemble method (also called committee method), RF creates multiple classification and regression trees (CARTs). The detailed process of RF can be described in the following steps: Step 1, a bootstrap sample of size n is randomly drawn with replacement from the original data. The remaining non-selected sample or "Out-of-Bag" sample (OOB) is about 30 % on average. Step 2, a classification tree is grown on the bootstrap sample without trimming, by recursively splitting data into distinct subsets with one parent node branched into two child nodes. At each node, a fixed number of features is randomly chosen without replacement from all original features, with "mtry" pre-specifying how many features are chosen. The best split is based on minimizing the mean square prediction error. Step 3, previous two steps are repeated to grow a pre-specified number of trees and make a decision based on the majority vote of all trees (classification) or average results over all trees (regression). Step 4, the prediction accuracy is computed using OOB samples [53].

As an output of the RF, the permutation PVIM, considering the difference in prediction accuracy before and after permuting the jth ($j = 1, \ldots, p$) feature X_j is defined as

$$PVIM_t(X_j) = \frac{\sum_{i \in B_t} \left(Y_i - \hat{Y}_{ti}\right)^2 - \sum_{i \in B_t} \left(Y_i - \hat{Y}_{ti}^*\right)^2}{|B_t|}.$$

Here B_t is the OOB sample for tree t, $t = 1, \ldots, ntree$. $\hat{Y}_{ti}$ is the predicted class for observation i got from tree t before permuting X_j and $\hat{Y}_{ti}^*$ is the predicted class after permuting X_j. The final importance measure is averaged over all trees

$$PVIM(X_j) = \sum_{t=1}^{ntree} PVIM_t(X_j)/ntree.$$

If one feature is randomly permuted, its original association with the response will be broken. Therefore, the idea of PVIM is this: if one feature is an important factor for response, the prediction accuracy should decrease substantially when using its permuted version and all other non-permuted features to predict the OOB sample.

According to the asymptotic theory of RF, RF is sparse when sample size approaches to infinity with a fixed number of features p (i.e. only a small number of causal features is truly associated with the response) [55], which matches the goal of feature screening. The PVIM gives an important measure for each feature, based on their level of associations with response, and hence can be used for feature screening [56]. The PVIM assess each variable's overall impacts by counting not only marginal effects, but also all other complex correlative, interactive, and joint effects, without requiring model structures or explicitly putting interactive terms into the model [32]. The overall effects of each feature are assessed implicitly by the multiple features in the same tree and also by the permuting process when all other features are left unchanged but kept in the same model. Therefore, the variable with weak marginal but strong overall effects will be assigned a high PVIM value [31, 32].

Availability of supporting data

The data set that we analyzed was freely download from http://cgd.jax.org/datasets/datasets.html) [39].

Abbreviations

GWAS: Genome-wide association studies; RF: Random forest; PVIM: Permutation variable importance measure (PVIM); SNPs: Single nucleotide polymorphisms; MDR: Multifactor-dimensionality reduction; CPM: Combinatorial partitioning method; SIS: Sure independence screening; ISIS: Iterated sure independence screening; DC-SIS: Distance correlation sure independence screening; CC-SIS: Conditional correlation sure independence screening; ICC-SIS: CC-SIS: Iterated conditional correlation sure independence screening; HDL: High density lipoprotein; NMRI: Naval Medical Research Institute; Chr: Chromosome; Dcorr: Distance correlation; OOB: "Out-of-Bag" sample; CART: Classification and regression trees.

Competing interests

The authors declare that there is no conflict of interest.

Authors' contributions

GF conceived the research and wrote the manuscript; GW performed the programming and data analysis; CC participated in idea discussions and manuscript revisions; All authors have read and approved the final version of the manuscript.

Acknowledgements

This work was supported by a grant from the National Science Foundation (DMS-1413366) to GF (http://www.nsf.gov).

References

1. Altshuler D, Daly MJ, Lander ES. Genetic mapping in human disease. Science. 2008;322(5903):881–8.
2. Marchini J, Donnelly P, Cardon LR. Genome-wide strategies for detecting multiple loci that influence complex diseases. Nat Genet. 2005;37(4): 413–7.
3. Balding DJ. A tutorial on statistical methods for population association studies. Nat Rev Genet. 2006;7(10):781–91.
4. Yoo W, Ference BA, Cote ML, Schwartz A. A comparison of logistic regression, logic regression, classification tree, and random forests to identify effective gene-gene and gene-environmental interactions. Int J Appl Sci Technol. 2012;2(7):268.
5. Carlson CS, Eberle MA, Kruglyak L, Nickerson DA. Mapping complex disease loci in whole-genome association studies. Nature. 2004;429(6990):446–52.
6. Schwender H, Bowers K, Fallin MD, Ruczinski I. Importance measures for epistatic interactions in case-parent trios. Ann Hum Genet. 2011;75(1): 122–32.
7. Phillips PC. Epistasis—the essential role of gene interactions in the structure and evolution of genetic systems. Nat Rev Genet. 2008;9(11): 855–67.
8. Moore JH. A global view of epistasis. Nat Genet. 2005;37(1):13–14.
9. Culverhouse R, Suarez BK, Lin J, Reich T. A perspective on epistasis: limits of models displaying no main effect. Am J Hum Genet. 2002;70(2): 461–71.
10. Glazier AM, Nadeau JH, Aitman TJ. Finding genes that underlie complex traits. Science. 2002;298(5602):2345–349.
11. Zhang Y, Liu JS. Bayesian inference of epistatic interactions in case-control studies. Nat Genet. 2007;39(9):1167–1173.
12. Cordell HJ. Detecting gene–gene interactions that underlie human diseases. Nat Rev Genet. 2009;10(6):392–404.
13. Chen X, Liu CT, Zhang M, Zhang H. A forest-based approach to identifying gene and gene–gene interactions. Proc Natl Acad Sci. 2007;104(49):19199–19203.
14. Manolio TA, Collins FS, Cox NJ, Goldstein DB, Hindorff LA, Hunter DJ, et al. Finding the missing heritability of complex diseases. Nature. 2009;461(7265):747–53.
15. Zuk O, Hechter E, Sunyaev SR, Lander ES. The mystery of missing heritability: Genetic interactions create phantom heritability. Proc Natl Acad Sci. 2012;109(4):1193–1198.
16. Gibson G. Hints of hidden heritability in GWAS. Nat Genet. 2010;42(7): 558–560.
17. Ritchie MD, Hahn LW, Moore JH. Power of multifactor dimensionality reduction for detecting gene-gene interactions in the presence of genotyping error, missing data, phenocopy, and genetic heterogeneity. Genet Epidemiol. 2003;24(2):150–7.
18. Hahn LW, Ritchie MD, Moore JH. Multifactor dimensionality reduction software for detecting gene–gene and gene–environment interactions. Bioinformatics. 2003;19(3):376–82.
19. Hoh J, Wille A, Ott J. Trimming, weighting, and grouping snps in human case-control association studies. Genome Res. 2001;11(12): 2115–119.
20. Nelson M, Kardia S, Ferrell R, Sing C. A combinatorial partitioning method to identify multilocus genotypic partitions that predict quantitative trait variation. Genome Res. 2001;11(3):458–70.
21. Fan J, Han F, Liu H. Challenges of big data analysis. Natl Sci Rev. 2014;1(2): 293–314.
22. Fan J, Samworth R, Wu Y. Ultrahigh dimensional feature selection: beyond the linear model. J Mach Learn Res. 2009;10:2013–038.
23. Fan J, Li R. Statistical challenges with high dimensionality: Feature selection in knowledge discovery. 2006. arXiv preprint math/0602133, http://arxiv.org/abs/math/0602133.
24. Wang L, Zheng W, Zhao H, Deng M. Statistical analysis reveals co-expression patterns of many pairs of genes in yeast are jointly regulated by interacting loci. PLoS Genet. 2013;9(3):1003414.
25. He Q, Lin DY. A variable selection method for genome-wide association studies. Bioinformatics. 2011;27(1):1–8.
26. Fan J, Lv J. Sure independence screening for ultrahigh dimensional feature space. J R Stat Soc Ser B Stat Methodol. 2008;70(5):849–911.
27. Li R, Zhong W, Zhu L. Feature screening via distance correlation learning. J Am Stat Assoc. 2012;107(499):1129–1139.
28. Fan J, Song R, et al. Sure independence screening in generalized linear models with np-dimensionality. Ann Stat. 2010;38(6):3567–604.
29. Székely GJ, Rizzo ML, Bakirov NK. Measuring and testing dependence by correlation of distances. Ann Stat. 2007;35(6):2769–794.
30. Liu J, Li R, Wu R. Feature selection for varying coefficient models with ultrahigh-dimensional covariates. J Am Stat Assoc. 2014;109(505):266–74.
31. Cook NR, Zee RY, Ridker PM. Tree and spline based association analysis of gene–gene interaction models for ischemic stroke. Stat Med. 2004;23(9): 1439–1453.
32. Lunetta KL, Hayward LB, Segal J, Van Eerdewegh P. Screening large-scale association study data: exploiting interactions using random forests. BMC Genet. 2004;5(1):32.

33. Jiang R, Tang W, Wu X, Fu W. A random forest approach to the detection of epistatic interactions in case-control studies. BMC Bioinforma. 2009;10(Suppl 1):65.

34. Winham SJ, Colby CL, Freimuth RR, Wang X, de Andrade M, Huebner M, et al. Snp interaction detection with random forests in high-dimensional genetic data. BMC Bioinforma. 2012;13(1):164.

35. Schwarz DF, König IR, Ziegler A. On safari to random jungle: a fast implementation of random forests for high-dimensional data. Bioinformatics. 2010;26(14):1752–1758.

36. Wang X, Le Roy I, Nicodeme E, Li R, Wagner R, Petros C, et al. Using advanced intercross lines for high-resolution mapping of HDL cholesterol quantitative trait loci. Genome Res. 2003;13:1654–1664.

37. Wang X, Korstanje R, Higgins D, Beverly P. Haplotype analysis in multiple crosses to identify a QTL gene. Genome Res. 2004;14:1767–1772.

38. Su Z, Ishimori N, Chen Y, Leiter EH, Churchill GA, Paigen B, Stylianou IM. Four additional mouse crosses improve the lipid QTL landscape and identify Lipg as a QTL gene. J Lipid Res. 2009;50(10):2083–094.

39. Zhang W, Korstanje R, Thaisz J, Staedtler F, Harttman N, Xu L, et al. Genome-wide association mapping of quantitative traits in outbred mice. G3 (Bethesda). 2012;14:167–74.

40. Yang H, Ding Y, Hutchins LN, Szatkiewicz J, Bell TA, Paigen BJ, et al. A customized and versatile high-density genotyping array for the mouse. Nat Methods. 2009;6(9):663–6.

41. Yalcin B, Nicod J, Bhomra A, Davidson S, Cleak J, Farinelli L, et al. Commercially available outbred mice for genome-wide association studies. PLoS Genet. 2010;6(9):e1001085.

42. Knijnenburg TA, Wessels LF, Reinders MJ, Shmulevich I. Fewer permutations, more accurate p-values. Bioinformatics. 2009;25(12):161–8.

43. Churchill GA, Doerge RW. Empirical threshold values for quantitative trait mapping. Genetics. 1994;138:963–71.

44. Flint J, Eskin E. Genome-wide association studies in mice. Nat Rev Genet. 2012;13(11):807–17.

45. Su Z, Wang X, Tsaih SW, Zhang A, Cox A, Sheehan S, Paigen B. Genetic basis of HDL variation in 129/SvImJ and C57BL/6J mice: Importance of testing candidate genes in targeted mutant mice. J Lipid Res. 2009;50(1):116–25.

46. Korstanje R, Li R, Howard T, Kelmenson P, Marshall J, Paige B, Churchill G. Influence of sex and diet on quantitative trait loci for HDL cholesterol levels in an SM/J by NZB/BINJ intercross population. J Lipid Res. 2004;45:881–8.

47. Wergedal JE, Ackert-Bicknell CL, Beamer WG, Mohan S, Baylink DJ. Mapping genetic loci that regulate lipid levels in a NZB/B1NJ*RF/J intercross and a combined intercross involving NZB/B1NJ, RF/J, MRL/MpJ, and SJL/J mouse strains. J Lipid Res. 2007;48:1724–1734.

48. Su Z, Leduc MS, Korstanje R, Paigen B. Untangling HDL quantitative trait loci on mouse chromosome 5 and identifying Scarb1 and Acads as the underlying genes. J Lipid Res. 2010;51:2706–713.

49. Cha PC, Takahashi A, Hosono N, Low SK, Kamatani N, Kubo M, et al. A genome-wide association study identifies three loci associated with susceptibility to uterine fibroids. Nat Genet. 2011;43(5):447–50.

50. Ripke S, Sanders A, Kendler K, Levinson D, Sklar P, Holmans P, et al. Genome-wide association study identifies five new schizophrenia loci. Nat Genet. 2011;43(10):969–76.

51. Bis JC, DeCarli C, Smith AV, van der Lijn F, Crivello F, Fornage M, et al. Common variants at 12q14 and 12q24 are associated with hippocampal volume. Nat Genet. 2012;44(5):545–51.

52. Morrison AC, Voorman A, Johnson AD, Liu X, Yu J, Li A, et al. Whole genome sequence-based analysis of a model complex trait, high density lipoprotein cholesterol. Nat Genet. 2013;45(8):899.

53. Breiman L. Random forests. Mach Learn. 2001;45(1):5–32.

54. Goldstein BA, Hubbard AE, Cutler A, Barcellos LF. An application of random forests to a genome-wide association dataset: methodological considerations & new findings. BMC Genet. 2010;11(1):49.

55. Biau G, Devroye L, Lugosi G. Consistency of random forests and other averaging classifiers. J Mach Learn Res. 2008;9:2015–033.

56. Qi Y, Bar-Joseph Z, Klein-Seetharaman J. Evaluation of different biological data and computational classification methods for use in protein interaction prediction. Proteins Struct Funct Bioinforma. 2006;63(3):490–500.

14

Characterization of the biological processes shaping the genetic structure of the Italian population

Silvia Parolo[1], Antonella Lisa[1], Davide Gentilini[2], Anna Maria Di Blasio[2], Simona Barlera[3], Enrico B. Nicolis[3], Giorgio B. Boncoraglio[4], Eugenio A. Parati[4] and Silvia Bione[1*]

Abstract

Background: The genetic structure of human populations is the outcome of the combined action of different processes such as demographic dynamics and natural selection. Several efforts toward the characterization of population genetic architectures and the identification of adaptation signatures were recently made. In this study, we provide a genome-wide depiction of the Italian population structure and the analysis of the major determinants of the current existing genetic variation.

Results: We defined and characterized 210 genomic loci associated with the first Principal Component calculated on the Italian genotypic data and correlated to the North–south genetic gradient. Using a gene-enrichment approach we identified the immune function as primarily involved in the Italian population differentiation and we described a locus on chromosome 13 showing combined evidence of North–south diversification in allele frequencies and signs of recent positive selection. In this region our bioinformatics analysis pinpointed an uncharacterized long intergenic non-coding (lincRNA), whose expression appeared specific for immune-related tissues suggesting its relevance for the immune function.

Conclusions: Our study, combining population genetic analyses with biological insights provides a description of the Italian genetic structure that in future could contribute to the evaluation of complex diseases risk in the population context.

Keywords: Latitude, Immunity, Pathogen, LincRNA

Background

Understanding the genetic structure of human populations is crucial to reconstruct their history and to elucidate the genetic predisposition to diseases. In fact, the genetic structure of human populations was shaped by several demographic events and selective forces, which have contributed to the current diversification and to the differences in diseases prevalence and predisposition [1, 2]. Some relevant examples highlighting the relationship between migration, selection and disease were recently reported, like the gradient in type 2 diabetes genetic risk moving out of Africa [3] or the demonstration that common risk alleles for inflammatory diseases are targets of recent positive selection [4]. Therefore, the study of the genetic architecture of common disorders requires a deep knowledge of the dynamics affecting the population under investigation.

In recent years, the genetic structure of several human populations has been characterized both at worldwide and regional level using genome-wide markers. In Europe, the genetic variation pattern showed a southeast-northwest gradient with a strict correspondence between genetic and geographic distances [5–7]. Along the European latitudinal gradient, Italy plays a major role due to its central position and its geographical conformation extended in the Mediterranean area. The genetic structure of the Italian population has been explored since a long time, starting from pioneering studies based on classic

---
* Correspondence: bione@igm.cnr.it
[1]Computational Biology Unit, Institute of Molecular Genetics-National Research Council, Pavia, Italy
Full list of author information is available at the end of the article

genetic markers [8], to recent works involving genome-wide approaches [9]. Altogether these studies demonstrated the presence of a North–South gradient in allele frequencies along the peninsula and the differentiation of Sardinia from the mainland. The observed European latitudinal cline in allele frequencies has been interpreted as the consequence of human migrations since Paleolithic [10].

In addition to demographic processes, several evidence of positive selection differentially shaping the genome of human populations have been described [11, 12]. In the European population, the best known signature of adaptation is represented by the lactase gene (*LCT*) which confers ability to digest lactose in adulthood. The lactase persistence shows a latitudinal cline with particularly high rates among Northern Europeans and it was demonstrated to be a target of natural selection [13]. Moreover, weak polygenic adaptation acting on many loci at the same time and slightly modifying allele frequencies has been also described as a shaper of human diversity [14]. As an example, human height, a polygenic highly heritable trait, has been proposed as a target of widespread selection on standing variation resulting in differences in adult height between northern and southern European populations [15].

Although the genetic structure of different populations has been deeply characterized, the underlining biological processes are still poorly understood thus requiring further investigations, both at worldwide and regional level.

In this paper, we exploited genome-wide genotypic data to recapitulate the genetic structure of the Italian population in its geographic context, refining the picture of the North–South gradient in genetic variation. A total of 210 genomic loci, sufficient to explain the latitudinal cline in genetic variation, were identified and characterized by different bioinformatics approaches.

Results

The genetic structure of the Italian population

To investigate the genetic structure of the Italian population we assembled a genome-wide genotype dataset of 1736 Italian individuals, as detailed in the Methods section.

After a quality-control procedure, the Italian genetic diversity was summarized by Principal Component Analysis (PCA) using the *smartpca* tool of the EIGENSOFT package [16]. To gain insight into the observed differentiation and test the existence of positive correlation with geography we assigned a geographic place of origin to the individuals through the analysis of their surnames (see Methods and Additional file 1). We observed that the clustering of individuals obtained from the plot of the first two Principal Components (PCs) reflected the geographical origin of each individual obtained from the surname analysis (Fig. 1). In particular, the first

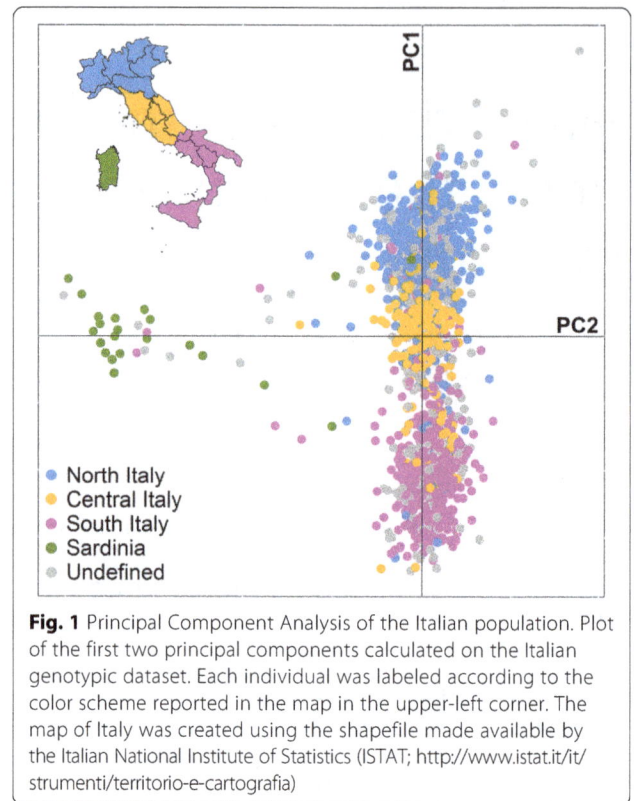

Fig. 1 Principal Component Analysis of the Italian population. Plot of the first two principal components calculated on the Italian genotypic dataset. Each individual was labeled according to the color scheme reported in the map in the upper-left corner. The map of Italy was created using the shapefile made available by the Italian National Institute of Statistics (ISTAT; http://www.istat.it/it/strumenti/territorio-e-cartografia)

principal component (0.17 % of total variance explained) showed a North–South gradient that well correlated with latitude (Pearson's correlation coefficient $r = 0.876$, $p = 8.805 \times 10^{-7}$). The regional subdivision of the Italian population was also evaluated using the pairwise F_{ST} parameter as a measure of genetic distance. A significant correlation between the matrix of F_{ST} and the matrix of the kilometric distances between regional capitals was found (Mantel test, $z = 59.7$, $p = 3.499 \times 10^{-5}$). The second principal component (0.09 % of total variance explained) differentiated Sardinian individuals from the others, reflecting their known genetic diversity. The other PCs did not show any correlation with the Italian geography.

We also evaluated the Italian genetic diversity in the surrounding geographic context through the analysis of available genotype data from populations of the European and Mediterranean area (Additional file 2). The PCA and the ancestry estimation method implemented in ADMIXTURE [17] revealed that Italy stood at the crossroad between continental Europe and the Mediterranean region thus confirming the North–South gradient previously described (Additional file 3, 4, 5).

Genomic loci contributing to latitudinal cline in the Italian population

To evaluate the involvement of specific biological processes in the North–South differentiation of the Italian population, we investigated the genetic variants contributing to PC1.

Through a linear regression analysis, after applying a genome-wide p-value threshold of 1×10^{-7}, we identified a total of 270 SNPs significantly associated with PC1 and sufficient to recapitulate the Italian latitudinal cline (Additional file 6). On the basis of linkage disequilibrium (LD) features of the genomic regions where the single nucleotide polymorphisms (SNPs) were located, we defined a total of 210 loci contributing to the North–South gradient (Fig. 2 and Additional file 7). The identified loci covered a total of 74.5 Mb, they were on average 355 kb wide (range: 10 kb–2.4 Mb) and were distributed along all autosomes. Thirteen loci appeared devoid of any transcribed regions whereas the remaining contained 702 RefSeq genes, with an average of 3.3 gene/locus. According to the HUGO Gene Nomenclature Committee (HGNC) classification [18], 82 % of genes were protein coding ($n = 578$), 14 % were non-coding RNAs ($n = 99$) and the remaining 4 % were pseudogenes ($n = 25$). When we tested the enrichment in gene content of the 210 genomic intervals, a slight overrepresentation was observed ($p = 0.0595$) and it resulted statistically significant considering only the protein-coding genes ($p = 0.0014$). Moreover, the enrichment in genes causing Mendelian diseases resulted significant ($p = 0.0223$). Using the National Human Genome Research Institute (NHGRI) Genome Wide Association Study (GWAS) catalogue [19], we found that 475 genetic variants involved in the predisposition to common disorders were located in the Italian PC1 loci ($p = 0.0126$).

To evaluate the involvement of the 702 genes in specific biological functions, we performed a gene-sets enrichment analysis. The human leukocyte antigen (HLA) region was excluded from the analysis, since it harbors several genes with known immune functions. The overrepresentation of Gene Ontology (GO) terms was evaluated using MSigDB of the Gene Set Enrichment Analysis (GSEA) package (see Methods) [20]. 8 GO terms resulted significantly enriched in the "Biological process" category (Table 1 and Additional file 8). Among them, the GO term "Signal transduction" resulted as the most enriched, indicating the presence of an high number of genes involved in cell function regulation. The second most significant GO term was "Regulation of cellular metabolic process". Moreover, the GO term "Immune system process" resulted significantly overrepresented and it contained seven genes (*IL6, CHUK, CXCR4, CD79A, CNR2, FCGR2B* and *MAL*) in common with the "Signal transduction" term and four genes (*IL6, CHUK, HDAC4* and *CEBPB)* shared with the "Regulation of cellular metabolic process" clade pointing out an overall interconnection among these biological processes. Taking into account the "Cellular Component" ontology, "Membrane" resulted as the most significantly enriched term together with 7 other terms referring to the membrane portion of the cell (Table 2 and Additional file 8). None of the "Molecular function" GO terms resulted enriched below the defined threshold. The analysis of canonical pathways, performed with the Ingenuity Pathway Analysis (IPA) tool, identified "Role of NFAT in Regulation of the Immune Response" as the most enriched pathway. Interestingly, other pathways were related to the immune response processes, underlining the relevance of this biological function in the Italian population PC1-associated gene list (Table 3 and Additional file 8).

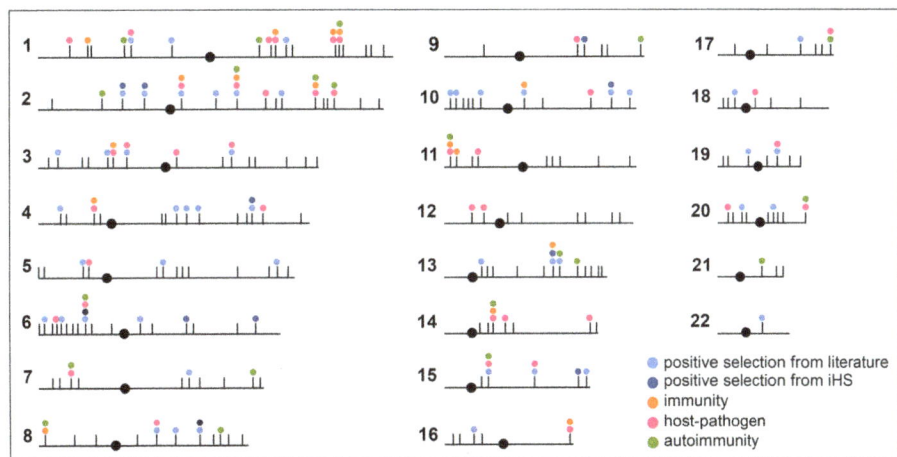

Fig. 2 Genomic distribution of the 210 Italian PC1-associated loci. Chromosomes were represented as horizontal straight lines with centromeres represented as black circles. The vertical dashes correspond to the 210 loci. The circles above the loci were colored based on positive selection features and functional annotation of the contained genes according to the legend in the lower-right corner

Table 1 Significantly enriched GO Biological Processes

GO term ID	GO term name	# Genes	FDR q-value
GO:0007165	Signal transduction	53	2.66E-07
GO:0031323	Regulation of cellular metabolic process	31	1.00E-05
GO:0019222	Regulation of metabolic process	31	1.00E-05
GO:0006139	Nucleobase-containing compound metabolic process	40	1.36E-05
GO:0002376	Immune system process	18	4.10E-05
GO:0043283	Biopolymer metabolic process	47	4.10E-05
GO:0019219	Regulation of nucleobase-containing compound metabolic process	25	4.20E-05
GO:0010468	Regulation of gene expression	26	5.14E-05

Impact of natural positive selection on the 210 Italian PC1 loci

Signals of positive selection were identified using the integrated haplotype score (iHS) statistics [21], which was calculated for each of the autosomal SNPs. We grouped SNPs into non-overlapping genomic intervals of 200 kb: the proportion of SNPs with an |iHS| greater than or equal to 2 was calculated for each interval and those lying in the 5 % tail of the resulting distribution were considered as significant. This approach resulted in the selection of 509 genomic intervals. The intersection between the PC1-associated loci and the iHS significant windows resulted in the identification of 17 loci harboring both signals, thus highlighting the contribution of selection in the Italian North–South differentiation.

Additional insights into the contribution of positive selection as a mechanism involved in the determination of the Italian North–South genetic gradient were obtained by comparison with literature data and testing the enrichment in gene lists for biological functions known to be target of positive selection. Taking into account recent publications based on genome-wide genotypic data and performed on different populations [11, 22, 23], a total of 47 Italian PC1-associated loci was found to overlap at least one genomic interval for which evidence of

positive selection were demonstrated (Fig. 2). Eight of the 17 loci identified using the iHS parameter were previously described as targets of positive selection by different studies (Fig. 2 and Additional file 7). When we evaluated loci enrichment for sets of gene involved in skin pigmentation, immunity, response to infectious disease, sensory perception and metabolism, previously defined by Grossman et al. [23], we found a significant result for immunity and pigmentation (respectively INRICH target-test $p = 0.017$ and $p = 0.004$).

In particular, the locus on chromosome 13 at nucleotide position 74,690,999–75,337,499 (locus 155 in Additional file 7) resulted to contain two intervals with a significant proportion of SNPs with |iHS| > = 2 and to overlap to selection signals previously identified by analyses performed on the HapMap and Human Genome Diversity Projects (HGDP) populations [11, 22, 23]. The fine mapping of SNPs showing significant association with Italian PC1 gradient and of SNPs showing |iHS| values exceeding the threshold, together with the genomic intervals reported to be adaptation targets by previous studies, allowed us to define a core-region of 209 kb (chromosome 13, position 74,863,339–75,072,592) in which North–South differentiation and positive selection signatures were clustered (Fig. 3a). The core-region contained a single validated RefSeq gene encoding for a lincRNA (*LINC00381*) with no functional information available. According to UCSC [24] annotation, a second gene (AX747962) transcribed from the opposite strand was present (Fig. 3b). The annotation of lincRNAs based on the work by Cabili et al. (2011) [25] confirmed the existence of this transcript and suggested the presence of a third transcriptional unit (*TCONS_00022202*) giving rise to two alternative spliced isoforms with high expression levels in white blood cells and in lymph nodes. Data from the ENCODE project [26] supported the transcriptional activity of this region in a Normal Human Epidermal Keratinocyte (NHEK) cell line, where an RNAseq peak, an enrichment of histone H3 acetylation on lysine 27 (H3K27Ac) and histone H3 mono-methylation on lysine 4 (H3K4Me1), together with

Table 2 Significantly enriched GO Cellular Components

GO term ID	GO term name	# Genes	FDR q-value
GO:0016020	Membrane	72	1.78E-13
GO:0005886	Plasma membrane	55	2.27E-11
GO:0005737	Cytoplasm	65	2.85E-09
GO:0044425	Membrane part	55	4.23E-09
GO:0044459	Plasma membrane part	44	4.23E-09
GO:0031226	Intrinsic to plasma membrane	38	5.48E-08
GO:0031224	Intrinsic to membrane	45	9.45E-08
GO:0005887	Integral to plasma membrane	37	9.45E-08
GO:0016021	Integral to membrane	44	1.61E-07

Table 3 Significantly enriched IPA Canonical Pathways

IPA pathway	# Genes	Ratio	p-value
Role of NFAT in Regulation of the Immune Response	14	0.082	2.40E-04
Glutamate Receptor Signaling	7	0.123	8.32E-04
Phagosome formation	10	0.09	8.71E-04
Dendritic Cell Maturation	13	0.073	1.20E-03
Synaptic Long Term Depression	11	0.077	1.70E-03
Protein Kinase A Signaling	21	0.054	2.04E-03
JAK/Stat Signaling	7	0.097	3.31E-03
TREM1 Signaling	7	0.093	4.17E-03
CREB Signaling in Neurons	11	0.064	6.92E-03
Chondroitin Sulfate Degradation (Metazoan)	3	0.2	7.08E-03

a cluster of DNaseI hypersensitive sites were demonstrated in the region. The 209 kb core-region also contained a SNP (rs17714988; position 74,995,660) reported as associated with cytokine responses in smallpox vaccine recipients [27]. When analyzed at haplotypic level, the rs17714988 allele, correlated with a higher level of secreted IFNα, was found on the haplotype containing the alleles for which we demonstrated both positive selection and association with the Italian latitudinal gradient.

Discussion

In this study, we investigated the genomic loci contributing to the genetic latitudinal gradient of the Italian population at genome-wide level. By Principal Component Analysis we identified the North–South gradient as the main axis of the Italian genetic variation in agreement with other studies [5, 7, 9]. Our results are slightly different from the previous work of Di Gaetano and co-authors that investigated the genetic structure of the Italian population using genome-wide markers because it identified the PC1 as the one separating Sardinia from the rest of Italy and the PC2 as latitude-related [9]. However, while in our study the proportion of Sardinian samples reflected that observed in the Italian population, in the study by Di Gaetano et al. Sardinian samples were over-represented, probably causing the differences in the results. In our study, we used the analysis of surnames to establish a correlation between PC1 and latitude. We are aware that our approach has some limitations. In particular, the use of surnames to define the geographic origin of samples can misclassify some individuals because it does not take into account the maternal contribution. However, since we used the origin information after the PCA to interpret this result our choice did not alter the subsequent analyses, which were only based on genetic data. Furthermore, similar results would have been obtained using the place of birth or the place of residence but they were only partly available for our samples and we could not make comparisons about their usefulness.

On the basis of the SNPs significantly contributing to the first Italian principal component, we identified 210 genomic loci that we considered as the main contributors to the North–South gradient. The evaluation of the loci by an interval-based enrichment approach revealed us that they were not randomly located in the genome but preferentially spanned genic regions and, in particular, regions containing protein coding genes. Moreover, the identified loci resulted enriched in disease-associated genes and risk-variants underlining the functional relevance of these regions. Within the most associated loci, genomic regions known for their contribution to the European genetic diversity were contained such as the *LCT, HERC2/OCA2* and *HLA* regions [28, 29]. The human pigmentation genetic diversity, showing a latitudinal gradient shaped by natural selection due to light exposure, was largely contributing to the Italian North–South gradient. Indeed, in addition to *HERC2/OCA2*, other pigmentation genes as *SLC45A2* [30], *HPS5* [31] and *EXOC2-IRF4* [32] were found in PC1-associated loci. Interestingly, it was recently reported that the positively selected gene SLC45A2 was also associated with melanoma susceptibility in a South European population, thus underlining the important link between selection and diseases [33]. In addition to these specific examples, the important role of pigmentation emerged also from the gene-set enrichment analysis together with the immune response, another biological function known to be target of recent selection [23]. The Gene Ontology analysis for Biological Process showed an enrichment of genes involved in signal transduction and in particular of membrane receptors triggering the immune cascade like the Toll-like receptors (*TLR1, TLR6, TLR10*) or *CD79A*, a subunit of the B-cell antigen receptor. In agreement with this result, the Gene Ontology analysis for Cellular Component revealed an enrichment of genes acting at the plasma membrane level, thus modulating cell behavior in response to external stimuli. The enrichment of genes with a role in the immune system emerged even more clearly from the analysis of canonical pathways. The

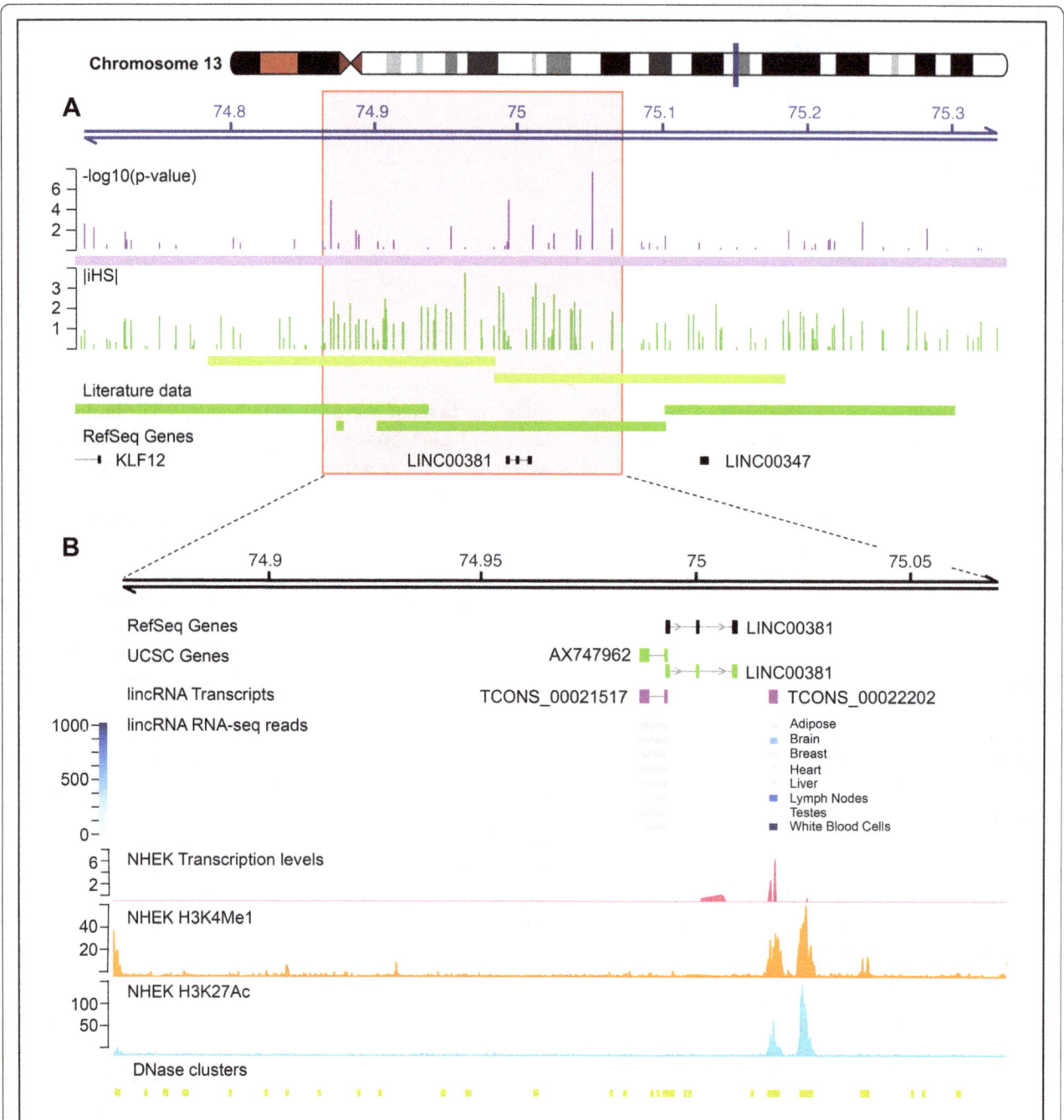

Fig. 3 Characterization of the newly identified locus at 13q22.1. **a** Below the line representing base positions in Mb, different features were represented: the PC1 association signals (*vertical dark violet lines*), the |iHS| value for each SNP tested (*vertical dark green*), the 200 kb intervals defined as positively selected according to the iHS analysis (*light green bars*), the genomic intervals with evidence of positive selection from the literature (*darker green bars*) and RefSeq genes (*black lines*); **b** detail of the core region defined showing: RefSeq genes (*black*), UCSC genes (*green*), lincRNA transcripts (*purple*) and lincRNA RNAseq reads (*blue scale*) according to Cabili et al. 2011 [25], transcription levels and epigenetic features in NHEK cell line from the ENCODE project, the DNase hypersensitivity clusters from 125 ENCODE cell types

pathways identified by the IPA and MSignDB analyses highlighted different aspects of the immune response. The majority of them converged to the Nf-kB signaling as previously suggested [23, 34, 35] and several genes encoding for its components (i.e. *CHUK*, *NFKBIA*) or regulators (i.e. *IGF1R*, *UBE2V*, *HRAS*, *PIK32C2G*, *TLRs*) were located in the Italian PC1-associated loci. Taken together, these data pointed out the immune response as the biological process mainly differentiated along the Italian peninsula, probably as a preferential target of natural selection.

The most likely explanation for the contribution of immunity to population differentiation is its function in host defense against pathogens [36, 37]. In fact, several of the genes that contribute to the Italian population structure were described as involved in infectious disease susceptibility or resistance. For example, malaria, which was endemic in the Mediterranean area and especially in Italy [38, 39] emerged as an infectious disease which had a great impact on the Italian genetic diversity. *HBB*, a gene known to harbor alleles conferring protection against malaria and to be a target of balancing selection [40], is among the genes showing strong differentiation in our dataset. Moreover, the complement factor 1 (*CR1*) gene, suggested to be involved in malaria susceptibility [41], and the *FCGR2B* gene, demonstrated to harbor malaria protective alleles [42, 43], were also identified by our analysis, thus strengthening the mark of malaria in the Italian genome.

Malaria was not the only pathology for which we recognized traces in the Italian population. The Toll-like receptor gene cluster, shown to modulate the response to *Yersinia pestis* [44] and its member *TLR1* involved in leprosy susceptibility [45], were also identified as well as the *IFITM3* gene, whose expression was demonstrated to protect against influenza A infection [46]. Furthermore, a region on chromosome 2 (locus 22 in Additional file 7), recently described as positively selected as a consequence of adaptation to *Vibrio cholera* [35], resulted linked to the PC1 trait and subjected to positive selection in our analysis (Additional file 9).

Recent studies highlighted the presence of adaptation signals in non-coding regions likely owing regulatory functions [37, 47]. In this regard, the locus on chromosome 13, which we demonstrated correlated to the Italian PC1 trait and subjected to positive selection, appeared particularly interesting as it contains only three lincRNAs transcripts. Among them, the *TCONS_00022202* transcript appeared as the best candidate to exert its role in the immune system, because it is mainly expressed in lymph nodes and white blood cells. The transcriptional activity of the *TCONS_00022202* locus was further supported by recent data provided by the ENCODE project demonstrating that it is enriched in an enrichment in modifications typical of active chromatin and is highly transcribed in the NHEK cell line. The NHEK cell line derived from primary epidermal keratinocytes which represent an effective barrier to the entry of infectious agents and play an active role in the initiation of the immune response. These cells produce a variety of cytokines, growth factors, interleukins and antimicrobial peptides thus representing a cell model to investigate inflammation and immune response. Given that non-coding RNAs are emerging as important regulators of gene expression in the immune response [48, 49], we suggested that the *TCONS_00022202*

transcript may represent a new immune-related molecule deserving further investigations.

Intriguingly, a polymorphism located about 22 kb upstream to the *TCONS_00022202* lincRNA was associated with the IFNα response in smallpox vaccine recipients, a phenotype that resembles the host response to the virus. Since the allele correlating with higher level of interferon expression was on the positively selected haplotype, we proposed that the observed signature of positive selection is the effect of adaptation to *Variola virus*. Moreover, the observation that 2 other SNPs (rs17070309 and rs12256830) associated with smallpox-induced cytokine response [27] are located within the Italian PC1-associated loci (loci 104 and 128 in Additional file 7), reinforced the hypothesis that smallpox virus could have shaped the Italian genome diversity.

Conclusions

In conclusion, our study provides new insights into the Italian population structure by characterizing the main determinants of the current genetic diversity and results in the identification of immunity as the main biological process responsible for genetic differentiation in Italy positive selection target, likely triggered by infective agents. Interestingly, recent studies suggested an important role of loci involved in host defense against pathogens also in autoimmune disease susceptibility. For example, it was proposed that the genetic architecture of inflammatory bowel disease was shaped by pathogen-driven selection [50, 51]. Further investigations are required for a better comprehension of evolutionary processes and their relationship with disease predisposition.

Methods

All the reported genomic coordinates were based on the February 2009 assembly of the human genome (hg19/GRCh37). The statistical analyses, unless otherwise specified, were performed with R, version 2.15.3 [52].

Study samples and genotyping

Before the quality control procedure a total of 1736 individuals was available for this study. In particular, 1648 individuals of self-reported Italian origin, recruited in North Italy had surname information accessible. Their genotype data were assembled from a study of cerebrovascular disease including 697 cases and 951 controls. Controls were recruited among blood donors and volunteer healthy people, 409 already analyzed in a study on obesity and 392 in the PROCARDIS study [53]. All individuals were enrolled in the study following written informed consent and ethical approval from the institutional review boards for each sample collection, namely Ethics Committee of the Fondazione IRCCS

Istituto Neurologico Carlo Besta, Istituto Auxologico Italiano and Lombardy Region. 88 samples from the Tuscan cohort (TSI) genotyped in the HapMap project phase III [54] were added to the study cohort, for a total of 1736 Italian individuals. For the evaluation of the genetic variability of the Italian population in the context of European and Mediterranean populations, we analyzed genotypic data of 303 individuals drawn from the Human Genome Diversity Project (HGDP [55]), 186 individuals from the Behar et al., 2010 study [56], 50 individuals from McEvoy et al., 2009 [57] and 25 individuals from the Wellcome Trust Case Control Consortium—WTCCC [58] (Additional file 2).

Surname-based definition of individual's geographical origin

The geographic origin of individuals was defined through the analysis of their surnames. The birth place and the place of residence were not available for all the individuals and previous analyses demonstrated that they are not suitable to infer the individual's geographic origin because of recent migrations [59]. In Italy surnames are transmitted patrilineally and can be considered as Y-chromosome genetic markers. For this reason we used the surname analysis as a tool to infer the place of origin. In particular, our surname analysis was based on the Italian Surnames database that was established extracting data from the complete national telephone directory of year 1993 (18,554,688 subscribers corresponding of about 33 % of the whole Italian population) and includes a total of 332,525 different surnames together with their frequencies in the different Italian administrative zones [60]. A supervised frequency-based approach combined with linguistic and historical records was used to analyze surnames and to determine their putative geographical origin. For the purposes of this study, the Italian territory was subdivided into four main areas: North (comprising 8 administrative regions, namely: Piedmont, Aosta Valley, Lombardy, Liguria, Veneto, Trentino Alto Adige, Friuli Venezia Giulia and Emilia Romagna), Central (comprising 5 administrative regions, namely: Tuscany, Marche, Umbria, Lazio and Abruzzo), South (comprising 6 administrative regions, namely: Molise, Campania, Apulia, Basilicata, Calabria and Sicily) and Sardinia. The analysis of surnames frequency distribution combined in the four main areas allowed to assign a geographical origin to a total of 1238 individuals. The remaining 410 individuals (25 %) had a surnames whose geographical origin could not be unambiguously assigned [60]. The surnames analysis was conducted independently and anonymously from the genotypic analyses and the match of data was conducted by authorized personnel. The 88 individuals from the TSI cohort of HapMap were assigned to Central Italy based on their reported origin.

Genotype data analysis and quality control procedures

Managing of genotype data and quality control procedures were performed with PLINK 1.0.7 [61]. For the Italian dataset a total of 487,999 SNPs was initially available for the analyses. The quality control procedure resulted in the exclusion of 35,003 markers with minor allele frequency below 0.05, 4100 markers for genotyping rate below 0.97 and 21 individuals for genotype call below 0.97. Because LD features could distort the PCA analysis, one member of each pair of SNPs with r^2 greater than 0.4, in windows of 200 SNPs (sliding window of 25 SNPs), was removed using the indep-pairwise command in PLINK. After the quality control procedure, a total of 1715 individuals and 172,111 SNPs was considered for the analysis. Finally, 1000 SNPs from the 8p23.1 genetic region, known to harbor a large inversion polymorphism [62, 63], were excluded because they could distort the subsequent analyses. On the dataset used to calculate the iHS statistics we did not exclude the SNPs highly correlated. The quality control procedure applied to the Mediterranean dataset is described in the Additional file 3.

Principal component analysis and ancestry estimation

Principal Component Analysis (PCA) was carried out using the *smartpca* tool of the EIGENSOFT package version 3.0 using the default parameter and no outlier exclusion [16]. The correlation between PC1 and latitude was tested using the R cor.test function. To each individual we attributed the latitude value corresponding to the capital of the administrative region identified as individual place of origin.

The SNPs significantly associated with the first principal component were identified through a linear regression model in PLINK. PC1 was used as a response variable and the SNP as the explanatory one. The analysis of SNPs associated with Italian PC2 identified a small number of significant SNPs, likely because samples from Sardinia were too few. For this reason PC2 was not further examined. To infer the ancestry proportions in the European/Mediterranean dataset we applied the unsupervised clustering algorithm ADMIXTURE [17]. The analysis was repeated from $K = 2$ to $K = 7$. The optimal number of K was estimated through the cross-validation procedure using the −cv = 10 option.

Test for selection

The presence of signal of selection was tested using the iHS statistics [21], calculated using the R package *rehh* [64]. This test detected the presence of extended haplotypes surrounding each core SNP to identify candidate alleles for selective sweeps. Before running the analysis, the genotypes, not LD-pruned, were phased using *fastPHASE* [65]. For each SNP the ancestral state was identified from NCBI dbSNP (build 139) and the genetic position along the chromosomes was taken from the HapMap Consortium (release 22, B36). To determine the significant regions the SNP's iHS scores were grouped in non overlapping 200 kb windows and for each window we calculated the fraction of SNPs with an |iHS| > = 2. The windows with a total number of SNPs less than 20 were excluded from the analysis. The fraction of windows in the top 5 % tail of iHS distribution was considered as significant.

The comparison with literature data was performed selecting articles reporting genome-wide analyses of positive natural selection carried out on reference populations belonging to the HapMap project or to the Human Genome Diversity Project, published up to 2013.

Loci definition

The genomic intervals corresponding to each PC1-associated SNP were defined on the basis of the LD feature of the genome through the Gene Relationships Across Implicated Loci (GRAIL) tool [66]. The tool was run using the list of 270 significant SNPs and the HapMap CEU (release 22) as a reference population. The genes overlapping the regions were defined from the UCSC RefSeq Genes track [67].

Enrichment analyses

The interval-based enrichment tests were performed with INRICH v.1.0 [68] using 1,000,000 permutations both in the first and in the second phase of the analysis. Specifically, protein coding genes, ncRNAs and pseudo-genes were defined according to the NCBI Gene database (http://www.ncbi.nlm.nih.gov/gene/) limiting the query to RefSeq records. The list of genes involved in Mendelian diseases was defined filtering the Online Mendelian Inheritance in Man (http://omim.org/) catalogue to exclude unconfirmed diseases, traits not involved in disorders and inconsistent or tentative records. The list of SNPs associated with complex disorders was retrieved from the NHGRI GWAS catalogue (http://www.genome.gov/gwastudies/; date accessed on April, 3rd 2014) [19]. The manually curated list of genes involved in pathways known to be target of positive selection was downloaded from the Composite of Multiple Signals website (http://www.broadinstitute.org/mpg/cms) [23].

The gene-set enrichment analysis was performed using the MSigDB tool of the GSEA package (http://www.broadinstitute.org/gsea/msigdb/index.jsp) querying Gene Ontology as source annotation database and considering the categories with a corrected p-value less than 1×10^{-4}. The canonical pathways were investigated using QIAGEN's Ingenuity® Pathway Analysis (IPA®, QIAGEN Redwood City, www.qiagen.com/ingenuity) and the first 10 significant canonical pathways were reported.

The functional categories reported in Fig. 2 were generated as follows. The immune category was defined combining evidence of genes involved in the immune system from the Immune System term of Gene Ontology Biological Process, the immune-related IPA and Reactome canonical pathways. The autoimmunity category was defined from the presence of association signals with autoimmune diseases. The host-pathogen category was manually defined exploiting the information from NCBI Gene database and literature confirmation.

Availability of data and materials

The list of the 210 loci associated with Italian population PC1 is available in Additional file 7. For each of the 210 loci, the genomic coordinates and the identifiers of the associated SNPs are provided.

Additional files

Additional file 1: Geographical distribution of Italian samples based on surnames. (DOCX 70 kb)

Additional file 2: The European/Mediterranean dataset. (DOCX 97 kb)

Additional file 3: Analysis of the Italian dataset with European and Mediterranean populations. (DOCX 149 kb)

Additional file 4: PCA of European and Mediterranean populations. (A) Plot of the first two principal components of the Italian population combined with populations from continental Europe and Mediterranean area; (B) geographical localization of the analyzed samples. Legend of symbols and colors used is reported below. The map of European/Mediterranean area was obtained plotting a suitable portion of the spatial world data downloaded from http://thematicmapping.org/. (PDF 1212 kb)

Additional file 5: Graphical representation of ADMIXTURE analysis results. The analysis was performed assuming 4 ancestral populations ($K = 4$) and including all the samples used for Principal Component Analysis in the European/Mediterranean dataset. The populations with an asterisk (*) are those of the present study. (PDF 147 kb)

Additional file 6: PCA of the Italian dataset using the 270 PC1-associated SNPs. Plot of the first two principal components showing that the 270 SNPs recreate the genetic latitudinal gradient observed in Italy. (PDF 73 kb)

Additional file 7: Features of the 210 Italian PC1-associated loci. (XLSX 47 kb)

Additional file 8: Gene-enrichment analyses. The genes present in the enriched GO Biological Process categories, Cellular Component categories and IPA canonical pathways are reported. (XLSX 57 kb)

Additional file 9: UCSC genome browser view of the locus 22 of Additional file 7. (A) In this panel is showed the genomic region chr2:95965626–97094126. In particular are reported: the PC1 association signals (vertical dark violet lines), the |iHS| value for each tested SNP (vertical dark green), the genomic region previously identified as positively selected by Karlsson et al. (2013) [35] (horizontal green bar) and the two genomic intervals resulted positively selected in our analysis (horizontal red bars); (B) detail of the identified positively selected region showing the RefSeq genes contained. (PDF 113 kb)

Abbreviations

GO: Gene Ontology; GRAIL: Gene Relationships Across Implicated Loci; GSEA: Gene Set Enrichment Analysis; GWAS: Genome Wide Association Study; H3K4Me1: Histone H3 mono-methylation on lysine 4; H3K27Ac: Histone H3 acetylation on lysine 27; HGDP: Human Genome Diversity Projects; HGNC: HUGO

Gene Nomenclature Committee; HLA: Human leukocyte antigen; iHS: Integrated haplotype score; IPA: Ingenuity Pathway Analysis; LD: Linkage disequilibrium; lincRNA: Long intergenic non-coding RNA; NHEK: Normal Human Epidermal Keratinocyte; NHGRI: National Human Genome Research Institute; PCA: Principal Component Analysis; PCs: Principal Components; SNPs: Single nucleotide polymorphisms; WTCCC: Wellcome Trust Case Control Consortium.

Competing interests
The authors declare that they have no competing interests.

Authors' contributions
AL, GBB, EAP and SBi conceived the research and developed the study design. AMDB, SBa, GBB and EAP provided DNA samples and genotypic data. SP, DG and EBN performed the quality control and merging procedures of genotypic data. SP performed all the other analyses. AL did the supervision of the statistical analysis and provided data from the Italian Surnames Database. SP, AMDB, SBa, GBB and SBi contributed to the interpretation and discussion of the results. SP and SBi wrote the manuscript. All authors read and approved the final manuscript.

Acknowledgments
We want to greatly thank Prof. Luigi Luca Cavalli-Sforza and Prof. Gianna Zei for their contribution without which this work would not have been able to even begin. We also thank Dr. Chiara Mondello for her critical reading of the manuscript. This study makes use of data generated by the Wellcome Trust Case–control Consortium. A full list of the investigators who contributed to the generation of the data is available from www.wtccc.org.uk. Funding for the project was provided by the Wellcome Trust under award 076,113 and 085,475. The population allele and genotype frequencies of the Finnish and Swedish sample were obtained from the data source funded by the Nordic Center of Excellence in Disease Genetics based on samples regionally selected from Finland, Sweden and Denmark. This work was supported by Cariplo Foundation Grant n. 2010/0253, Italian Ministry of Health Grant n. RC 2009/LR8 and RC 2010/LR8, and European Community, Sixth Framework Program Grant n. LSHM-CT-2007-037273. Silvia Parolo was supported by a fellowship of the PhD program in Genetic and Biomolecular Sciences of the University of Pavia.

Author details
[1]Computational Biology Unit, Institute of Molecular Genetics-National Research Council, Pavia, Italy. [2]Molecular Biology Laboratory, Istituto Auxologico Italiano, Milan, Italy. [3]Department of Cardiovascular Research, IRCCS Mario Negri Institute for Pharmacological Research, Milan, Italy. [4]Department of Cerebrovascular Diseases, IRCCS Istituto Neurologico Carlo Besta, Milan, Italy.

References
1. Myles S, Tang K, Somel M, Green RE, Kelso J, Stoneking M. Identification and analysis of genomic regions with large between-population differentiation in humans. Ann Hum Genet. 2008;72(Pt 1):99–110.
2. Moonesinghe R, Ioannidis JP, Flanders WD, Yang Q, Truman BI, Khoury MJ. Estimating the contribution of genetic variants to difference in incidence of disease between population groups. Eur J Hum Genet. 2012;20(8):831–6.
3. Corona E, Chen R, Sikora M, Morgan AA, Patel CJ, Ramesh A, et al. Analysis of the genetic basis of disease in the context of worldwide human relationships and migration. PLoS Genet. 2013;9(5):e1003447.
4. Raj T, Kuchroo M, Replogle JM, Raychaudhuri S, Stranger BE, De Jager PL. Common risk alleles for inflammatory diseases are targets of recent positive selection. Am J Hum Genet. 2013;92(4):517–29.
5. Novembre J, Johnson T, Bryc K, Kutalik Z, Boyko AR, Auton A, et al. Genes mirror geography within Europe. Nature. 2008;456(7218):98–101.
6. Lao O, Lu TT, Nothnagel M, Junge O, Freitag-Wolf S, Caliebe A, et al. Correlation between genetic and geographic structure in Europe. Curr Biol. 2008;18(16):1241–8.
7. Nelis M, Esko T, Mägi R, Zimprich F, Zimprich A, Toncheva D, et al. Genetic structure of Europeans: A view from the North-East. PLoS One. 2009;4(5):e5472.
8. Cavalli-Sforza LL, Menozzi P, Piazza A. The history and geography of human genes. Princeton: Princeton University Press; 1994.
9. Di Gaetano C, Voglino F, Guarrera S, Fiorito G, Rosa F, Di Blasio AM, et al. An overview of the genetic structure within the Italian population from genome-wide data. PLoS One. 2012;7(9):e43759.
10. Soares P, Achilli A, Semino O, Davies W, Macaulay V, Bandelt HJ, et al. The archaeogenetics of Europe. Curr Biol. 2010;20(4):R174–183.
11. Akey JM. Constructing genomic maps of positive selection in humans: Where do we go from here? Genome Res. 2009;19(5):711–22.
12. Scheinfeldt LB, Tishkoff SA. Recent human adaptation: Genomic approaches, interpretation and insights. Nat Rev Genet. 2013;14(10):692–702.
13. Bersaglieri T, Sabeti PC, Patterson N, Vanderploeg T, Schaffner SF, Drake JA, et al. Genetic signatures of strong recent positive selection at the lactase gene. Am J Hum Genet. 2004;74(6):1111–20.
14. Pritchard JK, Pickrell JK, Coop G. The genetics of human adaptation: Hard sweeps, soft sweeps, and polygenic adaptation. Curr Biol. 2010;20(4):R208–215.
15. Turchin MC, Chiang CW, Palmer CD, Sankararaman S, Reich D, Hirschhorn JN, et al. Evidence of widespread selection on standing variation in Europe at height-associated SNPs. Nat Genet. 2012;44(9):1015–9.
16. Patterson N, Price AL, Reich D. Population structure and eigenanalysis. PLoS Genet. 2006;2(12):e190.
17. Alexander DH, Novembre J, Lange K. Fast model-based estimation of ancestry in unrelated individuals. Genome Res. 2009;19(9):1655–64.
18. Gray K, Daugherty L, Gordon S, Seal R, Wright M, Bruford E. Genenames.org: The HGNC resources in. Nucleic Acids Res. 2013;41(D1):D545–52.
19. Welter D, MacArthur J, Morales J, Burdett T, Hall P, Junkins H, et al. The NHGRI GWAS Catalog, a curated resource of SNP-trait associations. Nucleic Acids Res. 2014;42(D1):D1001–6.
20. Subramanian A, Tamayo P, Mootha V, Mukherjee S, Ebert B, Gillette M, et al. Gene set enrichment analysis: A knowledge-based approach for interpreting genome-wide expression profiles. Proc Natl Acad Sci U S A. 2005;102(43):15545–50.
21. Voight BF, Kudaravalli S, Wen X, Pritchard JK. A map of recent positive selection in the human genome. PLoS Biol. 2006;4(3):e72.
22. Pickrell JK, Coop G, Novembre J, Kudaravalli S, Li JZ, Absher D, et al. Signals of recent positive selection in a worldwide sample of human populations. Genome Res. 2009;19(5):826–37.
23. Grossman SR, Andersen KG, Shlyakhter I, Tabrizi S, Winnicki S, Yen A, et al. Identifying recent adaptations in large-scale genomic data. Cell. 2013;152(4):703–13.
24. Kent W, Sugnet C, Furey T, Roskin K, Pringle T, Zahler A, et al. The human genome browser at UCSC. Genome Res. 2002;12(6):996–1006.
25. Cabili M, Trapnell C, Goff L, Koziol M, Tazon-Vega B, Regev A, et al. Integrative annotation of human large intergenic noncoding RNAs reveals global properties and specific subclasses. Genes Dev. 2011;25(18):1915–27.
26. Dunham I, Kundaje A, Aldred S, Collins P, Davis C, Doyle F, et al. An integrated encyclopedia of DNA elements in the human genome. Nature. 2012;489(7414):57–74.
27. Kennedy RB, Ovsyannikova IG, Pankratz VS, Haralambieva IH, Vierkant RA, Poland GA. Genome-wide analysis of polymorphisms associated with cytokine responses in smallpox vaccine recipients. Hum Genet. 2012;131(9):1403–21.
28. Heath S, Gut I, Brennan P, McKay J, Bencko V, Fabianova E, et al. Investigation of the fine structure of European populations with applications to disease association studies. Eur J Hum Genet. 2008;16(12):1413–29.
29. Donnelly MP, Paschou P, Grigorenko E, Gurwitz D, Barta C, Lu RB, et al. A global view of the OCA2-HERC2 region and pigmentation. Hum Genet. 2012;131(5):683–96.
30. Lucotte G, Mercier G, Dieterlen F, Yuasa I. A Decreasing Gradient of 374 F Allele Frequencies in the Skin Pigmentation Gene SLC45A2, from the North of West Europe to North Africa. Biochem Genet. 2010;48(1–2):26–33.
31. Zhang Q, Zhao B, Li W, Oiso N, Novak E, Rusiniak M, et al. Ru2 and Ru encode mouse orthologs of the genes mutated in human Hermansky-Pudlak syndrome types 5 and 6. Nat Gen. 2003;33(2):145–53.
32. Praetorius C, Grill C, Stacey SN, Metcalf AM, Gorkin DU, Robinson KC, et al. A polymorphism in IRF4 affects human pigmentation through a tyrosinase-dependent MITF/TFAP2A pathway. Cell. 2013;155(5):1022–33.
33. López S, García O, Yurrebaso I, Flores C, Acosta-Herrera M, Chen H, et al. The interplay between natural selection and susceptibility to melanoma on

allele 374 F of SLC45A2 gene in a South European population. PLoS One. 2014;9(8):e104367.

34. Kamberov Y, Wang S, Tan J, Gerbault P, Wark A, Tan L, et al. Modeling recent human evolution in mice by expression of a selected EDAR variant. Cell. 2013;152(4):691–702.

35. Karlsson E, Harris J, Tabrizi S, Rahman A, Shlyakhter I, Patterson N, et al. Natural selection in a Bangladeshi population from the Cholera-Endemic Ganges River Delta. Sci Transl Med. 2013;5(192):192ra86.

36. Fumagalli M, Sironi M, Pozzoli U, Ferrer-Admettla A, Pattini L, Nielsen R. Signatures of environmental genetic adaptation pinpoint pathogens as the main selective pressure through human evolution. Plos Genet. 2011;7(11):e1002355.

37. Karlsson E, Kwiatkowski D, Sabeti P. Natural selection and infectious disease in human populations. Nat Rev Gen. 2014;15(6):379–93.

38. Cavalli-Sforza LL, Bodmer WF. The Genetics of Human populations. San Francisco: W. H. Freeman and Company; 1971.

39. Majori G. Short history of malaria and its eradication in Italy with short notes on the fight against the infection in the mediterranean basin. Mediterr J Hematol Infect Dis. 2012;4(1):e2012016.

40. Mangano VD, Modiano D. An evolutionary perspective of how infection drives human genome diversity: The case of malaria. Curr Opin Immunol. 2014;30C:39–47.

41. Stoute J. Complement receptor 1 and malaria. Cell Microbiol. 2011;13(10):1441–50.

42. Clatworthy M, Willcocks L, Urban B, Langhorne J, Williams T, Peshu N, et al. Systemic lupus erythematosus-associated defects in the inhibitory receptor Fc gamma RIIb reduce susceptibility to malaria. Proc Natl Acad Sci U S A. 2007;104(17):7169–74.

43. Willcocks L, Carr E, Niederer H, Rayner T, Williams T, Yang W, et al. A defunctioning polymorphism in FCGR2B is associated with protection against malaria but susceptibility to systemic lupus erythematosus. Proc Natl Acad Sci U S A. 2010;107(17):7881–5.

44. Laayouni H, Oosting M, Luisi P, Ioana M, Alonso S, Ricano-Ponce I, et al. Convergent evolution in European and Rroma populations reveals pressure exerted by plague on Toll-like receptors. Proc Natl Acad Sci U S A. 2014;111(7):2668–73.

45. Wong S, Gochhait S, Malhotra D, Pettersson F, Teo Y, Khor C, et al. Leprosy and the adaptation of human toll-like receptor 1. Plos Pathog. 2010;6(7):e1000979.

46. Everitt A, Clare S, Pertel T, John S, Wash R, Smith S, et al. IFITM3 restricts the morbidity and mortality associated with influenza. Nature. 2012;484(7395):519–U146.

47. Fumagalli M, Sironi M. Human genome variability, natural selection and infectious diseases. Curr Opin Immunol. 2014;30C:9–16.

48. Atianand MK, Fitzgerald KA. Long non-coding RNAs and control of gene expression in the immune system. Trends Mol Med. 2014;20(11):623–31.

49. Heward JA, Lindsay MA. Long non-coding RNAs in the regulation of the immune response. Trends Immunol. 2014;35(9):408–19.

50. Jostins L, Ripke S, Weersma R, Duerr R, McGovern D, Hui K, et al. Host-microbe interactions have shaped the genetic architecture of inflammatory bowel disease. Nature. 2012;491(7422):119–24.

51. Quintana-Murci L, Clark A. Population genetic tools for dissecting innate immunity in humans. Nat Rev Immunol. 2013;13(4):280–93.

52. Team RC. R: A language and environment for statistical computing. Vienna: R Foundation for Statistical Computing; 2013.

53. Peden J, Hopewell J, Saleheen D, Chambers J, Hager J, Soranzo N, et al. A genome-wide association study in Europeans and South Asians identifies five new loci for coronary artery disease. Nat Gen. 2011;43(4):339–U389.

54. Altshuler D, Gibbs R, Peltonen L, Dermitzakis E, Schaffner S, Yu F, et al. Integrating common and rare genetic variation in diverse human populations. Nature. 2010;467(7311):52–8.

55. Li J, Absher D, Tang H, Southwick A, Casto A, Ramachandran S, et al. Worldwide human relationships inferred from genome-wide patterns of variation. Science. 2008;319(5866):1100–4.

56. Behar D, Yunusbayev B, Metspalu M, Metspalu E, Rosset S, Parik J, et al. The genome-wide structure of the Jewish people. Nature. 2010;466(7303):238–42.

57. McEvoy B, Montgomery G, McRae A, Ripatti S, Perola M, Spector T, et al. Geographical structure and differential natural selection among North European populations. Genome Res. 2009;19(5):804–14.

58. Burton P, Clayton D, Cardon L, Craddock N, Deloukas P, Duncanson A, et al. Genome-wide association study of 14,000 cases of seven common diseases and 3,000 shared controls. Nature. 2007;447(7145):661–78.

59. Zei G, Lisa A, Fiorani O, Magri C, Quintana-Murci L, Semino O, et al. From surnames to the history of Y chromosomes: The Sardinian population as a paradigm. Eur J Hum Genet. 2003;11(10):802–7.

60. Boattini A, Lisa A, Fiorani O, Zei G, Pettener D, Manni F. General Method to Unravel Ancient Population Structures through Surnames, Final Validation on Italian Data. Hum Biol. 2012;84(3):235–70.

61. Purcell S, Neale B, Todd-Brown K, Thomas L, Ferreira MA, Bender D, et al. PLINK: A tool set for whole-genome association and population-based linkage analyses. Am J Hum Genet. 2007;81(3):559–75.

62. Deng L, Zhang Y, Kang J, Liu T, Zhao H, Gao Y, et al. An unusual haplotype structure on human chromosome 8p23 derived from the inversion polymorphism. Hum Mutat. 2008;29(10):1209–16.

63. Tian C, Plenge RM, Ransom M, Lee A, Villoslada P, Selmi C, et al. Analysis and application of European genetic substructure using 300 K SNP information. PLoS Genet. 2008;4(1):e4.

64. Gautier M, Vitalis R. rehh: An R package to detect footprints of selection in genome-wide SNP data from haplotype structure. Bioinformatics. 2012;28(8):1176–7.

65. Scheet P, Stephens M. A fast and flexible statistical model for large-scale population genotype data: Applications to inferring missing genotypes and haplotypic phase. Am J Hum Genet. 2006;78(4):629–44.

66. Raychaudhuri S, Plenge R, Rossin E, Ng A, Purcell S, Sklar P, et al. Identifying relationships among genomic disease regions: Predicting genes at pathogenic snp associations and rare deletions. Plos Genet. 2009;5(6):e1000534.

67. Pruitt K, Tatusova T, Maglott D. NCBI reference sequences (RefSeq): A curated non-redundant sequence database of genomes, transcripts and proteins. Nucleic Acids Res. 2007;35:D61–5.

68. Lee PH, O'Dushlaine C, Thomas B, Purcell SM. INRICH: Interval-based enrichment analysis for genome-wide association studies. Bioinformatics. 2012;28(13):1797–9.

15

Freeze–thaw *Caenorhabditis elegans* freeze–thaw stress response is regulated by the insulin/IGF-1 receptor *daf-2*

Jian-Ping Hu[1†], Xiao-Ying Xu[1†], Li-Ying Huang[1], Li-shun Wang[2*] and Ning-Yuan Fang[1*]

Abstract

Background: Adaption to cold temperatures, especially those below freezing, is essential for animal survival in cold environments. Freezing is also used for many medical, scientific, and industrial purposes. Natural freezing survival in animals has been extensively studied. However, the underlying mechanisms remain unclear. Previous studies demonstrated that animals survive in extremely cold weather by avoiding freezing or controlling the rate of ice-crystal formation in their bodies, which indicates that freezing survival is a passive thermodynamic process.

Results: Here, we showed that genetic programming actively promotes freezing survival in *Caenorhabditis elegans*. We found that *daf-2*, an insulin/IGF-1 receptor homologue, and loss-of-function enhanced survival during freeze–thaw stress, which required the transcription factor *daf-16*/FOXO and age-independent target genes. In particular, the freeze–thaw resistance of *daf-2*(rf) is highly allele-specific and has no correlation with lifespan, dauer formation, or hypoxia stress resistance.

Conclusions: Our results reveal a new function for *daf-2* signaling, and, most importantly, demonstrate that genetic programming contributes to freezing survival.

Keywords: *C. elegans*, Freeze–thaw stress response, Insulin/IGF-1 receptor *daf-2*, Transcription factor *daf-16*/FOXO

Background

Cold temperature is a critical environment stimulus to animals. Subzero temperatures especially may adversely affect animals by direct lethal effects and damage caused by ice formation [1]. The ability of animals to sense and respond to cold temperatures, even below freezing, is essential for survival is cold environments. Freezing is also widely used for many medical, scientific, and industrial purposes, such as strain preservation and organ preservation. Understanding how to enhance survival and maintain normal physiological functions in the presence of freeze stress is critical for animals in nature and human research.

Natural freezing survival in animals has been extensively studied. Previous studies demonstrated that animals survive in extremely cold weather by avoiding freezing or controlling the rate of ice-crystal formation in their bodies [2], which indicates that freezing survival is a passive thermodynamic process. However, response to freeze–thaw stress and other biological phenomena, such as longevity and hypoxia resistance, may be genetically programmed. Because it is a powerful model in molecular genetics, the nematode *Caenorhabditis elegans* is suitable for studying genetic response to freeze–thaw stress.

Insulin/insulin-like growth factor 1 (IGF-1)-like signaling is the best-characterized pathway that regulates the lifespan and other stress-resistance traits of *C. elegans*. The insulin/IGF-1 receptor homologue *daf-2* activates a conserved phosphatidylinositol-3-OH kinase (PI(3)K)/3-phosphoinositide-dependent kinase-1 (PDK1)/Akt signal transduction pathway, which prevents FOXO transcription factor *daf-16* entry into the nucleus [3–5]. *daf-2* reduction/loss-of-function (rf) produces a longer lifespan [6]. Conversely, the PTEN phosphatase homologue, *daf-18*(rf), suppresses life-span extension induced by *daf-2*(rf) [7, 8].

---
* Correspondence: lishunwang@fudan.edu.cn; fangnysh@126.com
†Equal contributors
²The Division of Translational Medicine, Minhang Hospital, Fudan University, Shanghai, China
¹The Department of Geriatrics, Ren-Ji Hospital, Shanghai Jiao-Tong University School of Medicine, Shanghai, China

Insulin/IGF-1 signaling is also involved in formation of dauer larvae, which have an alternative, developmentally arrested third larval stage (L3) [9]. In unfavorable environments, such as crowding or food shortage, insulin/IGF-1 signaling or transforming growth factor-beta (TGF-β) is suppressed; unliganded nuclear receptor DAF-12 regulates dauer diapause [10]. Insulin/IGF-1 signaling is also involved in other stress tolerances, including oxidative stress, ultraviolet light, heat shock, or hypoxia stress. The longevity and stress tolerance produced by insulin/IGF-1 signaling mutants require nuclear translocation or nuclear activity of *daf-16* [11]. Savory et al. [12] showed that *daf-16* is important for delta-9 desaturase gene expression, which is important for survival at low temperatures. Ohta [13] showed that the insulin-signaling pathway in the intestines and neurons is essential for temperature experience-dependent cold tolerance in animals. Animal survival in freezing conditions is a more complicated phenomenon than in cold temperatures without ice formation. Organs can be injured during freezing by physical factors, such as ice-crystal formation, dehydration, and cold [14]. Moreover, animals can also suffer biochemical damage, such as oxidative stress or hypoxia stress. [15]

To date, few studies have addressed the genes that regulate freezing tolerance or survival. We investigated the roles of *daf-2*(rf) in freeze–thaw stress-response regulation. We exposed *daf-2(e1370)* and other rf strains to freeze–thaw stress to identify *daf-2*(rf)-improved freeze-induced mortality and cell damage. Then, we tested other *daf-2*(rf) alleles with various phenotypic severities and performed molecular analysis to determine which signaling is involved. We performed the experiments through both genetic and morphological analyses with different *C. elegans* mutants.

Results

Insulin/IGF-1 receptor homologue *daf-2*(rf) regulates freeze–thaw stress survival

To investigate the role of *daf-2*(rf) in freezing tolerance, we evaluated the survival rate of *daf-2(e1370)* and wild-type (*N2*) strains exposed to freeze–thaw stress. Two-day-old adults were exposed to -80 °C for 8 min and then thawed in a water bath at 30 °C; results showed that the *daf-2(e1370)* strain had a significantly increased survival rate compared with the *N2* strain (p < 0.01; Fig. 1a). To confirm the results of enhanced freezing survival produced by the reduction of *daf-2*/insulin-like signaling, we evaluated the survival rate of *N2* animals with *daf-2* RNAi interference or IGF$_1$R inhibitor treatment. Animals with both *daf-2* RNAi inactivation and IGF$_1$R inhibitor treatment had enhanced survival rates after freeze–thaw treatment (p < 0.01; Fig. 1a).

At high freezing rates, intracellular freezing occurs, which can lead to cell damage, mainly by ice-crystal formation [15, 16]. To reduce ice-crystal formation and cell damage, cells and organisms can be cryopreserved by slowly lowering the temperature until deep freezing temperature. *daf-2(e1370)* and *N2* animals were gradually cooled at a rate of about –1 °C/min (Additional file 1: Figure S1). Moreover, we also found that *daf-2(e1370)* animals had higher survival than *N2* at any growth stage. This result indicates that survival of *daf-2*(rf) animals following freeze–thaw damage is not stage- or age -specific (Fig. 1, Additional file 1: Figure S1A–C). We also found that the survival rate of L3/L4-stage worms was dramatically decreased compared with L1/L2-stage worms (Fig. 1b, Additional file 1: Figure S1A–B). This result indicates that freezing tolerance changed with development and age.

Freezing-induced behavioral defects and cell defects blocked by *daf-2*(rf)

Maintaining physiological functions in freezing temperatures is challenging; therefore, evaluation of response to different freezing conditions is important. Under freeze–thaw conditions, animals frozen at 0 °C and thawed in 30 °C water baths did not differ in survival rate (Additional file 2: Figure S2.), which indicates that the animals' physiological function loss is mainly caused by freezing rather than thawing. *daf-2(e1370)* animals survived and fully recovered locomotion after freeze–thaw stress. *N2* animals displayed significant locomotion defects after recovery from some stress (Fig. 2).

To investigate cell defects from freezing and protection by *daf-2*(rf), we examined the muscle cell morphology of animals exposed to freeze–thaw stress. Freezing caused striking nuclear fragmentation in muscle myocytes. RNAi inactivation of *daf-2* and downstream genes *age*, *pdk-1* maintained intact nuclei and protected myocytes from both nuclear fragmentation and death [Fig. 3].

daf-2 allelic specification for freeze–thaw stress survival is not a consequence of lifespan, dauer formation, or other stress resistance mechanisms

daf-2(rf) alleles tend to promote a prolonged lifespan and form dauer larvae. Consequently, *daf-2*(rf) mutants are always resistant to harsh environments [17]. To analyze whether freeze–thaw survival is correlated with lifespan, we tested freezing survival rates associated with 11 *daf-2* alleles' (Fig. 4). The freeze–thaw stress survival phenotypes were not well correlated with lifespan (r = 0.538; p = 0.088).

We found that *daf-2(e1370)* worms, which did not have the longest lifespan, had the highest freeze–thaw stress survival, followed by *e1391*, *e1371*, *e979*, and *m579*. Five alleles that were weak or produced no increased freezing survival had significantly increased lifespans as long as or longer than *e1370* [17]. Similarly,

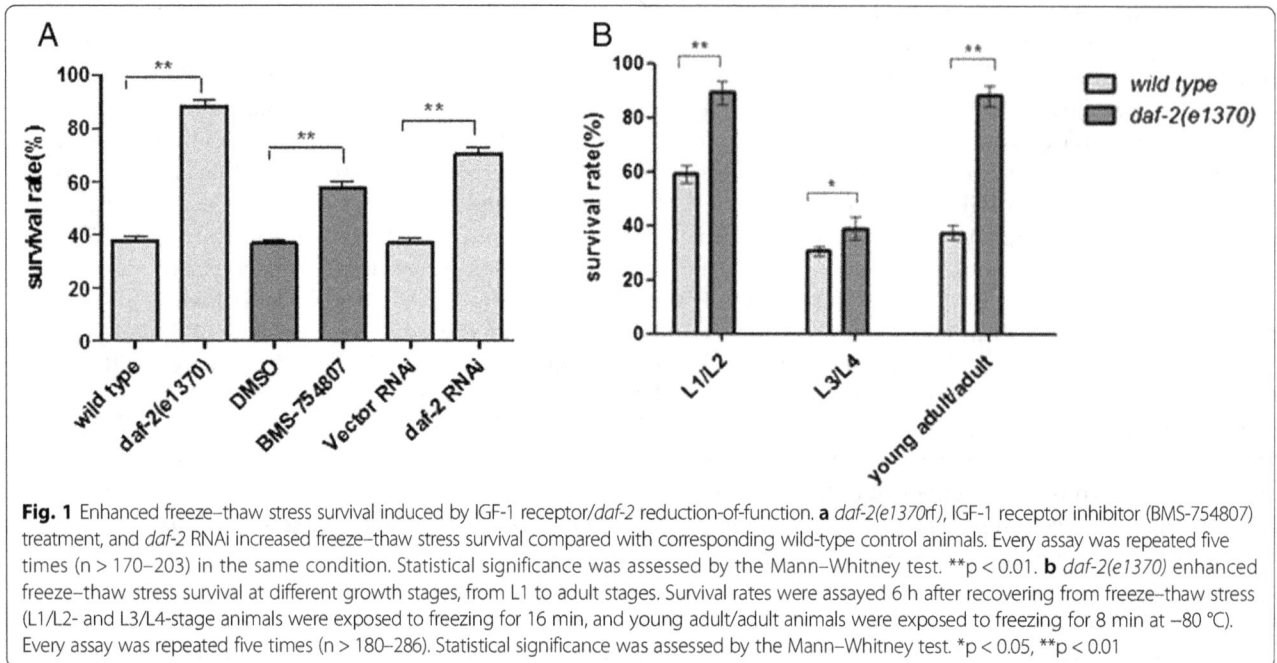

Fig. 1 Enhanced freeze–thaw stress survival induced by IGF-1 receptor/*daf-2* reduction-of-function. **a** *daf-2(e1370*rf*)*, IGF-1 receptor inhibitor (BMS-754807) treatment, and *daf-2* RNAi increased freeze–thaw stress survival compared with corresponding wild-type control animals. Every assay was repeated five times (n > 170–203) in the same condition. Statistical significance was assessed by the Mann–Whitney test. **p < 0.01. **b** *daf-2(e1370)* enhanced freeze–thaw stress survival at different growth stages, from L1 to adult stages. Survival rates were assayed 6 h after recovering from freeze–thaw stress (L1/L2- and L3/L4-stage animals were exposed to freezing for 16 min, and young adult/adult animals were exposed to freezing for 8 min at −80 °C). Every assay was repeated five times (n > 180–286). Statistical significance was assessed by the Mann–Whitney test. *p < 0.05, **p < 0.01

one allele that did not increase survival (*m41*) and two weaker alleles (*e1391* and *e979*) produced stronger Daf-c phenotypes than *e1370*, and one allele that did not increase survival (*e1368*) produced the same Daf-c phenotype as *e1370* [18, 19]. With regard to stress resistance, four *daf-2* alleles (*e1391*, *e979*, *e579*, and *m596*) were significantly resistant to thermal stress [18] but had weaker or no increased freeze–thaw survival. *daf-2* allele *e579* was significantly resistant to hypoxia stress but had weaker freeze–thaw survival. Three *daf-2* alleles (*e1371*, *e1391*, and *e979*) were weak or non-resistant to hypoxia stress but had higher freeze–thaw survival. *daf-2*(rf) freeze–thaw stress survival was highly allele-specific and did not appear to be a consequence of mechanisms that regulate lifespan, dauer formation, or other stress-resistance traits (Additional file 3: Figure S3).

The insulin-signaling pathway, but not dauer signaling, longevity genes, or *trpa-1*, is essential for freeze–thaw stress survival

To determine the molecules downstream of the insulin receptor involved freeze–thaw stress survival, we tested various mutants defective in the known insulin-signaling pathway (Fig. 5). Phenotypic analysis showed that mutants defective in the *daf-2*/insulin receptor or its downstream molecules had abnormal enhancement or reduction of freeze–thaw stress survival (Fig. 5). Abnormal increments of freeze–thaw stress survival in *daf-2* and downstream molecules mutants were suppressed by mutation or RNAi in the *daf-16*/FOXO-type transcriptional factor (Fig. 6).

Because the insulin-signaling pathway and other signaling molecules are essential for dauer larva formation, we tested other molecular components, including TGF-β and steroid hormonal signaling (Fig. 5) [19, 20]. However, we did not observe a considerable increase in these mutants (Fig. 5). The results indicate that the *daf-2*/insulin-signaling pathway, but not other dauer larva formation signaling, is essential for freezing survival.

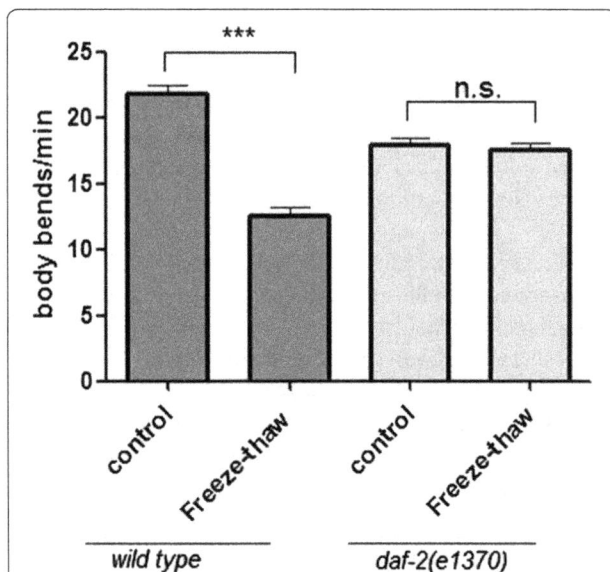

Fig. 2 Behavioral effects of freeze–thaw stress in wild-type (*N2*) and *daf-2(e1370*rf*)* strains. Locomotion rate was quantified as the number of body bends per min after recovery from freeze–thaw stress. Locomotion of N2 (n = 10) was significantly reduced after recovery from freeze–thaw stress; *daf-2(e1370)* remained unchanged (n = 10). Statistical significance was assessed by the Mann–Whitney test. ***p < 0.001, n.s. - not significant

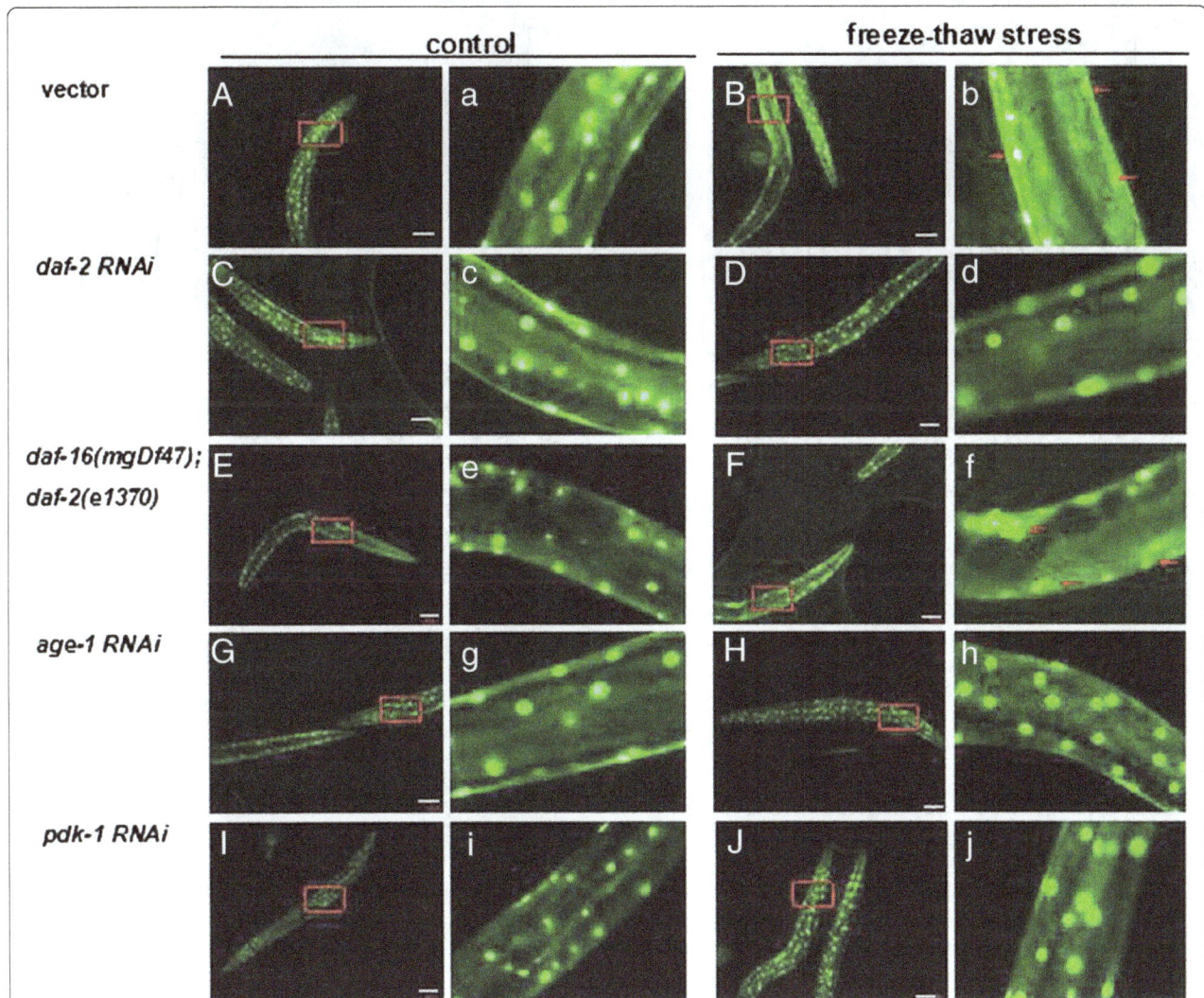

Fig. 3 Freeze–thaw stress-induced morphologic cell defects blocked by *daf-2*(rf). Adult animals were treated with freeze–thaw stress as described in the Methods; after 6 h of recovery, surviving animals were scored for cell morphology. All GFP reporter genes [40] were stably integrated. Every assay was repeated three times (n > 12–21) in the same conditions. Nuclear-localized *pmyo-3:gfp* reporter gene expression in body wall muscle nuclei. *PD4251* (**A, a**), *daf-2* RNAi (**C, c**), [*daf-16(mgDf47)*; *daf-2(e1370)*] (**E, e**), *age-1* RNAi (**G, g**), or *pdk-1* RNAi (**I, i**) animals not exposed to freeze–thaw stress treatment were observed. *PD4251* (**B, b**) and [*daf-16(mgDf47)*; *daf-2(e1370)*] (**F, f**) animals' nuclear GFP expression was fragmented (arrow) with freeze–thaw stress. Additionally, *daf-2* RNAi (**D, d**), *age-1* RNAi (**H, h**), and *pdk-1* RNAi (**J, j**) animals had preserved nuclear morphology with freeze–thaw stress

daf-2 is the best-characterized longevity gene in *C. elegans*. We tested other longevity genes mutants, including *eat-2*, *clk-1*, and *isp-1* mutants, and found that these animals did not have enhanced freezing survival compared with the *N2* strain. Recently, it was reported that the cold receptor transient receptor potential (TRP) channel, which is encoded by *trpa-1*, plays a central role in ageing and stress response to cold temperatures [21]. However, no reduced freeze–thaw survival was observed in *trpa-1* loss-of-function mutants. These results indicate that longevity genes and *trpa-1* are not essential for freeze–thaw stress survival.

daf-16 nuclear tranlsocation is not responsible for *daf-2*(rf)-enhanced freeze–thaw survival

To elucidate how *daf-16* affects *daf-2*(rf) enhanced freeze–thaw stress survival, cellular distributions of a *daf-16*::GFP fusion protein were first categorized on a scale of 1 (unlocalized) to 3 (fully nuclear localized) (Fig. 7). Distributions were scored in *daf-2* RNAi animals and compared with controls maintained at 20 °C. As a positive control, *daf-2* inactivation resulted in a marked translocation of *daf-16* to the nucleus (Fig. 7a–d, left histogram). With freeze–thaw stress, *daf-16* nuclear translocation significantly increased in both *daf-2* RNAi and control animals (Fig. 7e–h). However, the difference

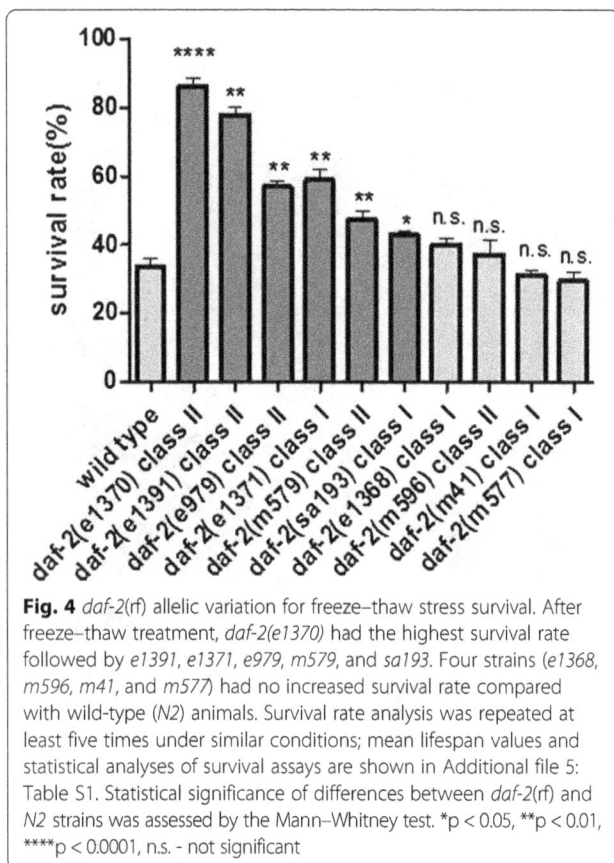

Fig. 4 *daf-2*(rf) allelic variation for freeze–thaw stress survival. After freeze–thaw treatment, *daf-2(e1370)* had the highest survival rate followed by *e1391, e1371, e979, m579,* and *sa193*. Four strains (*e1368, m596, m41,* and *m577*) had no increased survival rate compared with wild-type (*N2*) animals. Survival rate analysis was repeated at least five times under similar conditions; mean lifespan values and statistical analyses of survival assays are shown in Additional file 5: Table S1. Statistical significance of differences between *daf-2*(rf) and *N2* strains was assessed by the Mann–Whitney test. *p < 0.05, **p < 0.01, ****p < 0.0001, n.s. - not significant

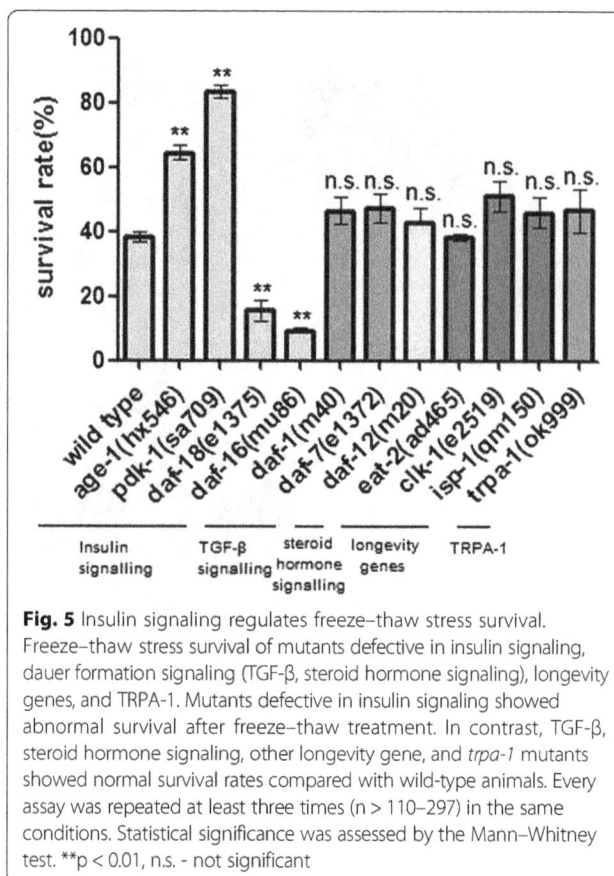

Fig. 5 Insulin signaling regulates freeze–thaw stress survival. Freeze–thaw stress survival of mutants defective in insulin signaling, dauer formation signaling (TGF-β, steroid hormone signaling), longevity genes, and TRPA-1. Mutants defective in insulin signaling showed abnormal survival after freeze–thaw treatment. In contrast, TGF-β, steroid hormone signaling, other longevity gene, and *trpa-1* mutants showed normal survival rates compared with wild-type animals. Every assay was repeated at least three times (n > 110–297) in the same conditions. Statistical significance was assessed by the Mann–Whitney test. **p < 0.01, n.s. - not significant

in *daf-16* nuclear translocation between *daf-2*(rf) and control animals disappeared, because there was no significant discrepancy in the subcellular localization of *daf-16*::GFP (Fig. 7, right histogram). These observations indicate that altering the subcellular localization of *daf-16* does not explain why *daf-2*(rf) has higher freeze–thaw stress survival than wild-type animals. We found that other mechanisms responsible for *daf-2*(rf) increased freezing survival.

To gather additional evidence to elucidate the nuclear activity of *daf-16* responsible for *daf-2*(rf) increased freeze–thaw stress survival, we assayed *daf-16* target genes. If the insulin pathway promotes *daf-16* nuclear activity, this pathway should regulate the expression level of *daf-16* target genes. We first examined the expression level of 21 direct transcriptional target genes of *daf-16* that are also involved in ageing, larval arrest, or fat formation. We found that mRNA levels of *C36A4.9* (*acs-19*, acetyl-CoA synthetase) and *C46A10.7* (*srh-99*, class H chemoreceptor/olfactory receptor) were markedly upregulated in *daf-2(e1370)* worms (Additional file 4: Figure 4). Mice deficient in the homologous AceCS2 (acetyl-CoA synthetase 2) gene cannot maintain normal body temperatures when starved or fed a LC/HF diet [22]. In *Drosophila*, central and peripheral elements of the

olfactory receptor system are responsible for temperature adaptation [23].

We infer that *acs-19* or *srh-99* may be required for freeze–thaw adaptation in *C. elegans*. For further elucidation, we assayed freeze–thaw survival rates with *acs-19* and *srh-99* RNAi in *daf-2(e1370)* and wild-type animals. We found that *acs-19* and *srh-99* RNAi significantly reduced *daf-2(e1370)* freeze–thaw survival, although there was still higher survival than in the *daf-16;daf-2* double mutant (Fig. 8). This result indicates that the *daf-16* target genes *acs-19* and *srh-99* are involved in the observed *daf-2*(rf) increased freeze–thaw stress survival. While, there is no evidence indicating other daf-16 target genes *sup-37* or *lig-1* involved in *daf-2(rf)* increased freezing survival (Fig. 8).

Discussion

In this study, we revealed genetic regulation of freeze–thaw stress responses in *C. elegans*. Phenotypic analysis of genetic deletion strains revealed that insulin/IGF-1 receptor *daf-2* controls both survival and behavior during freeze–thaw stress. *daf-2* reduction improved freeze–thaw stress survival, locomotion, and muscle cell protection. *daf-2(rf)* freeze–thaw response is highly allele-specific and not a consequence of lifespan, dauer formation, or other

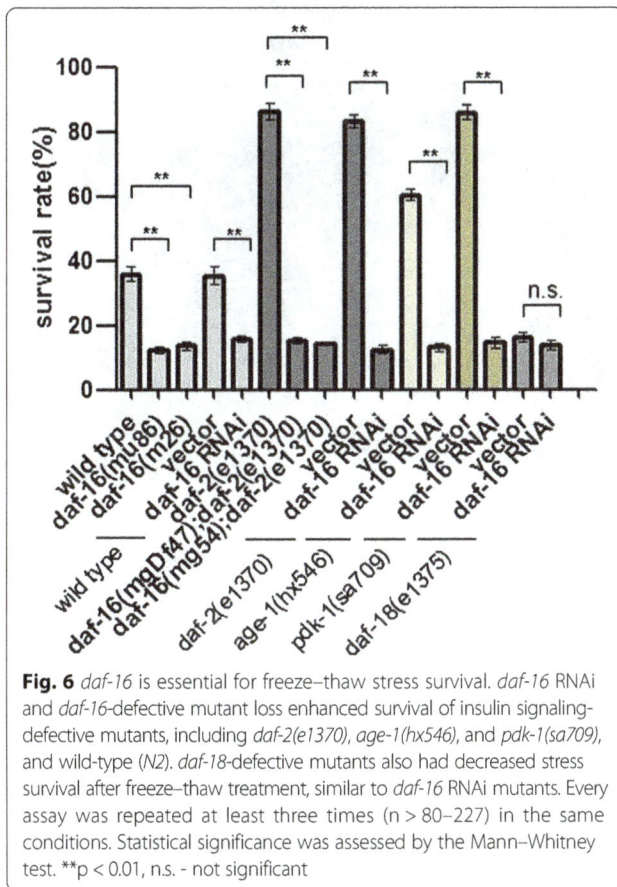

Fig. 6 *daf-16* is essential for freeze–thaw stress survival. *daf-16* RNAi and *daf-16*-defective mutant loss enhanced survival of insulin signaling-defective mutants, including *daf-2(e1370)*, *age-1(hx546)*, and *pdk-1(sa709)*, and wild-type (N2). *daf-18*-defective mutants also had decreased stress survival after freeze–thaw treatment, similar to *daf-16* RNAi mutants. Every assay was repeated at least three times (n > 80–227) in the same conditions. Statistical significance was assessed by the Mann–Whitney test. **p < 0.01, n.s. - not significant

stress-resistance trait regulation. We also revealed that insulin signaling, but not TGF-β, are related to freeze–thaw survival. Steroid hormone signaling participated in *daf-2*(rf) enhanced freeze–thaw stress regulation. *daf-16*::GFP cellular localization analysis and *daf-16* target gene screening revealed that *daf-16* regulated the target genes *acs-19* and *srh-99* but not *daf-16* nuclear translocation or *daf-2*(rf) enhanced freeze–thaw stress survival.

In the past few decades, there has been rapid progress in our understanding of how physiological mechanisms can protect freeze injury [24]. By contrast, very little is known about how genetic regulation promotes freeze resistance. At least in *C. elegans*, freeze resistance is not purely a passive thermodynamic process. In this study, we characterized insulin/IGF-1 receptor *daf-2*, which regulates freeze–thaw stress survival improvement in *C. elegans*. Our results indicate that genetic programming actively contribute to enhanced survival and physiological function recovery from freezing conditions.

Previous work revealed that insulin signaling is required for lifespan regulation, dauer formation, and stress tolerance. Recent studies demonstrated that the insulin-signaling pathway or *daf-16* is required for temperature experience-dependent cold tolerance of animals [12, 13]. To ascertain whether insulin signaling plays roles in

regulating freeze resistance in very stressful conditions, we performed survival experiments with freeze–thaw stress. In these experiments, we confirmed that insulin signaling via the *daf-16*/FOXO pathway is essential for freeze–thaw survival. Reduction-of-function of insulin/IGF-R *daf-2* protected muscle cell damage and promoted physiological activity recovery from freeze–thaw stress.

We found that freeze–thaw survival was *daf-2*(rf) allele-specific and is not a consequence of ageing, larval arrest, or other stress-resistance traits. Insulin signaling is also involved in dauer formation. To determine whether other dauer signaling participated in the freeze–thaw stress resistance, we assayed freeze–thaw survival of mutants. We found that dauer formation pathways, including TGF-β and steroid receptor signaling, were not essential for freeze–thaw stress survival. In addition, *daf-2* is the best-characterized longevity-regulated gene. Therefore, we also tested other freeze survival longevity gene mutants, including *eat-2* [25], *clk-1* [26], and *isp-1* [27]. We found that these other longevity genes were not required for freeze-stress resistance. Low temperatures led to increased longevity, in which the cold receptor TRP channel encoded by *trpa-1* is essential [21, 28]. Therefore, we tested the function of *trpa-1* and found that loss-of-function mutant *trpa-1* produced a normal phenotype compared with the wild-type strain. These results indicate that freeze–thaw survival is independent on cold receptor TRPA-1 signaling. These results are consistent with the *daf-2*(rf) allele relationship analysis between freeze–thaw survival and lifespan, larval arrest, and other stress–resistance traits. Freezing resistance may have an independent biological mechanism that differs from mechanisms that control ageing, dauer formation, and response to other stresses.

Finally, our results indicate that the *daf-16* target genes *C36A4.9* (*acs-19*) and *C46A10.7* (*srh-99*) are required for freezing survival of *daf-2(e1370)*. Previous studies demonstrated that *acs-19* maintained core body temperatures of mice in fasting conditions [22], and olfactory receptor be acclimated to the environmental temperature of *D. melanogaster* [23]. Inactivation by RNAi of *acs-19* or *srh-99* contributed to both decreased fat storage and enhanced dauer formation of *C. elegans* [29]. It is possible that an unrecognized underlying mechanism contributes to both fat storage and dauer formation phenomena, which are related to freezing tolerance and survival. Nevertheless, our analyses did not exclude the contribution of additional mechanisms to freezing survival. *acs-19* and *srh-99* RNAi reduced freezing survival of *daf-2(e1370)*; however, *daf-2(e1370) acs-19* RNAi and *daf-2(e1370) srh-99* RNAi mutants exhibited significantly higher survival compared with the *daf-16* mutant, which indicates that additional *daf-16* target genes or mechanisms must be involved. For example,

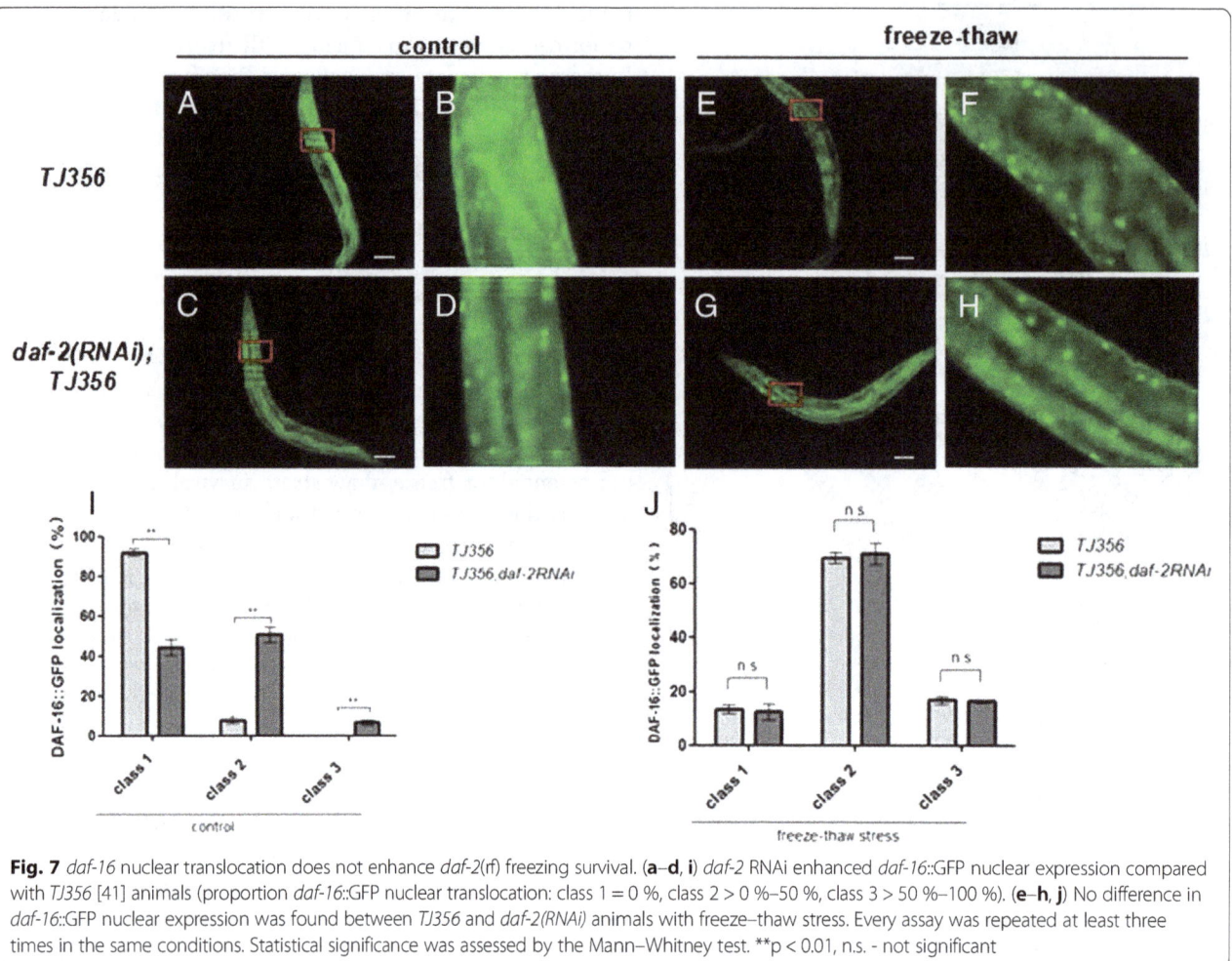

Fig. 7 *daf-16* nuclear translocation does not enhance *daf-2*(rf) freezing survival. (**a–d**, **i**) *daf-2* RNAi enhanced *daf-16*::GFP nuclear expression compared with *TJ356* [41] animals (proportion *daf-16*::GFP nuclear translocation: class 1 = 0 %, class 2 > 0 %–50 %, class 3 > 50 %–100 %). (**e–h**, **j**) No difference in *daf-16*::GFP nuclear expression was found between *TJ356* and *daf-2(RNAi)* animals with freeze–thaw stress. Every assay was repeated at least three times in the same conditions. Statistical significance was assessed by the Mann–Whitney test. **p < 0.01, n.s. - not significant

genes involved in the synthesis of trehalose [30–34], glycerol [30], heat-shock proteins [32, 35, 36], antioxidant enzymes [37], and Δ9 desaturase enzymes [12] could participate in freeze–thaw stress survival by enhancing cold tolerance of *C. elegans* and other species. Future studies are needed to identify other unknown genetic programming and additional components of the *daf-2*(rf)/*daf-16* pathway that ultimately may lead to a thorough understanding of how *daf-2*(rf) promotes freezing survival in *C. elegans*.

Conclusion

In conclusion, our mutant survival assay revealed that insulin/IGF-1 receptor *daf-2* played important roles in freeze–thaw stress responses in *C. elegans*. Freezing resistance is *daf-2* allele-specific and not a consequence of ageing, dauer formation, and other stress-regulation traits. Reduction-of-function of *daf-2* enhanced freeze–thaw survival, because it is dependent on insulin signaling pathway. The *daf-16*/FOXO-regulating target genes *acs-19* are *srh-99* essential for *daf-2*(rf) enhanced freezing

resistance. Considering that the insulin/IGF-1 receptor showed striking conservation across phylogeny [38, 39], our work indicates that a similar phenomenon may also occur in other organisms.

Methods
Strain selection and maintenance

Caenorhabditis elegans were maintained on nematode growth medium (NGM) agar plates seeded with OP50, which is a slow-growing *Escherichia coli* mutant. The strains used included *N2*, *eat-2(ad465)*, *isp-1(qm150)*, *clk-1(qm30)*, *daf-1(m40)*, *daf-7(e1372)*, *daf-12(rh286)*, *daf-2(e1370)*, *daf-2(e979)*, *daf-2(e1391)*, *daf-2(e1368)*, *daf-2(e1371)*, *daf-2(m596)*, *daf-2(m577)*, *daf-2(m579)*, *daf-2(Sa193)*, *daf-2(m41)*, *[daf-16(mgDf47); daf-2(e1370)]*, *TJ356(daf-16::GFP)*, and *PD4251(pmyo-3::GFP)*. All strains were provided by the Caenorhabditis Genetics Center funded by the National Institutes of Health National Center for Research Resources. Unless otherwise stated, all strains were cultured at 15 °C to the L4 stage and then transferred to animals at 20 °C for 2 d.

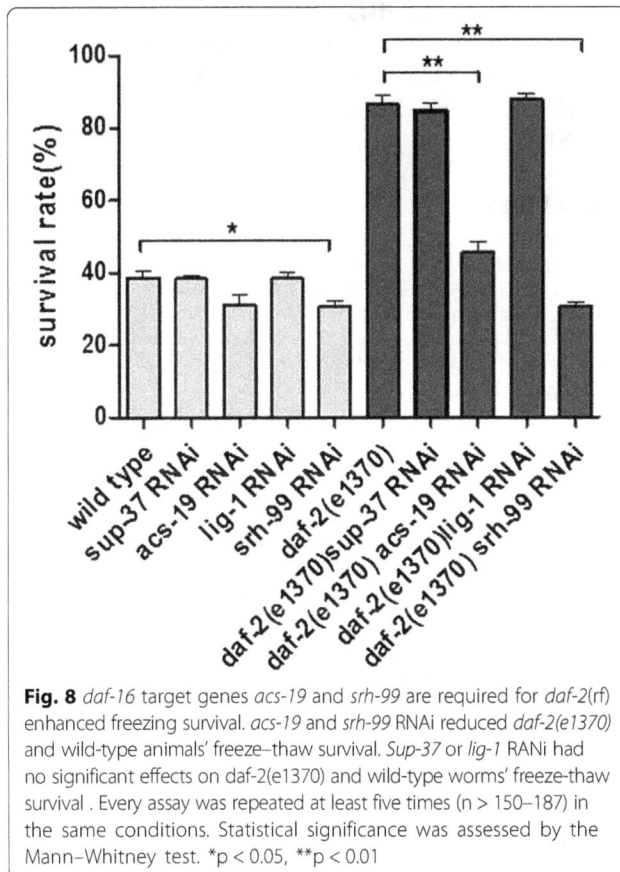

Fig. 8 *daf-16* target genes *acs-19* and *srh-99* are required for *daf-2*(rf) enhanced freezing survival. *acs-19* and *srh-99* RNAi reduced daf-2(e1370) and wild-type animals' freeze–thaw survival. *Sup-37* or *lig-1* RANi had no significant effects on daf-2(e1370) and wild-type worms' freeze-thaw survival . Every assay was repeated at least five times (n > 150–187) in the same conditions. Statistical significance was assessed by the Mann–Whitney test. *p < 0.05, **p < 0.01

Freeze–thaw stress conditions

Adult animals (2 d after L4) on NGM plates were washed with M9 buffer and transferred into FACS tube with 1 ml M9 buffer. *Caenorhabditis elegans* does not survive freezing very well, so to determine optimal freezing length for analyses, we placed animals from room temperature into −80 °C for different amounts of times (4 min, 5 min,6 min, 8 min, 12 min, or 16 min). At 4 min, there was no survival difference between *N2* and *daf-2(e1370)*. Ice began to form at 5 min, and all water froze in the tube at 6–7 min; however, all animals, including *N2* and *daf-2*(rf), died at 16 min. At 8 min, there were significant freezing survival differences among mutants. Therefore, animals were treated with freezing stress at −80 °C for 8 min for subsequent analyses. This experiment revealed that freezing stress damaged animals; it is possible that in the body, some cells did not completely freeze, but the freezing stress still injured animals.

Animals were exposed to freezing stress and then thawed at different temperatures (0 °C and 30 °C) for different lengths of time (10 min for 0 °C and approximately 1 min for 30 °C). We removed the tubes at the end of thawing process (when the ice was completely thawed), when the temperature in the tube is still 0 °C.

All treated animals were placed on dry NGM plates for 6 h before determining mortality score with an optical microscope. L1–adult animals were used for the *daf-2*(rf) survival test for different stages.

Programmed freezing conditions

L1–adult animals on NGM plates were gently washed with M9 buffer. For the freezing procedure, a Cryo 1 °C freezing container was used with a gradient cooling rate of −1 °C/min. Animals were transferred into freezing tubes with 0.9 ml buffer and 0.9 ml 30 % glycerin. At different time points (every 20 or 40 min), animals were thawed in a 30 °C water bath, and the survival rate was then assayed.

Feeding RNAi

dsRNA-expressing *E. coli* were streaked onto LB agar plates that contained ampicillin (50 μg ml^{-1}) and tetracycline (12.5 μg ml^{-1}), and then incubated at 37 °C overnight. Bacteria were inoculated in 3 ml LB liquid medium that contained only ampicillin (100 μg ml^{-1}) and then incubated at 37 °C overnight. All 3 ml of culture was spun down, the supernatant poured off until 150 μl was left (20× concentrated culture), and pellets were resuspended. Then, 50 μl of cells were resuspended to the center of RNAi plate (NGM/IPTG/ampicillin), allowed to dry (wrapped in aluminum foil), and induced overnight at room temperature (RNAi-seeded plates can be stored at room temperature for 2–3 d before use). Synchronized L1 worms were placed on each plate and incubated at 15 or 20 °C until they reached the desired stage for further experiments.

Scoring mortality

Worms were scored on NGM plates after 6 h of recovery from the thawing procedure. Worms were prodded with a pick at least three times over approximately 10 s; any that failed to move were counted as dead and removed from the plate.

Morphologic cell defects

The *PD4251* reporter gene *pmyo-3::gfp*, which is located in body wall muscle nuclei, was assayed with a fluorescence microscope. *pmyo-3::gfp* expression of strain *PD4251* by RNAi treatment with freezing and thawing was assayed.

Quantitation of locomotion rate by counting body bends

Ten animals were selected from fresh plates with fairly thin lawns, and one worm each was placed new plates. Assays started 24 h later (±1 h). A 3-min timer was used to count number of body bends. Every time the part of the worm just behind the pharynx reached a maximum bend in the opposite direction from the bend last counted was considered one body bend.

daf-16 nucleus translocation

TJ356 nuclear protein *daf-16*::GFP expression was assayed with a fluorescence microscope. *daf-16*::GFP expression of [*daf-2*(RNAi); *TJ356*] treated with freezing and thawing was assayed compared with that of *TJ356*.

RNA isolation for qRT-PCR

Total mRNA was extracted by TRIzol reagent (Invitrogen, Carlsbad, CA) from certain *C. elegans* and treated with RNase-free DNase (Promega, Madison, WI). Then, reverse transcription (RT) was performed with a TaKaRa RNA PCR kit (Takara, Dalian, China) following the manufacturer's instructions. The primers used are described in Additional file 5: Table S1 and sequenced in the DNA Sequencing Department of Biosune Systems Biology (Shanghai, China).

mRNA reverse transcription and qRT-PCR

One nanogram in vitro transcribed RNA was added to the RNA sample (500–1000 ng). DEPC H_2O was added to the RNA sample to 29.5 µl. Then, 0.5 µl each of 1 µg/µl random hexamer and 1 µg/µl poly dT were added. The mixture was incubated at 65 °C for 10 min, immediately put on ice for 5 min, and let stand at room temperature for 10 min. Then, 18.5 µl pre-mixture, which contained 2.5 µl 10 mM dNTP mix, 10 µl 5× first-strand buffer, 5 µl 0.1 M DTT, and 1 µl RNase OUT, was added. The mixture was then mixed and spun, and placed at room temperature for 2 min. Then, 1 µl Superscript II RT was added and gently mixed. The mixture was then spun down and let stand at room temperature for 10 min, incubated at 42 °C for 50 min, and heat inactivated at 70 °C for 15 min. Then, 1 µl RNase H was added to the solution, which was gently mixed, spun down, and blocked at 37 °C for 30 min.

SYBR Green qRT-PCR was performed on the LightCycler® 480 II System (Roche, Pleasanton, CA) using 5 µl 2× SYBR Green master mix (Roche), 2 µl RNASe free water, 1 µl of forward, 1 µl of reverse primer (10 µM), and 1 µl cDNA per reaction. Primer efficiency was assessed by a dilution series, and a dissociation curve was used to assess primer specificity. qRT-PCR data analysis was performed using genEx software (MultiD).

Statistical analyses

Percent survival was reported as mean ± SEM per trial. Every test was repeated at least three times under the same conditions. Survival rate was analyzed by the nonparametric Mann–Whitney test. Correlation analysis between *daf-2(rf)* allele freeze–thaw survival and lifespan/other stress traits were conducted by the nonparametric Spearman correlation test.

Availability of supporting data

Primer sequences,data are accessible through this link: https://mynotebook.labarchives.com/share/fangny/MjMuNHwxMzY4MDYvMTgvVHJlZU5vZGUvMzYxNTQ3NzkyNXw1OS40.

Additional files

Additional file 1: Figure S1. *daf-2(e1370)* enhanced freezing survival under programmed freezing conditions. Reduction-of-function mutant *daf-2(e1370)* (squares) exhibited significantly increased survival compared with wild-type (*N2*) animals under the programmed cooling conditions (-1 °C/min) at different stages. (JPG 259 kb)

Additional file 2: Figure S2. Survival rates of wild-type (*N2*) and *daf-2(e1370rf)* under different thawing processes. With different thawing treatments (30 °C water bath for 1 min or ice for 30 min), wild-type (*N2*) and *daf-2(e1370rf)* animals had unchanged survival rates after freeze–thaw stress. (JPG 94 kb)

Additional file 3: Figure S3. Correlated analysis of *daf-2(rf)* freeze–thaw survival and lifespan, larval arrest, and other stress-resistance traits. Freeze–thaw stress survival phenotypes were not correlated with lifespan, larval arrest, or hypoxia resistance , but was moderately correlated with heat shock survival . (JPG 41 kb)

Additional file 4: Figure S4. Screening for *daf-16* target genes required for freezing survival by QF-PCR. *daf-16* target genes C01B7.1(*sup-37*), C36A4.9 (*acs-19*) , C29A12.3(*lig-1*) and C46A10.7 (*srh-99*) have higher expression in *daf-2(e1370)* compared with wild-type (*N2*) animals. (JPG 92 kb)

Additional file 5: Table S1. *daf-2(rf)* allelic variation influence on freeze–thaw stress survival. (JPG 169 kb)

Competing interests

Authors have no financial and nonfinancial competing interests.

Authors' contributions

Hu JP assayed the freezing survival and drafted the manuscript. Xu XY performed the target gene analysis and feeding RNAi . Huang LY measured the daf-16::GFP and muscle GFP analysis. Wang LS and Fang NY contributed to design the experiment ,Wang LS analyzed the data and Fang NY revised the manuscript. All authors read and approved the final manuscript.

Acknowledgements

This work was supported by grants from the National Natural Science Foundation of China (NSFC81370360,81472758,31170783,U1302225) and Science and Technology Commission of Shanghai (11ZR1421500). All strains were provided by the Caenorhabditis Genetics Center.

References

1. Ramløv H. Aspects of natural cold tolerance in ectothermic animals. Hum Reprod. 2000;15:26–46.
2. Wharton D. A in Molecular and Physiological Basis of Nematode Survival (eds Perry, R.N. et al.) Ch. 8, 222-223 (Wallingford, 2011).
3. Morris JZ, Tissenbaum HA, Ruvkun G. A phosphatidylinositol-3-OH kinase family member regulating longevity and diapause in Caenorhabditis elegans. Nature. 1996;382:536–9.
4. Paradis S, Ruvkun G. Caenorhabditis elegansAkt/PKB transduces insulin receptor-like signals from age-1 PI3 kinase to the daf-16 transcription factor. Genes Dev. 1998;12:2488–98.
5. Paradis S, Ailion M, Toker A, et al. A PDK1 homolog is necessary and sufficient to transduce age-1 PI3 kinase signals that regulate diapause in Caenorhabditis elegans. Genes Dev. 1999;13:1438–52.
6. Kenyon C, Chang J, Gensch E, et al. C. Elegans mutant that lives twice as long as wild type. Nature. 1993;366:461–4.
7. Ogg S, Ruvkun G. The C. elegans PTEN homolog, DAF-18, acts in the insulin receptor-like metabolic signaling pathway. Mol Cell. 1998;2:887–93.

8. Gil EB, Link EM, Liu LX, et al. Regulation of the insulin-like developmental pathway of Caenorhabditis elegansby a homolog of the PTEN tumor suppressor gene. Proc Natl Acad Sci U S A. 1999;96:2925–30.

9. Patrick J. Hu, Dauer-wormbook. Available at: http://www.wormbook.org/chapters/www_dauer/dauer.html

10. Adam Antebi, Nuclear hormone receptors in C. Elegans-wormbook. Available at: http://www.wormbook.org/chapters/www_nuclearhormonerecep/nuclearhormonerecep.html

11. Ogg S, et al. The Fork head transcription factor daf-16 transduces insulin-like metabolic and longevity signals in C. elegans. Nature. 1997;389:994–9.

12. Savory FR, Sait SM, Hope I. A. daf-16 and Delta9 desaturase genes promote cold tolerance in long-lived Caenorhabditis elegans age-1 mutants. Plos One. 2011;6:e24550.

13. Ohta A, Ujisawa T, Sonoda S, et al. Light and pheromone-sensing neurons regulates cold habituation through insulin signalling in Caenorhabditis elegans. Nat Commun. 2014;22:4412.

14. Hermes-Lima M, Storey KB. Antioxidant defenses in the tolerance of freezing and anoxia by garter snakes. Am J Physiol. 1993;265:R646–52.

15. Mazur P. Cryobiology: the freezing of biological systems. Science. 1970;168:939–49.

16. Toner M, Cravalho EG, Karel M. Cellular response of mouse oocytes to freezing stress: prediction of intracellula ice formation. J Biomech Eng. 1993;115:169–74.

17. Finkel T, Holbrook NJ. Oxidants, oxidative stress and the biology of ageing. Nature. 2000;408(6809):239–47.

18. Gems D, Sutton AJ, Sundermeyer ML, et al. Two pleiotropic classes of daf-2mutation affect larval arrest, adult behavior, reproduction and longevity in Caenorhabditis elegans. Genetics. 1998;150(1):129–55.

19. Kimura KD, Tissenbaum HA, Liu Y, et al. daf-2, an insulin receptor-like gene that regulates longevity and diapause in Caenorhabditis elegans. Science. 1997;277(5328):942–6.

20. Fielenbach N, Antebi A. C. elegans dauer formation and the molecular basis of plasticity. Genes Dev. 2008;22:2149–65.

21. Xiao R, Zhang B, Dong Y, et al. A genetic program promotes C. elegans longevity at cold temperatures via a thermosensitive TRP channel. Cell. 2013;152(4):806–17.

22. Sakakibara I, Fujino T, Ishii M, et al. Fasting-induced hypothermia and reduced energy production in mice lacking acetyl-CoA synthetase 2. Cell Metab. 2009;9(2):191–202.

23. Riveron J, Boto T, Alcorta E.Transcriptional basis of the acclimation to high environmental temperature at the olfactory receptor organs of Drosophila melanogaster. BMC Genomics.4. doi:10.1186/1471-2164-14-259(2013).

24. Luis H. Toledo-Pereyra. in Organ Preservation for Transplantation 3rd edn, (eds Luis H. Toledo-Pereyra. et al.), Ch.4 45-57, (Landes Bioscience.2010).

25. Lakowski B, Hekimi S. The genetics of caloric restriction in Caenorhabditis elegans. Proc Natl Acad Sci U S A. 1998;95:13091–6.

26. Branicky R, Benard C, Hekimi S. clk-1, mitochondria, and physiological rates. Bioessays. 2000;22:48–56.

27. Feng J, Bussière F, Hekimi S. Mitochondrial electron transport is a key determinant of life span in Caenorhabditis elegans. Dev Cell. 2001;1(5):633–44.

28. Lee SJ, Kenyon C. Regulation of the longevity response to temperature by thermosensory neurons in Caenorhabditis elegans. Curr Biol. 2009;19(9):715–22.

29. Oh SW, Mukhopadhyay A, Dixit BL, et al. Identification of direct daf-16 targets controlling longevity, metabolism and diapause by chromatin immunoprecipitation. Nat Genet. 2005;38:251–7.

30. Lamitina ST, Strange K. Transcriptional targets of daf-16 insulin signaling pathway protect C. elegans from extreme hypertonic stress. Am J Physiol Cell Physiol. 2005;288:467–74.

31. McElwee J, Schuster E, Blanc E, Thornton J, Gems D. Diapause-associated metabolic traits reiterated in long-lived daf-2mutants in the nematode Caenorhabditis elegans. Mech Ageing Dev. 2006;127:458–72.

32. Wang J, Kim SK. Global analysis of dauer gene expression in Caenorhabditis elegans. Development. 2003;130:1621–34.

33. Jagdale GB, Grewal PS, Salminen SO. Both heat-shock and cold-shock influence trehalose metabolism in an entomopathogenic nematode. J Parasitol. 2005;91:988–94.

34. Honda Y, Tanaka M, Honda S. Trehalose extends longevity in the nematode Caenorhabditis elegans. Aging Cell. 2010;9:558–69.

35. Murphy CT, McCarroll SA, Bargmann CI, Fraser A, Kamath RS, et al. Genes that act downstream of daf-16 to influence the lifespan of Caenorhabditis elegans. Nature. 2003;424:277–83.

36. McElwee JJ, Schuster E, Blanc E, Thomas JH, Gems D. Shared transcriptional signature in Caenorhabditis elegans dauer larvae and long-lived daf-2mutants implicates detoxification system in longevity assurance. J Biol Chem. 2004;279:44533–43.

37. Larsen PL. Aging and resistance to oxidative damage in Caenorhabditis elegans. Proc Natl Acad Sci U S A. 1993;90:8905–9.

38. Fontana L, Partridge L, Longo VD. Extending healthy life span–from yeast to humans. Science. 2010;328:321–6.

39. Kenyon CJ. The genetics of ageing. Nature. 2010;464:504–12.

40. Fire A, Xu S, Montgomery MK, et al. Potent and specific genetic interference by double-stranded RNA in Caenorhabditis elegans. Nature. 1998;391(6669):806–11.

41. Henderson ST, Johnson TE. daf-16 integrates developmental and environmental inputs to mediate aging in the nematode Caenorhabditis elegans. Curr Biol. 2001;11:1975–80.

16

Comparative genome-wide association studies of a depressive symptom phenotype in a repeated measures setting by race/ethnicity in the multi-ethnic study of atherosclerosis

Erin B. Ware[1,2]*, Bhramar Mukherjee[3], Yan V. Sun[4], Ana V. Diez-Roux[5], Sharon L.R. Kardia[1] and Jennifer A. Smith[1]

Abstract

Background: Time-varying phenotypes have been studied less frequently in the context of genome-wide analyses across ethnicities, particularly for mood disorders. This study uses genome-wide association studies of depressive symptoms in a longitudinal framework and across multiple ethnicities to find common variants for depressive symptoms. Ethnicity-specific GWAS for depressive symptoms were conducted using three approaches: a baseline measure, longitudinal measures averaged over time, and a repeated measures analysis. We then used meta-analysis to jointly analyze the results across ethnicities within the Multi-ethnic Study of Atherosclerosis (MESA, n = 6,335), and then within ethnicity, across MESA and a sample from the Health and Retirement Study African- and European-Americans (HRS, n = 10,163).

Methods: This study uses genome-wide association studies of depressive symptoms in a longitudinal framework and across multiple ethnicities to find common variants for depressive symptoms. Ethnicity-specific GWAS for depressive symptoms were conducted using three approaches: a baseline measure, longitudinal measures averaged over time, and a repeated measures analysis. We then used meta-analysis to jointly analyze the results across ethnicities within the Multi-ethnic Study of Atherosclerosis (MESA, n = 6,335), and then within ethnicity, across MESA and a sample from the Health and Retirement Study African- and European-Americans (HRS, n = 10,163).

Results: Several novel variants were identified at the genome-wide suggestive level ($5 \times 10^{-8} < p$-value $\leq 5 \times 10^{-6}$) in each ethnicity for each approach to analyzing depressive symptoms. The repeated measures analyses resulted in typically smaller p-values and an increase in the number of single-nucleotide polymorphisms (SNP) reaching genome-wide suggestive level.

Conclusions: For phenotypes that vary over time, the detection of genetic predictors may be enhanced by repeated measures analyses.

Keywords: Depressive symptoms, Generalized estimating equations, Genome-wide association studies, Longitudinal, Psychogenetics

* Correspondence: ebakshis@umich.edu
[1]Department of Epidemiology, University of Michigan, Ann Arbor, MI, USA
[2]Institute of Social Research, University of Michigan, 1415 Washington Heights #4614, Ann Arbor, MI 48109, USA
Full list of author information is available at the end of the article

Background

With advances in the ability of statistical software to handle data with repeated measures, longitudinal data analysis is becoming more feasible in genetic association studies. While these analyses are more complicated and computationally intensive than analyses using only baseline measures, longitudinal data has been used to identify variants that influence complex traits above and beyond that of cross-sectional measurements [1]. Because depressive symptoms may vary over time in relation to a variety of circumstantial factors, repeated measures of depressive symptoms may provide a better characterization of an individual's phenotype than a single measure, thus increasing power to detect genetic susceptibility loci.

There are a number of circumstances where longitudinal data analysis may be more informative or powerful than cross-sectional analyses based on single or time averaged measures. If there is substantial variability over time in the outcome or interaction of other covariates or SNPs with time, a longitudinal analysis will clearly be more informative [2]. For a given fixed number of observations, cross sectional analyses will be more powerful than repeated measures in the presence of within-subject correlations (e.g. cross sectional n = 500; repeated measures n = 250 with two measures), but longitudinal analyses permits detection of factors associated with within person changes over time, which often allows stronger causal inferences [2]. A genetic association analysis with longitudinal data also follows these well-established properties, except for the fact that the analysis is repeated millions of times and tail behavior of the test statistics along with robustness issues become more critical since much smaller significance thresholds are used than traditional inference at a 5 % level of significance.

Depressive symptoms exist on a spectrum, varying in both severity and duration, and are often measured in population-based studies using the 20-item Center for Epidemiological Studies Depression scale (CES-D). Given the benefits of longitudinal analysis, the ability to detect genetic predictors of depression may be enhanced by analyzing depressive symptoms both over time and quantitatively [3], rather than applying cutoffs or defining disorders like Major Depressive Disorder (MDD) at the extreme of the continuum for a single time point [4].

The Multi-Ethnic Study of Atherosclerosis (MESA) European sub-sample was recently part of a discovery sample for a cross-sectional genome-wide association study (GWAS) of depressive symptoms conducted by the Cohorts for Heart and Aging Research in Genomic Epidemiology (CHARGE) consortium [5]. This GWAS focused on a single measure of depressive symptoms (as assessed by CES-D) in individuals of European descent. Though no loci reached genome-wide significance in the

discovery sample (composed of 34,549 individuals), one of the seven most significant SNPs had a suggestive association in the replication sample (rs161645, 5q21, $p = 9.19 \times 10^{-3}$). This SNP reached genome-wide significance ($p = 4.78 \times 10^{-8}$) in overall meta-analysis of the combined discovery and replication samples (n = 51,258) [5]. Important limitations of this GWAS include the reliance on a single measure of depressive symptoms and the focus on a single race/ethnic group.

In the present study, we use longitudinal data on a continuous measure of depressive symptoms collected over a 9 year period from three exams in MESA to conduct GWAS on depressive symptoms in four race/ethnicities. We also contrast different approaches of incorporating the repeated measures into the GWAS: (1) analyzing a single time-point measure (baseline), (2) averaging measures over time, and (3) conducting a repeated measures outcome analyses. Finally, we jointly analyze repeated measures GWAS results from MESA and up to ten exams from the Health and Retirement Study. The MESA study includes a total of 650, 507, and 5,178 participants with one, two, and three measures, respectively, while the HRS sample consists of 34, 147, and 9,982 individuals with one, two, and three-plus measures, respectively) in an overall meta-analysis for European Americans and African Americans to increase power. To our knowledge, there have been no GWAS of repeated measures of depressive symptoms measured over time in individuals of multiple race/ethnicities.

Results

Descriptive statistics

Descriptive statistics for MESA and HRS are presented in Table 1. The MESA sample includes 6,335 individuals (48 % male). Mean age at baseline is 62.2 years and approximately 40 %, 25 %, 12 %, and 23 % are of European (EA), African (AA), Chinese (CA), and Hispanic (HA) American self-reported ethnicity, respectively.

In MESA, the mean baseline depressive symptom score ranged from 6.3 (standard deviation (SD): 6.6) in the CA subsample to 9.9 (SD: 9.2) in the HA subsample out of a possible score of 60. CES-D scores increased over time in the EA (linear trend model for exam: $\beta_{exam} = 0.25$, $p < 0.0001$), AA ($\beta_{exam} = 0.03$, $p = 0.67$), and HA ($\beta_{exam} = 0.13$, $p = 0.11$) sub-groups, but this increase in trend was only significant in EA. The CA sub-group showed a non-significant decrease in depressive symptom score over time ($\beta_{exam} = -0.04$, $p = 0.67$). The intraclass correlation (within-person correlation) across all exams for which an individual had a valid CES-D score (up to three time-points) ranged from 0.44 in AA to 0.60 in EA.

The HRS analysis sample contains 10,163 respondents (41 % male), with 8,652 EA (85 %) and 1,511 AA (15 %). Mean age at baseline was 58 years. The CES-D8

Table 1 Descriptive statistics

| | MESA[1] n = 6,335 | | | | | | | | HRS[2] n = 10,163 | | | |
| | European American n = 2,514 | | African American n = 1,603 | | Hispanic American n = 1,443 | | Chinese American n = 775 | | European American n = 8,652 | | African American n = 1,511 | |
	Mean	(SD)	Mean	(SD)	Mean	(SD)	Mean	(SD)	Mean	(SD)	Mean	(SD)
Depression score[3]												
Baseline CES-D[4]	8	(7.8)	7.6	(7.6)	9.9	(9.2)	6.3	(6.6)	1.2	(1.8)	1.2	(1.8)
Averaged CES-D	8.7	(7.4)	7.8	(6.7)	10.2	(8.5)	6.2	(5.6)	1.2	(1.3)	1.9	(1.6)
Age	62.6	(10.2)	62.2	(10.1)	61.4	(10.3)	62.4	(10.4)	58.4	(8.8)	56.8	(8.2)
Sex (%)	n	%	n	%	n	%	n	%	n	%	n	%
Male	1207	48.0	744	46.4	711	49.3	385	49.7	3,556	41.1	639	42.3
Site (%)												
Baltimore, MD	505	20.1	482	30.1	-	-	-	-	-	-	-	-
Chicago, IL	526	20.9	258	16.1	-	-	275	35.4	-	-	-	-
Forsyth County, NC	548	21.8	425	26.5	3	0.2	-	-	-	-	-	-
Los Angeles, CA	133	5.3	143	8.9	554	38.4	498	64.3	-	-	-	-
New York, NY	209	8.3	295	18.4	431	29.9	2	0.3	-	-	-	-
St. Paul, MN	593	23.6	-	-	455	31.5	-	-	-	-	-	-
Anti-depressant Use (%)	307	12.2	61	3.8	84	5.8	19	2.5	-	-	-	-
Intraclass correlation												
Repeated Measures CES-D	0.60		0.44		0.57		0.57		0.48		0.51	

[1]Multi-Ethnic Study of Atherosclerosis, [2]Health and Retirement Study, [3]CES-D measured as 20-item sum in MESA and as 8-item sum in HRS, [4]Center for Epidemiologic Studies - Depression

depressive symptom score in HRS EA increased significantly over study waves ($\beta_{exam} = 0.03$, $p < 0.0001$) and decreased significantly in AA participants over time ($\beta_{exam} = -0.01$, $p = 0.04$). The intraclass correlation for the HRS participants across exams was 0.48 for EA participants and 0.51 for AA participants.

Ethnicity-specific association analysis in MESA
Table 2 shows the number of SNPs, minimum p-value of the adjusted association between SNP dosage and outcome,

and the genomic-control inflation factor, lambda, for each ethnicity in MESA and HRS. QQ plots are available in Additional file 1. The inflation factor, the extent to which the chi-square statistic is inflated due to confounding by ethnicity [6], is very close to 1.0 for all analyses, indicating adequate adjustment for population structure. One SNP reached the genome-wide significant threshold in the HA subset in the baseline CES-D approach in the intronic region of the *MUC13* gene (rs1127233, 3q22.1, $\beta = 0.2382$, p-value = 3.85×10^{-8}; averaged $\beta = 0.1598$, p-value = 9.23×10^{-6};

Table 2 Minimum p-value from GWAS of baseline, averaged, and repeated measures of CES-D[1] across ethnicities, MESA[2] and HRS[3]

| Study | Ethnicity | # of SNPS | Baseline CES-D score | | | Averaged CES-D score | | | Repeated measures CES-D score | | |
			Min p-value	# of unique SNPs[4]	λ[5]	Min p-value	# of unique SNPs	λ	Min p-value	# of unique SNPs	λ
MESA											
African American		2380122	2.05×10^{-7}	7	1.01	6.64×10^{-7}	9	1.00	1.63×10^{-7}	11	1.01
European American		2166730	1.33×10^{-7}	9	1.01	8.26×10^{-7}	6	1.00	6.04×10^{-7}	11	1.01
Chinese American		1801470	2.48×10^{-6}	1	0.99	1.42×10^{-6}	2	1.00	2.71×10^{-7}	4	1.02
Hispanic American		2148331	3.85×10^{-8}	10	1.00	1.61×10^{-6}	4	1.00	9.25×10^{-7}	11	1.01
HRS											
African American		2446939	-	-	-	-	-	-	2.07×10^{-6}	-	1.01
European American		2177692	-	-	-	-	-	-	6.54×10^{-7}	-	1.04

[1]Center for Epidemiological Studies – Depression, [2]Multi-Ethnic Study of Atherosclerosis, [3]Health and Retirement Study, [4]Number of unique (independent) SNPs, linkage disequilibrium R[2] < 0.80, INFO > 0.80, with ethnicity-specific minor allele frequency > 5 % and p-values < 1×10^{-5}, [5]genomic control lambda

repeat measures $\beta = 0.1753$, p-value $= 2.06\times10^{-6}$). This gene has previously been associated with cancer pathogenesis (e.g. [7–16]) but has not been implicated in any psychiatric disorders. This SNP was not associated with CES-D in the other race/ethnicities nor did it show consistent direction across ethnicity in the baseline CES-D analyses (AA: $\beta = -0.0112$, p-value $= 0.7707$; EA: $\beta = -0.0228$, p-value $= 0.4527$; CA: $\beta = 0.0562$, p-value $= 0.4351$). There were no other genome-wide significant SNPs in any of the ethnicities for any of the baseline, average, and repeated-measures modeling approaches though there were many suggestive $p < 10^{-6}$ findings.

Comparison of results across approaches

To compare association results between the different versions of the CES-D scores, we assessed scatter plots for the p-values ($p < 5\times10^{-4}$) from each pair of SNPs for the baseline CES-D score compared to the averaged CES-D score phenotype (Additional file 2), the baseline CES-D score compared to the repeated measures CES-D score (Additional file 3), and the averaged CES-D score against the repeated measures CES-D score (Additional file 4) within each of the four ethnicities in MESA. For all four ethnicities, the Spearman's rank correlations between the baseline versus averaged CES-D phenotype and between the baseline and repeated measures CES-D phenotypes ranged between 0.46 and 0.57. The correlations between p-values for the averaged versus repeated measures CES-D phenotype ranged between 0.85 and 0.92 (Table 3). We observed an increase in the number of unique (LD $R^2 < 0.8$) genome-wide suggestive SNPs from baseline to repeated measures for each ethnicity (EA: eight to nine; AA: four to 11; CA: one to four; HA: six to ten), with some (at least two SNPs appearing in multiple approaches as genome-wide suggestive within each ethnicity) consistency in the SNPs across approach (Additional file 5).

Meta-analysis across ethnicities in MESA

The results from the three meta-analyses performed within MESA across ethnicities for the baseline, averaged, and repeated measures CES-D scores are presented in Table 4. In the table, we present every unique

(LD $R^2 < 80$ %) SNP with $p < 1\times10^{-6}$. The meta-analysis only included SNPs with ethnicity-specific minor allele frequency (MAF) > 5 % calculated within ethnicity using only MESA participants. These meta-analyses showed no genome-wide significant results. Thirteen SNPs reached a genome-wide suggestive threshold in these meta-analyses. The smallest p-value was in the repeated measures meta-analysis on chromosome 2, (rs41379347, 2q32.2, p-value $= 1.81\times10^{-7}$). This SNP was only present (with MAF > 5 %) in the CA and HA subsamples. This SNP is in the intronic region of the STAT1 gene, IFN-γ transcription factor signal transducer and activator of transcription 1, previously implicated as a tumor suppressor [17, 18]. This SNP has not been previously associated with depressive symptoms.

Joint-analysis across studies for EA and AA

Results from the joint-analyses (MESA + HRS) for EA and AA, separately, are presented in Table 5. While no SNP reached the genome-wide level, eight SNPs (EA n = 3; AA n = 5) satisfied the suggestive threshold for significance. In EA the smallest p-value (rs6842756, 4q35.1, p-value $= 6.54\times10^{-7}$) was located within the ENPP6 gene, which is expressed primarily in the kidney and brain and has not been implicated in any disorders or diseases [http://omim.org/]. In AA the smallest observed p-value (rs2426733, 20q13.31, p-value $= 2.07\times10^{-6}$) was located downstream of the RBM38 oncogene. RBM38 encodes an RNA binding protein found to regulate MDM2 (12q14.3-q15) gene expression through mRNA stability [19, 20], but has not been identified in genetic studies of psychiatric disorders [17] (http://omim.org/).

Meta-analysis across all ethnicities in MESA and HRS

For the meta-analysis across all ethnicities in both HRS and MESA, we found no SNPs reaching genome-wide significance, though we found seven SNPs reaching genome-wide suggestive thresholds (Table 5). The most strongly associated SNPs in the meta-analysis, rs41379347 (p-value $= 1.81\times10^{-7}$) is located on chromosome 2 (in the STAT1 gene). The SNP rs41379347 was found previously in the MESA

Table 3 Spearman's correlation coefficients and 95 % confidence intervals for paired p-values in Multi-Ethnic Study of Atherosclerosis

		Baseline vs averaged CES-D score	Baseline vs repeated measures CES-D score	Averaged vs repeated measures CES-D score
		r, (95 % Confidence interval)	r, (95 % Confidence interval)	r, (95 % Confidence interval)
MESA	African American	0.53, (0.53, 0.53)	0.54, (0.54, 0.54)	0.88, (0.88, 0.88)
	European American	0.54, (0.54, 0.54)	0.57, (0.57, 0.57)	0.92, (0.92, 0.92)
	Chinese American	0.48, (0.48, 0.48)	0.46, (0.46, 0.47)	0.85, (0.85, 0.85)
	Hispanic American	0.54, (0.54, 0.54)	0.56, (0.55, 0.56)	0.88, (0.88, 0.88)

Table 4 Meta-analysis results[1] across ethnicities in MESA[2] (p-values $< 1\times10^{-5}$) for each depressive symptom score modeling approach

Approach	CHR	SNP	Location	Coded allele	Coded allele frequency	Z-score	P-value	Direction[3]	Closest gene[4] within ±50kB
Baseline									
	8	rs2440212	97270629	A	0.66	4.47	7.73×10^{-6}	++++	(GDF6)
	9	rs13440434	131953827	A	0.87	−4.50	6.79×10^{-6}	————	(GPR107)
	10	rs7087469	54339854	A	0.13	4.76	1.93×10^{-6}	++?+	-
	13	rs9560521	89457392	A	0.13	4.69	2.69×10^{-6}	++++	(LINC00559)
	16	rs8046816	71863525	A	0.47	4.53	5.92×10^{-6}	++++	-
	20	rs17215529	3923402	A	0.85	4.79	1.66×10^{-6}	++?+	RNF24
Averaged									
	1	rs3100865	2795967	T	0.49	4.44	9.02×10^{-6}	++++	-
	2	rs41379347	191577187	T	0.89	−4.58	4.57×10^{-6}	??–	STAT1
	2	rs7602149	114357038	T	0.84	−4.57	4.78×10^{-6}	–?–	LOC728055
	2	rs13001068	182706602	A	0.92	4.50	6.95×10^{-6}	?+?+	(PDE1A)
	7	rs697521	16730681	T	0.13	−4.74	2.12×10^{-6}	–?–	BZW2
	8	rs7350109	60753909	A	0.81	−4.5	6.88×10^{-6}	–?–	-
	11	rs1448128	121291660	C	0.24	−4.58	4.61×10^{-6}	————	-
	22	rs5760767	23696411	T	0.51	4.58	4.62×10^{-6}	++++	(TMEM211)
Repeated measures									
	1	rs11590206	145665933	A	0.16	−4.72	2.33×10^{-6}	————	(GJA5)
	2	rs41379347	191577187	T	0.89	−5.22	1.81×10^{-7}	??–	STAT1
	2	rs7602149	114357038	T	0.84	−4.62	3.83×10^{-6}	–?–	LOC728055
	4	rs13139186	96637940	T	0.90	−4.48	7.44×10^{-6}	————	UNC5C
	4	rs233976	104823918	A	0.21	4.47	7.75×10^{-6}	?+++	TACR3
	7	rs11771332	86539742	A	0.81	−4.48	7.45×10^{-6}	?-?-	(KIAA1324L)
	9	rs2211185	1332721	T	0.77	4.55	5.42×10^{-6}	++++	-
	18	rs2728505	21474070	A	0.55	−4.47	7.84×10^{-6}	————	-
	22	rs5760767	23696411	T	0.51	4.54	5.68×10^{-6}	++++	(TMEM211)

[1]filtered at ethnicity-specific minor allele frequency 5 %, where the SNP was present in at least two ethnicities, linkage disequilibrium $R^2 < 80$ %, and heterogeneity p-value ≥ 0.1; [2]Multi-Ethnic Study of Atherosclerosis; [3]Order corresponding to direction positions: African, European, Chinese, Hispanic American; [4]parentheses indicate location outside of gene

meta-analysis across ethnicity. This SNP was only present (with MAF > 5 %) in the MESA CA and HA samples, and thus, no new information was gained in the joint analysis across MESA and HRS.

Consistency with previous GWAS on depressive symptom scores

There has been one published GWAS conducted on depressive symptom scores [5], for which MESA EA were part of the discovery sample. This GWAS found one genome-wide significant SNP in overall meta-analysis of 51,258 European-ancestry individuals (rs161645, 5q21, $p = 4.78\times10^{-8}$). In our EA subsample, p-values for this SNP in our baseline and repeated measures analysis were 0.116 and 0.055, respectively, with consistent effect directions (+) as the Hek, *et al.* [5] finding.

Additionally, this SNP had a cross-ethnicity, within MESA meta-analysis p-value of 0.067 in the baseline analysis, 0.006 in the averaged CES-D analysis, and 0.008 in the repeated measures analysis. The overall direction of effect was consistent with the published GWAS for EA, AA, and HA, though the direction of effect was opposite for CA. This SNP had p-values of 0.951 and 0.113 for the cross-study (i.e. combining MESA and HRS) EA and AA analyses, respectively.

Discussion

This is the first set of GWASs to the authors' knowledge, to investigate common genetic variants for depressive symptoms in a longitudinal setting across four different ethnicities. We performed GWASs within each ethnicity for three different longitudinal approaches to a depressive symptom phenotype (baseline, averaged, and

Table 5 Meta-analysis results[1] between MESA[2] and HRS[3] (p-values $< 1\times10^{-5}$) for repeated measures depressive symptom score GEE analyses

Race	CHR	SNP	Location	Coded allele	Coded allele frequency	Z-score	P-value	Direction[4,5]	Closest gene[6] within ±50kB
African American									
	1	rs10776776	114384683	T	0.55	4.73	2.30×10^{-6}	++	(SYT6)
	1	rs1417303	235193008	T	0.59	−4.43	9.46×10^{-6}	−	LOC440737
	2	rs4629180	101454802	A	0.83	−4.51	6.41×10^{-6}	−	(LOC731220)
	2	rs6711630	126534599	T	0.93	4.58	4.70×10^{-6}	++	
	7	rs10249133	12514004	T	0.39	−4.47	7.67×10^{-6}	−	(LOC100133035)
	8	rs17067630	3661853	A	0.85	4.70	2.57×10^{-6}	++	CSMD1
	11	rs11036016	40661316	A	0.80	4.68	2.94×10^{-6}	++	LRRC4C
	15	rs4551976	49264445	T	0.63	−4.45	8.48×10^{-6}	−	(CYP19A1)
	16	rs365962	85267450	C	0.69	−4.53	5.83×10^{-6}	−	(LOC101928614)
	20	rs2426733	55454729	A	0.40	−4.75	2.07×10^{-6}	−	(RBM38)
European American									
	1	rs12031875	71357685	A	0.82	4.81	1.54×10^{-6}	++	ZRANB2-AS2
	4	rs6842756	185341452	A	0.92	4.98	6.54×10^{-7}	++	ENPP6
	6	rs6941340	16145531	T	0.48	−4.47	7.95×10^{-6}	−	
	9	rs11794102	111772109	A	0.91	4.54	5.70×10^{-6}	++	PALM2-AKAP2
	13	rs6492314	110267411	C	0.28	−4.75	2.00×10^{-6}	−	
	16	rs12921740	20219533	T	0.51	−4.55	5.44×10^{-6}	−	(GP2)
	18	rs2612547	41290709	A	0.83	4.47	7.94×10^{-6}	++	SLC14A2
All samples									
	1	rs2300177	71270950	T	0.19	−4.56	5.09×10^{-6}	? − +−−?	PTGER3
	1	rs1539418	145676734	A	0.16	−4.53	5.87×10^{-6}	−−−−??	
	1	rs7415169	168997688	A	0.85	4.70	2.58×10^{-6}	++++++	
	2	rs6711630	126534599	T	0.93	4.58	4.70×10^{-6}	+????+	
	2	rs41379347	191577187	T	0.89	−5.22	1.81×10^{-7}	??−??	STAT1
	4	rs6552764	185341497	A	0.86	4.64	3.41×10^{-6}	− + ? − ++	ENPP6
	8	rs12541177	124869454	T	0.31	−4.48	7.43×10^{-6}	−−−−−−−	FAM91A1
	9	rs9408311	38704682	T	0.84	4.92	8.87×10^{-7}	++ − +++	
	15	rs16975781	94245464	T	0.90	4.62	3.83×10^{-6}	+++++−	
	16	rs12921740	20219533	T	0.59	−4.70	2.61×10^{-6}	−−− + −−	(GP2)

[1]Filtered at ethnicity-specific minor allele frequency of 5 %, where the SNP was present in at least two ethnicities, linkage disequilibrium $R^2 < 80$ %, and heterogeneity p-value ≥ 0.1; [2]Multi-Ethnic Study of Atherosclerosis; [3]Health and Retirement Study [4]Order corresponding to direction positions: African, European, Chinese, Hispanic American; [5]For all samples analyses, order corresponding to direction position: MESA African American, MESA European American, MESA Chinese American, MESA Hispanic American, HRS European American, HRS African American; [6]parentheses indicate location outside of gene

repeated measures) and meta-analyzed them across ethnicity and across study. Though our joint meta-analysis of all ethnicities in both studies comprises 16,498 individuals, and the power to detect genetic variants of depression has been shown to increase when assessing depression quantitatively — as opposed to using a dichotomous definition or cutoff point [21] — we did not find any variants that reached genome-wide significant levels in the European-, African-, Hispanic-, or Chinese-American, race/ethnicity-specific GWAS, in meta-analyses across ethnicity in MESA, or in joint analyses across study for the European and African Americans with any evidence

of replication. However, we did find several novel variants at a genome-wide suggestive level and we observed an increase in the number of unique (LD $R^2 < 0.8$) genome-wide suggestive SNPs from baseline to repeated measures for each ethnicity (Additional file 5). We have taken the single SNP that has been credibly associated with depressive symptoms from Hek et al., [5] and presented evidence that a longitudinal framework may improve upon findings for depressive symptoms.

Hek, et al. [5] identified a SNP (rs161645) associated with a large sample of European-ancestry participants measured at a single time point. It is important to note that European Americans from MESA were used in the discovery sample for the previously published GWAS. We found that in the EA subsample, repeated measures better characterized depressive symptoms and the longitudinal analysis resulted in a repeated measures p-value for rs161645 ($p = 0.055$) less than half that of the baseline measures model ($p = 0.116$). If we consider this SNP a true signal (or proxy for a true signal), we indeed demonstrate that the p-value has decreased from the baseline to the repeated measures analysis.

A repeated measures analysis makes use of the full information content in the outcome and exposure/covariates for longitudinal data. For example, in an analysis with repeated measures data, if there is drop-out in the study and we use subject level averages, the homoscedasticity assumption of linear models is violated as different averages will be based on different number of observations and the ones with more observation will have higher precision. Averaging the exposure data may also lead to substantial loss in power. If there is a time trend or interaction of covariates (or SNPs) with time, a longitudinal model is expected to have larger power than a cross-sectional or averaged model. Longitudinal modeling is a better general framework as it allows incorporation of time-varying covariates (instead of averaging them) and allows exploration of G × E interaction in follow-up analysis with cumulative exposure trajectory. Although we saw an increase in the number of unique genome-wide suggestive SNPs for repeated measures compared to baseline, we note that since most of the SNPs are non-significant, this may be simply a comparison of false positives. However, in view of the existing literature one can argue that a longitudinal analysis is generally more efficient than using a summary quantity in the presence of repeated measures data.

For repeated measures, there are multiple modeling approaches. GEE produces unbiased and consistent estimates of the fixed effect parameters, even under misspecification of the correlation structure. Also, if the correlation structure is correctly specified, there is gain in terms of efficiency. GEE can be argued as a better framework than a linear regression model in terms of its robust estimates of the standard error and behavior of QQ plots as it protects under model misspecification [22]. That is why we chose the GEE framework for this large-scale association analysis instead of an alternative linear mixed model analysis.

Though GWAS have been used for over a decade, most variants identified for diseases have had very modest effect sizes, often explaining less than 1 % of the variance of quantitative traits [23]. Because of the small effect sizes, very large sample sizes are required to reach adequate power to detect genetic effects and produce reliable inferences [24]. Preliminary steps have been taken to increase power in our study through the characterization of a longitudinal phenotype. Most individual studies, including this one, are underpowered to detect these variants and often collaboration across many studies, involving meta-analysis, are used to increase sample size, and thus power [23, 25]. Though this framework is frequently used for common traits with standard measures, it is exceedingly difficult to find studies measuring depressive symptoms using the CES-D in multiple ethnicities, across time.

The depressive symptom GWAS literature to date includes one GWAS, with only one genome-wide significant result [5]. The literature for similar phenotypes, such as Major Depressive Disorder (MDD), has nine GWAS studies [26–34], a mega-analysis of the nine GWAS that included almost 19,000 European unrelated individuals [35], and a recent low-coverage, whole-genome sequencing analysis in the Chinese ethnicity [36]. Only two loci reached genome wide significance in individual studies [28, 37], but these loci were not significantly associated with MDD in the meta-analysis [35]. The whole-genome sequencing analysis, using a joint discovery-replication analysis and linear mixed models including a genetic relatedness matrix as a random effect, identified two loci on chromosome 10, one near the *SIRT1* gene ($p = 2.53 \times 10^{-10}$) and the other in an intron of the *LHPP* gene ($p = 6.45 \times 10^{-12}$) [36]. Meta-analyses of genetic predictors of MDD (up to early 2015) are currently consistent with chance findings and hypothesized candidate genes identified from physiological pathways (such as *TPH2, HTR2A, MAOA, COMT*) have rarely been identified/replicated as predictors of MDD in GWAS [34, 38–40]. Accordingly, we did not find a significant association with depressive symptoms for the SNPs that reached genome-wide significance in MDD GWAS nor those in hypothesized candidate genes. However, whole-genome sequencing and statistical modeling alternatives to traditional linear regression provide a promising avenue for discovering new genes that influence depressive illness, and follow-up of these new regions will be imperative.

One potentially important reason that SNPs detected through GWAS and biological candidate genes rarely replicate is because despite the CES-D correlating strongly with depression and having been used in hundreds of studies, the CES-D is not a diagnostic tool. The CES-D only measures depressive symptoms over the past week. The MESA study exams were spaced approximately 12 – 24 months apart (the HRS surveys 24 months apart). It is possible that failure to capture changes in depressive symptoms between the assessments introduced measurement error in the phenotype. Additionally, in the baseline and repeated measures analyses, though log-transformed to improve normality, the distribution of CES-D still deviated from the normal distribution. This is a consistent limitation of CES-D scores in the literature, and it should be noted that the p-values from our baseline and repeated measures models may reflect the non-normal distribution of the phenotype.

We included only common variants (those with ethnicity-specific MAF > 5 %) in our analysis. One reason we may not have found any significant genetic variants of depressive symptoms is that we did not look at rare variants or copy number variants. New methods for analyzing rare variants or SNP sets, such as Sequence Kernel Association Testing (SKAT), are being developed and applied and may help to further elucidate genetic predictors of depressive symptoms at a gene-level and across ethnicities [41]. Additionally, it is possible that multiple SNPs with small effects, working in concert, could affect individual susceptibility to depression and depressive symptoms [42]. Further, no interactions (gene-gene or gene-environment) were evaluated in these analyses, which may play an important role in revealing the pathogenesis of depression and depressive symptoms.

Conclusion

Since combining genetic information across ethnicities can result in false-positive findings from population stratification within genetically distinct populations, we conducted GWASs separately by ethnicity adjusting for ethnicity-specific principal components and filtered initial GWAS results by ethnicity-specific minor alleles to remove low frequency variants for more robust findings. The meta-analysis software accounts for both magnitude and direction of effect when combining information across studies (in this case different ethnicities) which is especially appropriate when studies contain differences in ethnicity, phenotype distribution, gender or constraints in sharing of individual level data [43].

Identifying genes that are associated with depression has tremendous potential to transform our understanding and treatment of depression. Utilizing longitudinal measures in GWA studies for depressive symptoms allows researchers to get a better picture of depression over the life-course. Though this study did not find any gene variants that reached genome-wide significance in the repeated measures approach, it provides a first step in examining depressive symptoms in different longitudinal settings and also across multiple ethnicities.

Methods
Discovery sample
MESA is a longitudinal study supported by NHLBI with the overall goal of identifying risk factors for subclinical atherosclerosis [44]. The MESA cohort (N = 6,814) was recruited in 2000–2002 from six Field Centers in Baltimore, MD; Chicago, IL; Forsyth County, NC; Los Angeles, CA; New York, NY; and St. Paul, MN. MESA participants were 45–84 years of age and free of clinical

cardiovascular disease at baseline. Participants attended a baseline examination and three additional follow-up examinations approximately 18–24 months apart. At each clinic visit, participants completed a series of demographic, personal history, medical history, access to care, behavioral, and psychosocial questionnaires in English, Spanish, or Chinese. Depressive symptoms were assessed using the Center for Epidemiologic Studies Depression scale (CES-D) at exams 1, 3 and 4. The total number of participants and the corresponding response rates (of participants alive) were: exam 1 (n = 6,814), exam 2 (n = 6,239, 92 %), exam 3 (n = 5,946, 89 %), exam 4 (n = 5,704, 87 %). After removing participants with missing genetic data, depressive symptom score, or covariates used for analysis, the final sample size was 6,335 individuals (European (EA): 2,514; African (AA): 1,603; Chinese (CA): 775; Hispanic (HA): 1,443). Data supporting the results of this article are available in the dbGaP repository, phs000209.v12.p3, http://www.ncbi.nlm.nih.gov/projects/gap/cgi-bin/study.cgi?study_id=phs000209.v12.p3. Written informed consent was obtained from participants after the procedure had been fully explained and institutional review boards at each site approved study protocol (University of Minnesota Human Subjects Committee Institutional Review Board (IRB), Johns Hopkins Office of Human Subjects Research IRB, University of California Los Angeles Office for the Protection of Research Subjects IRB, Northwestern University Office for the Protection of Research Subjects IRB, Wake Forest University Office of Research IRB, Columbia University IRB).

Depressive symptom score
Depressive symptom score was assessed using the 20-item CES-D Scale [45], which was for use in general population surveys [45, 46]. The CES-D has an excellent internal consistency (Cronbach's alpha = 0.90) [45], and assesses depressive symptoms at a specific period in time (over the past week). The outcome measure for this analysis is a sum of the 20 items, ranging from 0 to 60. If more than 5 items were missing, the CES-D score was not calculated. If 1–5 items were missing, the scores were summed for completed items, dividing the sum by the number of questions answered and then multiplying by 20. There were 5,178 (81.7 %) participants with three measures of CES-D, 507 (8.0 %) with two measures, and 650 (10.3 %) with only baseline CES-D measures, for a total of 17,198 observations. We corrected for anti-depressant use through a similar algorithm to adjusting blood pressure for persons taking anti-hypertensive medication [5]. Detailed methods are described in Additional file 6. After adjustment for anti-depressant use, CES-D scores were log-transformed to improve normality.

Genotyping

Approximately one million SNPs were genotyped using the Affymetrix Genome-Wide Human SNP Array 6.0. Imputation was performed using the IMPUTE 2.1.0 program in conjunction with HapMap Phase I and II reference panels (CEU + YRI + CHB + JPT, release 22 - NCBI Build 36 for AA, CA, and HA participants; CEU, release 24 - NCBI Build 36 for EA). Imputation SNPs were filtered at an INFO score of 0.80. We accounted for population substructure by including the top four ethnicity-specific principal components (estimated from genome-wide data) as adjustment covariates in all analyses, as proposed previously by MESA investigators and elsewhere [47, 48].

Joint sample

The Health and Retirement Study (HRS) was used as a joint sample to be combined with MESA GWAS results in a meta-analysis [49]. These two studies have comparable participants, and similar measures of phenotype. The HRS surveys a representative sample of more than 26,000 Americans over the age of 50 every two years starting in 1992. HRS data includes information on depressive symptoms measured with a short form of the CES-D, the CES-D8. The CES-D8 includes a subset of eight items from the full 20-item CES-D [45]. The depression score for each participant was composed of the total number of affirmative depression answers. The HRS depression symptom score ranges from 0 to 8. Participants missing two or more of the eight items were excluded from the analyses. Written informed consent was obtained and the IRB at the University of Michigan approved study protocol before data collection.

Over 12,000 HRS participants were genotyped for about 2.5 million SNPs using the Illumina Human Omni-2.5 Quad beadchip. Genotypes were imputed for EA and AA using MACH software (HapMap Phase II, release #22, CEU panel for EA and CEU + YRI panel for African Americans). Imputation SNPs were filtered at an INFO score of 0.80. We accounted for population substructure by including the top four ethnicity-specific principal components (estimated from genome-wide data) as adjustment covariates in all analyses. There were 10,163 HRS participants after removing those with missing outcome, covariate or genetic information. A total of 34 (0.3 %) had only one measure of CES-D8, 147 (1.4 %) had two measures, and 9,982 (98.2 %) had three or more CES-D8 measures, for a total of 72,273 observations.

Genome-wide association analysis

We contrasted GWAS results using different approaches to incorporate the time-varying phenotypic data: using a single (baseline) measure, taking the average across exams, or conducting a repeated measures analysis that accounts for correlation of responses within individuals.

Baseline and averaged GWA studies were analyzed using a one-step linear regression approach, adjusting for age, sex, site (in MESA) and the first four genome-wide principal components, stratified by race in PLINK v.1.07 [50, 51]. Each SNP was analyzed separately, using SNP dosages, in an additive genetic model.

For the repeated measures, we used generalized estimating equations (GEE) to account for within-individual correlations between repeated CES-D measures [52]. Within the 'geepack' package in the R software, we used an exchangeable (compound symmetric) correlation structure because empirical correlations for CES-D measures for exam 1, 3, and 4 were similar and we saw no significant trend in CES-D over time for any ethnicity except for the EA sub-sample [53, 54].

Comparison of p-values across phenotype approach

To examine whether p-values from GWAS in MESA were consistent in rank across the three analysis approaches (baseline, averaged across exams, repeated measures), we calculated Spearman's correlations between the ranks of p-values for SNP-phenotype associations within ethnic group.

Meta-analysis

To increase statistical power to detect SNP association, we performed a fixed-effects meta-analysis combining results across all four ethnicities within the MESA study for each of the three phenotype definitions (baseline, averaged, repeated measures), weighting by sample size. In order to further investigate consistency of associations across different studies we also conducted a meta-analysis for EA and AA (separately) across the MESA and HRS studies for the repeated measures phenotype. We use only the AA and EA samples due to the availability of a large enough sample size for these two ethnicities in HRS. Finally, we performed a meta-analysis across all ethnicities and all studies to further elucidate any genetic variants across ethnicity. For the analysis that includes both MESA and HRS, the repeated measures phenotype was selected to allow for maximum power. All meta-analyses were performed using METAL [43].

Availability of supporting data

Data supporting the results of this article are available in the dbGap repository, phs000209.v12.p3, http://www.ncbi.nlm.nih.gov/projects/gap/cgi-bin/study.cgi?study_id=phs000209.v12.p3.

Additional files

Additional file 1: QQ plot of *p*-values from GWA analyses adjusted for age, sex, study site and top four principal components, ethnicity-specific minor allele frequency greater than 5 %. (PDF 369 kb)

Additional file 2: Comparison of p-values (p-value < 5×10^{-4}) for genome-wide association studies for baseline CES-D score compared to averaged CES-D score. CES-D: Center for Epidemiological Studies – Depression, (a) African Americans, (b) European Americans, (c) Chinese Americans, (d) Hispanic Americans. (EPS 1757 kb)

Additional file 3: Comparison of p-values (p-value < 5×10^{-4}) for genome-wide association studies for baseline CES-D score compared to repeated measures CES-D score. CES-D: Center for Epidemiological Studies – Depression, (a) African Americans, (b) European Americans, (c) Chinese Americans, (d) Hispanic Americans. (EPS 1450 kb)

Additional file 4: Comparison of p-values (p-value < 5×10^{-4}) for genome-wide association studies for averaged CES-D score compared to repeated measures CES-D score. CES-D: Center for Epidemiological Studies – Depression, (a) African Americans, (b) European Americans, (c) Chinese Americans, (d) Hispanic Americans. (EPS 1424 kb)

Additional file 5: Individual SNP information for unique SNPs reaching genome-wide suggestive p-value threshold for MESA ethnicity-specific GWAS analyses for each methodological approach (MAF > 5 %, INFO > 0.8, LD R^2 < 0.80). (PDF 110 kb)

Additional file 6: Methodological information on anti-depressant adjustment. (PDF 269 kb)

Competing interests

Drs. Ware, Smith, Mukherjee, Sun, Diez-Roux, and Kardia declare no potential conflicts of interest.

Authors' contributions

EBW contributed to the design, data acquisition, analysis, interpretation of the data, and writing and revising of the manuscript; JAS, BM, YVS, ADR, and SLRK contributed to the design of the study, drafting of the manuscript, critical evaluation of intellectual content, and data acquisition. All authors have read and approved the final manuscript.

Authors' information

Not applicable.

Acknowledgements

MESA and the MESA SHARe project are conducted and supported by the National Heart, Lung, and Blood Institute (NHLBI) in collaboration with MESA investigators. Support for MESA is provided by contracts N01-HC-95159 through N01-HC-95169 and UL1-RR-024156. Funding for genotyping was provided by NHLBI Contract N02-HL-6-4278 and N01-HC-65226. Support for this study was also provided through R01-HL-101161.
HRS is supported by the National Institute on Aging (NIA U01AG009740). The genotyping was funded separately by the National Institute on Aging (RC2 AG036495, RC4 AG039029). Genotyping was conducted by the NIH Center for Inherited Disease Research (CIDR) at Johns Hopkins University. Genotyping quality control and final preparation of the data were performed by the Genetics Coordinating Center at the University of Washington.

Author details

[1]Department of Epidemiology, University of Michigan, Ann Arbor, MI, USA. [2]Institute of Social Research, University of Michigan, 1415 Washington Heights #4614, Ann Arbor, MI 48109, USA. [3]Department of Biostatistics, University of Michigan, Ann Arbor, MI, USA. [4]Department of Epidemiology, Emory University, Atlanta, GA, USA. [5]Department of Epidemiology and Biostatistics, Drexel University, Philadelphia, PA, USA.

References

1. Smith EN, Chen W, Kahonen M, Kettunen J, Lehtimaki T, Peltonen L, et al. Longitudinal genome-wide association of cardiovascular disease risk factors in the Bogalusa heart study. PLoS Genet. 2010;6(9):e1001094.
2. Diggle P, Heagery P, Kung-Yee L, Zeger S. Analysis of Longitudinal Data. Oxford, United Kingdom: Oxford University Press; 2002.
3. Hettema JM, Neale MC, Myers JM, Prescott CA, Kendler KS. A population-based twin study of the relationship between neuroticism and internalizing disorders. Am J Psychiatry. 2006;163(5):857–64.
4. Kendler KS, Gardner Jr CO. Boundaries of major depression: an evaluation of DSM-IV criteria. Am J Psychiatry. 1998;155(2):172–7.
5. Hek K, Demirkan A, Lahti J, Terracciano A, Teumer A, Cornelis MC, et al. A Genome-Wide Association Study of Depressive Symptoms. Biol Psychiatry. 2013;73(7):667–78.
6. Devlin B, Roeder K. Genomic control for association studies. Biometrics. 1999;55(4):997–1004.
7. Chauhan SC, Ebeling MC, Maher DM, Koch MD, Watanabe A, Aburatani H, et al. MUC13 mucin augments pancreatic tumorigenesis. Mol Cancer Ther. 2012;11(1):24–33.
8. Chauhan SC, Vannatta K, Ebeling MC, Vinayek N, Watanabe A, Pandey KK, et al. Expression and functions of transmembrane mucin MUC13 in ovarian cancer. Cancer Res. 2009;69(3):765–74.
9. Gupta BK, Maher DM, Ebeling MC, Sundram V, Koch MD, Lynch DW, et al. Increased expression and aberrant localization of mucin 13 in metastatic colon cancer. J Histochem Cytochem. 2012;60(11):822–31.
10. Maher DM, Gupta BK, Nagata S, Jaggi M, Chauhan SC. Mucin 13: structure, function, and potential roles in cancer pathogenesis. Mol Cancer Res. 2011;9(5):531–7.
11. Moehle C, Ackermann N, Langmann T, Aslanidis C, Kel A, Kel-Margoulis O, et al. Aberrant intestinal expression and allelic variants of mucin genes associated with inflammatory bowel disease. J Mol Med. 2006;84(12):1055–66.
12. Samuels TL, Handler E, Syring ML, Pajewski NM, Blumin JH, Kerschner JE, et al. Mucin gene expression in human laryngeal epithelia: effect of laryngopharyngeal reflux. Ann Otol Rhinol Laryngol. 2008;117(9):688–95.
13. Shimamura T, Ito H, Shibahara J, Watanabe A, Hippo Y, Taniguchi H, et al. Overexpression of MUC13 is associated with intestinal-type gastric cancer. Cancer Sci. 2005;96(5):265–73.
14. Williams SJ, Wreschner DH, Tran M, Eyre HJ, Sutherland GR, McGuckin MA. Muc13, a novel human cell surface mucin expressed by epithelial and hemopoietic cells. J Biol Chem. 2001;276(21):18327–36.
15. Clark HF, Gurney AL, Abaya E, Baker K, Baldwin D, Brush J, et al. The secreted protein discovery initiative (SPDI), a large-scale effort to identify novel human secreted and transmembrane proteins: a bioinformatics assessment. Genome Res. 2003;13(10):2265–70.
16. Kimura K, Wakamatsu A, Suzuki Y, Ota T, Nishikawa T, Yamashita R, et al. Diversification of transcriptional modulation: large-scale identification and characterization of putative alternative promoters of human genes. Genome Res. 2006;16(1):55–65.
17. Sherry ST, Ward MH, Kholodov M, Baker J, Phan L, Smigielski EM, et al. dbSNP: the NCBI database of genetic variation. Nucleic Acids Res. 2001;29(1):308–11.
18. Hix LM, Karavitis J, Khan MW, Shi YH, Khazaie K, Zhang M. Tumor STAT1 transcription factor activity enhances breast tumor growth and immune suppression mediated by myeloid-derived suppressor cells. J Biol Chem. 2013;288(17):11676–88.
19. Xu E, Zhang J, Chen X. MDM2 expression is repressed by the RNA-binding protein RNPC1 via mRNA stability. Oncogene. 2013;32(17):2169–78.
20. Yan W, Zhang J, Zhang Y, Jung YS, Chen X. p73 expression is regulated by RNPC1, a target of the p53 family, via mRNA stability. Mol Cell Biol. 2012;32(13):2336–48.
21. van der Sluis S, Posthuma D, Nivard MG, Verhage M, Dolan CV. Power in GWAS: lifting the curse of the clinical cut-off. Mol Psychiatry. 2013;18(1):2–3.
22. Voorman A, Lumley T, McKnight B, Rice K. Behavior of QQ-plots and genomic control in studies of gene-environment interaction. PLoS One. 2011;6(5):e19416.
23. de Bakker PI, Ferreira MA, Jia X, Neale BM, Raychaudhuri S, Voight BF. Practical aspects of imputation-driven meta-analysis of genome-wide association studies. Hum Mol Genet. 2008;17(R2):R122–128.
24. Roberts R, Wells GA, Stewart AF, Dandona S, Chen L. The genome-wide association study–a new era for common polygenic disorders. J Cardiovasc Transl Res. 2010;3(3):173–82.
25. McCarthy MI, Hirschhorn JN. Genome-wide association studies: past, present and future. Hum Mol Genet. 2008;17(R2):R100–101.
26. Huang J, Perlis RH, Lee PH, Rush AJ, Fava M, Sachs GS, et al. Cross-disorder genomewide analysis of schizophrenia, bipolar disorder, and depression. Am J Psychiatry. 2010;167(10):1254–63.

27. Lewis CM, Ng MY, Butler AW, Cohen-Woods S, Uher R, Pirlo K, et al. Genome-wide association study of major recurrent depression in the U.K. population. Am J Psychiatry. 2010;167(8):949–57.

28. McMahon FJ, Akula N, Schulze TG, Muglia P, Tozzi F, Detera-Wadleigh SD, et al. Meta-analysis of genome-wide association data identifies a risk locus for major mood disorders on 3p21.1. Nat Genet. 2010;42(2):128–31.

29. Muglia P, Tozzi F, Galwey NW, Francks C, Upmanyu R, Kong XQ, et al. Genome-wide association study of recurrent major depressive disorder in two European case–control cohorts. Mol Psychiatry. 2010;15(6):589–601.

30. Rietschel M, Mattheisen M, Frank J, Treutlein J, Degenhardt F, Breuer R, et al. Genome-wide association-, replication-, and neuroimaging study implicates HOMER1 in the etiology of major depression. Biol Psychiatry. 2010;68(6):578–85.

31. Shi J, Potash JB, Knowles JA, Weissman MM, Coryell W, Scheftner WA, et al. Genome-wide association study of recurrent early-onset major depressive disorder. Mol Psychiatry. 2011;16(2):193–201.

32. Shyn SI, Shi J, Kraft JB, Potash JB, Knowles JA, Weissman MM, et al. Novel loci for major depression identified by genome-wide association study of Sequenced Treatment Alternatives to Relieve Depression and meta-analysis of three studies. Mol Psychiatry. 2011;16(2):202–15.

33. Sullivan PF, Neale MC, Kendler KS. Genetic epidemiology of major depression: Review and meta-analysis. Am J Psychiatr. 2000;157(10):1552–62.

34. Wray NR, Pergadia ML, Blackwood DH, Penninx BW, Gordon SD, Nyholt DR, et al. Genome-wide association study of major depressive disorder: new results, meta-analysis, and lessons learned. Mol Psychiatry. 2012;17(1):36–48.

35. Major Depressive Disorder Working Group of the Psychiatric GC. A mega-analysis of genome-wide association studies for major depressive disorder. Mol Psychiatry. 2013;18(4):497–511.

36. CONVERGE consortium. Sparse whole-genome sequencing identifies two loci for major depressive disorder. Nature. 2015;523(7562):588–91.

37. Hek K, Mulder CL, Luijendijk HJ, van Duijn CM, Hofman A, Uitterlinden AG, et al. The PCLO gene and depressive disorders: replication in a population-based study. Hum Mol Genet. 2010;19(4):731–4.

38. Bosker FJ, Hartman CA, Nolte IM, Prins BP, Terpstra P, Posthuma D, et al. Poor replication of candidate genes for major depressive disorder using genome-wide association data. Mol Psychiatry. 2011;16(5):516–32.

39. Sullivan PF, de Geus EJ, Willemsen G, James MR, Smit JH, Zandbelt T, et al. Genome-wide association for major depressive disorder: a possible role for the presynaptic protein piccolo. Mol Psychiatry. 2009;14(4):359–75.

40. Wray NR, Pergadia ML, Blackwood DH, Penninx BW, Gordon SD, Nyholt DR, et al. Genome-wide association study of major depressive disorder: new results, meta-analysis, and lessons learned. Mol Psychiatry. 2012;17(1):36–48.

41. Wu MC, Lee S, Cai T, Li Y, Boehnke M, Lin X. Rare-variant association testing for sequencing data with the sequence kernel association test. Am J Hum Genet. 2011;89(1):82–93.

42. Demirkan A, Penninx BW, Hek K, Wray NR, Amin N, Aulchenko YS, et al. Genetic risk profiles for depression and anxiety in adult and elderly cohorts. Mol Psychiatry. 2011;16(7):773–83.

43. Willer CJ, Li Y, Abecasis GR. METAL: fast and efficient meta-analysis of genomewide association scans. Bioinformatics. 2010;26(17):2190–1.

44. Bild DE, Bluemke DA, Burke GL, Detrano R, Diez Roux AV, Folsom AR, et al. Multi-ethnic study of atherosclerosis: objectives and design. Am J Epidemiol. 2002;156(9):871–81.

45. Radloff L. The CES-D scale: A self-report depression scale for research in the general population. Appl Psychol Meas. 1977;1:385–401.

46. Comstock GW, Helsing KJ. Symptoms of depression in two communities. Psychol Med. 1976;6(4):551–63.

47. Setiawan VW, Doherty JA, Shu XO, Akbari MR, Chen C, De Vivo I, et al. Two estrogen-related variants in CYP19A1 and endometrial cancer risk: a pooled analysis in the Epidemiology of Endometrial Cancer Consortium. Cancer Epidemiol Biomarkers Prev. 2009;18(1):242–7.

48. Sun YV, Peyser PA, Kardia SL. A common copy number variation on chromosome 6 association with the gene expression level of endothelin 1 in transformed B lymphocytes from three racial groups. Circ Cardiovasc Genet. 2009;2(5):483–8.

49. Juster FT, Suzman R. An Overview of the Health and Retirement Study. J Hum Resour. 1995;30:[S7] of S7–S56.

50. Purcell S. PLINK 1.07. http://pngu.mgh.harvard.edu/purcell/plink/

51. Purcell S, Neale B, Todd-Brown K, Thomas L, Ferreira MA, Bender D, et al. PLINK: a tool set for whole-genome association and population-based linkage analyses. Am J Hum Genet. 2007;81(3):559–75.

52. Zeger SL, Liang KY. Longitudinal data analysis for discrete and continuous outcomes. Biometrics. 1986;42(1):121–30.

53. R Core Team. R: A language and environment for statistical computing. Vienna, Austria: R Foundation for Statistical Computing. 2014.http://www.R-project.org/

54. Yan J, Hojsgaard S, Halekoh U. geepack: Generalized estimating equation package, 2012. URL http://CRAN.R-project.org/package=geepack. R package version 1.1-6.

17

Improving accuracy of genomic prediction by genetic architecture based priors in a Bayesian model

Ning Gao[1,2], Jiaqi Li[1], Jinlong He[1], Guang Xiao[1], Yuanyu Luo[1], Hao Zhang[1], Zanmou Chen[1] and Zhe Zhang[1,2*]

Abstract

Background: In recent years, with the development of high-throughput sequencing technology and the commercial availability of genotyping bead chips, more attention is being directed towards the utilization of abundant genetic markers in animal and plant breeding programs, human disease risk prediction and personal medicine. Several useful approaches to accomplish genomic prediction have been developed and used widely, but still have room for improvement to gain more accuracy. In this study, an improved Bayesian approach, termed BayesBπ, which differs from the original BayesB in priors assigning, is proposed. An effective method for calculating the locus-specific π by converting p-values from association between SNPs and traits' phenotypes is given and systemically validated using a German Holstein dairy cattle population. Furthermore, the new method is applied to a loblolly pine (*Pinus taeda*) dataset.

Results: Compared with the original BayesB, BayesBπ can improve the accuracy of genomic prediction up to 7.62 % for milk fat percentage, a trait which shows a large effect of quantitative trait loci (QTL). For milk yield, which is controlled by small to moderate effect genes, the accuracy of genomic prediction can be improved up to 4.94 %. For somatic cell score, of which no large effect QTL has been reported, GBLUP performs better than Bayesian methods. BayesBπ outperforms BayesCπ in 10 out of 12 scenarios in the dairy cattle population, especially in small to moderate population sizes where accuracy of BayesCπ are dramatically low. Results of the loblolly pine dataset show that BayesBπ outperforms BayesB in 14 out of 17 traits and BayesCπ in 8 out of 17 traits, respectively.

Conclusions: For traits controlled by large effect genes, BayesBπ can improve the accuracy of genomic prediction and unbiasedness of BayesB in moderate size populations. Knowledge of traits' genetic architectures can be integrated into practices of genomic prediction by assigning locus-specific priors to markers, which will help Bayesian approaches perform better in variable selection and marker effects shrinkage.

Keywords: Genomic selection, Bayesian approaches, Priors, Genetic Architecture

Background

In the field of medicine, risk prediction of major diseases such as cancer is essential for taking preventive measures early before worsening [1–4]. Similarly, it is important to predict genetic values of candidates for early selection, through which the production costs will be reduced immensely, in breeding programs both of domestic animal and economically important plants [5–9]. Therefore, developing prediction methods exploiting the availability of genomic big-data is a renewed hot topic in the scientific community nowadays.

With the development of high-throughput sequencing technology and the commercial availability of genotyping bead chips in recent years, large numbers of single nucleotide polymorphisms (SNPs) covering the whole genome can be obtained quickly and cheaply. The utilizations of these genomic data to accelerate genome wide association studies (GWAS), disease prediction and personal medicine of human beings, and breeding programs of animals and plants are attracting more and more attention [10]. The paradigm of involving dense genomic

* Correspondence: zhezhang@scau.edu.cn
[1]National Engineering Research Center for Breeding Swine Industry, Guangdong Provincial Key Lab of Agro-animal Genomics and Molecular Breeding, College of Animal Science, South China Agricultural University, Guangzhou 510642, China
[2]Department of Animal Sciences, Animal Breeding and Genetics Group, Georg-August-Universität Göttingen, Göttingen 37075, Germany

markers into genetic merit prediction, which was termed genomic selection (GS), was first proposed by Meuwissen et al. [11]. Nowadays, GS has been applied to genetic merit prediction in human beings [12], model organisms [13], dairy cattle [14–17] and other domestic animals [18–21], and has even been applied to the breeding programs of economically important crops [22–24], forest trees [25, 26], and aquaculture species [27]. Methods for GS keep developing rapidly [28] and can be divided into two categories, direct and indirect approaches, based on the manners in which they use the genetic markers [29]. Direct approaches are derived from best linear unbiased prediction (BLUP), and termed genomic best linear unbiased prediction (GBLUP) [30], which firstly construct a numerator relationship matrix with SNPs and then mixed model equations are solved to obtain genetic merit directly. Indirect approaches include ridge regression BLUP (RRBLUP) [11], Bayesian variable selection approaches [11, 31], and Bayesian shrinkage [32, 33] first estimate marker effects and then get genetic values by summing the effects of all relevant markers.

In Meuwissen et al. [11], least square (LS), RRBLUP, and Bayesian non-linear models (i.e., BayesA and BayesB) were compared for the accuracy of quantitative trait loci (QTL) detection and genetic value prediction. Results from literatures shown that Bayesian approaches, which integrate a priori that a large proportion of SNPs (with a high probability π) are non-effective, are more powerful than other models in both of QTL detection and genetic value prediction. Since 2001, many Bayesian methods for GS have been developed [11, 31, 34], which were reviewed by Meuwissen [28]. The concept of a "Bayesian alphabet", which denotes the growing number of Bayesian methods that differ in the priors while sharing a similar sampling model, was first proposed by Gianola [35].

Although widely used in the animal and plant breeding programs, the original Bayesian models have been shown to have some drawbacks [31, 35]. The first is the arbitrary assignment of the proportion of non-effective SNPs (π), which is treated as a constant close to 1 (i.e., 0.95 or 0.99) in most situations [31, 35]. The second is the data-independent prior degree of freedoms assigned to locus-specific variances [35]; The full-conditional posterior has only one additional degree of freedom compared to the prior distribution, regardless of the number of phenotypes and genotypes [31]. To overcome these two deficiencies, BayesCπ and BayesDπ [31] were developed, in both of which the non-informative parameter π and/or scale parameter S are treated as variable and sampled from relevant prior distributions. Additionally, changes to the distribution of marker effects and variances have been performed [36, 37]. BayesLASSO [38, 39] uses an exponential distribution as a prior of marker effects, different from the prior normal distribution in BayesA and

BayesB. In BayesR [15] and BayesRS [34], the prior of marker effects is treated as a serious normal distribution. All of these approaches show some advantages under different circumstances, but none of them can be considered as the golden rule.

Although Bayesian approaches outperform GBLUP under most circumstances, the priors assigned to the established Bayesian approaches still may have room for further improvement. It has been shown that genetic architectures of traits can influence genomic prediction accuracy [40]. Therefore, traits' genetic architectures should be taken into account by assigning locus- or trait-specific priors to genomic prediction models. By assigning different marker weights to build a trait-specific numerator relationship matrix, locus-specific priors have been utilized in methods derived from BLUP, such as TABLUP [41, 42], BLUP|GA [43], and iterated-GBLUP [44]. These approaches confirmed that locus-specific priors show benefits compared to common priors. Moreover, by converting p-values derived from GWAS into marker-specific weights, the locus-specific priors have been utilized in the genomic prediction of human traits via BLUP [12], through which a greater degree of accuracy was gained. All these previous studies indicated that locus-specific priors in genomic prediction show favorable features in BLUP models. However, it has not been tested whether more accuracy will be gained in Bayesian models with locus-specific priors. Based on the assumptions of BayesB and prior knowledge of traits' genetic architectures, we argue here that a locus-specific prior (π) is more appropriate for Bayesian methods for genomic prediction. With a locus-specific prior, the accuracy of genomic prediction may be improved due to a more appropriate marker effect shrinkage and variable selection. The aim of this study is to propose and validate a modified BayesB method which can utilize locus-specific priors. The performance of the modified Bayesian approach in genomic prediction is compared with that of GBLUP, the original BayesB and BayesCπ.

Results
Statistical summary for all traits
Two datasets, a German cattle population [45] and a loblolly pine (*Pinus taeda*) dataset [25] were analyzed in this study. The statistical summary of all traits in the two datasets are shown in Table 1. It should be noted that phenotypes in the German cattle population were rescaled to standard normal distribution, i.e., $y \sim N(0, 1)$, where y denotes the phenotypes. For these traits, the traditional estimated breeding values, with high reliability, were close to the true breeding values. The variation of the regressed phenotypes of the loblolly pine was dramatically large (Table 1), and their heritability are relatively low [25].

Table 1 Descriptive statistics of trait phenotypes

Datasets	Traits[a]	N	Min.	Mean	Max.	S.D.	CV%
Dairy cattle	MY	5024	−3.383	0.000	3.319	1.000	–
	MFP	5024	−3.569	0.000	4.281	1.000	–
	SCS	5024	−4.462	0.000	3.469	1.000	–
Loblolly Pine	HT	927	−287.700	20.300	226.10	73.315	361.158
	HTLC	927	−94.110	3.304	89.080	24.976	755.932
	BHLC	927	−1.578	0.092	1.573	0.507	551.087
	DBH	927	−5.439	0.294	1.349	4.150	1411.565
	CWAL	927	−91.190	2.443	130.800	27.326	1118.543
	CWAC	927	−140.600	2.276	157.000	42.033	1846.793
	BD	927	−0.608	−0.004	1.739	0.249	−6225.000
	BA	927	−24.560	−0.261	21.140	7.315	−2802.682
	Rootnum_bin	927	−0.779	0.107	0.602	0.258	241.121
	Rootnum	927	−2.422	0.321	4.368	0.960	299.065
	Rust_bin	927	−0.482	−0.014	0.822	0.399	−2850.000
	Rust_gall_vol	927	−1.175	−0.022	5.212	1.132	−5145.454
	Stiffness	927	−3.244	0.095	6.082	1.225	1289.474
	Lignin	927	−3.644	0.050	4.073	1.200	2400.000
	LateWood	927	−4.544	0.090	4.878	1.571	1745.556
	Density	927	−10.290	−0.053	17.610	2.498	−4713.208
	C5C6	927	−8.102	−0.049	9.057	2.649	−5406.122

[a]MY, milk yield; MFP, milk fat percentage; SCS, somatic cell score; HT, total stem height; HTLC, total height to the base of the live crown; BHLC, basal height of the live crown; DBH, traits stem diameter; CWAL, crown width along the planting beds; CWAC, crown width across the planting beds; BD, average branch diameter; BA, branch angle average; Rootnum_bin, presence or absence of roots; Rootnum, Root number; Rust_bin, presence or absence of rust; Rust_gall_vol, gall volume; lignin, lignin content; LateWood, latewood percentage; Density, wood specific gravity; C5C6, C5C6 content. In the dairy cattle population, phenotypes were rescaled to standard normal distributions

Capturing of genetic architecture

In order to capture genetic architectures of traits, analysis of variance (ANOVA) based on single markers is performed for three traits in the dairy cattle population. The logarithms of p-values from ANOVA reflect the genetic architecture of these traits (Fig. 1). For milk fat percentage, a set of SNPs with dramatically low p-values on chromosome 14 were detected via ANOVA (Fig. 1), which is consistent with our prior knowledge about the genetic architecture of this trait that 30 % of the genetic variation is due to segregation of the *DGAT1* gene [46, 47] located on chromosome 14. For milk yield, clusters of SNPs with low p-values were detected on chromosome 1, 5, 7, and 14, which is consistent with the prior knowledge that there is a major gene on chromosome 14 and some genes with moderate or small effects on other chromosomes. For the somatic cell score, none significant association between phenotypes and SNPs has been detected, which in agree with the prior knowledge that no major genes affect this trait.

Furthermore, we found that the p-values from ANOVA can be converted with formula (4) to a probability form, which can be used as a genetic-architecture-based π in

Bayesian methods. The distribution of locus-specific π for three traits in the dairy cattle population can reveal the genetic architecture of these traits at some extent (Fig. 1).

Validating BayesBπ with the German dairy cattle dataset

Results of genomic prediction for three traits in the German Holstein dairy cattle (Table 2) show that when the population size is 200, BayesB outperforms BayesBπ. However, when the population sizes are 500 and 1000, BayesBπ performs better than BayesB. For milk fat percentage, BayesBπ gives 6.25 and 7.62 % higher prediction accuracies than BayesB when the population sizes are 500 and 1000, respectively. For milk yield, the accuracy of BayesBπ is 4.94 % higher than that of BayesB when $N = 500$, while BayesBπ and BayesB performed similarly for $N = 1000$; When the population size reached 2000, BayesBπ performs not better than BayesB. For somatic cell score, improvement of BayesB with locus-specific π is only observed when the population size is 1000; In other population sizes, GBLUP performed better than the Bayesian methods. BayesBπ outperforms BayesCπ in 10 out of 12 scenarios in the dairy cattle population, especially in small to moderate population sizes where

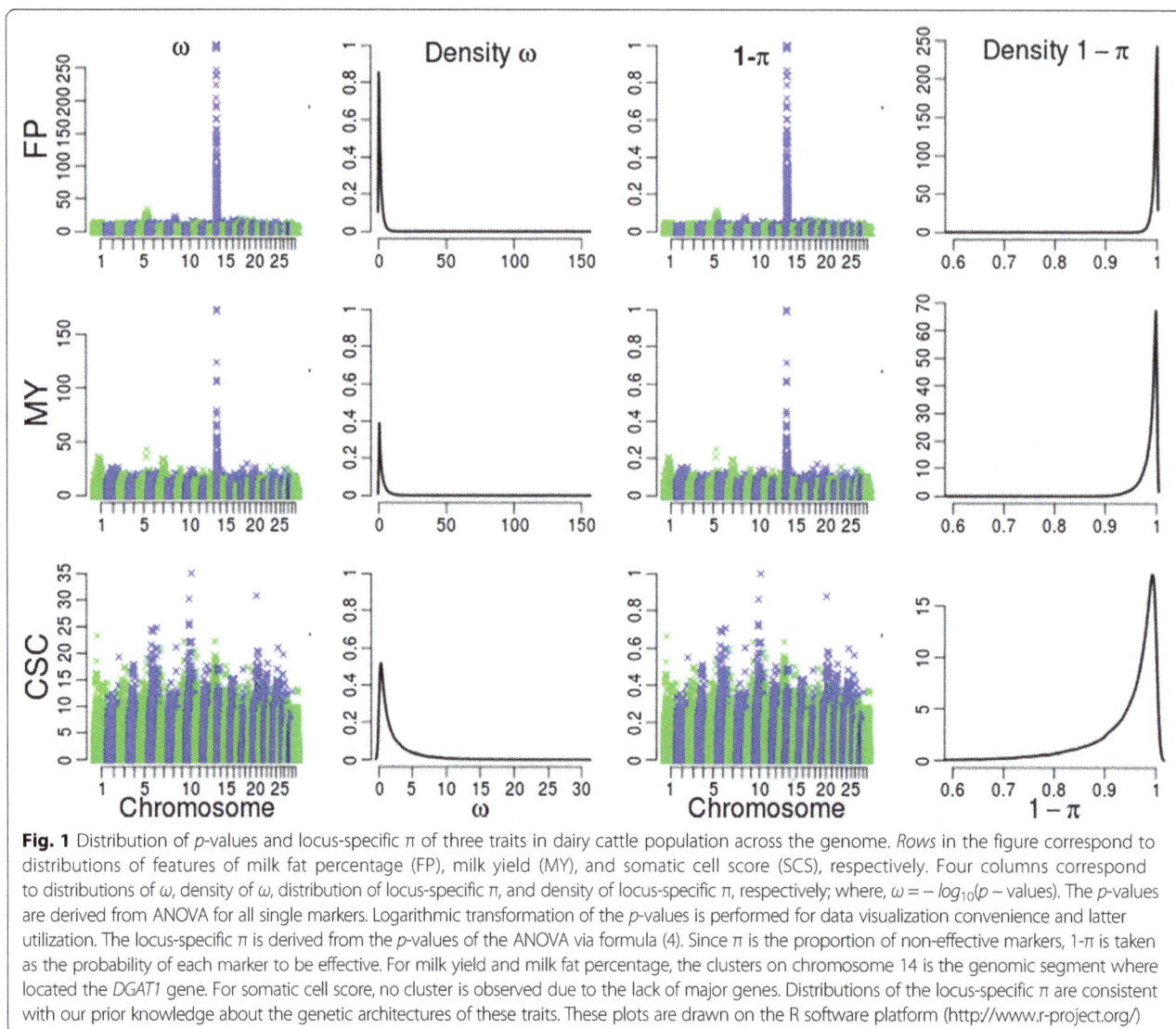

Fig. 1 Distribution of p-values and locus-specific π of three traits in dairy cattle population across the genome. *Rows* in the figure correspond to distributions of features of milk fat percentage (FP), milk yield (MY), and somatic cell score (SCS), respectively. Four columns correspond to distributions of ω, density of ω, distribution of locus-specific π, and density of locus-specific π, respectively; where, $ω = -log_{10}(p-values)$. The p-values are derived from ANOVA for all single markers. Logarithmic transformation of the p-values is performed for data visualization convenience and latter utilization. The locus-specific π is derived from the p-values of the ANOVA via formula (4). Since π is the proportion of non-effective markers, 1-π is taken as the probability of each marker to be effective. For milk yield and milk fat percentage, the clusters on chromosome 14 is the genomic segment where located the *DGAT1* gene. For somatic cell score, no cluster is observed due to the lack of major genes. Distributions of the locus-specific π are consistent with our prior knowledge about the genetic architectures of these traits. These plots are drawn on the R software platform (http://www.r-project.org/)

accuracies of BayesCπ are dramatically low (Fig. 2 & Table 2). For milk fat percentage, the prediction unbiasedness of BayesBπ is the best among four approaches (Additional file 1: Table S1), indicating that BayesBπ is suitable for genomic prediction of traits controlled by large effect genes.

Impacts of population sizes on accuracy of genomic selection are tested by averaging accuracies among traits in each subpopulation (Fig. 2). Similarly, impacts of traits' genetic architectures on genomic selection accuracy are detected by averaging accuracies among subpopulations for each trait (Table 2). Accuracies of BayesB and BayesBπ are higher than that of GBLUP for all population sizes. BayesBπ outperformed BayesB in moderate size populations, but the accuracies of BayesB and BayesBπ become similar when the population sizes become either smaller or larger (Fig. 2). When taking average genomic prediction accuracies of three traits in the

dairy cattle dataset across subpopulations, BayesBπ outperforms BayesB and BayesCπ in all three traits (Table 2). Moreover, BayesBπ outperforms GBLUP for both milk yield and milk fat percentage, but was not better than GBLUP for the somatic cell score.

Applying BayesBπ to the loblolly pine population

Results of loblolly pine dataset show that BayesBπ outperforms BayesB in 14 out of 17 traits and BayesCπ in 8 out of 17 traits, respectively (Table 3). The scale of accuracies for all traits are consistent with that reported by other scholars previously [25], although some differences exist due to random sampling in cross-validation. In four development related traits—CWAL, CWAC, BD, and Rootnum_bin, genomic prediction accuracies of BayesBπ are 0.52, 1.28, 0.76, and 1.84 % higher than that of BayesB; 1.84, 1.28, 1.14, and 0.72 % higher than that of GBLUP, respectively. In other traits, advances of

Table 2 Accuracy of genomic prediction of three traits in Germany cattle population r(EBVs, GEBVs)

Traits	N	GBLUP	BayesB	BayesBπ	BayesCπ
MY	200	**0.438 ± 0.010**	0.385 ± 0.018	0.382 ± 0.016	0.128 ± 0.016
	500	0.547 ± 0.007	0.547 ± 0.012	**0.574 ± 0.009**	0.324 ± 0.010
	1000	0.620 ± 0.005	**0.663 ± 0.005**	**0.663 ± 0.004**	0.560 ± 0.006
	2000	0.693 ± 0.003	**0.722 ± 0.002**	0.716 ± 0.002	0.718 ± 0.002
	Mean	0.574 ± 0.006	0.579 ± 0.009	**0.584 ± 0.008**	0.432 ± 0.008
MFP	200	0.353 ± 0.012	**0.558 ± 0.018**	0.544 ± 0.018	0.112 ± 0.012
	500	0.467 ± 0.008	0.629 ± 0.011	**0.670 ± 0.010**	0.332 ± 0.005
	1000	0.594 ± 0.004	0.709 ± 0.007	**0.763 ± 0.003**	0.709 ± 0.007
	2000	0.698 ± 0.003	**0.815 ± 0.002**	0.799 ± 0.002	0.799 ± 0.001
	Mean	0.528 ± 0.007	0.678 ± 0.010	**0.694 ± 0.008**	0.488 ± 0.006
SCS	200	**0.347 ± 0.017**	0.292 ± 0.015	0.290 ± 0.018	0.161 ± 0.017
	500	**0.469 ± 0.008**	0.440 ± 0.011	0.465 ± 0.009	0.265 ± 0.006
	1000	0.568 ± 0.004	0.570 ± 0.006	**0.572 ± 0.006**	0.535 ± 0.005
	2000	**0.650 ± 0.007**	0.647 ± 0.002	0.647 ± 0.002	0.646 ± 0.002
	Mean	0.508 ± 0.009	0.487 ± 0.008	**0.494 ± 0.009**	0.402 ± 0.008

The highest accuracies (Mean ± SE) among methods in different scenarios (subpopulations for different traits) are in bold faces. For each trait, accuracies among subpopulations are averaged to test the overall performances (i.e., the "Mean" accuracies here) of methods. For example, the overall performance of GBLUP in MY is the mean of its prediction accuracies for this trait among subpopulation 200, 500, 1000, and 2000

BayesBπ over BayesB range from 0.52 % for CWAL to 4.05 % for C5C6, with an average improvement of 2.13 %. The unbiasedness of BayesBπ shows a trend of larger than that of BayesB and GBLUP (Additional file 1: Table S2).

Discussion

Performance of BayesBπ in genomic prediction

Compared with original BayesB, BayesCπ, and GBLUP, the proposed new approach, which with a locus-specific prior instead of the common prior used in the other

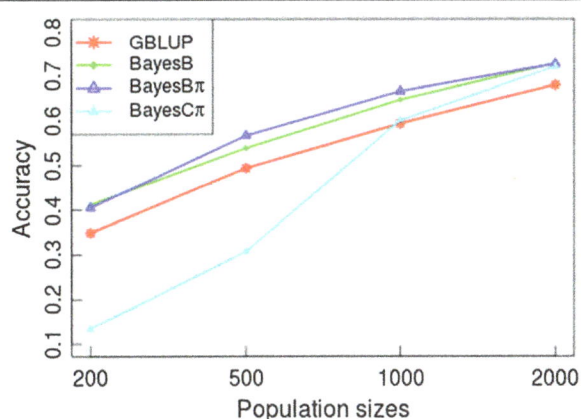

Fig. 2 Impact of population sizes on genomic prediction accuracy. Genomic prediction accuracies of each method in each subpopulation are averaged among three traits to test the overall performance of methods in different subpopulations. For example, accuracies of GBLUP in subpopulation 200 are averaged among three traits to gain its' overall performance in this population size

three methods, gives improved genomic prediction accuracies and unbiasednesses in both moderate size populations (Fig. 2) and traits controlled by large effect genes (i.e. milk fat percentage, Table 2). When the population sizes are small to moderate, the performance of BayesCπ will be dramatically decreased, while other three approaches gain relevantly reliable prediction accuracies. As expected and systematically tested previously, the accuracy of genomic prediction differs among approaches in small to moderate sample sizes but becomes similar when the reference population is large enough [24]. The accuracy of genomic prediction is also dependent on the consistency between the priors utilized and the true genetic architecture of the target traits [40]. It has been shown previously that methods given reasonable differential weights to markers outperformed those given common weights [12, 43]. Our results showed that the original BayesB can be improved by assigning locus-specific priors.

Priors in genomic prediction

When dealing with the problem of "p>>n" in the process of genomic prediction by Bayesian methods, the distributions of priors have a relatively large impact on posterior distributions of parameters to be estimated, for example marker effect and variance [48]. However, the sensitivity to priors differs among methods [49], while BayesA and BayesB are more sensitive because the assigned priors are not derived from real data [50]. Therefore, suitable priors are important for performing

Table 3 Accuracy of 17 traits in the loblolly pine population r(Deregressed Phenotypes, GEBVs)

Trait category	Traits	GBLUP	BayesB	BayesBπ	BayesCπ
Growth	HT	**0.376 ± 0.003**	0.351 ± 0.003	0.363 ± 0.002	0.374 ± 0.002
	HTLC	**0.451 ± 0.002**	0.449 ± 0.002	0.448 ± 0.002	0.449 ± 0.001
	BHLC	**0.487 ± 0.006**	0.468 ± 0.007	0.479 ± 0.007	0.487 ± 0.002
	DBH	**0.458 ± 0.002**	0.436 ± 0.003	0.446 ± 0.003	0.458 ± 0.002
Development	CWAL	0.381 ± 0.003	0.386 ± 0.003	**0.388 ± 0.003**	0.382 ± 0.002
	CWAC	0.468 ± 0.002	0.468 ± 0.002	**0.474 ± 0.002**	0.469 ± 0.002
	BD	0.262 ± 0.004	0.263 ± 0.004	**0.265 ± 0.004**	0.264 ± 0.003
	BA	**0.512 ± 0.003**	0.497 ± 0.002	0.500 ± 0.003	**0.512 ± 0.002**
	Rootnum_bin	0.277 ± 0.003	0.272 ± 0.004	**0.279 ± 0.003**	0.275 ± 0.002
	Rootnum	**0.262 ± 0.003**	0.245 ± 0.003	0.253 ± 0.003	0.261 ± 0.002
Disease resistance	Rust_bin	0.306 ± 0.004	**0.368 ± 0.004**	0.353 ± 0.004	0.32 ± 0.003
	Rust_gall_vol	0.259 ± 0.005	**0.325 ± 0.006**	0.292 ± 0.006	0.267 ± 0.004
Wood quality	Stiffness	**0.424 ± 0.003**	0.401 ± 0.003	0.410 ± 0.003	0.422 ± 0.002
	Lignin	**0.179 ± 0.005**	0.173 ± 0.005	0.176 ± 0.005	0.178 ± 0.003
	LateWood	0.254 ± 0.003	0.254 ± 0.003	**0.257 ± 0.003**	0.253 ± 0.002
	Density	**0.239 ± 0.003**	0.226 ± 0.003	0.234 ± 0.003	**0.239 ± 0.002**
	C5C6	**0.264 ± 0.004**	0.247 ± 0.004	0.257 ± 0.004	0.262 ± 0.003
Mean accuracy	–	0.345 ± 0.003	0.343 ± 0.004	**0.346 ± 0.004**	0.345 ± 0.002

The highest accuracies (Mean ± SE) among methods in relevant traits and subpopulations are in bold faces

genomic selection in plant and animal breeding practices or genomic prediction of human complex traits.

To the knowledge of the co-authors and pointed by other scholars, there is no golden rule for the assignment of priors in the paradigm of genomic prediction so far, especially when the biological meanings of priors are taken into account [51]. Nowadays, researchers are mostly focused on changing prior distribution of marker effect and variance for the purpose of gaining more prediction accuracy. The prior of marker effects was set to a normal distribution with a zero mean under most circumstances [11, 31]. In the work of Knurr et al. [36, 37], a spike-and-slab-shaped prior of marker effects was introduced, and the effects were limited to the interval of $-l \sim -b$, $-b \sim b$, and $b \sim l$. They concluded that an approach involving the mixture of uniform priors was suitable for genomic selection since through which different priors can be introduced into the prediction procedure [37]. Marker effects variance was usually set as an inverse chi-squared distribution with degree of freedom and scale parameters derived from the genetic architecture of traits [11, 31] or from a double exponential distribution [38, 39]. In this paper, we introduce one approach to obtain locus-specific prior by converting the p-values from associating phenotypes to single markers into probability form. The locus-specific prior here can be understood as a prior considering the probability of each marker to be effective or non-effective.

Locus-specific π in genomic prediction

Results from this study confirm that locus-specific priors can improve genomic prediction accuracy in domestic animals and plants. A locus-specific π calculated based on p-values from associating phenotypes to single markers outperforms other non-informative priors, especially for traits with large effect genes, such as the milk fat percentage of dairy cattle. The results of the loblolly pine show that genomic prediction accuracy is improved via a locus-specific π (Table 3).

In recent years, it has been shown that traits' genetic architectures impact genomic prediction accuracy at some extent [40]. Therefore, knowledge of traits' genetic architectures should be incorporated into practices of genomic prediction. Previously developed Bayesian approaches attempt to integrate prior knowledge of traits into genomic prediction by assigning different priors during the procedure of marker effects estimation. However, the priors of developed Bayesian methods are identical among all markers, which therefore cannot perfectly integrate prior knowledge into the paradigm of genomic prediction. Locus-specific priors at the level of variances of marker effects have been showed to be helpful in the estimation of phenotypes by combining genetic markers and records of phenotyped individuals [15, 34]. The difference between BayesBπ and BayesB is in the assigning of the proportion (π) of non-effective markers, which is assigned to be a fixed value close to 1 in BayesB while to be locus-specific constants calculated based on p-values from associating phenotypes to single markers in BayesBπ.

In the iteration of MCMC in BayesB [11], variable selection is based on π and the likelihood ratio, where π is identical among markers as discussed previously. The non-effective marker proportion π is estimated from data in BayesCπ and BayesDπ [31], but is same among markers within a single iteration. Assuming that the impact of the likelihood ratio on variable selection is identical for different methods, the non-effective marker proportion π is another important parameter that affects the decision whether markers are fitted in the model. The locus-specific π in our study assigns a more reasonable prior to the MCMC algorithm, and performs better variable selection. When genomic prediction is performed on extremely dense markers panels or full sequences [52], approaches with better variable selection will show advantages. Through formula (4), a constant close to zero would be assigned to markers with large effects, which would increases the probability of these markers being fitted in the model; however, a fixed value close to 1 would be assigned to markers with zero or small effects, thus decreasing the probability of these markers being fitted. It is by the new method that the sampling machine can perform more reasonable marker effect shrinkage and variable selection.

Methods for calculating locus-specific π

In this study, the locus-specific π was derived by converting p-values from associating phenotypes to single markers into probability form via formula (4) and then involved into the MCMC procedure directly. Our results show that priors derived through this strategy are consistent with prior knowledge about the genetic architectures of different traits in dairy cattle population. The magnitudes of π among these traits are highly variable, which reveals that the absolute values of π are at some extent impacted by the denominator of formula (4) (Fig. 1). Traits with large effect genes, such as the milk fat percentage, will return a relatively larger denominator; while traits with moderate to small effect genes, such as the milk yield and somatic cell score, will return a smaller denominator.

One way of dealing with such conflict may be to firstly dividing genetic markers into different classes based on p-values of ANOVA and then calculating π for each category of markers, which will involves efforts to find suitable thresholds as that for genome wide association studies [53]. Alternatively, effective loci that have been previously reported [54] can be considered during the calculation of locus-specific priors. With the development of sequencing technologies, more and more data from all levels of central dogma, termed multi-omics

data, is becoming available and can be involved in the paradigm of genomic prediction [55]. These data tend to be trait- or gene-specific, and thus can be integrated into genomic prediction by assigning locus-specific priors based on these data. In summary, as the publicly available information for commercially important traits increases, along with the development of suitable methods to integrate this information into genomic prediction, more genomic prediction accuracy can be gained in the near future.

Conclusions

In this study, we proposed and validated a modified BayesB method, BayesBπ, which can integrate prior knowledge into genomic prediction by assigning locus-specific priors to genetic markers. We conclude, based on the results of genomic prediction for three traits in German Holstein dairy cattle and 17 traits in a loblolly pine dataset, that firstly, for traits controlled by large effect genes, BayesBπ can improve the genomic prediction accuracy and unbiasedness of BayesB and BayesCπ. Secondly, knowledge of the genetic architecture can improve the performance of Bayesian models in genomic prediction by assigning locus-specific priors to markers. Thirdly, converting p-values of ANOVA to a locus-specific π is an efficient methodology for traits controlled by major genes in moderate size populations. Furthermore, BayesBπ may serves as a favorable method for variable selection when full sequences data are involved into genomic prediction.

Methods
Data sets

Two datasets, a dairy cattle and a loblolly pine dataset, are used to validate the new genomic prediction model. The cattle population consists of 5024 individuals [45]. Three traits, milk yield, milk fat percentage, and somatic cell score of this population are selected as model traits. After genotyping with Illumina BovineSNP50 [56] Bead chip, 42,551 SNPs were obtained for further study. Traditional estimated breeding values (EBVs) with high reliabilities for the three traits are used as the response variables of the statistical models in this study. For in detail description of this population see Zhang et al. [43], where this dataset was used to compare the accuracies of GS with GBLUP [30], TABLUP [42], BLUP|GA [43] and BayesB [11]. The dataset is online available with link http://www.g3journal.org/content/5/4/615/suppl/DC1.

The publicly available loblolly pine dataset consists of 927 lines from the United States, of which 17 traits related to growth, wood quality, disease resistance, and development were recorded [25]. For computational convenience, deregressed phenotypes given by Resende et, al. [25] are used as the response variables of GS models. The statistical summary of the deregressed phenotypes for all 17 traits is shown in Table 1. All trees were genotyped with an Illumina Infinium array [57], and 4853 SNPs were obtained. For more details about this loblolly pine dataset see Resende, et. al. [25]. The dataset is online available with link http://www.genetics.org/content/190/4/1503/suppl/DC1.

Whole genome prediction models

The statistical model for GBLUP in this study can be written as

$$\mathbf{y} = \mathbf{X}\mu + \mathbf{Z}\mathbf{u} + \mathbf{e}, \tag{1}$$

where $\mathbf{y}$ is a vector of phenotypic values; μ denotes the overall mean; $\mathbf{u}$ is a vector of additive genetic merits for all individuals, which is assumed to be multivariate normal $\mathbf{u} \sim N(0, \sigma_u^2 \mathbf{G})$; σ_u^2 denotes variance of additive genetic merits; $\mathbf{G}$ is a marker-derived numerator relationship matrix [30]; $\mathbf{e}$ is the model residuals, where $e \sim N(0, \sigma_e^2 \mathbf{I})$; σ_e^2 denotes the residual variance; and $\mathbf{X}$ and $\mathbf{Z}$ are incidence matrices linking the overall mean and additive genetic merits to the phenotypes, respectively. The original and modified BayesB are involved in the estimation of marker effects in the training population. The statistical model of both methods can be written as

$$\mathbf{y} = \mathbf{X}\mathbf{b} + \sum_{i=1}^{N} \mathbf{z}_i g_i + \mathbf{e}, \tag{2}$$

where $\mathbf{y}$ is a vector of phenotypic values; $\mathbf{b}$ is a vector of fixed effects (overall mean in this study); $g_i \sim N(0, \sigma_{g_i}^2)$ is the substitution effect of marker i; $\sigma_{g_i}^2$ is the variance of marker effects; N is the total number of markers; $\mathbf{e} \sim N(0, \mathbf{I}\,\sigma_e^2)$ is the vector of residuals; σ_e^2 is the residual variance; $\mathbf{X}$ is the design matrix for $\mathbf{b}$; and $\mathbf{z}_i$ is a vector of indicators for genotypes of marker i with values equal to 0, 1, and 2 to indicate the marker genotypes 11, 12, and 22, respectively. The marker effect variance $\sigma_{g_i}^2$ is assumed a priori to be 0 with a probability of π or to follow a scaled inverse χ-squared distribution (i.e., $\sigma_{g_i}^2 \sim x^{-2}(v, S)$) with a probability of $(1 - \pi)$, where the degree of freedom $v = 4.234$ and scale parameter $S = 0.0429$ [11]. The prior distribution of the error variance (i.e., σ_e^2) is a scaled inverse χ-squared distribution with parameters $v = -2$ and $S = 0$.

Gibbs sampling is used in the MCMC algorithm to obtain samples of each parameter from its full-conditional posterior distribution. Given a Gaussian response variable, the likelihood of which is $p(y|\mu, g, \sigma^2) = \prod_{i=1}^{n} N(y_i | \mu + \sum_{j=1}^{p} x_{ij} g_j, \sigma^2)$, where $N(y_i | \mu + \sum_{j=1}^{p} x_{ij} g_j, \sigma^2)$ is a normal density for the random variable y_i centered at $\mu + \sum_{j=1}^{p} x_{ij} g_j$ and with variance σ^2. According to Meuwissen et al. [11], the prior of unknowns in model (2) can be assigned as

$$p(\mu, g, \sigma^2 | df, S, \omega) \propto \left\{ \prod_{j=1}^{p} p\left(\beta_j | \theta_{g_j}, \sigma^2\right) p\left(\theta_{g_j} | \omega\right) \right\} x^{-2}(\sigma^2 | df, S).$$

Then the joint posterior density of all unknowns can be written as

$$p(\mu, g, \sigma^2 | y, df, S, \omega) \propto \prod_{i=1}^{n} N\left(y_i | \mu + \sum_{j=1}^{p} x_{ij} g_j, \sigma^2\right)$$

$$\times \left\{ \prod_{j=1}^{p} p(g_j | \theta_{g_j}, \sigma^2) p\left(\theta_{g_j} | \omega\right) \right\} x^{-2}(\sigma^2 | df, S).$$

Conditional posterior of each parameter can be deduced from the joint posterior density. However, we cannot use these conditional posterior distributions directly for estimating parameters because all of them are conditional on other unknowns. While we can introduce a MCMC procedure based on a Gibbs sampler to solve this problem. The general steps of Gibbs sampler (i.e., BayesA) are given below.

Step 1: Initialization of parameters. Initialize μ, g_i and $\sigma_{g_i}^2$ with small positive numbers.

Step 2: Update the $\sigma_{g_i}^2$. Sampling $\sigma_{g_i}^2$ from its' fully conditional distribution, $P\left(\sigma_{g_i}^2 | g_i\right) = x^{-2}\left(v + n_i, S + g_i' g_i\right)$, where $v = 4.234$, $S = 0.0429$, n_i is the number of haplotype effects at the ith segment.

Step 3: Update the σ_e^2. First adjust e with $e = y - Xg - 1_n'\mu$, then update σ_e^2 by drawing a single sample from $x^{-2}(n - 2, e_i' e_i)$.

Step 4: Update the overall mean μ by sample from $N\left(\frac{1}{n}(1_n' y - 1_n' Xg), \frac{\sigma_e^2}{n}\right)$.

Step 5: Update effects of all chromosome segments by sampling all effects from

$$N\left(\frac{X_{ij}' y - X_{ij}' Xg_{ij=0} - X_{ij}' 1_n \mu}{X_{ij}' X_{ij} + \sigma_e^2 / \sigma_i^2}, \ \sigma_e^2 / \left(X_{ij}' X_{ij} + \sigma_e^2 / \sigma_i^2\right)\right),$$

where, X_{ij} is the column of X of effect g_{ij}; $g_{ij=0}$ equal to g except that the effect of g_{ij} is set to zero.

Step 6: Repeat step 2 to step 5 for a large number of cycles.

BayesB uses a prior that a large proportion (π) of markers are non-effective and the prior distribution of $\sigma_{g_i}^2$ is

$$\begin{cases} \sigma_{g_i}^2 = 0 & \text{with probability } \pi \\ \sigma_{g_i}^2 \sim x^{-2}(v, S) & \text{with probability } (1-\pi) \end{cases},$$

where $v = 4.234$ and $S = 0.0429$. The Gibbs sampler of

BayesA will not move through the entire space of method BayesB, because the sampling of $\sigma_{g_i}^2 = 0$ is impossible, if $g_i'g_i > 0$. This problem is resolved by sampling $\sigma_{g_i}^2$ and g_i simultaneously using a Metropolis-Hasting (MH) algorithm. Thus, the Monte Carlo Markov Chain (MCMC) algorithm of BayesB consists of running a Gibbs chain as in BayesA, except that samples of $\sigma_{g_i}^2$ are obtained by running a Metropolis-Hasting (MH) algorithm for 100 cycles instead of simply sampling $\sigma_{g_i}^2$ from an inverse chi-square distribution. The parameter π is used at the beginning of the Metropolis-Hasting (MH) algorithm in the sampling model. Once the MH algorithm began, a random number α is sampled from a uniform distribution. If $\alpha \geq 1 - \pi$, the variance of marker effects is not resampled and set as 0 or not updated according to the likelihood ratio. However, the variance is sampled from an inverse χ-squared distribution and accepted according to the likelihood ratio when $\alpha < 1 - \pi$. If the variance is 0, the effect of current marker is set as 0, otherwise it is sampled from its posterior distribution. Therefore, the updating of marker effects is affected by the variance.

In this study, the MCMC in Bayesian methods are iterated 10,000 times with 100 cycles in Metropolis-Hastings algorithm, and the first 2000 iterations are discarded as burn-in. Samples from the remaining iterations are averaged to obtain estimates of marker effects. In BayesB, π is set to 0.95, while is calculated with formula (4) in BayesBπ. Our new method is termed BayesBπ because it is an improved version of the original BayesB by assigning genetic architecture based priors. Calculation of GBLUP, BayesB, and BayesBπ are conducted with our in house programs, while BayesCπ is conducted with R package "GBLR" [58].

Locus-specific priori

From the aspect of whole genome, π is the proportion of non-effective markers. However, from the aspect of single markers, π is an important parameter which decides the extent to which a marker is fitted in the model, and thus affects the estimation of marker effects. Therefore, π should be different among markers, which is consistent with our prior knowledge that some genome segments have large effects and others show moderate to zero effects across the whole genome. Here we propose a method to obtain the locus-specific π based on traits' genetic architecture. The locus-specific π is obtained by rescaling p-values derived from the analysis of variance (ANOVA) to a probability form. ANOVA is performed by the R software package (http://www.r-project.org/) on single markers in the reference population to get the p-values. The model for ANOVA can be written as

$$\mathbf{y} = \mathbf{Xb} + \mathbf{Zg} + \mathbf{e}, \tag{3}$$

where, $\mathbf{y}$ is a vector of phenotypes; $\mathbf{X}$ is a design matrix linking records to the fixed effects included in $\mathbf{b}$; $\mathbf{Z}$ is a design matrix indicating the genotypes of individual SNPs; g is the effect of single markers; and $\mathbf{e}$ is a vector of residuals. Then the p-values derived from ANOVA on single SNPs are transformed to the locus-specific π through the formula

$$\pi_i = \frac{max(\boldsymbol{\omega}) - \omega_i}{\max(\boldsymbol{\omega}) - min(\boldsymbol{\omega})}, \tag{4}$$

where, π_i is the locus-specific π of ith marker; $\boldsymbol{\omega} = -log_{10}(\boldsymbol{p})$; $\omega_i = -log_{10}(p_i)$; p_i is the p-value of ith marker; and $\boldsymbol{p}$ is the vector of p-values of all markers.

In BayesBπ, the locus-specific π of SNPs are obtained from the reference population through the method mentioned above. In the following MCMC algorithm, each marker uses its corresponding π to perform the estimation of variances and marker effects.

Model validation

The accuracy of genomic prediction is defined as the correlation between the GEBVs and the response variables (conventional EBVs in the dairy cattle population, and regressed phenotypes in the loblolly pine dataset). Regression of the GEBVs on the response variables are performed, and the regression coefficients are taken as the genomic prediction unbiasednesses. The accuracy and unbiasedness of BayesBπ are compared with that of GBLUP [30], the original BayesB [11], and BayesCπ. The dairy cattle population is used as a standard dataset for models validating. In order to investigate the impact of population sizes on genomic selection accuracy, subsets with sizes of 200, 500, 1000, and 2000 are randomly sampled from the complete dairy cattle dataset. For all subpopulations and traits, a 5-fold cross-validation is performed 20 times to get the mean accuracies and unbiasednesses for the three methods. Within the loblolly pine dataset, a 10-fold cross-validation is performed 10 times. Therefore, the mean accuracy and unbiasedness are obtained by averaging estimated values of 100 validations for both datasets. In the dairy cattle dataset, the mean accuracies of the subpopulations and traits are further averaged to show the impact of both population sizes and the genetic architectures of traits on the performance of different approaches. The extents of improvement with our new method compared to the original BayesB are calculated with the formula $\beta = \frac{acc_{B\pi} - acc_B}{acc_B} \times 100\%$, where, β is the extent of improvement with our new method compared to the original BayesB; $acc = \frac{cov(GEBVs, y)}{\sigma_{GEBVs}\sigma_y}$ is the Pearson's correlation coefficient between genomic estimated breeding values (GEBVs) and model response variables (i.e., y in the formula here, which is traditional EBVs in the dairy cattle population and deregressed phenotypes in the loblolly pine dataset), where σ_{GEBVs} and σ_y are the standard deviations of GEBVs and model response variables; $acc_{B\pi}$ and acc_B are accuracies of our new method and that of the original BayesB, respectively.

Availability of supporting data

The data used in this study are online available through http://www.g3journal.org/content/5/4/615/suppl/DC1 for dairy cattle dataset and http://www.genetics.org/content/190/4/1503/suppl/DC1 for loblolly pine dataset, respectively.

Abbreviations

SNP: Single nucleotide polymorphism; GWAS: Genome wide association study; GS: Genomic selection; BLUP: Best linear unbiased prediction; GBLUP: Genomic best linear unbiased prediction; RRBLUP: Ridge regression best linear unbiased prediction; TABLUP: Best linear unbiased prediction method including a trait-specific relationship matrix; BLUP|GA: Best linear unbiased prediction method including a genetic architecture (GA) based relationship matrix; LS: Least square; BayesA: Bayesian method A; BayesB: Bayesian method B; BayesCπ: Bayesian method Cπ; BayesDπ: Bayesian method Dπ; BayesR: Bayesian method R; BayesLASSO: Bayesian method with least absolute shrinkage and selection operator (LASSO); QTL: Quantitative trait loci; MCMC: Monte carlo markov chain; MH: Metropolis-hasting algorithm; ANOVA: Analysis of variance; DGAT1: Diacylglycerol acyltransferase 1 gene.

Competing interests

The authors declare no competing interests.

Authors' contributions

NG proposed the method for calculating loci-specific π and helped draft the manuscript. ZZ provided the dairy cattle dataset, participated in the design and helped draft the manuscript. JLH carried out the cross validation calculation of the dairy cattle dataset. YYL carried out the cross validation calculation of the loblolly pine dataset. GX helped the results visualization. HZ and ZMC participated in the design and contributed to the manuscript. JQL participated in the design and helped draft the manuscript. All authors read and approved the manuscript.

Acknowledgements

This work is supported by the National Natural Science Foundation of China (31200925, 31371258), the earmarked fund for China Agriculture Research System (CARS-36), the Guangdong Natural Science Foundation (2014A030313453), and the Key Scientific and Technological Projects of Guangzhou (2011Y2-00008). Prof. Dr. Henner Simianer provides the original analysis of the Germany cattle dataset and helpful suggestions to this work. We would like to thank the two reviewers and the associate editor for their essential suggestions and comments which improved the manuscript considerably.

References

1. Vachon CM, van Gils CH, Sellers TA, Ghosh K, Pruthi S, Brandt KR, et al. Mammographic density, breast cancer risk and risk prediction. Breast Cancer Res. 2007, 9(6):doi:10.1186/bcr1829.
2. Jostins L, Barrett JC. Genetic risk prediction in complex disease. Hum Mol Genet. 2011;20:R182–8.
3. Domchek SM, Eisen A, Calzone K, Stopfer J, Blackwood A, Weber BL. Application of breast cancer risk prediction models in clinical practice. J Clin Oncol. 2003;21(4):593–601.
4. Bonassi S, Au WW. Biomarkers in molecular epidemiology studies for health risk prediction. Mutat Res Rev Mutat Res. 2002;511(1):73–86.
5. Bouquet A, Juga J. Integrating genomic selection into dairy cattle breeding programmes: a review. Animal. 2013;7(5):705–13.
6. Lin Z, Hayes BJ, Daetwyler HD. Genomic selection in crops, trees and forages: a review. Crop & Pasture Science. 2014;65(11):1177–91.
7. Hayes BJ, Cogan NOI, Pembleton LW, Goddard ME, Wang J, Spangenberg GC, et al. Prospects for genomic selection in forage plant species. Plant Breed. 2013;132(2):133–43.
8. Zhao Y, Mette MF, Reif JC. Genomic selection in hybrid breeding. Plant Breed. 2015;134(1):1–10.
9. Stock KF, Reents R. Genomic Selection: Status in Different Species and Challenges for Breeding. Reprod Domest Anim. 2013;48:2–10.
10. de los Campos G, Hickey JM, Pong-Wong R, Daetwyler HD, Calus MPL. Whole-Genome Regression and Prediction Methods Applied to Plant and Animal Breeding. Genetics. 2013;193(2):327.
11. Meuwissen THE, Hayes BJ, Goddard ME. Prediction of total genetic value using genome-wide dense marker maps. Genetics. 2001;157(4):1819–29.
12. de los Campos G, Vazquez AI, Fernando R, Klimentidis YC, Sorensen D: Prediction of complex human traits using the genomic best linear unbiased predictor. Plos Genetics. 2013, 9(7):doi:10.1371/journal.pgen.1003608.
13. Ober U, Ayroles JF, Stone EA, Richards S, Zhu D, Gibbs RA, et al: Using whole-genome sequence data to predict quantitative trait phenotypes in Drosophila melanogaster. Plos Genetics 2012, 8(5):doi:10.1371/journal.pgen.1002685.
14. Luan T, Woolliams JA, Lien S, Kent M, Svendsen M, Meuwissen THE. The Accuracy of Genomic Selection in Norwegian Red Cattle Assessed by Cross-Validation. Genetics. 2009;183(3):1119–26.
15. Erbe M, Hayes BJ, Matukumalli LK, Goswami S, Bowman PJ, Reich CM, et al. Improving accuracy of genomic predictions within and between dairy cattle breeds with imputed high-density single nucleotide polymorphism panels. J Dairy Sci. 2012;95(7):4114–29.
16. Hayes BJ, Bowman PJ, Chamberlain AC, Verbyla K, Goddard ME. Accuracy of genomic breeding values in multi-breed dairy cattle populations. Genet Sel Evol. 2009, 41:doi:10.1186/1297-9686-1141-1151
17. Verbyla KL, Hayes BJ, Bowman PJ, Goddard ME. Accuracy of genomic selection using stochastic search variable selection in Australian Holstein Friesian dairy cattle. Genet Res. 2009;91(5):307–11.
18. Morota G, Abdollahi-Arpanahi R, Kranis A, Gianola D: Genome-enabled prediction of quantitative traits in chickens using genomic annotation. BMC Genomics 2014, 15:doi:10.1186/1471-2164-115-1109
19. Slack-Smith A, Kinghorn BP, van der Werf JHJ. Accuracy of genomic selection in predicting carcass traits in meat sheep. Anim Prod Sci. 2010;50(11-12):XIII-XIII.
20. Christensen OF, Madsen P, Nielsen B, Ostersen T, Su G. Single-step methods for genomic evaluation in pigs. Animal. 2012;6(10):1565–71.
21. Daetwyler HD, Hickey JM, Henshall JM, Dominik S, Gredler B, van der Werf JHJ, et al. Accuracy of estimated genomic breeding values for wool and meat traits in a multi-breed sheep population. Anim Prod Sci. 2010;50(11-12):1004–10.
22. Zhao Y, Gowda M, Liu W, Wuerschum T, Maurer HP, Longin FH, et al. Accuracy of genomic selection in European maize elite breeding populations. Theor Appl Genet. 2012;124(4):769–76.
23. Heffner EL, Jannink J-L, Iwata H, Souza E, Sorrells ME. Genomic Selection Accuracy for Grain Quality Traits in Biparental Wheat Populations. Crop Sci. 2011;51(6):2597–606.
24. Zhong S, Dekkers JCM, Fernando RL, Jannink J-L. Factors Affecting Accuracy From Genomic Selection in Populations Derived From Multiple Inbred Lines: A Barley Case Study. Genetics. 2009;182(1):355–64.
25. Resende Jr MFR, Munoz P, Resende MDV, Garrick DJ, Fernando RL, Davis JM, et al. Accuracy of Genomic Selection Methods in a Standard Data Set of Loblolly Pine (Pinus taeda L.). Genetics. 2012;190(4):1503.
26. Resende Jr MFR, Munoz P, Acosta JJ, Peter GF, Davis JM, Grattapaglia D, et al. Accelerating the domestication of trees using genomic selection: accuracy of prediction models across ages and environments. New Phytol. 2012;193(3):617–24.

27. Sonesson AK, Meuwissen THE: Testing strategies for genomic selection in aquaculture breeding programs. Genet Sel Evol. 2009; 41:doi:10.1186/1297-9686-1141-1137

28. Meuwissen T, Hayes B, Goddard M. Accelerating Improvement of Livestock with Genomic Selection. Annu Rev Anim Biosci. 2013;1:221–37.

29. Zhang Z, Zhang Q, Ding X. Advances in genomic selection in domestic animals. Chin Sci Bull. 2011;56(25):2655–63.

30. VanRaden PM. Efficient Methods to Compute Genomic Predictions. J Dairy Sci. 2008;91(11):4414–23.

31. Habier D, Fernando RL, Kizilkaya K, Garrick DJ. Extension of the bayesian alphabet for genomic selection. BMC Bioinformatics 2011; 12:doi:10.1186/1471-2105-1112-1186

32. Xu SZ. Estimating polygenic effects using markers of the entire genome. Genetics. 2003;163(2):789–801.

33. ter Braak CJF, Boer MP, Bink M. Extending Xu's Bayesian model for estimating polygenic effects using markers of the entire genome. Genetics. 2005;170(3):1435–8.

34. Brondum RF, Su G, Lund MS, Bowman PJ, Goddard ME, Hayes BJ. Genome position specific priors for genomic prediction. BMC Genomics. 2012, 13:doi:10.1186/1471-2164-1113-1543.

35. Gianola D, de los Campos G, Hill WG, Manfredi E, Fernando R. Additive Genetic Variability and the Bayesian Alphabet. Genetics. 2009;183(1):347–63.

36. Knurr T, Laara E, Sillanpaa MJ. Genetic analysis of complex traits via Bayesian variable selection: the utility of a mixture of uniform priors. Genet Res. 2011;93(4):303–18.

37. Knurr T, Laara E, Sillanpaa MJ. Impact of prior specifications in a shrinkage-inducing Bayesian model for quantitative trait mapping and genomic prediction. Genet Sel Evol. 2013, 45:doi:10.1186/1297-9686-1145-1124.

38. Legarra A, Robert-Granie C, Croiseau P, Guillaume F, Fritz S. Improved Lasso for genomic selection. Genet Res. 2011;93(1):77–87.

39. de los Campos G, Naya H, Gianola D, Crossa J, Legarra A, Manfredi E, et al. Predicting Quantitative Traits With Regression Models for Dense Molecular Markers and Pedigree. Genetics. 2009;182(1):375–85.

40. Daetwyler HD, Pong-Wong R, Villanueva B, Woolliams JA. The Impact of Genetic Architecture on Genome-Wide Evaluation Methods. Genetics. 2010;185(3):1021–31.

41. Zhang Z, Ding X, Liu J, Zhang Q, de Koning DJ. Accuracy of genomic prediction using low-density marker panels. J Dairy Sci. 2011;94(7):3642–50.

42. Zhang Z, Liu J, Ding X, Bijma P, de Koning D-J, Zhang Q. Best linear unbiased prediction of genomic breeding values using a trait-specific marker-derived relationship matrix. PLos One. 2010, 5(9):doi:10.1371/journal.pone.0012648.

43. Zhang Z, Ober U, Erbe M, Zhang H, Gao N, He J, et al. Improving the accuracy of whole genome prediction for complex traits using the results of genome wide association studies. PLos One. 2014, 9(3):doi:10.1371/journal.pone.0093017.

44. Wang H, Misztal I, Aguilar I, Legarra A, Fernando RL, Vitezica Z, et al. Genome-wide association mapping including phenotypes from relatives without genotypes in a single-step (ssGWAS) for 6-week body weight in broiler chickens. Front Genet. 2014, 5:doi:10.3389/fgene.2014.00134.

45. Zhang Z, Erbe M, He J, Ober U, Gao N, Zhang H, et al. Accuracy of whole-genome prediction using a genetic architecture-enhanced variance-covariance matrix. G3 (Bethesda, Md). 2015;5(4):615–27.

46. Grisart B, Farnir F, Karim L, Cambisano N, Kim JJ, Kvasz A, et al. Genetic and functional confirmation of the causality of the DGAT1 K232A quantitative trait nucleotide in affecting milk yield and composition. Proc Natl Acad Sci U S A. 2004;101(8):2398–403.

47. Drinkwater NR, Gould MN. The Long Path from QTL to Gene. Plos Genetics. 2012;8(9):e1002975.

48. Leon-Novelo L, Casella G. Prior influence in linear regression when the number of covariates increases to infinity. Stat Probab Lett. 2012;82(3):438–45.

49. Nadaf J, Riggio V, Yu T-P, Pong-Wong R. Effect of the prior distribution of SNP effects on the estimation of total breeding value. BMC Proc. 2012;6 Suppl 2:S6–6.

50. Lehermeier C, Wimmer V, Albrecht T, Auinger H-J, Gianola D, Schmid VJ, et al. Sensitivity to prior specification in Bayesian genome-based prediction models. Stat Appl Genet Mol Biol. 2013;12(3):375–91.

51. Gianola D. Priors in Whole-Genome Regression: The Bayesian Alphabet Returns. Genetics. 2013;194(3):573–96.

52. Meuwissen T, Goddard M. Accurate Prediction of Genetic Values for Complex Traits by Whole-Genome Resequencing. Genetics. 2010;185(2):623–U338.

53. Fernando RL, Nettleton D, Southey BR, Dekkers JCM, Rothschild MF, Soller M. Controlling the proportion of false positives in multiple dependent tests. Genetics. 2004;166(1):611–9.

54. Hu Z-L, Fritz ER, Reecy JM. AnimalQTLdb: a livestock QTL database tool set for positional QTL information mining and beyond. Nucleic Acids Res. 2007;35:D604–9.

55. Kadarmideen HN. Genomics to systems biology in animal and veterinary sciences: Progress, lessons and opportunities. Livest Sci. 2014;166:232–48.

56. Matukumalli LK, Lawley CT, Schnabel RD, Taylor JF, Allan MF, Heaton MP, et al: Development and characterization of a high density SNP genotyping assay for cattle. PLos One. 2009, 4(4):doi:10.1371/journal.pone.0005350.

57. Eckert AJ, van Heerwaarden J, Wegrzyn JL, Nelson CD, Ross-Ibarra J, Gonzalez-Martinez SC, et al. Patterns of Population Structure and Environmental Associations to Aridity Across the Range of Loblolly Pine (Pinus taeda L., Pinaceae). Genetics. 2010;185(3):969–82.

58. Perez P, de los Campos G. Genome-Wide Regression and Prediction with the BGLR Statistical Package. Genetics. 2014;198(2):483–U463.

18

Genetic association meta-analysis: a new classification to assess ethnicity using the association of MCP-1 -2518 polymorphism and tuberculosis susceptibility as a model

Tania Vásquez-Loarte, Milana Trubnykova and Heinner Guio[*]

Abstract

Background: In meta-analyses of genetic association studies, ancestry and ethnicity are not accurately investigated. Ethnicity is usually classified using conventional race/ethnic categories or continental groupings even though they could introduce bias increasing heterogeneity between and within studies; thus decreasing the external validity of the results. In this study, we performed a meta-analysis using a novel ethnic classification system to test the association between *MCP-1* -2518 polymorphism and pulmonary tuberculosis. Our new classification considers genetic distance, migration and linguistic origins, which will increase homogeneity within ethnic groups.

Methods: We included thirteen studies from three continents (Asia, Africa and Latin America) and considered seven ethnic groups (West Africa, South Africa, Saharan Africa, East Asia, South Asia, Persia and Latin America).

Results: The results were compared to the continental group classification. We found a significant association between *MCP-1* -2518 polymorphism and TB susceptibility only in the East Asian and Latin American groups (OR 3.47, $P = 0.08$; OR 2.73, $P = 0.02$). This association is not observed in other ethnic groups that are usually considered in the Asian group, such as India and Persia, or in the African group.

Conclusions: There is an association between *MCP-1* -2518 polymorphism and TB susceptibility only in the East Asian and Latin American groups. We suggest the use of our new ethnic classification in future meta-analysis of genetic association studies when ancestry markers are not available. This new classification increases homogeneity for certain ethnic groups compared to the continental classification. We recommend considering previous data about migration, linguistics and genetic distance when classifying ethnicity in further studies.

Keywords: Polymorphism, CCL2, MCP-1, Tuberculosis, Ethnicity

Background

Tuberculosis disease (TB) is a major public health problem worldwide. To create new strategies that will improve TB control, we need a better understanding of the biological, environmental, social, and ethnic factors [1]. One promising route is the study of polymorphisms involved in pulmonary TB susceptibility [2, 3]. Several human genes have been associated with TB development [4–6], including the *monocyte chemoattractant protein 1*

(*MCP-1*), also called *CCL2*. *MCP-1* belongs to a group of CC chemokines located in chromosome 17q11.2. MCP-1 protein interacts with chemokine C-C motif receptor 2 (CCR2) to activate and recruit monocytes, macrophages, CD4+ T cells and immature dendritic cells to the site of infection [7–9]. The presence of MCP-1 protein in an adequate concentration is important for granuloma formation and *M. tuberculosis* clearance [10, 11].

Although there are more than ten genetic polymorphisms in the *MCP-1* promoter and coding region, only the *MCP-1* -2518 A/G allele (reference sequence 1024611) is functional and affects gene expression [10]. A substitution from A to G in -2518 position of the promoter region

---
* Correspondence: heinnerguio@gmail.com
Laboratorio de Biotecnología y Biología Molecular, Instituto Nacional de Salud, Avenida Defensores del, Morro 2268, Lima 9, Peru

increases the levels of MCP-1. This action decreases the concentration of IL-12p40, which recruits and activates memory/effector Th1 cells, thus impairing long-term protection to intracellular pathogens [10]. Observational studies have shown that *MCP-1* -2518 A/G polymorphism is associated with the development of pulmonary tuberculosis (pTB) and could be a potential marker for latent TB and disease severity [3, 12]. However, this association is different among countries such as Persia, India, Korea and China, which share continental groups [13, 14].

Geographical distribution by continents is the conventional way to assess ethnicity in meta-analysis of genetic association studies. However, population genetics has demonstrated that ethnic composition is related more with genetic distance, migration and linguistic origins rather than continental groups. In terms of ancestry biomarkers, continental grouping relies on markers such as Y-DNA and mtDNA haplogroups and varies within continents [15, 16]. As a consequence, conventional classification might introduce bias and increase heterogeneity between and within studies, decreasing the external validity of the results. Thus, it is questionable if the conventional classification is an appropriate proxy for ethnicity.

In order to have a better understanding of the relationship between ethnicity and the susceptibility to infectious diseases such as TB, we evaluated the association between the *MCP-1* -2518 A/G polymorphism and pTB susceptibility using a new multi-factorial ethnic classification and compared it with the conventional approach of continental groups. This new classification is based on previous research on genetic distance, migration and linguistic origins [16–19], which improves the homogeneity of ethnic groups. We believe that our new classification for ethnicity offers a more robust approach to explain susceptibility to disease, and that it can increase the internal validity of genetic studies when ancestry markers are not available.

Methods
Search strategy
A literature search was carried out in NCBI database, Scielo and Lilacs to identify genetic association studies between MCP-1 polymorphism and pTB risk prior to December 2013. We used the following MESH terms: (("Polymorphism, Genetic"[Mesh]) AND "Chemokine CCL2"[Mesh]) AND "Tuberculosis"[Mesh]. Mesh term "MCP-1" gave the result CCL2. "Reference sequence 1024611 A/G" gave zero results. Our selection criteria included: 1) studies evaluating the association between MCP-1-2518 A/G and TB risk, 2) observational studies, 3) pulmonary TB, 4) studies performed in adults and children, 4) patients without HIV or cancer, 5) available allelic and the genotype frequencies to estimate an odds ratio (OR), 6) control groups that met

Hardy Weinberg Equilibrium, and 7) articles published until December 2013. Studies that did not meet these criteria were excluded. When original articles included more than one study population, we considered each as an independent study. In case of multiple publications on the same study, we included the study with the larger sample and/or the most recently published. The data search retrieved 23 articles. Ten studies were excluded because they were reviews or meta-analyses, or corresponded to pediatric populations, spinal TB, latent TB, HIV positive individuals or data from controls was inaccessible. At the end, 13 studies (7651 cases and 8056 controls) [3, 10, 20–30] (Fig. 1) were considered.

Data collection
All articles were separately extracted, reviewed and collated by two independent reviewers who checked for any discordance and reached a consensus in all items. Authors were contacted by email when we needed more information about an article. The following information was extracted for each study: author, year of publication, country of origin, ethnicity, sample size, type of study population, TB definition, allele and the genotype frequencies in cases, controls and methods. The information was systematically reviewed using STROBE and STREGA parameters [31, 32].

Ethnic classification
We proposed a new ethnic classification based on previous information about genetic distance, migration and linguistic origins [16, 17, 33–35] and compared it to the conventional classification. The new ethnic classification considered previous findings about genetic distance [17]. For this purpose, data such as country of origin was extracted from each study. Finally our new ethnic classification included: Middle East Asia (Persia), East Asia (Korea and China), South Asia (India), Saharan region (Morocco and Tunisia), South Africa (South Africa), West Africa (Guinea-Bissau, Gambia, and Ghana) and Latin America (Peru and Mexico). The conventional classification includes three groups: Africa (37 %), Asia (43.8 %) and Latin America (18.8 %). The characteristics of each study are listed in Table 1. We hypothesized that the new classification creates ethnic groups that have more homogeneity than the groups obtained by the conventional group classification.

Statistical analysis
For each study, the Hardy Weinberg Equilibrium (HWE) was calculated for the controls using X^2 statistic. Genotypes deviated from HWE if two-sided p values were <0.05. Begg funnel plot and Egger's test indicated publication bias if p value was <0.05. Sensitivity analysis was performed by

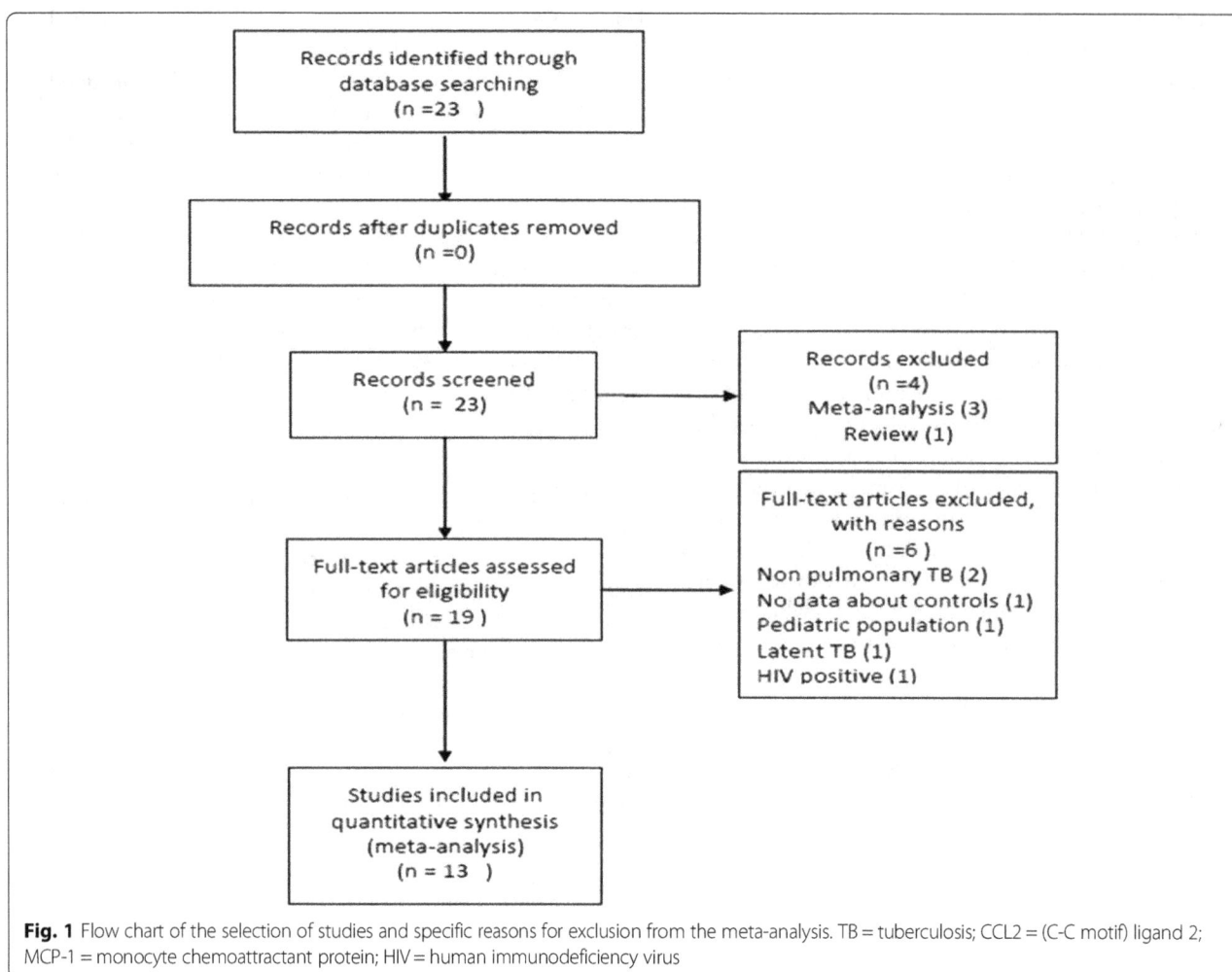

Fig. 1 Flow chart of the selection of studies and specific reasons for exclusion from the meta-analysis. TB = tuberculosis; CCL2 = (C-C motif) ligand 2; MCP-1 = monocyte chemoattractant protein; HIV = human immunodeficiency virus

removing one study at a time to assess the stability of the meta-analysis results.

To prove our hypothesis, we assessed heterogeneity and the magnitude of association for each ethnic group. We assessed heterogeneity by using the χ^2 based Q test and I^2 statistic. P values less than 0.01 were considered significant for heterogeneity. To assess the magnitude of association (pooled OR), in the presence of homogeneity, we used a fixed effects model (inverse variance weighted). Otherwise, we used a random effects model (DerSimonian and Laird, D + L). Pooled OR for the association between *MCP-1* - 2518 A/G polymorphism and pTB risk was determined in three steps. First, we did an allelic comparison (G vs. A) to determine the pooled OR in the overall data and by ethnic subgroups. Second, using our new ethnic classification, we analyzed four genotype models: a) recessive (GG vs. AG + AA), b) homogenous co-dominant (GG vs. AA), c) heterogeneous co-dominant GA vs. AA) and c) dominant (GG + GA vs. AA). Third, we compared these results to those obtained from the analysis using the conventional classification. Odds ratio estimates

were considered significant if P was <0.05, and were expressed using a 95 % confidence interval (CI). When analyzing by ethnicity, we used the groups that had ≥1 degree of freedom. For our analysis, the wild type allele was A, and the risk allele was G. We did not adjust our model for environmental effects. The statistical analysis was performed using STATA 11.0. (STATA Corp, College Station, TX, USA).

Results

The conventional ethnic analysis found heterogeneity in the three continental study groups for both allelic and genotype analysis. Using our new classification, we found homogeneity for South Asia (India) and West Africa, which were further analyzed with a fixed effects model. We could not improve homogeneity within the rest of the ethnic groups and used a random effects model for their analysis. According to the Begg's funnel plot and Egger's test, we did not find any bias in the analysis of the entire group (t = 1.76, P = 0.1; Fig. 2). Sensitivity analysis did not find any prominent effect of each individual

Table 1 Characteristics of studies included in the meta-analysis

Author, year, reference	Country	Male cases (%)	Age, mean (SD)	Diagnosis of cases	Control source and characteristics	Methods
Africa						
Ben-Selma et al., [23]	Tunisia	75	44(-)/-	Clinical and radiological pTB, BCG+	Healthy individuals, same community and ethnicity, BCG+	RFLP
Arji et al., [24]	Morocco	56	30(16)/38(17)	Clinical and radiological pTB, AFB+, HIV-, HBV-, HCV-	Healthy blood donors	RFLP
Möller et al., [22]	South Africa	-	-	Clinical and radiological pTB, AFB+, HIV-	Healthy individuals, same community, HIV-	SNPlex genotyping system
Thye et al., [20]	Ghana	-	-	Clinical and radiological pTB, AFB+, HIV-	Healthy individuals, TST-	Light type-based genotype
Velez et al., [21]	Guinea-Bissau	60	37(14)/36(12)	Clinical Pulmonary pTB, AFB+, HIV-	Healthy individuals, same community	Real-time PCR
Velez et al., [21]	Gambia	69	33(14)/ 29(13)	Clinical Pulmonary pTB, AFB+, HIV-	Neighbors, spouses	Real-time PCR
Asia						
Flores-Villanueva et al., [10]	Korea	67	38(-)/34(-)	Clinical and radiological pTB, AFB+, culture+, HIV-	Healthy blood donors	RFLP
Chu et al., [27]	Hong Kong	66	48(18)/31(9)	Clinical pTB, AFB+, HIV-	Healthy blood donors	RFLP
Xu et al., [29]	China	51	45(14)/42(13)	Clinical and radiological pTB, AFB+, in treatment	Healthy children	SSP-PCR
Yang et al., [28]	China	66	-	Clinical and radiological pTB, AFB+, in treatment	Surgery and Gynecology patients, no prior TB	RFLP
Naderi et al., [30]	Persia	22	50(21)/51(13)	Patients with confirmed pTB	Healthy individuals	Tetra-ARMS PCR
Mishra et al., [26]	India	69	37(7)/38(6)	AFB+ or patients under treatment	Healthy individuals, same ethnicity, AFB-	RFLP
Alagarasu et al., [25]	India	66	34(10)/31(9)	Clinical and radiological pTB, AFB+, HIV-	Healthy individuals	RFLP
Latin America						
Flores-Villanueva et al., [10]	Mexico	68	37(7)/36(7)	Clinical and radiological pTB, AFB+, culture+	Healthy neighbors, 334 TST+, 176 TST-	RFLP
Ganachari et al., [3]	Mexico	65	36(6)/37(3)	BCG+, clinical and radiological pTB, AFB+, HIV-	Healthy neighbors, TST+, HIV-	Tetra-ARMS
Ganachari et al., [3]	Peru	58	30(10)/34(9)	Clinical and radiological pTB, AFB+	Healthy individuals	Tetra-ARMS

pTB = pulmonary TB, AFB = acid fast bacilli, BCG_v = Bacillus Calmette-Guérin vaccine, HIV = human immune deficiency virus, HBV = Hepatitis B virus, HCV = Hepatitis C virus, TST = tuberculosis skin test, RFLP = restriction fragment length polymorphism, Tetra-ARMS = amplification refractory mutation system-PCR, PCR = polymerase chain reaction

Begg's funnel plot with pseudo 95% confidence limits

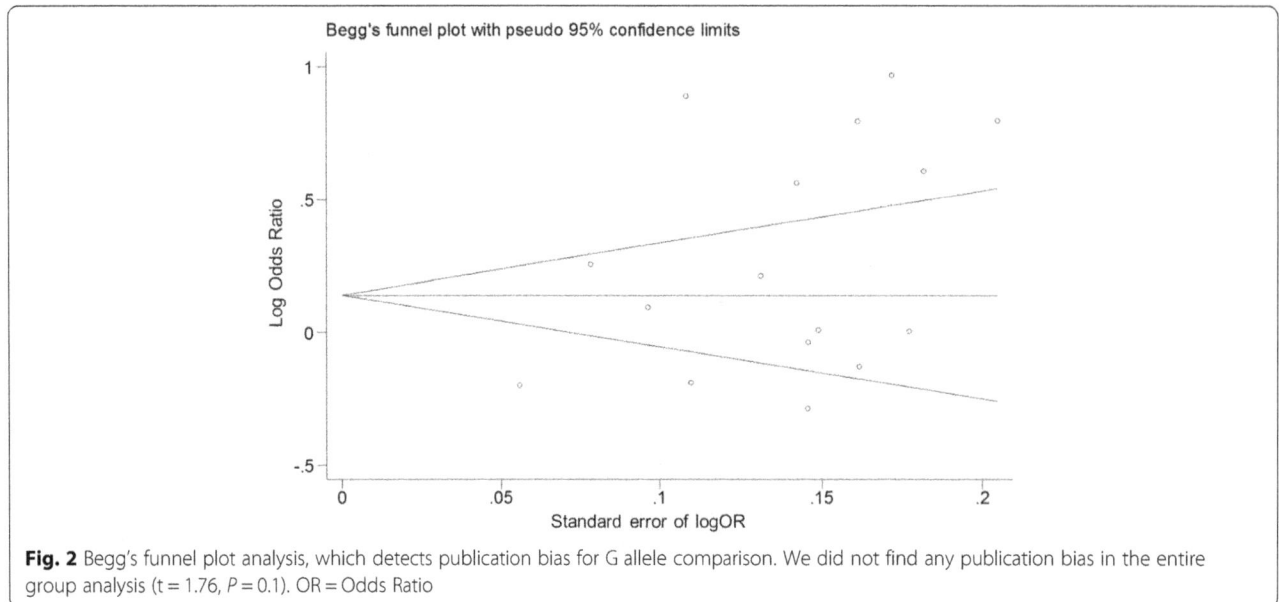

Fig. 2 Begg's funnel plot analysis, which detects publication bias for G allele comparison. We did not find any publication bias in the entire group analysis (t = 1.76, P = 0.1). OR = Odds Ratio

study when estimating the pooled OR. Characteristics of each study, allele and genotype distributions are shown in Tables 1 and 2 respectively.

G allele frequencies in conventional and new ethnic classification

The conventional classification showed that G allele is frequent in Asia and South America (45 % in cases vs. 39 % in controls and 70 % in cases vs. 59 % in controls, respectively) but not in Africa (22 % in cases vs. 20 % in controls). Our new ethnic classification showed that East Asia has the highest frequency of the G allele (57 % cases and 46 % controls) in the Asian group. In Latin America, this allele has a similar frequency in Mexico and Peru. In contrast, the African ethnic groups (Saharan, South and West Africa) have a low frequency of G allele in a similar proportion (Fig. 3). The pooled OR shows that presence of G allele increases the risk to develop pTB by 30 %. The conventional classification shows that this association is only significant for South America and Asia (OR 1.76, 95 % CI 1.7-2.6, P < 0.01; OR = 1.41,95 % CI 1.02-1.96, P = 0.03, respectively). Interestingly, our new ethnic classification showed that in the Asian continent, the G allele increases risk only in the East Asian ethnic group (OR 1.9, 95 % CI 1.2-3, P < 0.01), but not for South Asian and Persia (Fig. 4). We did not find any association for any of the African groups.

MCP-1 -2518 A/G genotypes and pTB susceptibility

The conventional classification showed that individuals from South America and Asia that carry GG genotype have 2.7 and 2.1 times the risk to develop pTB as compared to the ones with AA genotype (OR 2.72, 95 % CI 1.6–6.3, P = 0.02 and OR 2.09, 95 % CI 1.1–3.8, P = 0.01,

respectively). The recessive model also showed increased susceptibility in both continents but to a lesser extent (OR 2.12, 95 % CI 1.5–5.4, P < 0.01 in Asia and OR 1.76, 95 % CI 1.1–2.6, P < 0.01 in South America). Our new ethnic classification showed similar results for Latin America but not for Asian ethnic groups. Only the East Asian group that had the MCP-1 -2518 polymorphism in a homozygote co-dominant and recessive model had an increased risk to develop pTB (OR 3.47, 95 % CI 1.4–8.7, P < 0.01 and OR 2.34, 95 % CI 1.3-4.3, P < 0.01, respectively). The new classification did not find any association for the Persian and South Asian groups. Neither the new nor the continental group classifications found any association in Africa or any of its ethnic groups (Figs. 4, 5).

Ethnic groups and heterogeneity

We found heterogeneity within the ethnic groups from the conventional classification (I^2 73.1 %, P < 0.01, for Africa; I^2 83.8 %, P < 0.01, for Asia; I^2 90.7 %, P < 0.01, for South America). The new ethnic classification showed homogeneity for West Africa and South Asia (I^2 0 %, P = 0.3; I^2 0 %, P = 0.5, respectively). We found heterogeneity within Arabia, East Asia and Latin America (I^2 93.5 %, P < 0.01; I^2 88.7 %, P < 0.01; I^2 91.6 %, P < 0.01, respectively). We could not obtain results for South Africa and Persia, since there was only one study in each group.

Discussion

We found an association between the MCP-1 -2518 polymorphism and tuberculosis susceptibility in East Asian and Latin American populations [13, 14, 36]. Previous meta-analyses try to extrapolate this association to the Asian continent. However, since there is ethnic variability within each continent, we cannot generalize this conclusion to every

Table 2 *MCP-1* allele and genotype distribution in different ethnic groups

Author	Country	Continent	Ethnic group	Cases/Controls	G allele (%) cases/controls	Cases GG	Cases AG	Cases AA	Controls GG	Controls GG	Controls AG	Controls AA	P HWE
Ben-Selma et al., [23]	Tunisia	Africa	Saharian	168/150	33.6/21.7	25	63	80	8	49	93	0.6
Arji et al., [24]	Morocco	Africa	Saharian	337/204	21.7/27.0	9	128	200	15	80	109	0.8
Möller et al., [22]	South Africa	Africa	South Africa	431/482	22.5/26.0	26	142	263	39	173	270	0.2
Thye et al., [20]	Ghana	Africa	West Africa	1964/2312	17.1/20.2	63	546	1355	92	748	1472	0.8
Velez et al., [21]	Guinea- Bissau	Africa	West Africa	314/341	25.0/21.3	17	123	174	21	103	217	0.07
Velez et al., [21]	Gambia	Africa	West Africa	236/252	24.6/24.4	18	80	138	15	93	144	0.9
Flores-Villanueva et al., [10]	Korea	Asia	East Asia	129/162	60.1/36.4	46	63	20	22	74	66	0.5
Chu et al., [27]	China	Asia	East Asia	403/461	52.1/49.8	110	200	93	113	233	115	0.8
Xu et al., [29]	China	Asia	East Asia	100/100	55.5/36.0	29	53	18	13	46	41	0.7
Yang et al., [28]	China	Asia	East Asia	167/167	68.9/50.0	84	62	21	42	83	42	0.9
Naderi et al., [30]	Persia	Asia	Middle East	142/166	29.6/29.5	17	50	75	15	68	83	0.8
Mishra et al., [26]	India	Asia	South Asia	215/294	25.1/25.9	18	72	125	20	112	162	0.9
Alagarasu et al., [25]	India	Asia	South Asia	153/203	31.4/34.2	21	54	78	29	81	93	0.1
Flores-Villanueva et al., [10]	Mexico	South America	Latin America	435/334	72.0/51.3	229	168	38	91	161	82	0.8
Ganachari et al., [3]	Mexico	South America	Latin America	193/243	68.1/54.9	93	77	23	70	127	46	0.4
Ganachari et al., [3]	Peru	South America	Latin America	701/796	70.0/64.4	354	273	74	327	371	98	0.6

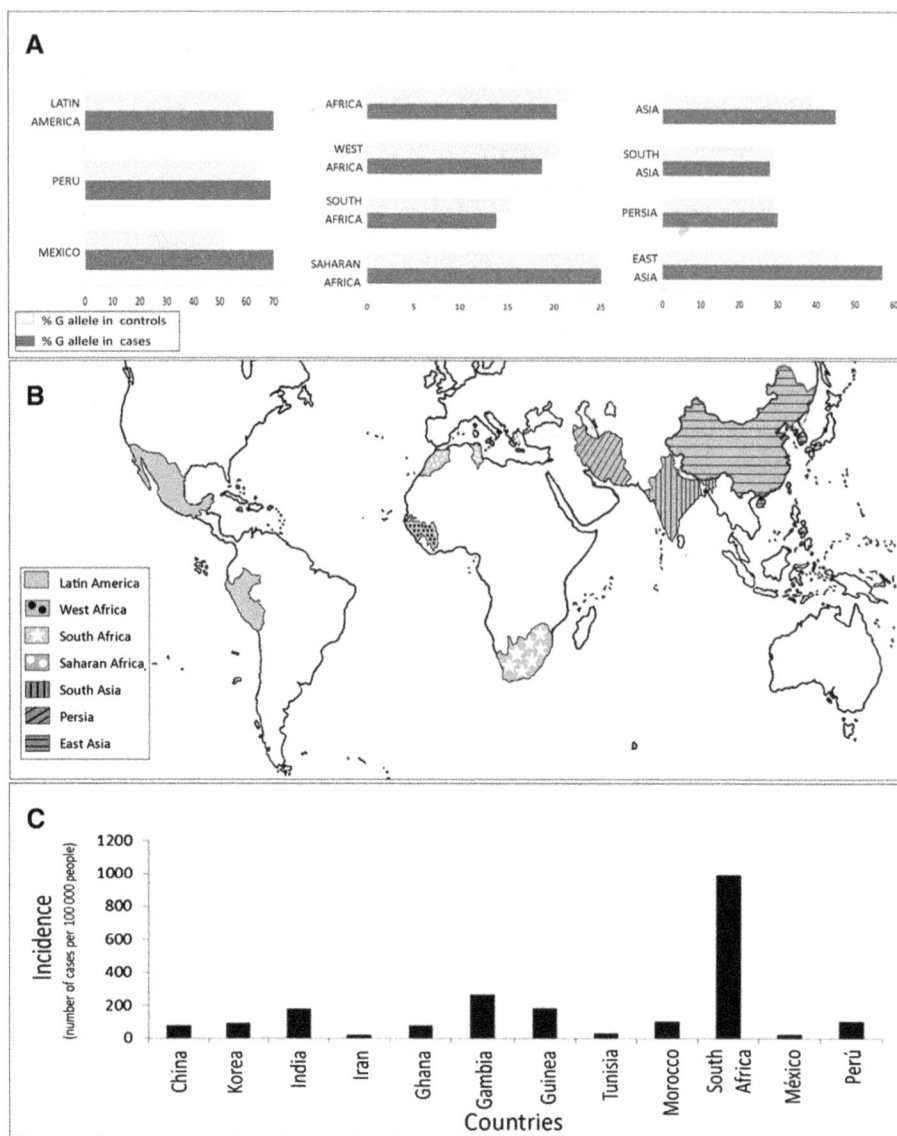

Fig. 3 G allele *MCP-1* -2518 polymorphism distribution in study populations and incidence of pulmonary TB (2010). **a** shows the frequency of G allele among ethnic countries. G allele is more frequent in individuals with pulmonary TB from East Asian and Latin American ethnic countries (* = P <0.01) and there is no difference within African subgroups, Persia and South Asia. **b** shows the ethnic countries considered in our new ethnic classification. **c** The chart shows the incidence of tuberculosis in the groups studied found at http://data.worldbank.org/indicator/SH.TBS.INCD. (this incidence includes HIV cases)

ethnic group. In this way, our meta-analysis groups study populations by using information about migration and linguistics to make ethnic groups more similar. Using this method, we found that an association does not apply to every country in the same continent.

Our new ethnic classification creates ethnic groups (e.g. West Africa and South Asia) with countries sharing similar characteristics. This new classification must be further evaluated with new studies related to genetic susceptibility for infectious and noninfectious diseases.

Regarding TB susceptibility, our new classification, in contrast to the conventional classification, helped to clarify that the association between MCP-1 -2518 A/G polymorphism and pTB is specific for certain populations such as East Asia and Latin America. To our knowledge, this is the first meta-analysis that uses a model of genetic susceptibility for pTB to assess if a new ethnic classification based on previous findings about genetic distance, migration and linguistic origins, improves homogeneity within each ethnic groups [13]. Thus, we propose our new classification as a good proxy when genetic markers are not available [17, 37–42].

Previous meta-analyses that use a continental group classification found an association between the *MCP-1-* 2518

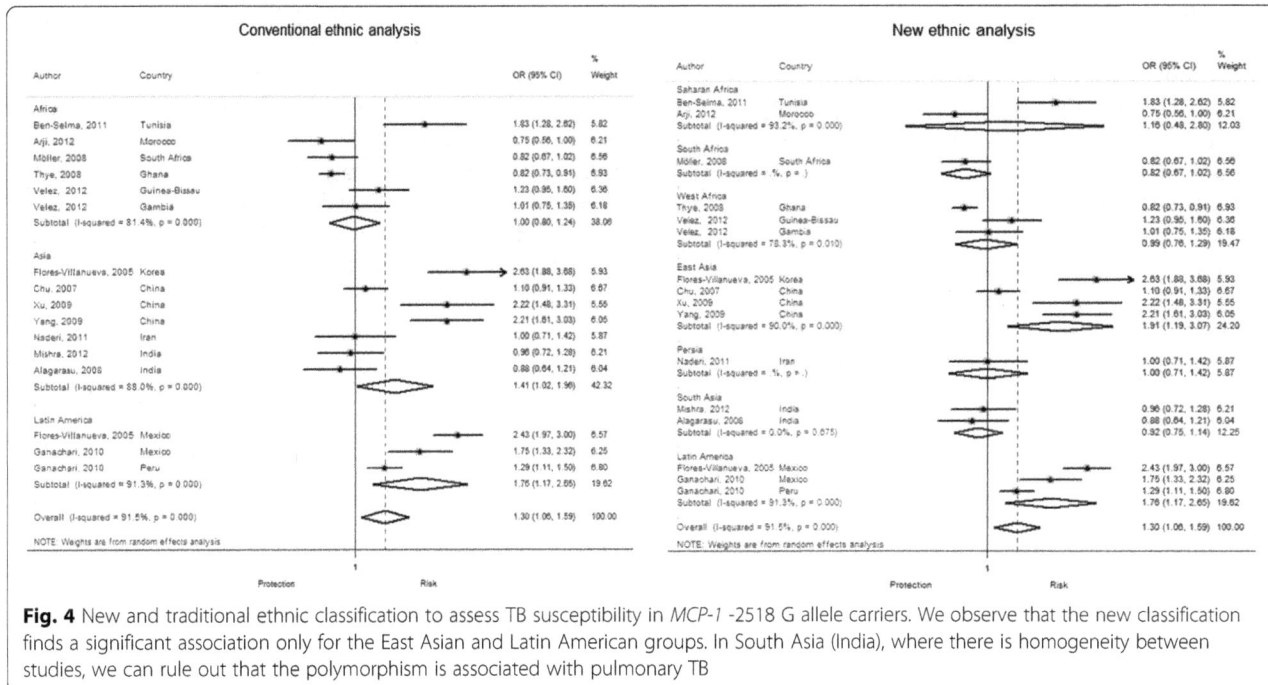

Fig. 4 New and traditional ethnic classification to assess TB susceptibility in *MCP-1* -2518 G allele carriers. We observe that the new classification finds a significant association only for the East Asian and Latin American groups. In South Asia (India), where there is homogeneity between studies, we can rule out that the polymorphism is associated with pulmonary TB

A/G polymorphism and pTB, which is significant for Asia and South America [13, 14]. However, these ethnic groups include countries that are different in terms of ancestry and therefore genetic susceptibility. Our new classification helps to improve homogeneity in South Asia and West Africa. However our new classification does not help us with homogeneity in East Asia and Latin America, where we found association between polymorphism *MCP-1* -2518 G

allele and pTB. Failure to reach homogeneity could be explained because of gene-gene or gene-environmental interactions. It has been reported that people carrying polymorphism *MCP-1* -2518 and *MMP-1* -1607 have a higher risk to develop severe TB [12]. The high frequency of G allele observed in cases compared to controls in both East Asia and Latin America might support the hypothesis of a similar ancestry between these two groups [43].

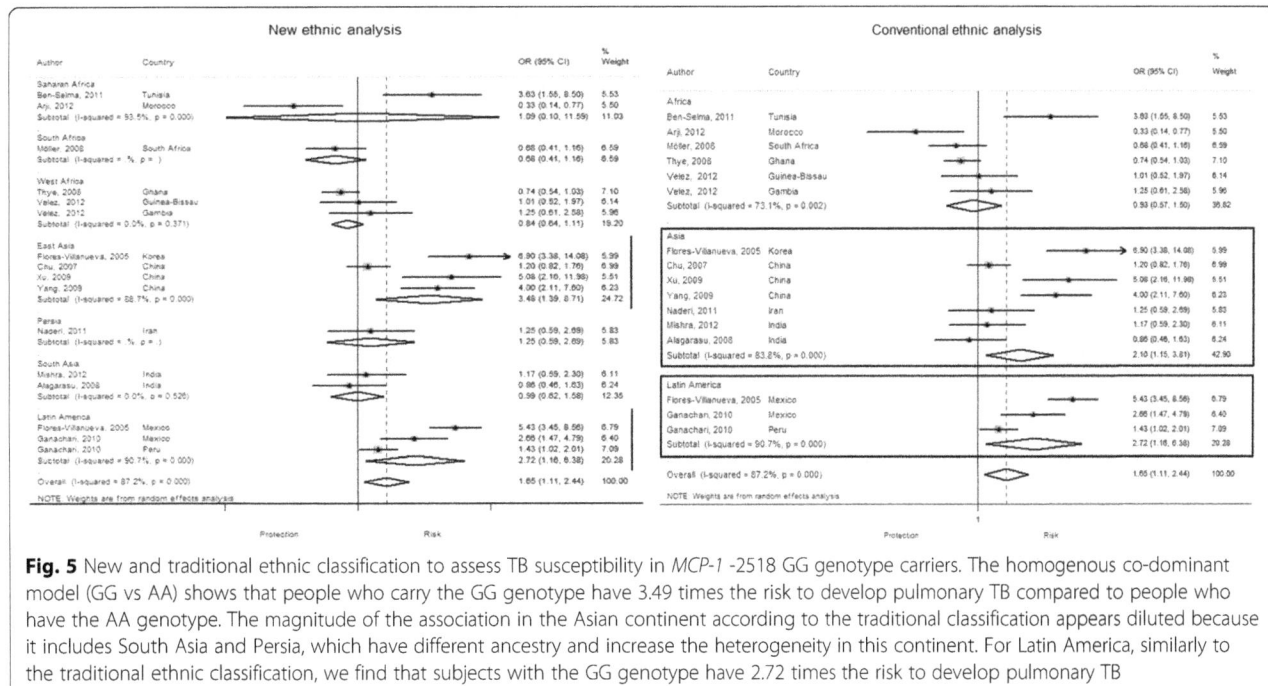

Fig. 5 New and traditional ethnic classification to assess TB susceptibility in *MCP-1* -2518 GG genotype carriers. The homogenous co-dominant model (GG vs AA) shows that people who carry the GG genotype have 3.49 times the risk to develop pulmonary TB compared to people who have the AA genotype. The magnitude of the association in the Asian continent according to the traditional classification appears diluted because it includes South Asia and Persia, which have different ancestry and increase the heterogeneity in this continent. For Latin America, similarly to the traditional ethnic classification, we find that subjects with the GG genotype have 2.72 times the risk to develop pulmonary TB

Recent studies from Africa show that the G polymorphism is not common among this population [21, 24]. Human population started in Africa, which means it is the oldest population, and therefore it has had the opportunity to accumulate genetic changes, such as the accumulation of -2518 MCP-1 A allele in its inhabitants that conferred protection and made it possible to adapt to hazardous environmental conditions [44]. We also have to consider other factors influencing TB susceptibility such as malnourishment, socioeconomic, environmental and health factors. The homogeneity found in West Africa cannot be completely explained in our study. We did not assess homogeneity in South Africa, since we only had one study population [22].

To deal with heterogeneity in Asia, we considered three groups: East Asia, South Asia and the Middle East [37–39]. The Asian population started from an "out of Africa" migration 50,000 years ago. It originated from two main migratory routes. The first one moved towards South Asia (India), and the second one to East Asia. Later, Central Asia was populated by Eurasian descendants. This is why we grouped Chinese and Korean populations under the East Asia group, India under the South Asia, and considered Persia under Persia. Also, these study populations have social, educational and mating habits that have that are particular to each group [16, 33, 34, 45]. Our classification groups two similar study populations from India under South Asia. Thus in this setting, it is unlikely that pTB susceptibility is due to the presence of *MCP-1* -2518 G polymorphism. In contrast to South Asia, we found an association between this polymorphism and pTB in East Asia where we also found heterogeneity. Interestingly, India (South Asia) and China (East Asia) accounted together for more than 40 % of TB cases worldwide in the last decade [46]. However the implementation of TB control strategies in China has helped decrease prevalence by 50 %, mortality rates by almost 80 % and TB incidence rates by 3.4 % per year between 1990 and 2010 [47]. Thus, even though there is a better control of TB in China, genetic factors might be playing an important role in the development of this disease. In contrast, India maintains its high incidence for the last 10 years, which might be due to a lack of social and public health control rather than genetic factors. It is difficult to assess homogeneity within Persia [48, 49] because we only had one study population. In further meta-analyses about genetic susceptibility we recommend to give a special consideration to Central Asian countries since they share European ancestry and therefore different genetic markers compared to East and South Asia [33].

Latin American ethnic groups that originated from a Han Chinese migration to South America 3000 years ago share HLA markers with this population [50]. This similarity might also explain similar frequency of *MCP-1* -2518 G allele and other genetic markers between East Asians and Latin Americans. Since Mexico and Peru share migration, common history, language routes and admixture indexes [40, 51–53], we decided to maintain them in the same ethnic group as previous meta-analyses. However, we consider that for Latin America, we should consider two groups: one of Andean and another of European origin.

The limitations in our study are very common to meta-analyses about genetic association studies. We could not consider environmental and genetic factors that influence this association because this information was not found in the original articles. Thus, for research in multifactorial diseases such as TB, we strongly recommend future studies to include information about malnourishment, socioeconomic factors, BCG and TST status, which could also help to control for heterogeneity.

We obtained homogeneity within South Asia and West Africa, where we can rule out that *MCP-1* -2518 polymorphism is associated to susceptibility. However, despite tuberculosis susceptibility found in Latin America and East Asia due to *MCP-1* -2518 polymorphism, the populations within each group are still genetically different.

Genetic association studies in populations from Persia, South Africa, South Asia, East Asia and the Americas, where infectious diseases represent a public health problem, will help assess heterogeneity in order to understand the role of ethnicity in genetic susceptibility to these diseases. In the absence of an adequate classification that groups similar genetic characteristics, suitable for understanding genetic susceptibility, our new classification might be a potential proxy for ethnic classification in meta-analysis of genetic association studies when genetic markers are not available.

Conclusions

In summary, using this novel approach, we found an association between the *MCP-1* -2518 polymorphism and pTB susceptibility, specifically in Latin American and East Asian populations not detected by using conventional classification. We encourage the use of our new ethnic classification in further genetic association studies for infectious and non-infectious diseases.

Abbreviations
CCL2: Chemokine (C-C motif) ligand 2; CCR2: Chemokine C-C motif receptor; HIV: Human immunodeficiency virus; HWE: Hardy weinberg equilibrium; MCP-1: Monocyte chemoattractant protein 1; MMP-1: Matrix metalloproteinase-1; mtDNA: Mitochondrial DNA; NCBI: National Center for Biotechnology Information; OR: Odds ratio; pTB: Pulmonary tuberculosis; STREGA: Strengthening the reporting of genetic association studies; STROBE: Strengthening the reporting of observational studies in epidemiology; TB: Tuberculosis.

Competing interests
The authors do not disclose any financial and non- financial competing interests.

Authors' contributions

TVL, MT, HG have equally contributed in the conception and design of the study, have been involved in drafting the manuscript, have given final approval of the version to be published and agree to be accountable for all aspects of the work in ensuring that questions related to the accuracy or integrity of any part of the work are appropriately investigated and resolved. TVL and MT have equally contributed in the acquisition of the data and interpretation of results. TVL has been involved in the data analysis. All authors read and approved the final manuscript.

Authors' information

This article was conceived when the authors' affiliation was Instituto Nacional de salud, Lima, Peru.

Acknowledgements

This work was supported by the Peruvian National Institute of Health. We thank Drs. Kim Hoffman and Cesar Sanchez for reviewing this manuscript.

References

1. Lönnroth K, Raviglione M. Global Epidemiology of Tuberculosis: Prospects for Control. Semin Respir Crit Care Med. 2008;29(05):481–91.
2. Qidwai T, Jamal F, Khan MY. DNA sequence variation and regulation of genes involved in pathogenesis of pulmonary tuberculosis. Scand J Immunol. 2012;75(6):568–87.
3. Ganachari M, Ruiz-Morales JA, Gomez de la Torre Pretell JC, Dinh J, Granados J, Flores-Villanueva PO. Joint effect of MCP-1 genotype GG and MMP-1 genotype 2G/2G increases the likelihood of developing pulmonary tuberculosis in BCG-vaccinated individuals. PloS one. 2010;5(1), e8881.
4. Pacheco AG, Cardoso CC, Moraes MO. IFNG +874 T/A, IL10 –1082G/A and TNF –308G/A polymorphisms in association with tuberculosis susceptibility: a meta-analysis study. Hum Genet. 2008;123(5):477–84.
5. Correa PA, Gomez LM, Cadena J, Anaya JM. Autoimmunity and Tuberculosis. Opposite Association with TNF Polymorphism. J Rheumatol. 2005;32(2):219–24.
6. Tian C, Zhang Y, Zhang J, Deng Y, Li X, Xu D, et al. The +874 T/A polymorphism in the interferon-γ gene and tuberculosis risk: An update by meta-analysis. Hum Immunol. 2011;72(11):1137–42.
7. Hodge DL, Reynolds D, Cerban FM, Correa SG, Baez NS, Young HA, et al. MCP-1/CCR2 interactions direct migration of peripheral B and T lymphocytes to the thymus during acute infectious/inflammatory processes. Eur J Immunol. 2012;42(10):2644–54.
8. Tanaka T, Terada M, Ariyoshi K, Morimoto K. Monocyte chemoattractant protein-1/CC chemokine ligand 2 enhances apoptotic cell removal by macrophages through Rac1 activation. Biochem Biophys Res Commun. 2010;399(4):677–82.
9. Sodhi A, Biswas SK. Monocyte chemoattractant protein-1-induced activation of p42/44 MAPK and c-Jun in murine peritoneal macrophages: a potential pathway for macrophage activation. J Interferon Cytokine Res. 2002;22(5):517–26.
10. Flores-Villanueva PO, Ruiz-Morales JA, Song CH, Flores LM, Jo EK, Montano M, et al. A functional promoter polymorphism in monocyte chemoattractant protein-1 is associated with increased susceptibility to pulmonary tuberculosis. J Exp Med. 2005;202(12):1649–58.
11. Hussain R, Ansari A, Talat N, Hasan Z, Dawood G. CCL2/MCP-I genotype-phenotype relationship in latent tuberculosis infection. PloS one. 2011;6(10), e25803.
12. Ganachari M, Guio H, Zhao N, Flores-Villanueva PO. Host gene-encoded severe lung TB: from genes to the potential pathways. Genes Immun. 2012;13(8):605–20.
13. Zhang Y, Zhang J, Zeng L, Huang H, Yang M, Fu X, et al. The -2518A/G Polymorphism in the MCP-1 Gene and Tuberculosis Risk: A Meta-Analysis. PLoS one. 2012;7(7), e38918.
14. Feng WX, Flores-Villanueva PO, Mokrousov I, Wu XR, Xiao J, Jiao WW, et al. CCL2-2518 (A/G) polymorphisms and tuberculosis susceptibility: a meta-analysis. Int J Tuberc Lung Dis. 2012;16(2):150–6.
15. Roewer L, Nothnagel M, Gusmao L, Gomes V, Gonzalez M, Corach D, et al. Continent-wide decoupling of chromosomal genetic variation from language and geography in native South Americans. PLoS Genet. 2013;9(4), e1003460.
16. Foster MW, Sharp RR. Race, ethnicity, and genomics: social classifications as proxies of biological heterogeneity. Genome Res. 2002;12(6):844–50.
17. Tishkoff SA, Reed FA, Friedlaender FR, Ehret C, Ranciaro A, Froment A, et al. The genetic structure and history of Africans and African Americans. Science. 2009;324(5930):1035–44.
18. Pagani L, Kivisild T, Tarekegn A, Ekong R, Plaster C, Gallego Romero I, et al. Ethiopian genetic diversity reveals linguistic stratification and complex influences on the Ethiopian gene pool. Am J Hum Genet. 2012;91(1):83–96.
19. Cooke GS, Hill AV. Genetics of susceptibility to human infectious disease. Nat Rev Genet. 2001;2(12):967–77.
20. Thye T, Nejentsev S, Intemann CD, Browne EN, Chinbuah MA, Gyapong J, et al. MCP-1 promoter variant -362C associated with protection from pulmonary tuberculosis in Ghana, West Africa. Hum Mol Genet. 2009;18(2):381–8.
21. Velez Edwards DR, Tacconelli A, Wejse C, Hill PC, Morris GA, Edwards TL, et al. MCP1 SNPs and pulmonary tuberculosis in cohorts from West Africa, the USA and Argentina: lack of association or epistasis with IL12B polymorphisms. PloS one. 2012;7(2), e32275.
22. Moller M, Nebel A, Valentonyte R, van Helden PD, Schreiber S, Hoal EG. Investigation of chromosome 17 candidate genes in susceptibility to TB in a South African population. Tuberculosis (Edinb). 2009;89(2):189–94.
23. Ben-Selma W, Harizi H, Boukadida J. MCP-1 -2518 A/G functional polymorphism is associated with increased susceptibility to active pulmonary tuberculosis in Tunisian patients. Mol Biol Rep. 2011;38(8):5413–9.
24. Arji N, Busson M, Iraqi G, Bourkadi JE, Benjouad A, Boukouaci W, et al. The MCP-1 (CCL2) -2518 GG genotype is associated with protection against pulmonary tuberculosis in Moroccan patients. J Infect Dev Ctries. 2012;6(1):73–8.
25. Alagarasu K, Selvaraj P, Swaminathan S, Raghavan S, Narendran G, Narayanan PR. CCR2, MCP-1, SDF-1a & DC-SIGN gene polymorphisms in HIV-1 infected patients with & without tuberculosis. Indian J Med Res. 2009;130(4):444–50.
26. Mishra G, Poojary SS, Raj P, Tiwari PK. Genetic polymorphisms of CCL2, CCL5, CCR2 and CCR5 genes in Sahariya tribe of North Central India: an association study with pulmonary tuberculosis. Infect Genet Evol. 2012;12(5):1120–7.
27. Chu SF, Tam CM, Wong HS, Kam KM, Lau YL, Chiang AK. Association between RANTES functional polymorphisms and tuberculosis in Hong Kong Chinese. Genes Immun. 2007;8(6):475–9.
28. Yang BF, Zhuang B, Li F, Zhang CZ, Song AQ. The relationship between monocyte chemoattractant protein-1 gene polymorphisms and the susceptibility to pulmonary tuberculosis. Chin J Tuberc Respir Dis. 2009;32(6):454–6.
29. Xu ZE, Xie YY, Chen JH, Xing LL, Zhang AH, Li BX, et al. Monocyte chemotactic protein-1 gene polymorphism and monocyte chemotactic protein-1 expression in Chongqing Han children with tuberculosis. Chin J Pediatr. 2009;47(3):200–3.
30. Naderi M, Hashemi M, Karami H, Moazeni-Roodi A, Sharifi-Mood B, Kouhpayeh H, et al. Lack of association between rs1024611 (-2581 A/G) polymorphism in CC-chemokine Ligand 2 and susceptibility to pulmonary Tuberculosis in Zahedan, Southeast Persia. Prague Med Rep. 2011;112(4):272–8.
31. Gallo V, Egger M, McCormack V, Farmer PB, Ioannidis JP, Kirsch-Volders M, et al. STrengthening the Reporting of observational studies in Epidemiology - Molecular Epidemiology (STROBE-ME): an extension of the STROBE statement. Eur J Clin Investig. 2012;42(1):1–16.
32. Little J, Higgins JP, Ioannidis JP, Moher D, Gagnon F, von Elm E, et al. STrengthening the REporting of Genetic Association studies (STREGA)–an extension of the STROBE statement. Eur J Clin Investig. 2009;39(4):247–66.
33. Ramsay M. Africa: continent of genome contrasts with implications for biomedical research and health. FEBS Lett. 2012;586(18):2813–9.
34. Heyer E, Balaresque P, Jobling MA, Quintana-Murci L, Chaix R, Segurel L, et al. Genetic diversity and the emergence of ethnic groups in Central Asia. BMC Genet. 2009;10:49.
35. Muro T, Iida R, Fujihara J, Yasuda T, Watanabe Y, Imamura S, et al. Simultaneous determination of seven informative Y chromosome SNPs to differentiate East Asian, European, and African populations. Legal Med. 2011;13(3):134–41.
36. Gong T, Yang M, Qi L, Shen M, Du Y. Association of MCP-1 -2518A/G and -362G/C variants and tuberculosis susceptibility: a meta-analysis. Infect Genet Evol. 2013;20:1–7.

37. Colonna V, Boattini A, Guardiano C, Dall'ara I, Pettener D, Longobardi G, et al. Long-range comparison between genes and languages based on syntactic distances. Hum Hered. 2010;70(4):245–54.

38. Dulik MC, Zhadanov SI, Osipova LP, Askapuli A, Gau L, Gokcumen O, et al. Mitochondrial DNA and Y chromosome variation provides evidence for a recent common ancestry between Native Americans and Indigenous Altaians. Am J Hum Genet. 2012;90(2):229–46.

39. Stoneking M, Delfin F. The human genetic history of East Asia: weaving a complex tapestry. Curr Biol. 2010;20(4):R188–93.

40. Majumder PP. The human genetic history of South Asia. Curr Biol. 2010;20(4):R184–7.

41. Abdulla MA, Ahmed I, Assawamakin A, Bhak J, Brahmachari SK, Calacal GC, et al. Mapping human genetic diversity in Asia. Science. 2009;326(5959):1541–5.

42. Bedoya G, Montoya P, Garcia J, Soto I, Bourgeois S, Carvajal L, et al. Admixture dynamics in Hispanics: a shift in the nuclear genetic ancestry of a South American population isolate. Proc Natl Acad Sci U S A. 2006;103(19):7234–9.

43. Qin H, Zhu X. Power comparison of admixture mapping and direct association analysis in genome-wide association studies. Genet Epidemiol. 2012;36(3):235–43.

44. Hill AV. Aspects of genetic susceptibility to human infectious diseases. Annu Rev Genet. 2006;40:469–86.

45. Tamang R, Singh L, Thangaraj K. Complex genetic origin of Indian populations and its implications. J Biosci. 2012;37(5):911–9.

46. Watts G. WHO annual report finds world at a crossroad on tuberculosis. BMJ. 2012;345, e7051.

47. de Colombani P, Dadu A, Dravniece G, Hoffner S, Ilyenkova V, Kovac Z, et al. Review of the national tuberculosis programme in Belarus, 10-21 October 2011. Copenhagen: WHO Regional Office for Europe; 2012. p. 76.

48. Fazeli Z, Vallian S. Phylogenetic relationship analysis of Persiaians and other world populations using allele frequencies at 12 polymorphic markers. Mol Biol Rep. 2012;39(12):11187–99.

49. Fazeli Z, Vallian S. Molecular phylogenetic study of the Persiaians based on polymorphic markers. Gene. 2013;512(1):123–6.

50. Tokunaga K, Ohashi J, Bannai M, Juji T. Genetic link between Asians and native Americans: evidence from HLA genes and haplotypes. Hum Immunol. 2001;62(9):1001–8.

51. Reich D, Patterson N, Campbell D, Tandon A, Mazieres S, Ray N, et al. Reconstructing Native American population history. Nature. 2012;488(7411):370–4.

52. Galanter JM, Fernandez-Lopez JC, Gignoux CR, Barnholtz-Sloan J, Fernandez-Rozadilla C, Via M, et al. Development of a panel of genome-wide ancestry informative markers to study admixture throughout the Americas. PLoS Genet. 2012;8(3), e1002554.

53. Johnson NA, Coram MA, Shriver MD, Romieu I, Barsh GS, London SJ, et al. Ancestral components of admixed genomes in a Mexican cohort. PLoS Genet. 2011;7(12), e1002410.

19

Identification and characterization of a high kernel weight mutant induced by gamma radiation in wheat (*Triticum aestivum* L.)

Xuejiao Cheng[1,2,3], Lingling Chai[1,2,3], Zhaoyan Chen[1,2,3], Lu Xu[1,2,3], Huijie Zhai[1,2,3], Aiju Zhao[4], Huiru Peng[1,2,3], Yingyin Yao[1,2,3], Mingshan You[1,2,3], Qixin Sun[1,2,3]* and Zhongfu Ni[1,2,3]*

Abstract

Background: Inducing mutations are considered to be an effective way to create novel genetic variations and hence novel agronomical traits in wheat. This study was conducted to assess the genetic differences between Shi4185 and its mutant line Fu4185, produced by gamma radiation with larger grain, and to identify quantitative trait loci (QTLs) for thousand kernel weight (TKW).

Results: Phenotypic analysis revealed that the TKW of Fu4185 was much higher than that of Shi4185 under five different environments. At the genomic level, 110 of 2019 (5.4 %) simple sequence repeats (SSR) markers showed polymorphism between Shi4185 and Fu4185. Notably, 30 % (33 out of 110) polymorphic SSR markers were located on the D-genome, which was higher than the percentage of polymorphisms among natural allohexaploid wheat genotypes, indicating that mutations induced by gamma radiation could be a potential resource to enrich the genetic diversity of wheat D-genome. Moreover, one QTL, *QTkw.cau-5D*, located on chromosome 5DL, with Fu4185 contributing favorable alleles, was detected under different environments, especially under high temperature conditions.

Conclusions: *QTkw.cau-5D* is an environmental stable QTL, which may be a desired target for genetic improvement of wheat kernel weight.

Keywords: Wheat, Kernel weight, Mutant, SSR marker, QTL

Background

Bread wheat (*Triticum aestivum* L.) is one of the most important food crops, which accounts for 20 % of the world's calorie consumption and growers (http://faostat3.fao.org/home/E). With the ever-growing population in the world, a pressing objective is to increase wheat productivity, which could be dissected by improving wheat yield potential and raising the yield ceiling [1]. Wheat yield can be attributed to the integration of the number of fertile spikes per unit area, grains per spike, and the kernel weight. Among these three yield components, kernel weight is a

highly heritable trait and has made significant contributions to yield potential in modern wheat breeding [2]. For example, TKW of Chinese wheat varieties has increased by 2.19 g every 10 years from 1940 to 2000 [3]. However, for the majority of the 20th century, TKW was mainly improved by applying selection techniques over phenotypic measurements. These years, detection of QTLs with molecular markers has drawn more attention. Up to date, QTLs for TKW have been mapped on all 21 wheat chromosomes [4–9]. In addition, the grain size-controlling genes in rice showed a significant association with the orthologs in wheat for the TKW trait [10–12].

Increasing temperature is a major element of observed global climate change (https://www.ncdc.noaa.gov/indicators/). For wheat, global production is estimated to fall by 6 % for each °C of further temperature increase [13]. Thus, high temperature (heat) stress during grain filling

* Correspondence: qxsun@cau.edu.cn; nizf@cau.edu.cn
[1]State Key Laboratory for Agrobiotechnology and Key Laboratory of Crop Heterosis and Utilization (MOE) and Key Laboratory of Crop Genomics and Genetic Improvement (MOA), Beijing Key Laboratory of Crop Genetic Improvement, China Agricultural University, Yuanmingyuan Xi Road No. 2, Haidian District, Beijing 100193, China
Full list of author information is available at the end of the article

becomes a major problem for almost all wheat production areas in temperate regions [14, 15] and developing heat-tolerant cultivars has become an important objective for breeders. Extensive research into the heat tolerance of wheat during grain filling revealed that kernel weight was more suited for screening than other traits evaluated [16–19]. Therefore, identification of DNA markers associated with kernel weight under post-anthesis high temperature stress would allow marker-assisted selection (MAS) and increase the efficiency for improving yield potential during breeding. Despite its importance, only a few QTL mapping studies of kernel weight have focused on heat stress.

A powerful approach for deciphering the biological functions of genes is to produce mutants with altered phenotypes or physiological responses. In wheat, chemical and ionizing radiation mutagenesis have been universally used to generate genetic variations for breeding researches and genetic studies. In total, 274 mutant varieties of wheat were developed by physical or chemical mutagens from 1930 to 2014 (http://www-naweb.iaea.org/nafa/pbg/). As a mutant induced by gamma radiation from Shi4185, Fu4185 showed higher kernel weight. This study aimed to analyze the genetic differences between Fu4185 and Shi4185 by SSR markers and to identify the genomic regions responsible for kernel weight in segregating populations derived from Shi4185 and Fu4185 under timely and late-sown conditions at different locations. This study will contribute to a better understanding of the genome-wide genetic variation and the stability of the QTL for kernel weight under heat stress.

Methods

Plant materials and field experiment

Shi4185 is an elite wheat cultivar in the Northern China winter wheat region. Fu4185 is a gamma radiation-induced mutant of advanced generations (M8), which was donated by Dr. Fengwu Zhao (Hebei Academy of Agricultural Sciences, China). F_1 seeds were generated by self-pollinating of the F_1 progeny of Shi4185 and Fu4185 in 2010. A total of 249 F_2 individuals and the two parental lines were planted in the experimental station at Shangzhuang, Beijing (40°06′N, 116°11′E) in the autumn of 2011 (BJ-2011), and their derived $F_{2:3}$ lines were grown at Gaoyi, Hebei (37°37′N, 114°35′E) (HB-2012) and Linfen, Shanxi (36°05′N, 111°30′E) (SX-2012) in the autumn of 2012 with three replicates. For spring sowing, the seeds of F_2 populations were surface sterilized with 70 % ethanol for 30 s and 10 % NaClO for 10 min, and vernalized for 40 days in Petri dishes on wet filter paper at 4 °C in the dark after germination. After that, the seeds of two F_2 populations were planted in BJ on February 17, 2014 (BJ-2014) and in SX on March 1, 2014 (SX-2014), respectively. The F_2 seeds

were space-planted (7.5 cm between plants), and each $F_{2:3}$ line was hand-sown using a randomized complete block design in two-row plots of 1.5 m long row with 0.3 m spacing between the rows.

Trait evaluation

Data of yield and yield contributing traits (plant height, spike length, spikelet number per spike, thousand kernel weight) of Shi4185 and Fu4185 that were planted at BJ in the autumn of 2011 were recorded using 30 plants (10 random plants from each replication) before harvest. After harvest, TKW of F_2 individuals were determined. For the 249 $F_{2:3}$ lines, bulked seeds of 10 random plants from the same line per replication were measured for TKW using a camera-assisted phenotyping system. This system was provided by Hangzhou Wanshen Detection Technology Co., Ltd. (Hangzhou, China).

DNA extraction and SSR marker analysis

Genomic DNA was extracted from seedling young leaves using the protocol as described by Sharp et al. [20]. The DNA was precipitated again with isopropanol, washed twice with 70 % ethanol and dissolved in TE buffer. DNA quality was checked using 1 % agarose to make sure no noticeable degradation.

The parental lines were screened with a total of 2019 SSR markers (*gwm, wmc, barc, cfd, ksm, gdm and pk*). Primer sequences for most SSR markers are available at http://wheat.pw.usda.gov/GG2/index.shtml. Primer sequences for SSR markers designed in our lab are partially listed in Table 1. PCR amplifications were carried out in a 10 μL volume containing 40 ng genomic DNA, 1 μL 10× reaction buffer, 0.2 μL 10 mmol L^{-1} dNTPs, 1.0 μL primer, 1 U rTaq DNA polymerase (Takara, Dalian) and 5.7 μL ddH$_2$O. The PCR was performed by initially denaturing the template DNA at 94 °C for 5 min, followed by 35 cycles at 94 °C for 45 s, 55 °C for 30 s, and 72 °C for 30 s, then terminated by a final extension for 10 min at 72 °C. PCR fragments were separated on 8 % non-denatured polyacrylamide gel electrophoresis (PAGE) and visualized by silver staining according to Marklund, Chaudhary et al. [21].

Marker development and QTL analysis

International Wheat Genome Sequencing Consortium (IWGSC) (http://www.wheatgenome.org/) has published survey sequence assemblies of the 21 individual chromosomes of Chinese Spring (*Triticum aestivum* L.). Firstly, the SSR markers were developed by using the survey sequence assembly of chromosome 5DL. SSR markers were designed under the conditions that dinucleotide and trinucleotide repeats were more than 30 and 21 times, respectively, in the SSR-flanking regions. Secondly, *Aegilops tauschii* is the diploid progenitor of the D-genome of

Table 1 Polymorphic SSR markers for genetic map construction

Markers	Forward primer (5'–3')	Reverse primer (5'–3')	Annealing temperature (°C)
Xcau1022	CCTAACCATCCAACCATAAGT	TTTCACGTACTCAAAAGTGG	55
Xcau1053	GGATGTACATTGAACAGTGCT	CCCACCCTACTCACTTCAA	55
Xcau1074	ACCTAAAATCTTCCCCCTACT	GCCAATTAATGCAGACTAGC	55
Xcau1087	GGAAAATCATGCACACATGG	CCCCACCCTACTCACTTCAA	58
Xcau1118	TGTAATCCGTCTCCTACCTTAT	GTCATACATGTCATCGGACTAC	55
Xcau1132	AGTCAATGAACAGAGCCATC	CTTGACCTAGATAGGGAAACAA	55

allohexaploid wheat, and its genome is an invaluable reference for wheat genomics [22]. Recently, the single nucleotide polymorphism (SNP) genetic map of *Aegilops tauschii*, was constructed, and the collinearity between *Brachypodium*, rice, and sorghum was established [23]. Therefore, comparative genomics approaches were employed to design SSR markers using the draft genome sequences, the extended SNP marker sequences and BAC scaffolds of *Aegilops tauschii* accession AL8/78. Information of polymorphic SSR markers for genetic map construction was listed in Table 1.

Linkage analysis of the marker was performed using JoinMap 4.0 software (http://www.kyazma.nl/index.php/mc.JoinMap/sc.General). According to the linkage distance between markers, a linkage map was drawn using MapChart v2.2 software (http://www.biometris.nl). Single marker analysis and QTL analysis by Composite interval mapping (CIM) method were performed by WinQTL Cart 2.5 software package [24]. The logarithm of the odds (LOD) threshold scores were calculated using 1,000 permutations [25], [26]. A QTL was declared when the LOD score was greater than 2.0.

Results

Phenotypic analysis

To characterize the phenotypic difference between Shi4185 and Fu4185, these two genotypes were firstly planted in BJ in the autumn of 2011 for analysis. As shown in Fig. 1, Shi4185 and Fu4185 showed significant differences in spike length and kernel size, whereas no difference was detected for plant height and spikelet number per spike. As expected, the TKW of Fu4185 was higher than that of Shi4185, which exhibited a good reproducibility in different environments (Table 2). In addition, there was a wide range of variations for both F_2 and $F_{2:3}$ populations, with coefficients of variation (CVs) ranging from 9.59 to 11.86 % in the F_2 populations and from 4.0 to 4.87 % in the $F_{2:3}$ populations, respectively. All populations showed continuous distributions with transgressive segregation on both sides for TKW (Additional file 1). Notably, the calculated broad-

Fig. 1 Phenotypic comparison of Shi4185 and Fu4185 planted at BJ-2011. **a** Plants at the filling stage; **b** Spikes at the filling stage; **c** Kernels after harvest

Table 2 TKW (g) of Shi4185 and Fu4185 in different environments

Environment	Sowing season	Shi4185	Fu4185
BJ-2011	Autumn	36.01 ± 0.86	42.82 ± 0.75**
HB-2012	Autumn	33.51 ± 1.29	37.14 ± 1.10*
SX-2012	Autumn	32.23 ± 1.26	39.44 ± 0.70**
BJ-2014	Spring	33.76 ± 1.09	38.79 ± 1.20**
SX-2014	Spring	29.90 ± 1.10	36.61 ± 1.41**

*Indicates a significance level of $P \leq 0.05$; **Indicates a significance level of $P \leq 0.01$. All data are given as mean ± SD

sense heritability (h_B^2) of TKW based on the data of $F_{2:3}$ populations in HB-2012 and SX-2012 were 88.5 % and 86.9 %, respectively, suggesting that TKW is suitable for QTL mapping (Table 3).

Genetic variations between Shi4185 and Fu4185

To investigate the genetic variations between Shi4185 and Fu4185, a total of 2019 SSR markers were used for analysis and 110 (5.4 %) markers were found to be polymorphic between Shi4185 and Fu4185. Two types of polymorphism were observed: presence or absence of the fragment (Type 1), and length differences of the fragment (Type 2). The numbers of the markers belong to Type 1 and 2 were 34 (30.9 %) and 76 (69.1 %), respectively. In addition, among the 110 polymorphic markers, 105 were located on 20 wheat chromosomes, with the exception of 4B. The numbers of polymorphic markers on each chromosome ranged from 1 (1A) to 14 (2B).

Single-marker analysis

To identify molecular markers associated with TKW, a total of 249 F_2 individuals, planted at BJ in the autumn of 2011, were genotyped using 58 polymorphic SSR markers. After the genotypic data of the F_2 population was merged with the phenotypic data of the F_2 and $F_{2:3}$ populations, respectively, single-marker analyses were conducted using WinQTLCart version 2.5. The results showed that marker Xbarc239 was significantly associated with TKW in both the F_2 and the $F_{2:3}$ populations at two different locations, whereas marker Xgwm68 was

only significantly linked with TKW in the $F_{2:3}$ populations (Table 4).

The genetic distance between Xbarc239 and Xgwm68 was 36.6 cM. Previous studies reported that Xbarc239 was located on chromosome 5DL [27]. To verify the chromosome location of Xbarc239, the amplified products of the parental lines were cloned into the pEASY-blunt Simple Cloning Vector (Transgen) for sequencing and blasted against the wheat chromosome survey sequence database (http://www.wheatgenome.org/). The results showed that the amplified products exhibited much higher similarity (98.6 %) to chr5DL_ab_k71_contigs_4520158 as compared with chr5AL_ab_k95_contigs_l_2810952 and chr5BL_ab_k71_contigs_10879865 (56.1 % and 66.7 %, respectively), suggesting that Xbarc239 was located on chromosome 5DL (Fig. 2, Additional file 2).

Genetic linkage map construction of chromosome 5DL and QTL mapping

To further explore polymorphic markers within the QTL region of TKW on chromosome 5DL, 137 SSR markers were developed using the wheat survey assembly sequence and the draft genome sequences of Aegilops tauschii, among which 26 markers showed polymorphism between Shi4185 and Fu4185. These polymorphic markers were used for linkage map construction by genotyping 249 individuals of the F_2 population (BJ-2011). Finally, one genetic linkage map of chromosome 5DL was constructed, including 8 SSR markers and spanning a total length of 84.96 cM (Fig. 3).

For QTL mapping, the genotype data of F_2 population were merged with the phenotype data of both F_2 and two $F_{2:3}$ populations, respectively. Though composite interval mapping, one QTL for TKW (QTkw.cau-5D) was detected besides Xbarc239, with Fu4185 contributing favorable alleles (Fig. 3). For F_2 population, the Fu4185 allele at QTkw.cau-5D increased TKW by 2.65 g, explaining 10.61 % of the phenotypic variation. For the two $F_{2:3}$ populations, the Fu4185 allele at QTkw.cau-5D increased TKW by 0.45 to 0.56 g, explaining 1.58 to 2.94 % of the phenotypic variation (Table 5).

To further assess the environmental stability of QTkw.cau-5D, another two F_2 populations derived from Fu4185 and Shi4185 were grown at two locations during the spring season of 2014. Four SSR markers (Xcau1053, Xcau1087, Xbarc239 and Xcau1118) were used to genotype the two populations. Genetic linkage map construction and composite interval mapping were carried out for each F_2 populations. QTL mapping analysis showed that QTkw.cau-5D could be detected in both the two populations, with Fu4185 contributing favorable alleles. The additive effects ranged from 1.43 to 1.70 g, explaining 9.51 to 14.87 % of the phenotypic variation (Table 5).

Table 3 TKW (g) of F_2 and $F_{2:3}$ populations in different environments

Environment	Population	Range	Mean	SD	h_B^2 (%)
BJ-2011	F_2	25.80–54.43	40.83	4.84	-
HB-2012	$F_{2:3}$	32.85–40.59	36.24	1.45	86.9
SX-2012	$F_{2:3}$	31.56–39.85	35.83	1.75	88.4
BJ-2014	F_2	20.55–47.07	37.56	3.67	-
SX-2014	F_2	23.64–44.05	34.60	3.32	-

-Indicates no data

Table 4 Single marker analysis of TKW in F_2 and $F_{2:3}$ populations

Environment	Population	marker	b0	b1	−2ln (L0/L1)	F (1, n-2)	Pr (F)
BJ-2011	F_2	*Xbarc239*	40.94	−2.0	21.27	22.03	0.000 ****
HB-2012	$F_{2:3}$	*Xbarc239*	36.28	−0.68	27.55	28.90	0.000****
		Xgwm68	36.26	−0.74	30.89	32.63	0.000****
SX-2012	$F_{2:3}$	*Xbarc239*	35.87	−0.73	21.68	22.47	0.000****
		Xgwm68	35.84	−0.55	11.33	11.49	0.001***

***Indicates a significance level of $P \leq 0.001$; ****Indicates a significance level of $P \leq 0.0001$

Discussions

Frequency, pattern and chromosome location of mutation at microsatellite loci between Shi4185 and Fu4185

Microsatellites or SSRs are composed of tandemly repeated, simple DNA sequence motifs of 1–6 nucleotide bases in length. Microsatellite markers are important tools for plant breeding, genetic and evolution studies [28]. A number of studies in animals and plants have calculated the natural mutation rate of microsatellites, and showed that the mutation rate varies greatly among species, ranging from 5×10^{-6} in Drosophila [29–31] to 10^{-3} in human [32, 33]. Gamma radiation can cause different types of mutations such as deletions, insertions, inversions and single-base substitution [34], but few studies have analyzed the frequency, polymorphism and pattern of microsatellite between the mutants and wild types in plants. In this study, SSR markers were employed to calculate the mutation frequencies induced by gamma radiation. Among the 2019 SSR markers, 110 (5.4 %) showed polymorphism between Shi4185 and Fu4185, which is lower than the percentage of polymorphism among natural allohexaploid wheat genotypes [35], but higher than the natural mutation rate of microsatellites per generation in different species [36]. In addition, two types of polymorphism were observed for SSR markers, including presence or absence of fragments (type 1) and fragment length differences for SSR markers (type 2). Multiple factors contribute to the observed variations. Type 1 might be produced by deletions, insertions,

inversions and point mutations in the primer binding regions, whereas type 2 might be caused by insertion or deletion of bases in the SSR repeat regions. Consistent with this hypothesis, we sequenced marker *Xbarc239* and this mutation proved to be a change in the number of TTA repeats.

The knowledge of chromosome location for polymorphic SSR markers between Shi4185 and Fu4185 is critical for the mutation research, because it could enable us to relate the SSR loci with the altered phenotypes. Our data showed that 105 of 110 polymorphic SSR markers were distributed on 20 chromosomes (except 4B). The numbers of polymorphic markers on each chromosome ranged from 1 (1A) to 14 (2B), suggesting that mutation frequency are not uniform across the genome. Although gamma radiation induced a large number of mutations on the genome, only a few agronomical traits showed differences between Fu4185and Shi4185. Moreover, only 2 of 110 polymorphic SSR markers were significantly associated with TKW. Collectively, we speculated that most mutation events may not result in noticeable phenotypic changes, indicating that the wheat genome could largely bear the mutations.

Hexaploid common wheat (*Triticum aestivum* L.; genome AABBDD) evolved by natural hybridization of emmer wheat (*Triticum turgidum* L.; genome AABB) and *Aegilops tauschii* Coss. (genome DD) [37, 38]. Only a few *Aegilops tauschii*'s intraspecific lineages contributed to the evolution of common wheat, which resulted in relatively narrow genetic variation on the D-genome in wheat [39–42].

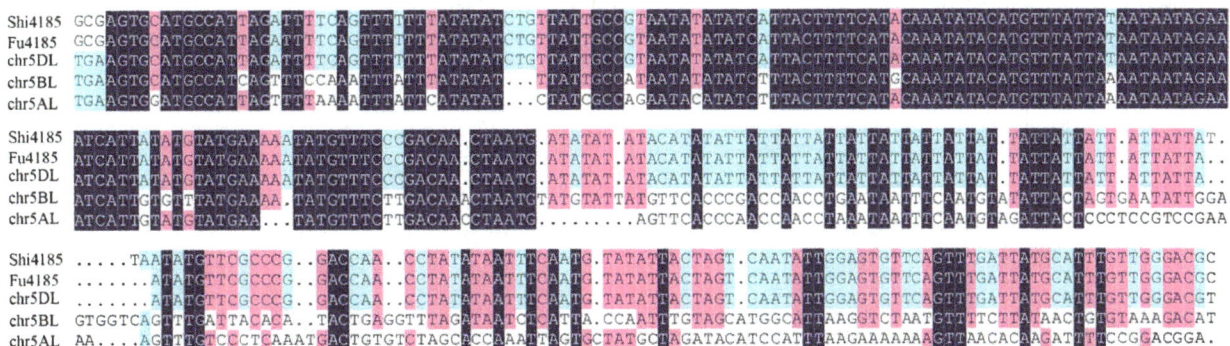

Fig. 2 Sequence alignment of *Xbarc239* of Shi4185 and Fu4185 with wheat chromosome survey sequence

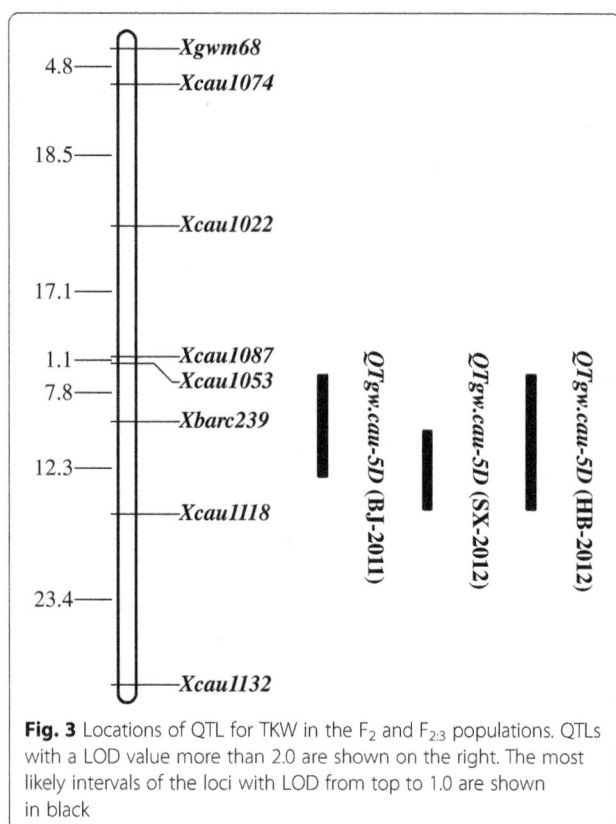

Fig. 3 Locations of QTL for TKW in the F$_2$ and F$_{2:3}$ populations. QTLs with a LOD value more than 2.0 are shown on the right. The most likely intervals of the loci with LOD from top to 1.0 are shown in black

weight in common wheat cultivars and the QTLs were assigned to various chromosomes [4–9]. For example, Simmonds et al. [45] reported that the effect of yield QTL, located on chromosome 6A, was driven primarily by increased kernel weight due to wider grains, indicating that the enhancement of kernel weight by MAS may benefit the genetic improvement of wheat yield. In this study, QTL mapping for TKW was conducted using F$_2$ and F$_{2:3}$ populations derived from Shi4185 and Fu4185. One major QTL (*QTkw.cau-5D*), with Fu4185 contributing favorable alleles, was consistently detected on chromosome 5DL under different environments, which was linked to SSR marker *Xbarc239*. However, phenotypic variation explained by *QTkw.cau-5D* was not very stable under different environments, with a range of 1.58 to 14.87 %. Moreover, the effect of *QTkw.cau-5D* in F$_{2:3}$ were smaller than in F$_2$ populations. These results may be partially explained by the following reasons: 1) Effects of other QTLs that were not detected; 2) Inaccurate phenotypic data of each genotype in F$_2$ populations for quantitative trait; 3) Impact of severe weather on TKW of F$_{2:3}$ populations. According to the weather record, we speculated that excessive rainfall at the late stage of wheat grain filling in HB-2012 and SX-2012 may result in the smaller effect of *QTkw.cau-5D* in F$_{2:3}$ as compared to F$_2$ populations.

Heat stress is one of key abiotic stress affecting wheat production in temperate regions [14, 15]. In traditional breeding, membrane thermostability and chlorophyll fluorescence were used as indicators of heat-stress tolerance in wheat, as they showed strong genetic correlation with grain yield, but the method was a time and labor-consuming process [46]. Mapping QTLs linked to heat stress tolerance traits will help to develop wheat cultivars suitable for high-temperature environments through MAS strategy. Recently, different traits like grain filling duration (GFD), thousand kernel weight (TKW) and yield have proved to be the preferable criteria to screen for heat-tolerant wheat in fields and many QTLs with significant effects on heat tolerance have been detected [17, 18, 47]. QTLs can be categorized according to the stability of their effects across environmental conditions: a "constitutive" QTL is consistently detected across most

Consistently, the average genetic diversity of D-genome in bread wheat was lower than that of *Aegilops tauschii* [43]. Interestingly, we found that 33 of 110 polymorphic SSR markers were located on D-genome. Moreover, the QTL of kernel weight was mapped on chromosome 5DL, with Fu4185 contributing favorable alleles. Taken together, we proposed that mutants induced by gamma radiation could be a potential resource to enrich the genetic diversity of wheat D-genome.

Environmental stability of QTL on chromosome 5DL controlling wheat kernel weight

Kernel weight is a complex quantitative and important agronomic trait. Mapping QTLs is the preliminary work for the genetic improvement by MAS [44]. To date, many studies have identified QTLs controlling kernel

Table 5 Additive effects of QTL for TKW in F$_2$ and F$_{2:3}$ populations

Environment	Population	Marker interval[a]	Position (cM)[b]	LOD	Additive effect[c]	Contribution (%)
BJ-2011	F$_2$	*Xcau1053-Xbarc239*	6.01	5.23	−2.65	10.61
HB-2012	F$_{2:3}$	*Xcau1053-Xbarc239*	7.81	2.28	−0.45	2.94
SX-2012	F$_{2:3}$	*Xbarc239-Xbarc1118*	5.98	2.27	−0.56	1.58
BJ-2014	F$_2$	*Xcau1087-Xbarc239*	5.05	4.97	−1.70	9.51
SX-2014	F$_2$	*Xcau1053-Xbarc239*	0.00	2.58	−1.43	14.87

[a]Marker interval is the interval containing the significant peak value of the QTL
[b]Position means the distance of the significant peak value for the QTL from the first marker in the marker interval
[c]A positive value indicates the Fu4185 allele having a positive effect on TKW, and negative value indicates Shi4185 allele having positive effect on TKW

Table 6 List of environment, flowering date, maturity date and No. of days for different daily maximum temperature [a] (Tmax) during grain filling

Environment	Flowering date	Maturity date	No. of days for different Tmax during grain filling			
			<30 °C	30–34 °C	>35 °C	Sum
BJ-2011	May 5, 2012	June 13, 2012	24	16	0	40
HB-2012	May 1, 2013	June 5, 2013	21	12	2	35
SX-2012	May 5, 2013	June 9, 2013	18	13	5	36
BJ-2014	May 11, 2014	June 20, 2014	16	21	4	41
SX-2014	May 15, 2014	June 25, 2014	9	29	4	42

[a]Data of Tmax from individual year of each environment (http://www.tianqi.com/)

environments; an "adaptive" QTL is detected only under specific environmental conditions [48, 49]. An important prerequisite for a successful MAS program aimed at improving heat tolerance is the identification of the "constitutive" QTLs. Recently, timely and late-sown were used for mapping QTLs associated with heat tolerance in wheat and two "constitutive" QTLs were detected on chromosome 2B and 7B [18]. Applying this method, we tested the effects of *QTkw.cau-5D* under different environments. Although the number of the days with the maximum temperature above 30 °C from flowering to harvesting at spring sowing season was much more than that at autumn sowing season (Table 6), *QTkw.cau-5D* was also detected under high temperature environments, which provide

further evidence that it is an environmental stable QTL. However, to determine the accurate effect of *QTkw.cau-5D* under heat stress, it is better to compare spring and autumn planted trials in the same year, which deserved for further investigation.

In this study, the marker *Xbarc239* linked to *QTkw.cau-5D* was physically located to the bin of 5DL1-0.60-0.74 using Chinese Spring deletion lines [27]. When comparing previous results [50–53], we found that *QTkw.cau-5D* was located to the same bin as the QTL for TKW between *Xswes340a* and *Xswes342b* (*Xgwm174*) [51] (Fig. 4). Although further analysis was needed to clarify the relationship of these two QTLs, *QTkw.cau-5D* detected in our study may be a more desired target for genetic

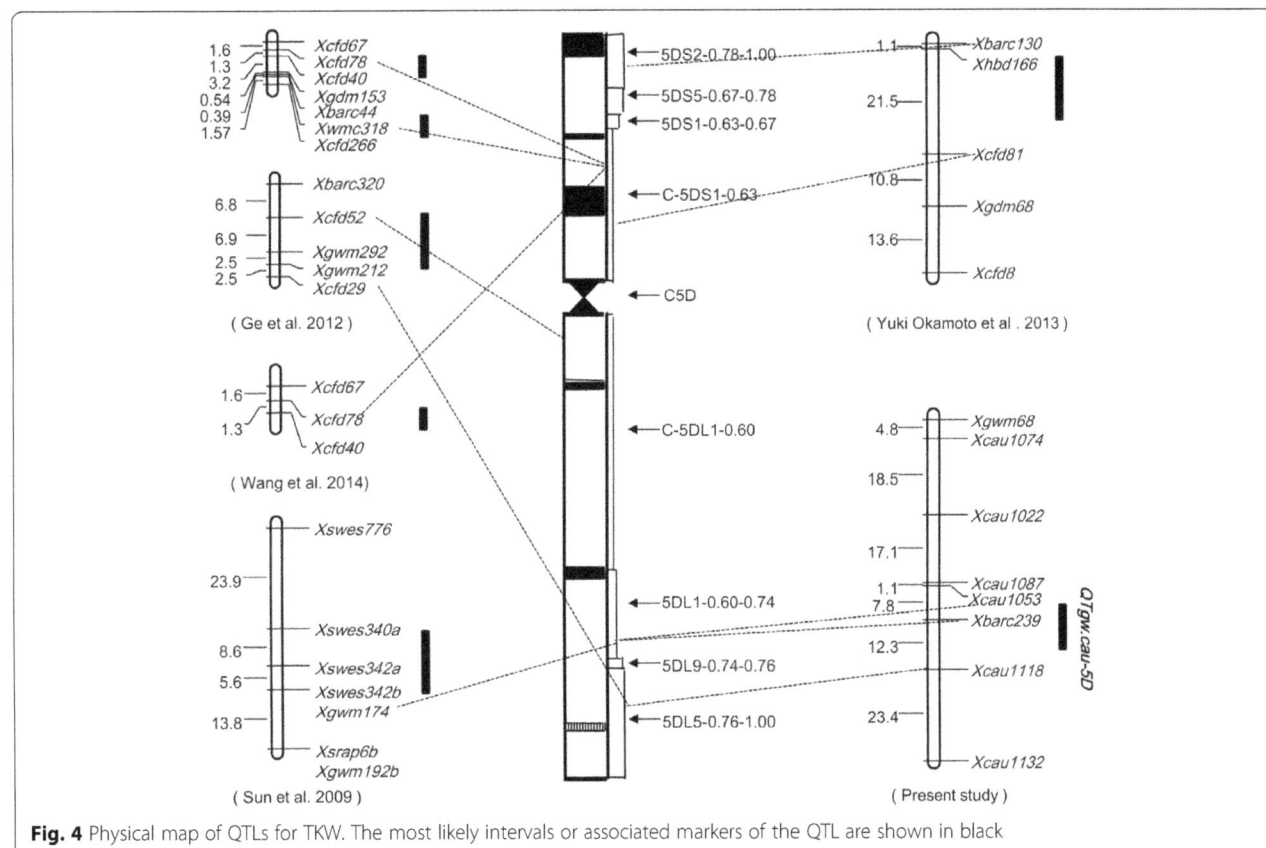

Fig. 4 Physical map of QTLs for TKW. The most likely intervals or associated markers of the QTL are shown in black

improvement of wheat kernel weight. Firstly, *QTkw.cau-5D* is an environmental stable QTL, whereas the QTL on chromosome 5D reported by Sun *et al.* [51] was only detected in 2 of 4 experimental environments; Secondly, the donor parent of Fu4185 (Shi4185) is an elite wheat cultivar in the Northern China winter wheat region. Moreover, a number of new cultivars derived from Shi4185 have been released in China, such as Henong130, Baomai10 and Nongda399. Theoretically, Fu4185 is a potential elite parent for genetic improvement of kernel weight in wheat breeding; Finally, considering the severe effect of heat stress on kernel weight in wheat production, our identified QTL associated with kernel weight under high temperature stress may increase the efficiency for improving wheat yield potential by MAS.

Conclusions
One favorable and environmental stable QTL allele on chromosome 5DL controlling kernel weight was identified in Fu4185, a mutant of an elite wheat cultivar Shi4185 induced by gamma radiation. Furthermore, 30 % (33 of 110) polymorphic SSR markers between Shi4185 and Fu4185 were located on the D-genome. Taken together, these data revealed that mutations induced by gamma radiation could be a potential resource to enrich the genetic diversity of wheat D-genome.

Abbreviations
QTL: Quantitative trait locus; TKW: Thousand kernel weight; SSR: Simple sequence repeats; MAS: Marker-assisted selection; PAGE: Polyacrylamide gel electrophoresis; IWGSC: International wheat genome sequencing consortium; SNP: Single nucleotide polymorphism; CIM: Composite interval mapping; LOD: Logarithm of the odds; CVs: Coefficients of variation; GFD: Grain filling duration; h_B^2: Broad-sense heritability; Tmax: Daily maximum temperature.

Competing interests
The authors declare that they have no conflict of interests.

Authors' contributions
ZFN, QXS and XJC conceived the project; XJC conducted the field evaluations and TKW phenotyping with assistances from HJZ, AJZ and LX; XJC carried out genotyping and sequencing experiments with assistances from LLC and ZYC. XJC analyzed the data with assistances from HRP, YYY and MSY; XJC and ZFN wrote the manuscript. All authors read and approved the final manuscript.

Acknowledgments
We thank Dr. Fengwu Zhao for providing Fu4185. This work was supported by the National Natural Science Foundation of China (91435204 and 31271710), the 863 Project Grant (2012AA10A309).

Author details
[1]State Key Laboratory for Agrobiotechnology and Key Laboratory of Crop Heterosis and Utilization (MOE) and Key Laboratory of Crop Genomics and Genetic Improvement (MOA), Beijing Key Laboratory of Crop Genetic Improvement, China Agricultural University, Yuanmingyuan Xi Road No. 2, Haidian District, Beijing 100193, China. [2]Department of Plant Genetics and Breeding, China Agricultural University, Beijing 100193, China. [3]National Plant Gene Research Centre, Beijing 100193, China. [4]Hebei Crop Genetic Breeding Laboratory Institute of Cereal and Oil Crops, Hebei Academy of Agriculture and Forestry Sciences, Shijiazhuang 050035, China.

References
1. Smith R. Increasing wheat yield requires genetics, management. Southwest Farm Press. http://southwestfarmpress.com/wheat/increasing-wheat-yield-requires-genetics-management. Accessed 6 February 2014.
2. Peng JH, Ronin Y, Fahima T, Roder MS, Li YC, Nevo E, et al. Domestication quantitative trait loci in Triticum dicoccoides, the progenitor of wheat. Proc Natl Acad Sci U S A. 2003;100:2489–94.
3. Wang LF, Ge HM, Hao CY, Dong YS, Zhang XY. Identifying loci influencing 1,000-kernel weight in wheat by microsatellite screening for evidence of selection during breeding. PLoS One. 2012;7(2):e29432.
4. Campbell KG, Bergmem CJ, Gualberto DG, Anderson JA, Giroux MJ, Hareland G, et al. Quantitative trait loci associated with kernel traits in a soft × hard wheat cross. Crop Sci. 1999;39:1184–95.
5. Campbell BT, Baenziger PS, Gill KS, Eskridge KM, Budak H, Erayman M, et al. Identification of QTLs and environmental interactions associated with agronomic traits on chromosome 3A wheat. Crop Sci. 2003;43:1493–505.
6. Kato K, Miura H, Sawada S. Mapping QTLs controlling grain yield and its components on chromosome 5A of wheat. Theor Appl Genet. 2000;101:1114–21.
7. Varshney RK, Prasad M, Roy JK, Kumar N, Singh H, Dhaliwal HS, et al. Identification of eight chromosomes and one microsatellite marker on 1AS associated with QTL for grain weight in bread wheat. Theor Appl Genet. 2000;100:1290–5.
8. Groos C, Robert N, Bervas E, Charmet G. Genetic analysis of grain protein-content, grain yield and thousand-kernel weight in bread wheat. Theor Appl Genet. 2003;106:1032–40.
9. Kumar N, Kulwal PL, Gaur A, Tyagi AK, Khurana JP, Khurana P, et al. QTL analysis for grain weight in common wheat. Euphytica. 2006;151:135–44.
10. Jiang QY, Hou J, Hao CY, Wang LF, Ge HM, Dong YS, et al. The wheat (*T. aestivum*) sucrose synthase 2 gene (*TaSus2*) active in endosperm development is associated with yield traits. Funct Integr Genomics. 2001;11(1):49–61.
11. Su ZQ, Hao CY, Wang LF, Dong YC, Zhang XY. Identification and development of a functional marker of *TaGW2* associated with grain weight in bread wheat (*Triticum aestivum* L.). Theor Appl Genet. 2011;122(1):211–23.
12. Zhang L, Zhao YL, Gao LF, Zhao GY, Zhou RH, Zhang BS, et al. *TaCKX6-D1*, the ortholog of rice *OsCKX2*, is associated with grain weight in hexaploid wheat. New Phytol. 2012;195(3):574–84.
13. Asseng S, Ewert F, Martre P, Rötter RP, Lobell DB, Cammarano D, et al. Rising temperatures reduce global wheat production. Nat Clim Change. 2015;5:143–7.
14. Reynolds M, Balota M, Delgado M, Amani I, Fischer R. Physiological and morphological traits associated with spring wheat yield under hot, irrigated conditions. Funct Plant Biol. 1994;21(6):717–30.
15. Wardlaw I, Wrigley C. Heat tolerance in temperate cereals: an overview. Funct Plant Biol. 1994;21(6):695–703.
16. Sharma RC, Tiwari AK, Ortiz-Ferrara G. Reduction in kernel weight as a potential indirect selection criterion for wheat grain yield under heat stress. Plant Breed. 2008;127:241–8.
17. Pinto RS, Reynolds MP, McIntyre CL, Olivares-Villegas JJ, Chapman SC. Heat and drought QTL in a wheat population designed to minimize confounding agronomic effects. Theor Appl Genet. 2010;121:1001–21.
18. Paliwal R, Röder MS, Kumar U, Srivastava JP, Joshi AK. QTL mapping of terminal heat tolerance in hexaploid wheat (*T. aestivum* L.). Theor Appl Genet. 2012;125(3):561–75.
19. Pandey GC, Sareen S, Siwach P, Tiwari R. Molecular characterization of heat tolerance in bread wheat (*Triticum aestivum* L.) using differences in thousand-grain weights (dTGW) as a potential indirect selection criterion. Cereal Res Commun. 2014;1:38–46.
20. Sharp PJ, Kreis M, Shewry PR, Gale MD. Location of β-amylase sequences in wheat and its relatives. Theor Appl Genet. 1988;75(2):286–90.

21. Marklund S, Chaudhary R, Marklund L, Sandberg K, Andersson L. Extensive mtDNA diversity in horses revealed by PCR-SSCP analysis. Anim Genet. 1995;26:193–6.

22. Brenchley R, Spannagl M, Pfeifer M, Barker GL, D'Amore R, Allen AM, et al. Analysis of the bread wheat genome using whole-genome shotgun sequencing. Nature. 2012;491(7426):705–10.

23. Luo MC, Gu YQ, You FM, Deal KR, Ma Y, Hu Y, et al. A 4-gigabase physical map unlocks the structure and evolution of the complex genome of Aegilops tauschii, the wheat D-genome progenitor. Proc Natl Acad Sci U S A. 2013;110(19):7940–5.

24. Wang S, Basten CJ, Zeng ZB. Windows QTL cartographer 2.5. Department of Statistics, North Carolina State University, Raleigh, NC.2012. http:// statgen.ncsu.edu/qtlcart/WQTLCart.htm. Accessed 01 Aug 2012.

25. Churchill GA, Doerge RW. Empirical threshold values for quantitative trait mapping. Genetics. 1994;138(3):963–71.

26. Doerge RW, Churchill GA. Permutation tests for multiple loci affecting a quantitative character. Genetics. 1996;142:285–94.

27. Yang J, Bai G, Shaner GE. Novel quantitative trait loci (QTL) for Fusarium head blight resistance in wheat cultivar Chokwang. Theor Appl Genet. 2005;111(8):1571–9.

28. Vigouroux Y, Jaqueth JS, Matsuoka Y, Smith OS, Beavis WD, Smith JS, et al. Rate and pattern of mutation at microsatellite loci in maize. Mol Biol Evol. 2002;19(8):1251–60.

29. Schug MD, Mackay TF, Aquadro CF. Low mutation rates of microsatellite loci in Drosophila melanogaster. Nat Genet. 1997;15(1):99–102.

30. Schlötterer C, Ritter R, Harr B, Brem G. High mutation rate of a long microsatellite allele in Drosophila melanogaster provides evidence for allele-specific mutation rates. Mol Biol Evol. 1998;1510:1269–74.

31. Schug MD, Hutter CM, Wetterstrand KA, Gaudette MS, Mackay TF, Aquadro CF. The mutation rates of di-, tri- and tetranucleotide repeats in Drosophila melanogaster. Mol Biol Evol. 1998;15(12):1751–60.

32. Brinkmann B, Klintschar M, Neuhuber F, Hühne J, Rolf B. Mutation rate in human microsatellites: influence of the structure and length of the tandem repeat. Am J Hum Genet. 1998;62(6):1408–15.

33. Xu X, Peng M, Fang Z. The direction of microsatellite mutations is dependent upon allele length. Nat Genet. 2000;24(4):396–9.

34. Morita R, Kusaba M, Iida S, Yamaguchi H, Nishio T, Nishimura M. Molecular characterization of mutations induced by gamma irradiation in rice. Genes Genet Syst. 2009;84(5):361–70.

35. Somers DJ, Isaac P, Edwards K. A high-density microsatellite consensus map for bread wheat (Triticum aestivum L.). Theor Appl Genet. 2004;109(6):1105–14.

36. Lynch M. Evolution of the mutation rate. Trends Genet. 2010;26(8):345–52.

37. Nesbitt M, Samuel D. From staple crop to extinction? The archaeology and history of the hulled wheats. In S. Padulosi, K. Hammer, J. Heller, editors. In Proceedings of the First International Workshop on Hulled Wheats: 21–22 July 1995. Italy: Tuscany, Castelvecchio Pascoli; 1995. pp. 41–102.

38. Petersen G, Seberg O, Yde M, Berthelsen K. Phylogenetic relationships of Triticum and Aegilops and evidence for the origin of the A, B, and D genomes of common wheat (Triticum aestivum). Mol Phylogenet Evol. 2006;39:70–82.

39. Tsunewaki K. Comparative gene analysis of common wheat and its ancestral species. II. Waxiness, growth habit and awnedness. Jpn J Bot. 1966;19:175–229.

40. Dvorak J, Luo MC, Yang ZL, Zhang HB. The structure of the Aegilops tauschii genepool and the evolution of hexaploid wheat. Theor Appl Genet. 1998;97:657–70.

41. Feldman M, Origin of cultivated wheat. In: Bonjean AP, Angus WJ, editors. The world wheat book: a history of wheat breeding. Paris: rue Lavoisier; 2001. p. 3–53.

42. Dubcovsky J, Dvorak J. Genome plasticity a key factor in the success of polyploid wheat under domestication. Science. 2007;316:1862–6.

43. Naghavi MR, Aghaei MJ, Taleei AR, Omidi M, Mozafari J, Hassani ME. Genetic diversity of the D-genome in T. aestivum and Aegilops species using SSR markers. Genet Resour Crop Evol. 2009;56:499–506.

44. Montaldo HH, Meza-Herrera CA. Use of molecular markers and major genes in the genetic improvement of livestock. Electron J Biotechnol. 1998;2:15–6.

45. Simmonds J, Scott P, Leverington-Waite M, Turner AS, Brinton J, Korzun V, et al. Identification and independent validation of a stable yield and thousand grain weight QTL on chromosome 6A of hexaploid wheat (Triticum aestivum L.). BMC Plant Biol. 2014;14(1):191.

46. Talukder SK, Babar MA, Vijayalakshmi K, Poland J, Prasad PVV, Bowden R, et al. Mapping QTL for the traits associated with heat tolerance in wheat (Triticum aestivum L.). BMC Genet. 2014;15(1):97.

47. Yang J, Sears RG, Gill BS, Paulsen GM. Quantitative and molecular characterization of heat tolerance in hexaploid wheat. Euphytica. 2002;126:275–82.

48. Vargas M, van Eeuwijk FA, Crossa J, Ribaut JM. Mapping QTLs and QTL x environment interaction for CIMMYT maize drought stress program using factorial regression and partial least squares methods. Theor Appl Genet. 2006;112(6):1009–23.

49. Collins NC, Tardieu F, Tuberosa R. Quantitative trait loci and crop performance under abiotic stress: where do we stand? Plant Physiol. 2008;147(2):469–86.

50. Ge H, You G, Wang L, Hao CY, Dong YS, Li ZS, et al. Genome selection sweep and association analysis shed light on future breeding by design in wheat. Crop Sci. 2012;52(3):1218–28.

51. Sun XY, Wu K, Zhao Y, Kong FM, Han GZ, Jiang HM, et al. QTL analysis of kernel shape and weight using recombinant inbred lines in wheat. Euphytica. 2009;165(3):615–24.

52. Wang Y, Hao C, Zheng J, Ge HM, Zhou Y, Ma ZQ et al. A haplotype block associated with thousand-kernel weight on chromosome 5DS in common wheat (Triticum aestivum L.). J Integr Plant Biol. 2014;57(8):662-72.

53. Okamoto Y, Nguyen AT, Yoshioka M, Iehisa JC, Takumi S. Identification of quantitative trait loci controlling grain size and shape in the D genome of synthetic hexaploid wheat lines. Breed Sci. 2013;63(4):423–9.

20

The association of CD40 polymorphisms with CD40 serum levels and risk of systemic lupus erythematosus

Jian-Ming Chen[1†], Jing Guo[2†], Chuan-Dong Wei[1], Chun-Fang Wang[1], Hong-Cheng Luo[1], Ye-Sheng Wei[1*] and Yan Lan[2*]

Abstract

Background: Current evidence shows that the CD40–CD40 ligand (CD40–CD40L) system plays a crucial role in the development, progression and outcome of systemic lupus erythematosus (SLE). The aim of this study was to investigate whether a CD40 gene single nucleotide polymorphism (SNP) is associated with SLE and CD40 expression in the Chinese population. We included controls ($n = 220$) and patients with either SLE ($n = 205$) in the study.

Methods: The gene polymorphism was measured using Snapshot SNP genotyping assays and confirmed by sequencing. We analyzed three single nucleotide polymorphisms of CD40 gene rs1883832C/T, rs1569723A/C and rs4810485G/T in 205 patients with SLE and 220 age-and sex-matched controls. Soluble CD40 (sCD40) levels were measured by ELISA.

Results: There were significant differences in the genotype and allele frequencies of CD40 gene rs1883832C/T polymorphism between the group of patients with SLE and the control group ($P < 0.05$). sCD40 levels were increased in patients with SLE compared with controls ($P < 0.01$). Moreover, genotypes carrying the CD40 rs1883832 C/T variant allele were associated with increased CD40 levels compared to the homozygous wild-type genotype in patients with SLE. The rs1883832C/T polymorphism of CD40 and its sCD40 levels were associated with SLE in the Chinese population.

Conclusions: Our results suggest that CD40 gene may play a role in the development of SLE in the Chinese population.

Keywords: CD40, Gene, Polymorphism, SLE

Background

Systemic lupus erythematosus (SLE) a kind of chronic autoimmune disease, leading to multiple organ damage, has the characteristics of various autoantibodies production. Although that the etiology and pathogenesis of SLE is not clear, it maybe immune regulation disorder caused by a complex interplay of genetic and environmental factors, hormones, antigen antibody and complement complex deposits lead to local or systemic tissue or organ damage [1–4]. Among them, genetic factors seem to play a key role in the susceptibility to SLE. In the past several years genome-wide association studies (GWAS) for SLE have identified literally hundreds of genetic loci

involved in the susceptibility conferred to complex inherited traits [5–7]. Even though this scenario represents an extraordinary advance in complex disease genetics, the modest effect sizes of the common polymorphisms found associated explain only a small fraction of the heritability in most of these multifactorial conditions, suggesting that many more loci remain to be discovered [8, 9]. One of the genes encoding a member of the tumor necrosis factor receptor family that plays a key role in adaptive immunity of SLE is CD40 [10].

CD40, a member of the tumor necrosis family of transmembrane glycoproteins, was identified on B cells, monocytes, dendritic cells, endothelial and epithelial cells, which is rapidly and transiently expressed on the surface of recently activated CD4[+]T cells and is a potent T-cell costimulatory molecule [11–13]. Interactions between CD40 and CD40L induce B cell immunoglobulin production as well as monocyte activation and

---
* Correspondence: wysh22@163.com; yylanyan@163.com
†Equal contributors
[1]Department of Laboratory Medicine, Affiliated hospital of Youjiang Medical University for Nationalities, Baise 533000, Guangxi, China
[2]Department of Dermatology, Affiliated Hospital of Youjiang Medical University for Nationalities, Baise 533000, Guangxi, China

dendrite cell differentiation [14, 15]. Some authors have demonstrated that the multipotent immunomodulator CD40, expressed on vascular endothelial cells, smooth muscle cells, mononuclear phagocytes, and platelets, promote awide array of pro-atherogenic functions in vitro [16–19]. The gene encoding CD40 is located on chromosome 20q11-13 in humans, which consists of nine exons and eight Introns. Recently, a number of polymorphisms in the gene encoding CD40 gene have been identified and a relationship between the CD40 gene polymorphisms and risk of different autoimmune and inflammatory diseases, such as multiple autoimmune diseases, Graves' disease and rheumatoid arthritis has been reported [20–22]. However, very little data has examined the association between rs1883832C/T, rs1569723A/C and rs4810485G/T polymorphisms in CD40 gene and SLE. Furthermore, the relationship between the CD40 gene polymorphisms and the plasma level of CD40 gene is unknown. In this study, we investigated the relationship of CD40 gene rs1883832C/T, rs1569723A/C and rs4810485G/T polymorphisms and their CD40 serum levels in a Chinese population.

Methods

Study population

Our study was designed as a retrospective study. The study consisted of 205 patients with SLE (36 males and 169 females, aged between 30 and 82 years). All patients with SLE were consecutively selected. They were recruited from the Department of Dermatology, Affiliated Hospital of Youjiang Medical University for Nationalities, Guangxi, China between October 2014 and November 2015. The 220 control subjects were matched to the patients on the basis of age and gender (42 males and 178 females, aged between 29 and 78 years). The control subjects underwent a routine medical check-up in the outpatient clinic of the Department of Internal Medicine, Affiliated Hospital of Youjiang Medical University for Nationalities, Guangxi, China between May 2013 and November 2014. According to the thorough clinical and laboratory evaluation, none of them were found to have any medical condition other than hypertension, autoimmune and inflammatory diseases. All study subjects were Chinese and resided in the same geographic area in China. The study was performed with the approval of the ethics committee of the Affiliated Hospital of Youjiang Medical University for Nationalities, and written informed consent was obtained from all the subjects.

DNA extraction

Genomic DNA was extracted from EDTA-anticoagulated peripheral blood leukocytes by the salting-out method [23]. Briefly, 3 ml of blood was mixed with Triton lysis buffer (0.32 M sucrose, 1 % Triton X100, 5 mM $MgCl_2$,

H_2O, 10 mM Tris–HCl, pH 7.5). Leucocytes were spun down and washed with H_2O. The pellet was incubated with proteinase K at 56 °C and subsequently salted out at 4 °C using a saturated NaCl solution. Precipitated proteins were removed by centrifugation. The DNA in the supernatant fluid was dissolved in 300 μlH_2O.

Determination of CD40 genotype

The CD40 gene rs1883832 C/T, rs1569723 A/C and rs4810485 G/T genotypes were determined by using a Snapshot SNP genotyping assay. The PCR primers were designed based on the GenBank reference sequence (accession no. NC_000020.11) (Table 1). To confirm the genotyping results, PCR-amplified DNA samples were examined by DNA sequencing, and the results were 100 % concordant.

Plasma CD40 determination

Plasma samples from the patients and healthy controls were separated from venous blood at room temperature, and stored at −70 °C until use. The quantity determination of plasma CD40 levels was performed by enzyme-linked immunosorbent assay (ELISA) kits (Fermentas, Lithuania), following the manufacturer's protocol. Developed color reaction was measured as OD450 units on an ELISA reader (RT-6000, China). The concentration of plasma CD40 was determined by using standard curve constructed with the kit's standards over the range of 0–1000 pg/ml.

Statistical analysis

Genotype and allele frequencies of CD40 were compared between SLE cases and controls using the $\chi2$ test and Fisher's exact test when appropriate, and odds ratios (OR) and 95 % confidence intervals (CIs) were calculated to assess the relative risk conferred by a particular allele and genotype. Demographic and clinical data between

Table 1 The primer sequences used for detecting the different CD40 SNPs

Reference SNP ID	PCR primers
rs1883832C/T	F:5'-GGACCTGGGGGCAAAGAAGA-3'
	R: 5'- CCCACTCCCAACTCCCGTCT -3'
	EF:5'-TTTTTTTTTTTTGCAGAGGCAGAC GAACCAT -3'
rs1569723 A/C	F: 5'- GGGATG GCCTGATCCAAAGG -3'
	R: 5'- CCCACAGTCCACCACCCATC -3'
	EF:5'-TTTTTTTTTTTTTTTTTTTTTTTTTTT TTTTTCGCTTTACACCCACAGCC-3'
rs4810485 G/T	F: 5'- ATCCCCCAAGTACCTGGCTCCT -3'
	R: 5'- CCTTGCTGCTTCCC TTGCTTTC -3'
	EF:5'- TTTTTTTTTTTTTTTTTTTTTTTTTTT TCCTACTTTAGAG GGCTGTAGATTCC -3'

groups were compared by $\chi 2$ test and by Student's t-test. Hardy–Weinberg equilibrium was tested for with a goodness of fit $\chi 2$-test with 1 ° of freedom to compare the observed genotype frequencies among the subjects with the expected genotype frequencies. The linkage disequilibrium (LD) between the polymorphisms was quantified using the Shi's standardized coefficient D' (|D'|) [24]. The haplotypes and their frequencies were estimated based on a Bayesian algorithm using the Phase program [25]. Statistical significance was assumed at the $P < 0.05$ level. The SPSS statistical software package version 11.5 was used for all of the statistical analysis.

Results

Clinical characteristics of the study participants

There were no significant differences in the age ($P > 0.05$) and percentage of males/females ($P > 0.05$) between the two groups. The serum CD40 levels were significantly higher in the group of patients with SLE than those in the control group [(mean +/− SD 58.5 +/− 22.8 pg/ml, $n = 205$) vs, (mean +/− SD 41.7 +/− 13.2 pg/ml, $n = 220$); $P < 0.001$] (Fig. 1).

The genotype and allele frequencies of CD40 gene

The genotype and allele frequencies of the CD40 gene rs1883832 C/T, rs1569723 A/C and rs4810485 G/T polymorphisms in the group of patients with SLE and in the control group are shown in Table 2. The genotype distributions of the three polymorphisms among the controls and the cases were in Hardy–Weinberg equilibrium, and the Hardy–Weinberg equilibrium p-values of the CC, CT and TT genotypes of rs1883832 C/T, rs1569723 A/C

Table 2 The genotype and allele frequencies of CD40 polymorphism in SLE patients and controls

Polymorphism	Control subjects $n = 220$ (%)	SLE patients $n = 205$ (%)	χ^2	P value
rs1883832 C/T				
CC	79 (35.9)	47 (22.9)	9.504	0.009
CT	101 (45.9)	105 (51.2)		
TT	40 (18.2)	53 (25.9)		
C	259 (58.9)	199 (48.5)	9.109	0.003
T	181 (41.1)	211 (51.5)		
rs1569723 A/C				
AA	54 (24.5)	51 (24.9)	0.284	0.868
AC	105 (47.7)	93 (45.4)		
CC	61 (27.7)	61 (29.8)		
A	213 (48.4)	195 (47.6)	0.061	0.805
C	227 (51.6)	215 (52.4)		
rs4810485 G/T				
GG	56 (25.5)	52 (25.4)	0.341	0.843
GT	107 (48.6)	95 (46.3)		
TT	57 (25.9)	58 (28.3)		
G	219 (49.8)	199 (48.5)	0.130	0.719
T	221 (50.2)	211 (51.5)		

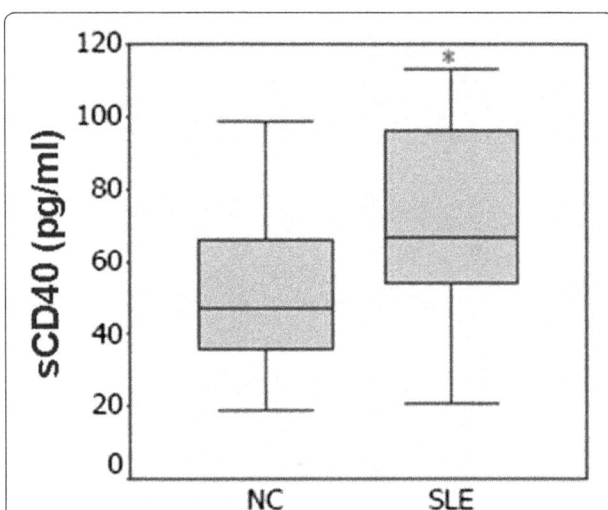

Fig. 1 The levels of CD40 in patients with SLE and normal control subjects. The expression of CD40 was significantly increased in patients with SLE compared to that in control subjects [(mean +/− SD 41.7+/− 13.2 pg/ml, $n = 205$) vs (mean +/− SD 58.5+/− 22.8 pg/ml, $n = 220$); $P < 0.001$]

and rs4810485 G/T were 0.440, 0.509 and 0.686 in controls, and were 0.718, 0.195 and 0.300 in cases, respectively. The frequencies of the CC, CT and TT genotypes of rs1883832 C/T were 35.9, 45.9 and 18.2 % in controls, and were 22.9, 51.2 and 25.9 % in cases, respectively. There were significant differences in the genotype and allele frequencies of the CD40 gene rs1883832 C/T polymorphism between the SLE and control groups. The rs1883832 T allele was associated with a significantly increased risk of SLE as compared with the rs1883832 C allele (OR = 1.517, 95 % CI, 1.157–1.990, $P = 0.003$). However, genotype and allele frequencies of the CD40 gene rs1569723 A/C and rs4810485 G/T polymorphisms in SLE patients were not significantly different than those in controls ($P > 0.05$).

Haplotype analysis of the CD40 gene

Haplotype analyses were performed and the possible six haplotype frequencies are shown in Table 3. Two major haplotypes (TCT and CAG) accounted for 51.5, 42.9 and 43.0, 47.0 % of these six haplotypes in both the cases and the controls, respectively. CD40 gene rs1883832 C/T polymorphism was in strong linkage disequilibrium with the rs1569723 A/C (|D'| = 0.867) and rs4810485 G/T (|D'| = 0.841). The rs1569723 A/C and rs4810485 G/T were in strong linkage disequilibrium (|D'| = 0.922). By haplotype analyses, we found T-C-T haplotype was associated with a significantly increased risk of SLE as

Table 3 Haplotype distribution in the patients with SLE and controls

CD40 gene (rs1883832/rs1569723/ rs4810485) haplotypes	SLE patients 2 n = 410 (%)	Controls 2 n = 440 (%)	OR (95 % CI)	P value
T-C-T	211 (51.5)	189 (43.0)	1.408 (1.074–1.845)	0.013
C-A-G	176 (42.9)	207 (47.0)	0.847 (0.646–1.110)	0.228
C-A-T	8 (2.0)	13 (3.0)	0.735 (0.253–2.138)	0.346
C-C-G	6 (1.5)	14 (3.2)	0.452 (0.172–1.187)	0.099
C-C-T	4 (1.0)	8 (1.8)	0.532 (0.159–1.780)	0.298
T-A-T	5 (1.2)	9 (2.0)	0.591 (0.196–1.779)	0.344

compared with the control group (OR = 1.408; 95 % CI, 1.074–1.845; $P = 0.013$).

Association between CD40 gene polymorphisms and sCD40 levels

Genotype at the rs1883832 C/T polymorphism was significantly associated with sCD40 levels in patients with SLE. The plasma CD40 levels were significantly higher in individuals with homozygous TT genotypes (62.6 +/− 23.3 pg/ml, $n = 53$) or heterozygous of CT genotypes (59.9 +/− 22.6 pg/ml, $n = 105$) than homozygous of CC genotypes (50.7 +/− 20.4 pg/ml, $n = 47$, $P < 0.01$, respectively). However, there were no significant differences in the plasma CD40 levels between TT and CT genotypes (Fig. 2). In addition, there were no significant associations of the CD40 rs1569723 A/C and rs4810485 G/T polymorphisms with plasma levels of CD40 (data not shown).

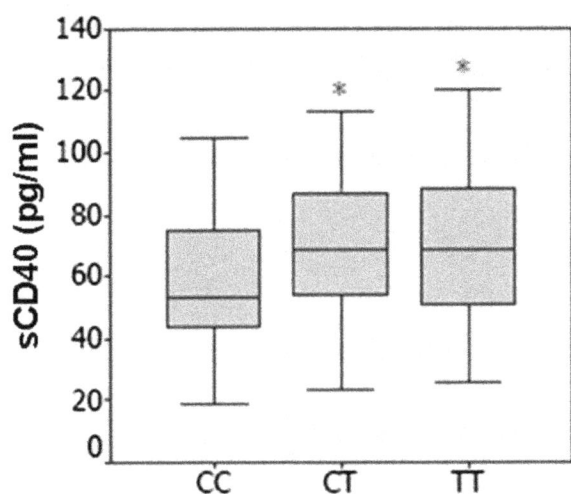

Fig. 2 Association between the levels of CD40 and the rs1883832 C/T polymorphism of CD40 gene was observed in patients with SLE. Plasma CD40 levels with CC homozygous were significantly lower than that of the TT homozygous or CT heterozygotes, respectively. However, there were no significant differences in the plasma CD40 levels between CT and TT genotypes

Discussion

CD40, the receptor for CD40L, is a 48-kDa transmembrane protein belonging to the TNF (tumor necrosis factor) superfamily, and is expressed on B cells, endothelial cells, macrophages, dendritic cells, T cells, and fibroblasts. Until now, little information has addressed the association between CD40 polymorphisms and its soluble level in Chinese patients. In this study, we focused on identifying a genetic marker that may help refine the SLE risk profile. We found that the rs1883832 C/T polymorphism of CD40 and the levels of sCD40 were significantly associated with the presence of SLE. The rs1883832 C/T polymorphism may affect the levels of sCD40. Moreover, we also found that the rs1883832 C/T polymorphism was in strong linkage disequilibrium with the rs1569723 A/C ($|D'| = 0.867$) and rs4810485 G/T ($|D'| = 0.841$). The rs1569723 A/C and rs4810485 G/T were in strong linkage disequilibrium ($|D'| = 0.922$). Major two haplotype frequencies of the TCT and CAG among the SLE in the present study were 0.515 and 0.429 respectively. By haplotype analyses, we found that TCT haplotype was associated with a significantly increased risk of SLE as compared with the control groups (OR = 1.408; 95 % CI, 1.074–1.845; $P = 0.013$). Our results suggest that the CD40 gene plays a central role in the mechanism of the SLE pathophysiology. Thus, CD40 gene rs1883832 C/T polymorphism may serve as novel genetic markers of susceptibility to SLE in the Chinese population.

SLE is a chronic inflammatory disease of collagen in the skin, of joints, and of internal organs, and is a complex disorder in which multiple genetic variants, together with environmental and hormonal factors, contribute to disease risk. The etiology of SLE remains unknown, and the pathological mechanisms underlying the related organ and tissue damage have not been fully elucidated [26]. Recently, increasing evidence showed that CD40 contributes to the pathogenesis of chronic inflammatory and autoimmune diseases due to its biological activity [27]. In several reports of SLE, CD40 has either been indirectly or directly shown to be a contributing factor to the disease. In one report, Zhang et al. presented that TT genotype carriers showed higher CD40 expression and serum soluble CD40 ncentration in male IS patients [28]. CD40 polymorphisms are also associated with SLE clinical manifestation, mainly nephritis and arthritis [29, 30]. However, Plasma levels of CD40 were significantly elevated in SLE patients in comparison with healthy controls. In the present study, our data also showed that the plasma sCD40 levels were significantly high in SLE patients compared to controls. The results of our study indirectly suggest that CD40 may play a role in patients with SLE. These observations make CD40 an interesting candidate gene for a role in human SLE.

Several studies have investigated associations between genetic variation in the CD40 gene and SLE, but results of these studies have been inconsistent. Vazgiourakis found that CD40 has been identified as a new susceptibility locus in Greek and Turkish patients with SLE. The rs4810485 minor allele T is under-represented in SLE and correlates with reduced CD40 expression in peripheral blood monocytes and B cells, with potential implications for the regulation of aberrant immune responses in the disease, the CD40 gene rs4810485 G/T polymorphisms between the group of patients with SLE and the control group in European-American population ($P < 0.05$) [31]. Meanwhile, Piotrowski reported that there was no apparent relationship in the genotype frequencies of CD40 gene rs4810485 G/T polymorphisms with the risk of SLE in Polish patients as compared to controls ($P > 0.05$) [32]. Our results showed that there were significant differences in the genotype and allele frequencies of CD40 gene rs1883832C/T polymorphism between the group of patients with SLE and the control group ($P < 0.05$). sCD40 levels were increased in patients with SLE compared with controls ($P < 0.01$). The rs1883832C/T polymorphism of CD40 and its sCD40 levels were associated with SLE in the Chinese population. However their findings suggest that the significant variation in prevalence of risk genetic locis among different populations may also explain some of the sizable geographic variation in disease prevalence. The reason for these discrepancies remains unclear, but several possibilities should be considered. First, it may be due to the genetic trait differences; CD40 gene polymorphisms were distinct in specific population, various ethnicities and geographic region. Furthermore, SLE is a multi-factorial disease and individual exposure to various environmental factors, and genetic susceptibility might have caused different results. In addition, the inadequate study design such as non-random sampling and a limited sample size should also be considered. The possible selection bias that might have been present in the hospital-based, case–control study is a relevant issue. Finally, we cannot exclude that the observed association depends on a gene in linkage disequilibrium with the CD40 gene or on the effect of CD40 on another peptide.

So far, investigations on the CD40 gene rs1883832 C/T polymorphism and its soluble level, which are associated with SLE, have not been performed. Our data demonstrated that CD40 gene rs1883832 C/T polymorphism was associated with SLE ($P < 0.05$). Also, the level of sCD40 was found to be elevated in SLE patients ($P < 0.01$). Moreover, genotypes carrying the CD40 rs1883832 C/T variant allele (TT or CT genotype) were associated with increased CD40 levels compared to the homozygous wild-type genotype (CC genotype) in patients with SLE ($P < 0.01$). Additionally, our results showed that sCD40 levels were not associated with the polymorphisms of the CD40 in healthy controls. A plausible explanation is that the sCD40 expression is inducible and its expression is upregulated after stimulation and such inflammatory stimulation in healthy controls should be missing. However, we found that individuals carrying the rs1883832 T allele of the CD40 gene rs1883832 C/T polymorphism, which has been associated with increased sCD40 production, were at a significantly increased risk of SLE. This finding suggests an association between CD40 genotypes and its soluble form. We speculate that CD40 gene rs1883832 C/T polymorphismmay exert an impact on its protein metabolism and stability.

Conclusion
On the basis of these findings, we conclude that the rs1883832 C/T polymorphism of CD40 and the levels of sCD40 were significantly associated with the risk of SLE in the Chinese population. These results suggest that further studies with larger cohorts of patients should be performed to illustrate the correlation of the CD40 gene polymorphism with SLE susceptibility, independently or in combination with other CD40 SNPs and other genes. Because genetic polymorphisms were often vary different between ethnic groups, further studies are also needed to clarify the association of the CD40 polymorphism with the risk of SLE in diverse ethnic populations.

Abbreviations
SLE: Systemic lupus erythematosus; CI: Confidence interval; OR: Odds ratio; SNPs: Single nucleotide polymorphisms; ELISA: Enzyme linked immunosorbent assay.

Competing interests
The authors declare that they have no competing interests.

Authors' contributions
JMC designed the study, was involved in data collection, analysis and interpretation of data, and was involved in drafting and critically revising the manuscript. JG was involved in designing the study, was involved in data collection, analysis and was involved in drafting and critically revising the manuscript. CDW was involved in designing the study and revising the manuscript. CFW was involved in analysis and interpretation of data, as well as revising the manuscript. HCL was involved in study design, data collection, as well as revising the manuscript. YSW was involved in study design, data collection, as well as revising the manuscript. YL was involved in data collection and revising the manuscript. All authors read and approved the final manuscript.

Acknowledgments

This study was supported by the National Natural Science Foundation (No. 81260234). This work was supported by Key Programs of Guangxi health department (No. Z2012086).

References

1. Okamura T, Morita K, Fujio K, Yamamoto K. Regulatory T cells in systemic lupus erythematosus. Nihon Rinsho Meneki Gakkai Kaishi. 2015;38:69–77.
2. Belot A, Kasher PR, Trotter EW, Foray AP, Debaud AL, Rice GI, et al. Protein kinase cδ deficiency causes mendelian systemic lupus erythematosus with B cell-defective apoptosis and hyperproliferation. Arthritis Rheum. 2013;65:2161–71.
3. O'Neill S, Cervera R. Systemic lupus erythematosus. Best Pract Res Clin Rheumatol. 2010;24:841–55.
4. Hawro T, Bogucki A, Krupińska-Kun M, Maurer M, Woźniacka A. Intractable Headaches, Ischemic Stroke, and Seizures Are Linked to the Presence of Anti-β2GPI Antibodies in Patients with Systemic Lupus Erythematosus. PLoS One. 2015;10:e0119911.
5. Finno CJ, Aleman M, Higgins RJ, Madigan JE, Bannasch DL. Risk of false positive genetic associations in complex traits with underlying population structure: a case study. Vet J. 2014;202:543–9.
6. Ciccacci C, Perricone C, Ceccarelli F, Rufini S, Di Fusco D, Alessandri C, et al. A multilocus genetic study in a cohort of Italian SLE patients confirms the association with STAT4 gene and describes a new association with HCP5 gene. PLoS One. 2014;9:e111991.
7. Mak A, Tay SH. Environmental factors, toxicants and systemic lupus erythematosus. Int J Mol Sci. 2014;15:16043–56.
8. Marks SD, Tullus K. Autoantibodies in systemic lupus erythematosus. Pediatr Nephrol. 2012;27:1855–68.
9. Belot A, Cochat P. Monogenic systemic lupus erythematosus. Nephrol Ther. 2012;8:1–4.
10. Gandhi KS, McKay FC, Cox M, Riveros C, Armstrong N, Heard RN, et al. The multiple sclerosis whole blood mRNA transcriptome and genetic associations indicate dysregulation of specific T cell pathways in pathogenesis. Hum Mol Genet. 2010;19:2134–43.
11. Norbert D, Kathrin P, Martin H, Harrer T, Schuster P, Ries M, et al. Chronic Immune Activation in HIV-1 Infection Contributes to Reduced Interferon Alpha Production via Enhanced CD40:CD40 Ligand Interaction. PLoS One. 2012;7:e33925.
12. Karimi MH, Marzban S, Hajiyan MR, Geramizadeh B, Pourfathollah AA, Rajabiyan MH, et al. Effect of CD40 silenced dendritic cells by RNA interference on mice skin allograft rejection. Immunotherapy. 2015;7:111–8.
13. Gao Y, Kazama H, Yonehara S. Bim regulates B-cell receptor-mediated apoptosis in the presence of CD40 signaling in CD40-pre-activated splenic B cells differentiating into plasma cells. Int Immunol. 2012;24:283–92.
14. Gorbacheva V, Fan R, Wang X, Baldwin 3rd WM, Fairchild RL, Valujskikh A. IFN-γ production by memory helper T cells is required for CD40-independent alloantibody responses. J Immunol. 2015;194:1347–56.
15. Rabant M, Gorbacheva V, Fan R, Yu H, Valujskikh A. CD40-independent help by memory CD4 T cells induces pathogenic alloantibody but does not lead to long-lasting humoral immunity. Am J Transplant. 2013;13:2831–41.
16. Portillo JA, Greene JA, Schwartz I, Subauste MC, Subauste CS. Blockade of CD40-TRAF2,3 or CD40-TRAF6 is sufficient to inhibit pro-inflammatory responses in non-haematopoietic cells. Immunology. 2015;144:21–33.
17. Yuan M, Ohishi M, Wang L, Raguki H, Wang H, Tao L, et al. Association between serum levels of soluble CD40/CD40 ligand and organ damage in hypertensive patients. Clin Exp Pharmacol Physiol. 2010;37:848–51.
18. Gerdes N, Zirlik A. Co-stimulatory molecules in and beyond co-stimulation-tipping the balance in atherosclerosis. Thromb Haemost. 2011;106:804–13.
19. Wu T, Guo R, Zhang B. Developments in the study of CD40/ CD40L gene and its polymorphism in atherosclerosis. Zhong Nan Da Xue Xue Bao Yi Xue Ban. 2012;37:413–8.
20. Wagner M, Wisniewski A, Bilinska M, Pokryszko-Dragan A, Cyrul M, Kusnierczyk P, et al. Investigation of gene-gene interactions between CD40 and CD40L in Polish multiple sclerosis patients. Hum Immunol. 2014;75:796–801.
21. Huber AK, Finkelman FD, Li CW, Concepcion E, Smith E, Jacobson E, et al. Genetically driven target tissue over expression of CD40:a novel mechanism in autoimmune disease. J Immunol. 2012;189:3043–53.
22. Li G, Diogo D, Wu D, Spoonamore J, Dancik V, Franke L, et al. Human genetics in rheumatoid arthritis guides a high-throughput drug screen of the CD40 signaling pathway. PLoS Genet. 2013;9:e1003487.
23. Wang DM, Tang S, Li Z, Cheng X, Gao SQ, Deng ZH. High through-put genomic DNA isolation technique and its application in HLA genotyping for samples from bone marrow donor program. Zhongguo Shi Yan Xue Ye Xue Za Zhi. 2009;17:1265–8.
24. ShiYY HL. SHEsis, a powerful software platform for analyses of inkage disequilibrium, haplotype construction, and genetic association at polymorphism loci. Cell Res. 2005;15:97–8.
25. Stephens M, Smith NJ, Donnelly P. A new statistical method for haplotype reconstruction from population data. Am J Hum Genet. 2001;68:978–89.
26. Comte D, Karampetsou MP, Tsokos GC. T cells as a therapeutic target in SLE. Lupus. 2015;24:351–63.
27. Bankert KC, Oxley KL, Smith SM, Graham JP, de Boer M, Thewissen M, et al. Induction of an Altered CD40 Signaling Complex by an Antagonistic Human Monoclonal Antibody to CD40. J Immunol. 2015;194:4319–27.
28. Zhang B, Wu T, Song C, Chen M, Li H, Guo R. Association of CD40-1 C/T polymorphism with cerebral infarction susceptibility and its effect on sCD40L in Chinese population. Int Immunophar-macol. 2013;16:461–5.
29. Joo YB, Park BL, Shin HD, Park SY, Kim I, Bae SC. Association of genetic polymorphisms in CD40 with susceptibility to SLE in the Korean population. Rheumatology (Oxford). 2013;52:623–30.
30. Pau E, Chang NH, Loh C, Lajoie G, Wither JE. Abrogation of pathogenic IgG autoantibody production in CD40L gene-deleted lupus-prone New Zealand Black mice. Clin Immunol. 2011;139:215–27.
31. Vazgiourakis VM, Zervou MI, Choulaki C, Bertsias G, Melissourgaki M, Yilmaz N, et al. A common SNP in the CD40 region is associated with systemic lupus erythematosus and correlates with alteredCD40 expression: implications for the pathogenesis. Ann Rheum Dis. 2011;70:2184–90.
32. Piotrowski P, Lianeri M, Wudarski M, Olesinska M, Jagodzinski PP. Single nucleotide polymorphism of CD40 region and the risk of systemic lupus erythematosus. Lupus. 2013;22:233–7.

21

Genetic linkage of hyperglycemia and dyslipidemia in an intercross between BALB/cJ and SM/J *Apoe*-deficient mouse strains

Qian Wang[1,2†], Andrew T. Grainger[3,4†], Ani Manichaikul[5], Emily Farber[5], Suna Onengut-Gumuscu[5] and Weibin Shi[1,2*]

Abstract

Background: Individuals with dyslipidemia often develop type 2 diabetes, and diabetic patients often have dyslipidemia. It remains to be determined whether there are genetic connections between the 2 disorders.

Methods: A female F_2 cohort, generated from BALB/cJ (BALB) and SM/J (SM) *Apoe*-deficient ($Apoe^{-/-}$) strains, was started on a Western diet at 6 weeks of age and maintained on the diet for 12 weeks. Fasting plasma glucose and lipid levels were measured before and after 12 weeks of Western diet. 144 genetic markers across the entire genome were used for quantitative trait locus (QTL) analysis.

Results: One significant QTL on chromosome 9, named *Bglu17* [26.4 cM, logarithm of odds ratio (LOD): 5.4], and 3 suggestive QTLs were identified for fasting glucose levels. The suggestive QTL near the proximal end of chromosome 9 (2.4 cM, LOD: 3.12) was replicated at both time points and named *Bglu16*. *Bglu17* coincided with a significant QTL for HDL (high-density lipoprotein) and a suggestive QTL for non-HDL cholesterol levels. Plasma glucose levels were inversely correlated with HDL but positively correlated with non-HDL cholesterol levels in F_2 mice on either chow or Western diet. A significant correlation between fasting glucose and triglyceride levels was also observed on the Western diet. Haplotype analysis revealed that "lipid genes" *Sik3*, *Apoa1*, and *Apoc3* were probable candidates for *Bglu17*.

Conclusions: We have identified multiple QTLs for fasting glucose and lipid levels. The colocalization of QTLs for both phenotypes and the sharing of potential candidate genes demonstrate genetic connections between dyslipidemia and type 2 diabetes.

Keywords: Dyslipidemia, Hyperglycemia, Type 2 diabetes, Quantitative trait locus, Genetic linkage

Background

Individuals with dyslipidemia have an increased risk of developing type 2 diabetes (T2D), and diabetic patients often have dyslipidemia, which includes elevations in plasma triglyceride and low-density lipoprotein (LDL) cholesterol levels and reductions in high-density lipoprotein (HDL) cholesterol levels [1]. Part of the increased diabetic risk associated with dyslipidemia is due to genetic variations that influence both lipoprotein homeostasis and the development of T2D. Indeed, a few rare gene mutations result in both dyslipidemia and T2D, which include *ABCA1* [2], *LIPE* [3], *LPL* [4], and *LRP6* [5]. Genome-wide association studies (GWAS) have identified >150 loci associated to variation in plasma lipids [6, 7] and >70 loci associated with T2D, fasting plasma glucose, glycated hemoglobin (HbA1c), or insulin resistance [8–10]. Nearly a dozen of the loci detected are associated

* Correspondence: ws4v@Virginia.EDU
†Equal contributors
[1]Department of Radiology & Medical Imaging, University of Virginia, Snyder Bldg Rm 266, 480 Ray C. Hunt Dr., P.O. Box 801339, Fontaine Research Park, Charlottesville, VA 22908, USA
[2]University of Virginia, Snyder Bldg Rm 266, 480 Ray C. Hunt Dr., P.O. Box 801339, Fontaine Research Park, Charlottesville, VA 22908, USA
Full list of author information is available at the end of the article

with both lipid and T2D-related traits at the genome-wide significance level, including *GCKR*, *FADS1*, *IRS1*, *KLF14*, and *HFE* (http://www.genome.gov/GWAStudies/). Surprisingly, half of them have shown opposite allelic effect on dyslipidemia and glucose levels [11], and this is in contrary to the positive correlations observed at the clinical level. Furthermore, it is challenging to establish causality between genetic variants and complex traits in humans due to small gene effects, complex genetic structure, and environmental influences.

A complementary approach to finding genetic components in human disease is to use animal models. Apolipoprotein E-deficient ($Apoe^{-/-}$) mice are a commonly used mouse model of dyslipidemia, with elevations in non-HDL cholesterol levels and reductions in HDL levels, even when fed a low fat chow diet [12, 13]. High fat diet feeding aggravates dyslipidemia. Moreover, these mice develop all phases of atherosclerotic lesions seen in humans [14] and are extensively used for atherosclerosis research [15–18]. We have found that $Apoe^{-/-}$ mice with certain genetic backgrounds develop significant hyperglycemia and T2D when fed a Western-type diet but become resistant with some other genetic backgrounds [16, 19, 20]. BALB/cJ (BALB) and SM/J (SM) $Apoe^{-/-}$ mice exhibit differences in dyslipidemia and T2D-related phenotypes [16]. The objective of the present study was to explore potential genetic connections between dyslipidemia and T2D through quantitative trait locus (QTL) analysis of a female cohort derived from an intercross between BALB-$Apoe^{-/-}$ and SM-$Apoe^{-/-}$ mice.

Methods

Ethics statement

All procedures were in accordance with current National Institutes of Health guidelines (https://grants.nih.gov/grants/olaw/Guide-for-the-Care-and-use-of-laboratory-animals.pdf) and approved by the institutional Animal Care and Use Committee (protocol #: 3109). Blood was drawn from the retro-orbital plexus of overnight fasted mice with the animals under isoflurane anesthesia.

Animals, experimental design and procedures

BALB and SM $Apoe^{-/-}$ mice were created using the classic congenic breeding strategy, as described [16]. BALB-$Apoe^{-/-}$ mice were crossed with SM-$Apoe^{-/-}$ mice to generate F_1s, which were intercrossed by brother-sister mating to generate a female F_2 cohort. Mice were weaned at 3 weeks of age onto a rodent chow diet. At 6 weeks of age, female F_2 mice were started on a Western diet containing 21 % fat, 34.1 % sucrose, 0.15 % cholesterol, and 19.5 % casein *by weight* (Harlan Laboratories, TD 88137) and maintained on the diet for 12 weeks. Mice were bled twice: once before initiation of the Western diet and once at the end of the 12-week feeding period.

Overnight fasted mice were bled into tubes containing 8 μL of 0.5 mol/L ethylenediaminetetraacetic acid. Plasma was prepared and stored at –80 °C before use.

Housing and husbandry

Breeding pairs were housed in a cage of 1 adult male and 2 females, and litters were weaned at 3 weeks of age onto a rodent chow diet in a cage of 5 or less. At 6 weeks of age, F_2 mice were switched onto the Western diet and maintained on the diet for 12 weeks. All mice were housed under a 12-h light/dark cycle at an ambient temperature of 23 °C and allowed free access to water and drinking food. Mice were fasted overnight before blood samples were collected.

Measurements of plasma glucose and lipid levels

Plasma glucose was measured with a Sigma glucose (HK) assay kit, as reported with modification to a longer incubation time [21]. Briefly, 6 μl of plasma samples were incubated with 150 μl of assay reagent in a 96-well plate for 30 min at 30 °C. The absorbance at 340 nm was read on a Molecular Devices (Menlo Park, CA) plate reader. The measurements of total cholesterol, HDL cholesterol, and triglyceride were performed as reported previously [13]. Non-HDL cholesterol was calculated as the difference between total and HDL cholesterol.

Genotyping

Genomic DNA was isolated from the tails of mice by using the phenol/chloroform extraction and ethanol precipitation method. The Illumina LD linkage panel consisting of 377 SNP loci was used to genotype the F_2 cohort. Microsatellite markers were typed for chromosome 8 where SNP markers were uninformative in distinguishing the parental origin of alleles. DNA samples from the two parental strains and their F_1s served as controls. Uninformative SNPs were excluded from QTL analysis. SNP markers were also filtered based on the expected pattern in the control samples, and F_2 mice were filtered based on 95 % call rates in genotype calls. After filtration, 228 F_2s and 144 markers were included in genome-wide QTL analysis.

Statistical analysis

QTL analysis was performed using J/qtl and Map Manager QTX software as previously reported [19, 22, 23]. One thousand permutations of trait values were run to define the genome-wide LOD (logarithm of odds) score threshold needed for significant or suggestive linkage of each trait. Loci that exceeded the 95th percentile of the permutation distribution were defined as significant ($P < 0.05$) and those exceeding the 37th percentile were suggestive ($P < 0.63$).

Prioritization of positional candidate genes

The Sanger SNP database (http://www.sanger.ac.uk/sanger/Mouse_SnpViewer/rel-1410) was used to prioritize candidate genes for overlapping QTLs affecting plasma glucose and HDL cholesterol levels on chromosome (Chr) 9, which were mapped in two or more crosses derived from different parental strains for either phenotype. We converted the original mapping positions in cM for the confidence interval to physical positions in Mb and then examined SNPs within the confidence interval. Probable candidate genes were defined as those with one or more SNPs in coding or upstream promoter regions that were shared by the parental strains carrying the "high" allele but were different from the parental strains carrying the "low" allele at a QTL, as previously reported [24].

Results

Trait value distributions

Fasting plasma glucose and lipid levels of F_2 mice were measured before and after 12-weeks of Western diet. Values of fasting plasma glucose, non-HDL cholesterol and triglyceride levels of F_2 mice on both chow and Western diets and of HDL cholesterol level on the chow diet were normally or approximately normally

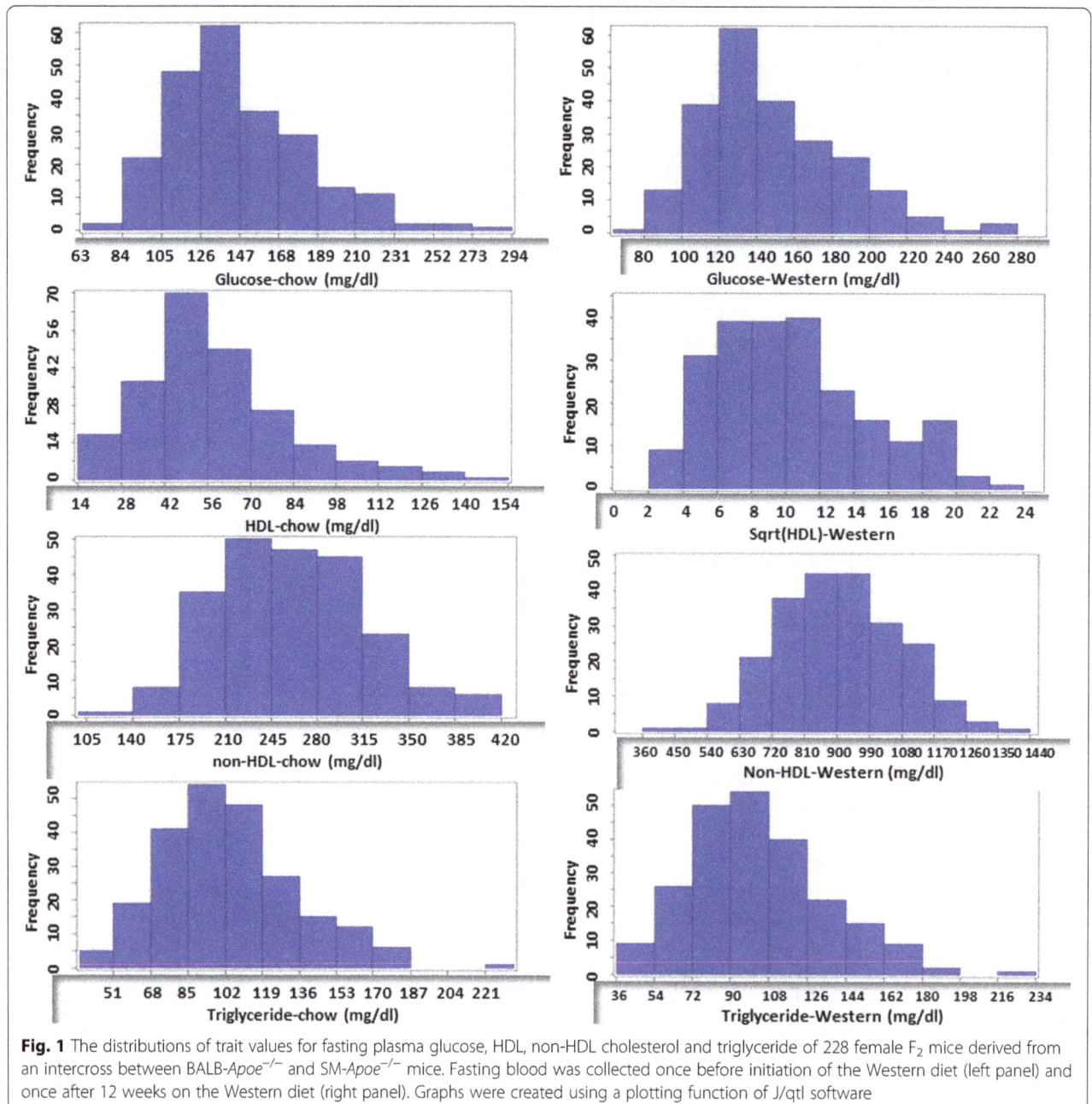

Fig. 1 The distributions of trait values for fasting plasma glucose, HDL, non-HDL cholesterol and triglyceride of 228 female F_2 mice derived from an intercross between BALB-*Apoe*$^{-/-}$ and SM-*Apoe*$^{-/-}$ mice. Fasting blood was collected once before initiation of the Western diet (left panel) and once after 12 weeks on the Western diet (right panel). Graphs were created using a plotting function of J/qtl software

distributed (Fig. 1). Values of square root-transformed HDL cholesterol levels on the Western diet showed a normal distribution. These data were then analyzed to search for QTLs affecting the traits. Loci with a genome-wide suggestive or significant P value are presented in Table 1.

Fasting glucose levels

A genome-wide scan for main effect QTL revealed a suggestive QTL near the proximal end of Chr9 for fasting glucose when mice were fed the chow diet (2.37 cM, LOD: 2.21) (Fig. 2 and Table 1). As this QTL was replicated on the Western diet, it was named $Bglu16$. For fasting glucose levels on the Western diet, a significant QTL on Chr9 and 3 suggestive QTLs, including $Bglu16$ on Chr9, were identified. The significant QTL on Chr9

peaked at 26.37 cM and had a LOD score of 5.425. It was named $Bglu17$. The suggestive QTL near the middle portion of Chr5 (67.4 cM, LOD 2.18) replicated $Bglu13$, initially mapped in a B6 x BALB $Apoe^{-/-}$ intercross [21]. The suggestive QTL on distal Chr5 (101.24 cM, LOD 3.198) was novel. The BALB allele conferred an increased glucose level for both of the Chr9 QTLs while the SM allele conferred increased glucose levels for the 2 Chr5 QTLs (Table 2).

Fasting lipid levels

Genome-wide scans for main effect QTLs showed that HDL, non-HDL cholesterol, and triglyceride levels were each controlled by multiple QTLs (Figs. 3, 4 and 5; Table 1). For HDL, 3 significant QTLs, located on Chr1, Chr7 and Chr9, and 1 suggestive QTL on Chr10, were

Table 1 Significant and suggestive QTLs for plasma glucose and lipid levels in female F_2 mice derived from BALB-$Apoe^{-/-}$ and SM-$Apoe^{-/-}$ mice

Locus	Chr	Trait	LOD[a]	p-value[b]	Peak (cM)	95 % CI[c]	High allele	Mode of inheritance[d]
Bglu16	9	Glucose-C	2.214	0.549	2.37	0.37–30.37	B	Additive
Bglu13	5	Glucose-W	2.1.8	<0.63	67.4	45.4–80.03	S	Recessive
-	5	Glucose-W	3.198	0.097	101.24	29.40–101.24	S	Additive
Bglu16	9	Glucose-W	3.12	<0.63	2.37	0–10.37	B	Additive
Bglu17	9	Glucose-W	**5.425**	**0.001**	26.37	16.37–40.37	B	Additive
Hdlq5	1	HDL-C	**8.64**	**0.000**	93.52	87.52–97.02	S	Additive
Hdlcl1	7	HDL-C	2.668	0.321	61.33	35.57–89.57	S	Dominant
Hdlq17	9	HDL-C	**4.614**	**0.014**	30.37	16.37–32.37	S	Additive
Hdlq26	10	HDL-C	2.181	0.591	61.22	25.03–61.22	S	Dominant
Hdlq5	1	HDL-W	**13.944**	**0.000**	87.52	83.52–93.52	S	Additive
Hdlcl1	7	HDL-W	**3.658**	**0.034**	85.57	77.57–89.67	S	Additive
Hdlq17	9	HDL-W	**10.625**	**0.000**	30.42	24.37–30.53	S	Additive
Chol7	1	non-HDL-C	2.093	0.626	66.95	9.52–74.56	B	Recessive
Nhdlq15	2	non-HDL-C	2.56	0.321	23.86	8.73–38.73	B	Additive
Hdlq34	5	non-HDL-C	2.106	0.614	19.4	19.4–30.5	S	Additive
Pnhdlc1	6	non-HDL-C	2.489	0.362	57.53	1.53–77.53	B	Recessive
Nhdlq1	8	non-HDL-C	2.221	0.537	44.14	10.14–60.14	B	Additive
Nhdlq12	12	non-HDL-C	2.73	0.245	39.41	15.41–59.41	B	Additive
Nhdlq15	**2**	**non-HDL-W**	**4.79**	**0.002**	31.80	22.73–40.73	B	Dominant
Nhdlq11	9	non-HDL-W	2.136	0.585	32.37	0.37–75.33	B	Additive
-	11	non-HDL-W	2.332	0.436	1.99	1.99–17.99	B	Dominant
Nhdlq16	**16**	**non-HDL-W**	**3.99**	**0.011**	46.66	35.43–46.66	S	Dominant
Tgq11	2	Triglyceride-C	2.952	0.169	26.73	12.73–60.83	B	Additive
-	5	Triglyceride-C	2.759	0.234	80.03	73.40–93.40	S	Heterosis
Trglyd	1	Triglyceride-W	3.291	0.091	97.02	79.24–97.02	S	Additive

[a]LOD scores were obtained from genome-wide QTL analysis using J/qtl software. The significant LOD scores were highlighted in bold. The suggestive and significant LOD score thresholds were determined by 1,000 permutation tests for each trait. Suggestive and significant LOD scores were 2.116 and 3.429, respectively, for glucose on the chow diet; 2.056 and 3.569 for glucose on the Western diet; 2.127 and 3.725 for HDL cholesterol, 2.09 and 3.662 for non-HDL cholesterol, and 2.102 and 3.522 for triglyceride on the chow diet; 2.10 and 3.486 for HDL, 2.123 and 3.628 for non-HDL, and 2.123 and 3.628 for triglyceride on the Western diet
[b]The p-values reported represent the level of genome-wide significance
[c]95 % Confidence interval in cM defined by a whole genome QTL scan
[d]Mode of inheritance was defined according to allelic effect at the nearest marker of a QTL

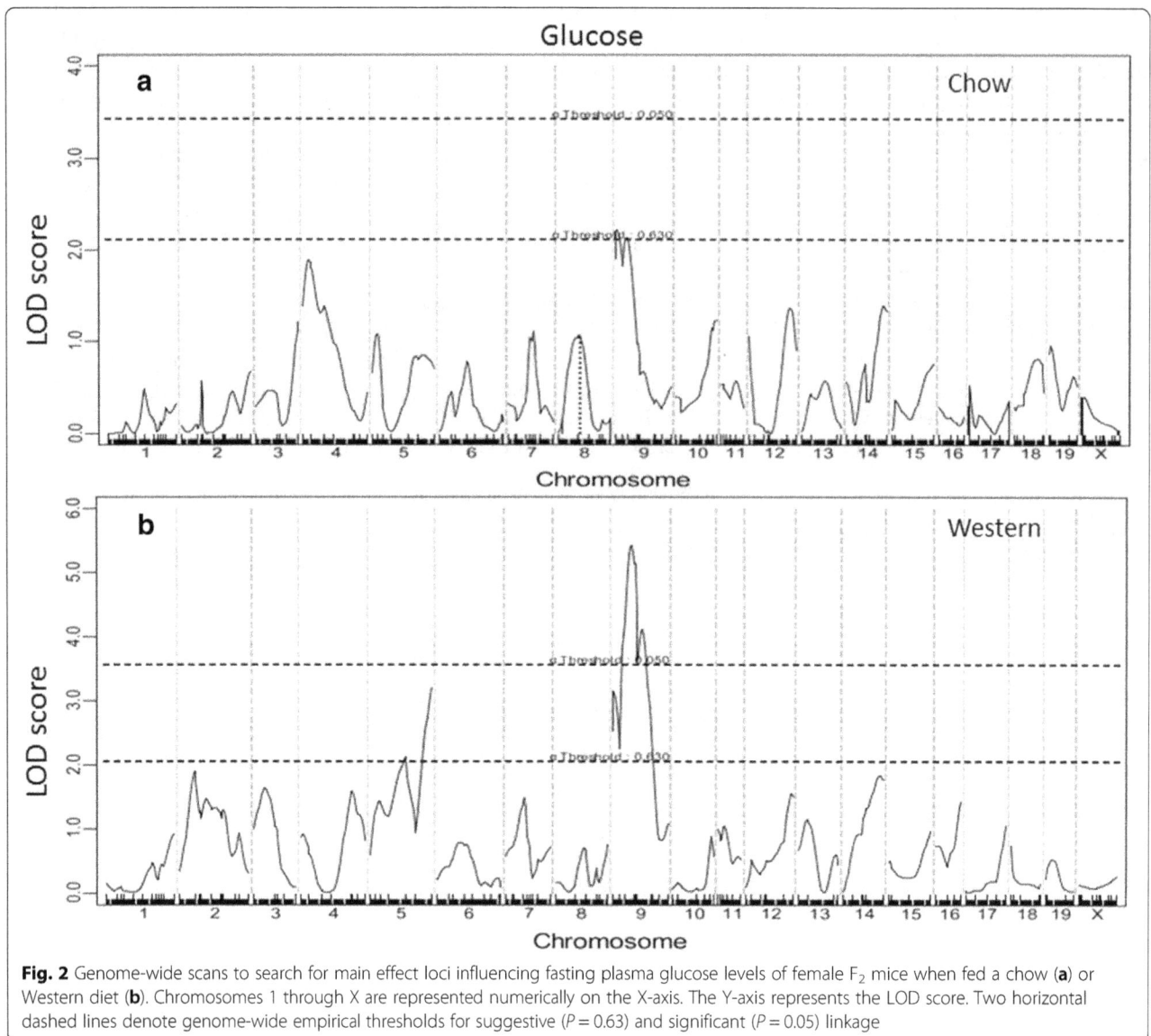

Fig. 2 Genome-wide scans to search for main effect loci influencing fasting plasma glucose levels of female F_2 mice when fed a chow (**a**) or Western diet (**b**). Chromosomes 1 through X are represented numerically on the X-axis. The Y-axis represents the LOD score. Two horizontal dashed lines denote genome-wide empirical thresholds for suggestive ($P = 0.63$) and significant ($P = 0.05$) linkage

identified. All 3 significant QTLs for HDL were detected when mice were fed either chow or Western diet, while the suggestive QTL on Chr10 was found when mice were on the chow diet. The significant QTL on Chr1 replicated *Hdlq5*, which had been mapped in numerous crosses [25]. The Chr7 QTL replicated *Hdlcl1*, initially mapped in (PERA/EiJ x B6-*Ldlr*)) x B6-*Ldlr* backcross [26]. The Chr9 QTL replicated *Hdlq17*, previously mapped in B6 x 129S1/SvImJ F_2 mice [27]. The suggestive QTL on Chr10 overlapped with *Hdlq26* mapped in a SM/J x NZB/BlNJ intercross [28]. For all 4 HDL QTLs, F_2 mice homozygous for the SS allele had higher HDL levels than those homozygous for the BB allele (Table 2).

For non-HDL cholesterol levels, 6 suggestive QTLs were detected when F_2 mice were fed the chow diet, and 2 significant and 2 suggestive QTLs were detected on the Western diet (Fig. 4). The 2 significant QTLs on

Chr2 and Chr16 and the suggestive QTL on Chr11 were novel. The former 2 QTLs were named *Nhdlq15* and *Nhdlq16*, respectively. *Nhdlq15* peaked at 31.8 cM on Chr2 and affected non-HDL levels in a dominant mode from the BB allele while *Nhdlq16* peaked at 46.66 cM on Chr16 and affected non-HDL levels in a dominant mode from the SS allele. The rest replicated previously identified ones in other mouse crosses: The Chr1 QTL peaked at 66.95 cM, overlapping with *Chol7* mapped in an intercross of 129S1/SvImJ and CAST/Ei mice [29]. The Chr5 QTL overlapped with *Hdlq34* mapped in PERA/EiJ x I/LnJ and PERA/EiJ x DBA/2 J intercrosses [30]. The Chr6 QTL overlapped with *Pnhdlc1*, initially mapped in a B6 x CASA/Rk intercross and then replicated in B6 x C3H $Apoe^{-/-}$ F_2 mice [31, 32]. The Chr8 QTL replicated *Nhdlq1*, initially mapped in B6 x 129S1/SvImJ F_2 mice [33]. The Chr9 QTL replicated *Nhdlq11*, initially mapped

Table 2 Allelic effects in different QTLs on plasma glucose and lipids of female F_2 mice derived from BALB and SM $Apoe^{-/-}$ mice

Locus name	Chr	Trait	LOD	Peak (cM)	Closest marker	BB	SS	SB
Bglu16	9	Glucose-C	2.214	2.37	rs13480073	109.0 ± 28.7 (n = 44)	93.9 ± 22.9 (n = 43)	97.4 ± 22.6 (n = 141)
Bglu13	5	Glucose-W	2.1.8	67.4	rs3726547	144.4 ± 30.6 (n = 43)	153.5 ± 40.4 (n = 88)	142.9 ± 35.3 (n = 97)
-	5	Glucose-W	3.198	101.24	rs13478578	132.7 ± 31.3 (n = 51)	158.8 ± 41.8 (n = 63)	147.5 ± 34.0 (n = 113)
Bglu16	9	Glucose-W	3.12	2.37	rs13480073	165.3 ± 40.9 (n = 54)	138.3 ± 30.0 (n = 43)	144.4 ± 35.7 (n = 141)
Bglu17	9	Glucose-W	**5.425**	26.37	CEL.9_49183636	168.0 ± 39.8 (n = 42)	134.7 ± 26.5 (n = 62)	146.1 ± 37.1 (n = 124)
Hdlq5	1	HDL-C	**8.64**	93.52	rs13476259	49.5 ± 20.9 (n = 60)	73.2 ± 26.5 (n = 62)	55.1 ± 19.4 (n = 106)
Hdlcl1	7	HDL-C	2.668	61.33	rs3724711	49.0 ± 20.0 (n = 50)	58.3 ± 21.0 (n = 63)	62.9 ± 25.5 (n = 115)
Hdlq17	9	HDL-C	**4.614**	30.37	CEL.9_49183636	49.5 ± 15.9 (n = 42)	69.4 ± 26.8 (n = 62)	56.2 ± 22.5 (n = 124)
Hdlq26	10	HDL-C	2.181	61.22	rs3688351	50.7 ± 19.2 (n = 60)	59.4 ± 21.9 (n = 53)	62.4 ± 25.8 (n = 114)
Hdlq5	1	sqrtHDL-W	**13.944**	87.52	rs3685643	66.6 ± 50.2 (n = 57)	201.1 ± 118.8 (n = 62)	117.0 ± 97.1 (n = 109)
Hdlcl1	7	sqrtHDL-W	**3.658**	85.57	rs6216320	95.5 ± 87.6 (n = 63)	173.3 ± 126.7 (n = 55)	122.4 ± 98.4 (n = 110)
Hdlq17	9	sqrtHDL-W	**10.625**	30.42	CEL.9_49183636	57.6 ± 49.3 (n = 42)	183.3 ± 115.0 (n = 62)	122.8 ± 101.7 (n = 124)
Chol7	1	non-HDL-C	2.093	66.95	rs6354736	279.5 ± 62.8 (n = 56)	257.8 ± 56.9 (n = 57)	251.2 ± 52.1 (n = 114)
Nhdlq15	2	non-HDL-C	2.56	23.86	mCV23209429	273.9 ± 56.1 (n = 55)	238.0 ± 47.0 (n = 53)	262.7 ± 59.0 (n = 120)
Hdlq34	5	non-HDL-C	2.106	19.4	rs3658401	244.5 ± 54.7 (n = 63)	276.0 ± 53.2 (n = 61)	259.3 ± 58.3 (n = 104)
Pnhdlc1	6	non-HDL-C	2.489	57.53	rs13478909	279.6 ± 51.3 (n = 51)	252.0 ± 65.2 (n = 57)	254.8 ± 53.5 (n = 120)
Nhdlq1	8	non-HDL-C	2.221	44.14	D8Mit50	275.0 ± 54.5 (n = 60)	242.4 ± 57.4 (n = 57)	262.9 ± 55.6 (n = 96)
Nhdlq12	12	non-HDL-C	2.73	39.41	rs6195664	278.6 ± 52.3 (n = 62)	243.8 ± 57.8 (n = 59)	257.4 ± 56.5 (n = 107)
Nhdlq15	**2**	non-HDL-W	**4.79**	31.8	rs13476507	954.1 ± 156.0 (n = 56)	806.9 ± 158.2 (n = 47)	915.6 ± 166.1 (n = 125)
Nhdlq11	9	non-HDL-W	2.136	32.37	rs3709825	958.4 ± 211.4 (n = 42)	856.8 ± 165.3 (n = 62)	906.6 ± 149.5 (n = 124)
-	11	non-HDL-W	2.332	1.99	rs4222040	927.3 ± 149.8 (n = 67)	849.0 ± 165.1 (n = 69)	917.0 ± 170.5 (n = 85)
Nhdlq16	**16**	non-HDL-W	**3.99**	46.66	rs3721202	820.2 ± 152.7 (n = 56)	931.9 ± 146.4 (n = 52)	928.4 ± 174.6 (n = 120)
Tgq11	2	Triglyceride-C	2.952	26.73	mCV23209429	123.7 ± 35.7 (n = 55)	101.9 ± 34.6 (n = 53)	107.3 ± 31.8 (n = 120)
-	5	Triglyceride-C	2.759	80.03	gnf05.120.578	110.2 ± 33.2 (n = 43)	119.3 ± 35.8 (n = 88)	101.6 ± 31.3 (n = 97)
Trglyd	1	Triglyceride-W	3.291	97.02	rs13476259	94.0 ± 28.6 (n = 59)	115.3 ± 33.0 (n = 62)	100.7 ± 30.8 (n = 106)
Bglu16	9	Glucose-C	2.214	2.37	rs13480073	109.0 ± 28.7 (n = 44)	93.9 ± 22.9 (n = 43)	97.4 ± 22.6 (n = 141)
Bglu13	5	Glucose-W	2.1.8	67.4	rs3726547	144.4 ± 30.6 (n = 43)	153.5 ± 40.4 (n = 88)	142.9 ± 35.3 (n = 97)
-	5	Glucose-W	3.198	101.24	rs13478578	132.7 ± 31.3 (n = 51)	158.8 ± 41.8 (n = 63)	147.5 ± 34.0 (n = 113)
Bglu16	9	Glucose-W	3.12	2.37	rs13480073	165.3 ± 40.9 (n = 54)	138.3 ± 30.0 (n = 43)	144.4 ± 35.7 (n = 141)
Bglu17	9	Glucose-W	**5.425**	26.37	CEL.9_49183636	168.0 ± 39.8 (n = 42)	134.7 ± 26.5 (n = 62)	146.1 ± 37.1 (n = 124)
Hdlq5	1	HDL-C	**8.64**	93.52	rs13476259	49.5 ± 20.9 (n = 60)	73.2 ± 26.5 (n = 62)	55.1 ± 19.4 (n = 106)
Hdlcl1	7	HDL-C	2.668	61.33	rs3724711	49.0 ± 20.0 (n = 50)	58.3 ± 21.0 (n = 63)	62.9 ± 25.5 (n = 115)
Hdlq17	9	HDL-C	**4.614**	30.37	CEL.9_49183636	49.5 ± 15.9 (n = 42)	69.4 ± 26.8 (n = 62)	56.2 ± 22.5 (n = 124)
Hdlq26	10	HDL-C	2.181	61.22	rs3688351	50.7 ± 19.2 (n = 60)	59.4 ± 21.9 (n = 53)	62.4 ± 25.8 (n = 114)
Hdlq5	1	sqrtHDL-W	**13.944**	87.52	r[55]s3685643	66.6 ± 50.2 (n = 57)	201.1 ± 118.8 (n = 62)	117.0 ± 97.1 (n = 109)
Hdlcl1	7	sqrtHDL-W	**3.658**	85.57	rs6216320	95.5 ± 87.6 (n = 63)	173.3 ± 126.7 (n = 55)	122.4 ± 98.4 (n = 110)
Hdlq17	9	sqrtHDL-W	**10.625**	30.42	CEL.9_49183636	57.6 ± 49.3 (n = 42)	183.3 ± 115.0 (n = 62)	122.8 ± 101.7 (n = 124)
Chol7	1	non-HDL-C	2.093	66.95	rs6354736	279.5 ± 62.8 (n = 56)	257.8 ± 56.9 (n = 57)	251.2 ± 52.1 (n = 114)
Nhdlq15	2	non-HDL-C	2.56	23.86	mCV23209429	273.9 ± 56.1 (n = 55)	238.0 ± 47.0 (n = 53)	262.7 ± 59.0 (n = 120)
Hdlq34	5	non-HDL-C	2.106	19.4	rs3658401	244.5 ± 54.7 (n = 63)	276.0 ± 53.2 (n = 61)	259.3 ± 58.3 (n = 104)
Pnhdlc1	6	non-HDL-C	2.489	57.53	rs13478909	279.6 ± 51.3 (n = 51)	252.0 ± 65.2 (n = 57)	254.8 ± 53.5 (n = 120)
Nhdlq1	8	non-HDL-C	2.221	44.14	D8Mit50	275.0 ± 54.5 (n = 60)	242.4 ± 57.4 (n = 57)	262.9 ± 55.6 (n = 96)
Nhdlq12	12	non-HDL-C	2.73	39.41	rs6195664	278.6 ± 52.3 (n = 62)	243.8 ± 57.8 (n = 59)	257.4 ± 56.5 (n = 107)
Nhdlq15	**2**	non-HDL-W	**4.79**	31.8	rs13476507	954.1 ± 156.0 (n = 56)	806.9 ± 158.2 (n = 47)	915.6 ± 166.1 (n = 125)

Table 2 Allelic effects in different QTLs on plasma glucose and lipids of female F_2 mice derived from BALB and SM $Apoe^{-/-}$ mice (Continued)

Nhdlq11	9	non-HDL-W	2.136	32.37	rs3709825	958.4 ± 211.4 (n = 42)	856.8 ± 165.3 (n = 62)	906.6 ± 149.5 (n = 124)
-	11	non-HDL-W	2.332	1.99	rs4222040	927.3 ± 149.8 (n = 67)	849.0 ± 165.1 (n = 69)	917.0 ± 170.5 (n = 85)
Nhdlq16	**16**	non-HDL-W	**3.99**	46.66	rs3721202	820.2 ± 152.7 (n = 56)	931.9 ± 146.4 (n = 52)	928.4 ± 174.6 (n = 120)
Tgq11	2	Triglyceride-C	2.952	26.73	mCV23209429	123.7 ± 35.7 (n = 55)	101.9 ± 34.6 (n = 53)	107.3 ± 31.8 (n = 120)
-	5	Triglyceride-C	2.759	80.03	gnf05.120.578	110.2 ± 33.2 (n = 43)	119.3 ± 35.8 (n = 88)	101.6 ± 31.3 (n = 97)
Trglyd	1	Triglyceride-W	3.291	97.02	rs13476259	94.0 ± 28.6 (n = 59)	115.3 ± 33.0 (n = 62)	100.7 ± 30.8 (n = 106)

Chr chromosome, LOD logarithm of odds, C chow diet, W Western diet, BB homozygous BALB allele, SS homozygous SM allele, SM heterozygous allele
Data are mean ± SD. The units for these measurements are mg/dL for plasma glucose or lipid levels. The number in the brackets represents the number of progeny with a specific genotype at a peak marker. The significant QTLs and their LOD scores were highlighted in bold

in B6 x C3H $Apoe^{-/-}$ F_2 mice [32]. The Chr12 QTL peaked at 44.14 cM, overlapping with Nhdlq12 mapped in a B6 x C3H $Apoe^{-/-}$ F_2 intercross [32].

For triglyceride levels, 3 suggestive QTLs, located on Chr1, 2, and 5, respectively, were identified (Fig. 5). The Chr1 QTL peaked at 97 cM, 17 cM distal the Apoa2 gene (80 cM). The Chr2 QTL replicated Tgq11, mapped

in an intercross between DBA/1J and DBA/2J [34]. The Chr5 QTL was novel.

Coincident QTLs for fasting glucose and lipids
LOD score plots for Chr9 showed that the QTL for fasting glucose (Bglu17) coincided precisely with the QTLs for HDL (Hdlq17) and non-HDL (Nhdlq11) in the

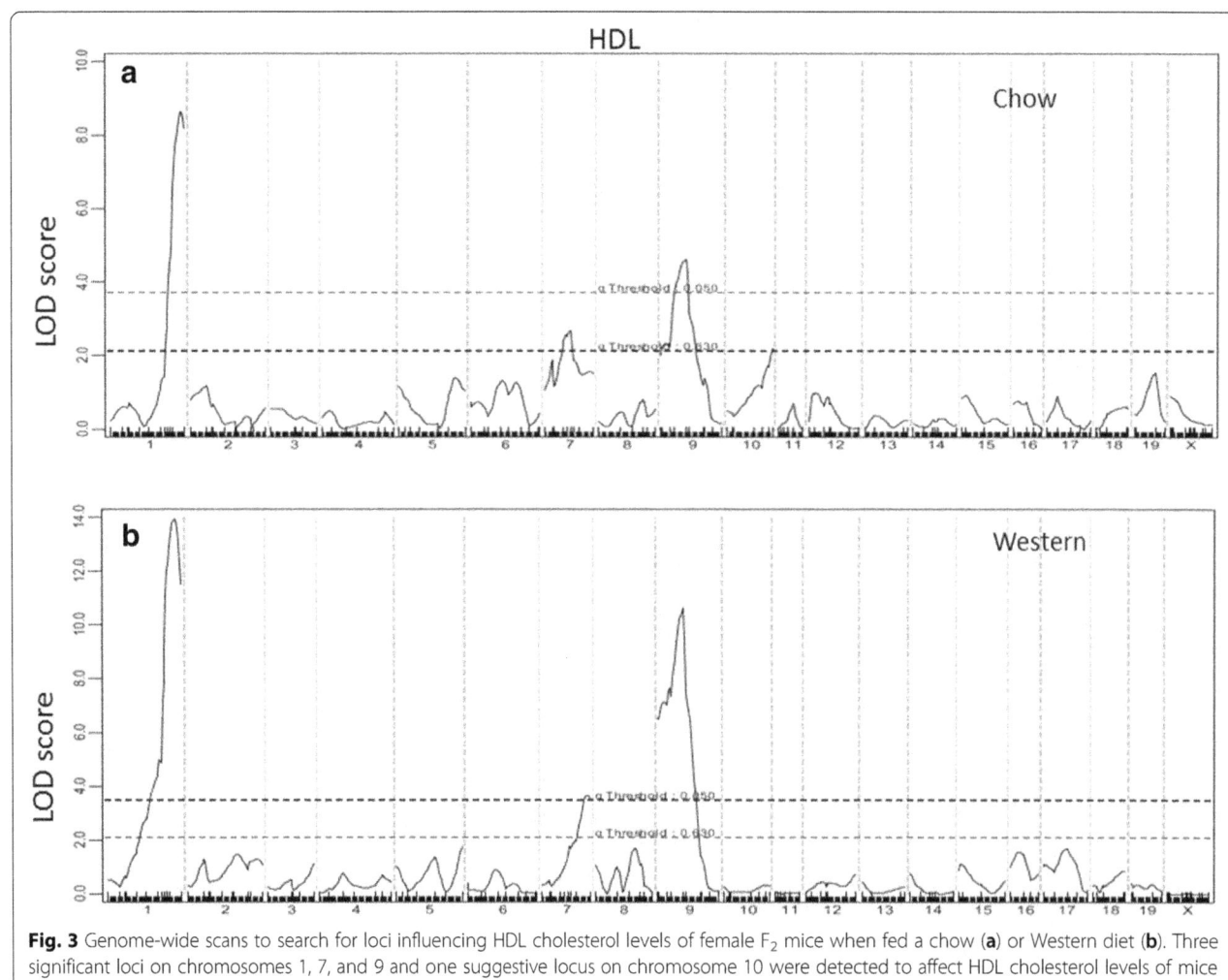

Fig. 3 Genome-wide scans to search for loci influencing HDL cholesterol levels of female F_2 mice when fed a chow (**a**) or Western diet (**b**). Three significant loci on chromosomes 1, 7, and 9 and one suggestive locus on chromosome 10 were detected to affect HDL cholesterol levels of mice

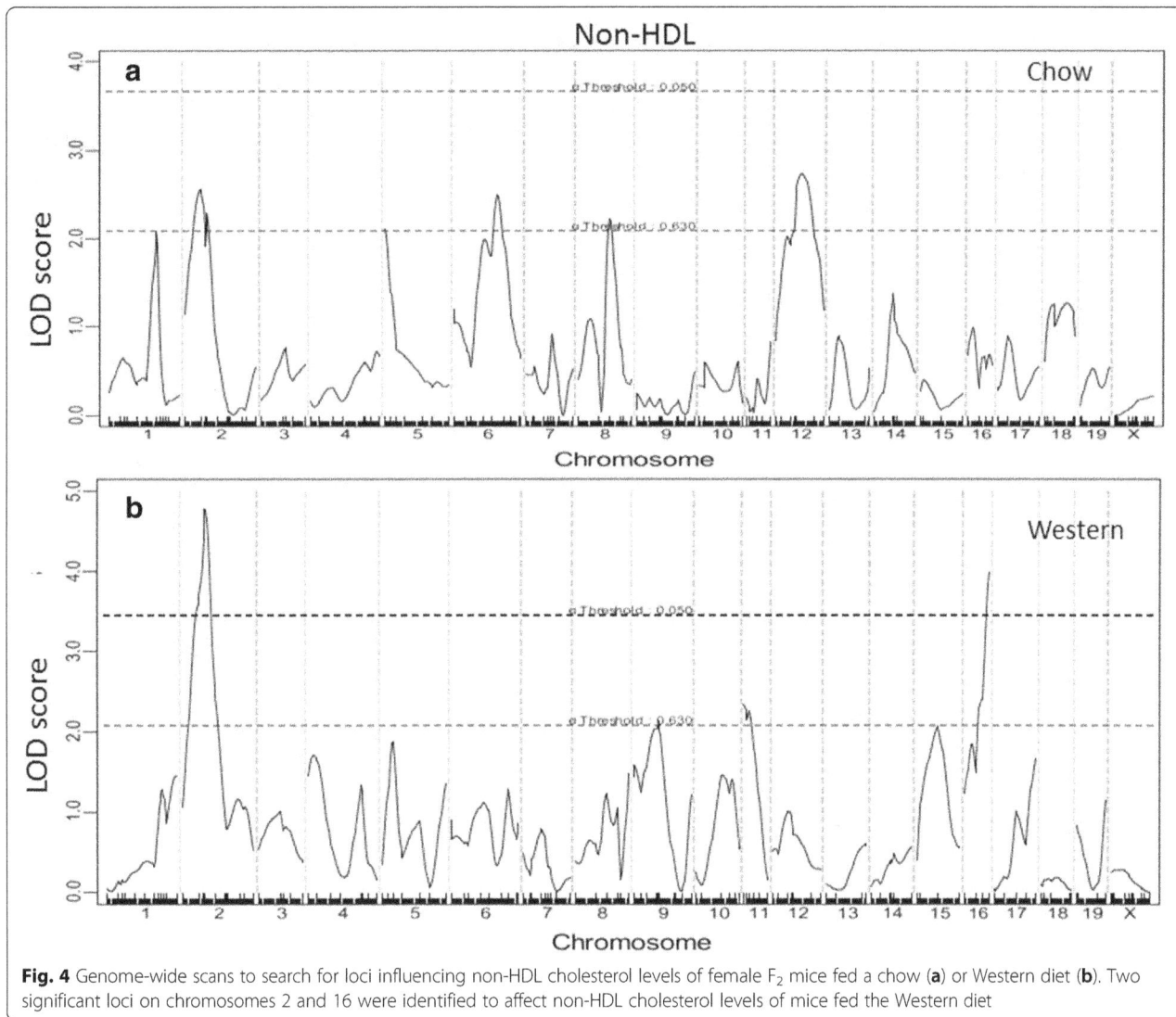

Fig. 4 Genome-wide scans to search for loci influencing non-HDL cholesterol levels of female F_2 mice fed a chow (**a**) or Western diet (**b**). Two significant loci on chromosomes 2 and 16 were identified to affect non-HDL cholesterol levels of mice fed the Western diet

confidence interval (Fig. 6). F_2 mice homozygous for the BB allele exhibited elevated levels of fasting glucose and non-HDL but decreased levels of HDL, compared to those homozygous for the SS allele (Table 2). These QTLs affected their respective trait values in an additive manner.

Correlations between plasma glucose and lipid levels

The correlations of fasting glucose levels with plasma levels of HDL, non-HDL cholesterol, or triglyceride were analyzed with the F_2 population (Fig. 7). A significant inverse correlation between fasting glucose and HDL cholesterol levels was observed when the mice were fed a chow ($R = -0.220$; $P = 8.1E-4$) or Western diet ($R = -0.257$; $P = 8.5E-5$). F_2 mice with higher HDL cholesterol levels had lower fasting glucose levels. Conversely, significant positive correlations between fasting glucose and non-

HDL cholesterol levels were observed when mice were fed either chow ($R = 0.194$; $P = 3.31E-3$) or Western diet ($R = 0.558$; $P = 4.7E-20$). F_2 mice with higher non-HDL cholesterol levels also had higher fasting glucose levels, especially on the Western diet. A significant positive correlation between plasma levels of fasting glucose and triglyceride was observed when mice were fed the Western diet ($R = 0.377$; $P = 3.9E-9$) but not the chow diet ($R = 0.065$; $P = 0.330$).

Prioritization of positional candidate genes for Chr9 coincident QTLs

Bglu17 on Chr9 has been mapped in 3 separate intercrosses, including previously reported C57BLKS x DBA/2 [35] and B6-*Apoe*[-/-] x BALB-*Apoe*[-/-] crosses [21]. *Hdlq17* on Chr9 has been mapped in multiple crosses, including B6 x 129, B6 x CAST/EiJ, B6-*Apoe*[-/-] x C3H-*Apoe*[-/-], and B6-*Apoe*[-/-] x BALB-*Apoe*[-/-] crosses

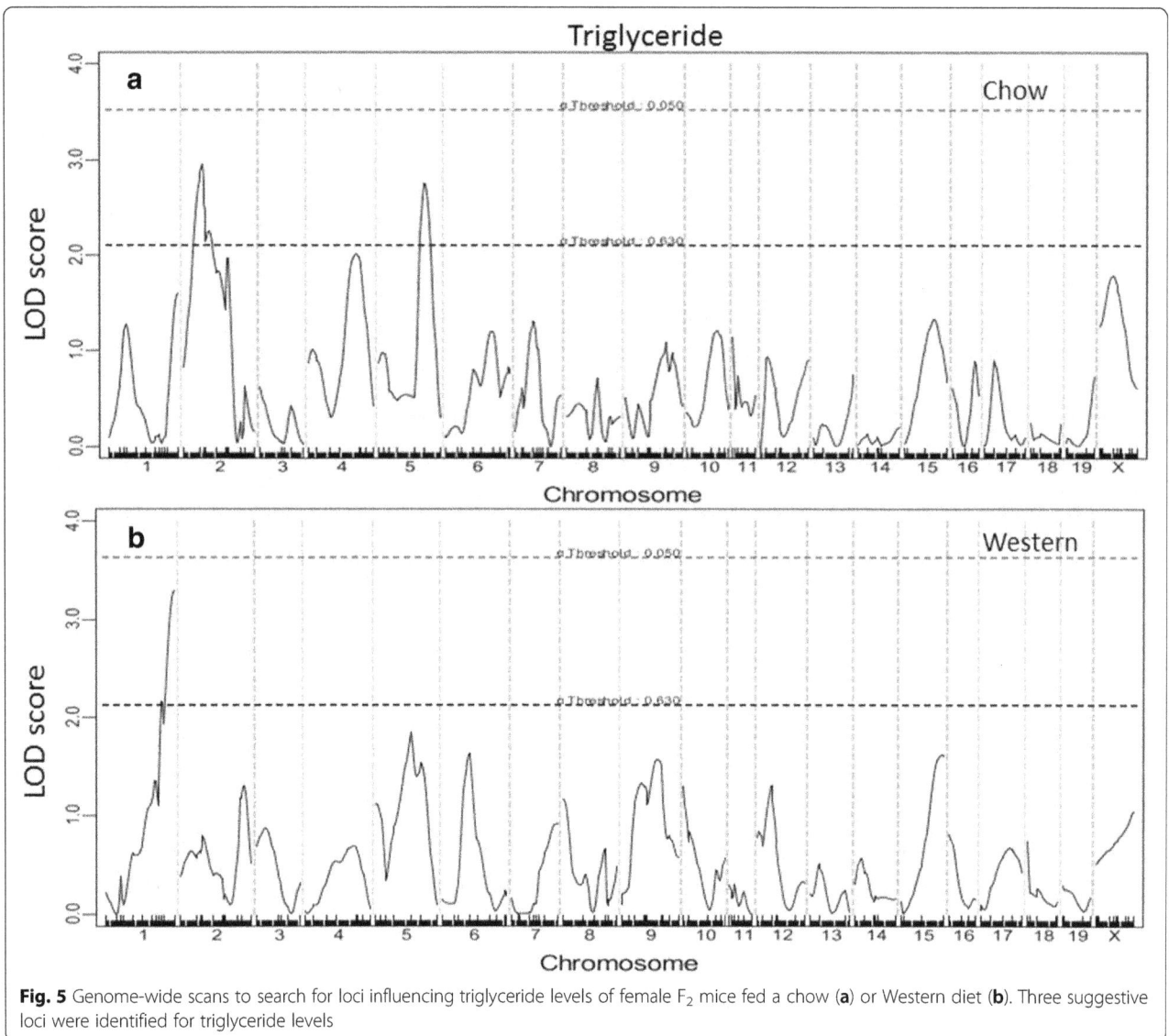

Fig. 5 Genome-wide scans to search for loci influencing triglyceride levels of female F_2 mice fed a chow (**a**) or Western diet (**b**). Three suggestive loci were identified for triglyceride levels

[24, 27, 31, 32, 36–38]. We conducted haplotype analyses using Sanger SNP database to prioritize positional candidate genes for both QTLs. Prioritized candidate genes for *Hdlq17* are shown in Additional file 1: Table S1, and candidate genes for *Bglu17* are shown in Additional file 2: Table S2. Most candidates for *Hdlq17* are also candidate genes for *Bglu17*. These candidates contain one or more non-synonymous SNPs in the coding regions or SNPs in the upstream regulatory region that are shared by the high allele strains but are different from the low allele strains at the QTL. All candidate genes were further examined for associations with relevant human diseases using the NIH GWAS database (http://www.genome.gov/GWAStudies/). *Sik3, Apoa1,* and *Apoc3* have been shown to be associated with variations in total, HDL, LDL-cholesterol or triglyceride levels [6, 7, 39], and *Cadm1* with obesity-related traits [40].

Discussion

BALB and SM are two mouse strains that exhibit distinct differences in HDL, non-HDL cholesterol, and type 2 diabetes-related traits when deficient in *Apoe* [16]. BALB-*Apoe*$^{-/-}$ mice have higher HDL, lower non-HDL cholesterol, and lower glucose levels than SM-*Apoe*$^{-/-}$ mice when they are fed a Western diet. To identify the genetic factors responsible for these differences, we performed QTL analysis on a female cohort derived from an intercross between the two *Apoe*$^{-/-}$ strains. We have identified four loci contributing to fasting glucose levels, four loci contributing to HDL cholesterol levels, nine loci for non-HDL cholesterol levels, and three loci for triglyceride levels. Moreover, we have observed genetic connections between dyslipidemia and type 2 diabetes in that the QTL for fasting glucose is colocalized with the QTLs for HDL and non-HDL cholesterol on chromosome

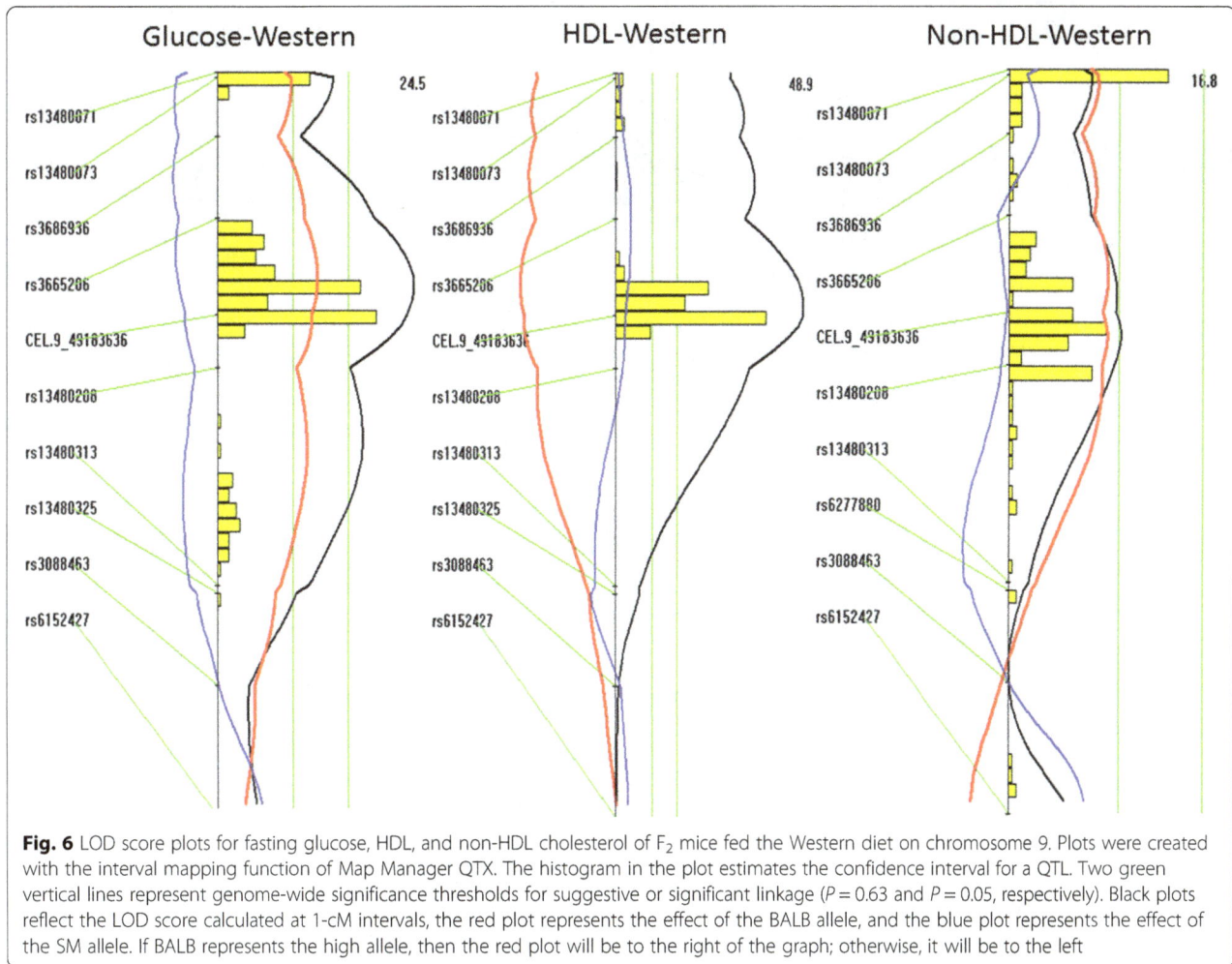

Fig. 6 LOD score plots for fasting glucose, HDL, and non-HDL cholesterol of F_2 mice fed the Western diet on chromosome 9. Plots were created with the interval mapping function of Map Manager QTX. The histogram in the plot estimates the confidence interval for a QTL. Two green vertical lines represent genome-wide significance thresholds for suggestive or significant linkage ($P = 0.63$ and $P = 0.05$, respectively). Black plots reflect the LOD score calculated at 1-cM intervals, the red plot represents the effect of the BALB allele, and the blue plot represents the effect of the SM allele. If BALB represents the high allele, then the red plot will be to the right of the graph; otherwise, it will be to the left

9 and these coincident QTLs share a large fraction of potential candidate genes.

We identified a significant QTL on chromosome 9, peaked at 26 cM, which affected fasting plasma glucose levels when mice were fed a chow or Western diet. We named it *Bglu17* to represent a novel locus regulating fasting glucose levels in the mouse. This locus is overlapping with a significant QTL (not named) for blood glucose levels on the intraperitoneal glucose tolerance test identified in a BKS-Cg-*Leprdb*$^{+/+m}$ x DBA/2 intercross and a suggested QTL identified in a B6-*Apoe*$^{-/-}$ x BALB-*Apoe*$^{-/-}$ intercross [21, 35]. Interestingly, we found that *Bglu17* coincided precisely with *Hdlq17*, a QTL for HDL cholesterol levels, and *Nhdlq11*, a QTL for non-HDL cholesterol levels. The colocalization of two or more QTLs for different traits suggests that these traits are controlled either by the same gene(s) or closely linked but different individual genes. *Hdlq17* has been mapped in multiple crosses derived from inbred mouse strains whose genomes have been resequenced by Sanger, including B6, 129, BALB, C3H/HeJ, and CAST/EiJ

[24, 27, 31, 32, 36–38]. *Nhdlq11* was previously mapped in a NZB/BINJ x SM/JF2 cross and a B6-*Apoe*$^{-/-}$ x C3H-*Apoe*$^{-/-}$ intercross [32, 41]. To determine whether *Bglu17* and *Hdlq17* share the same underlying candidate genes, we performed haplotype analyses on those crosses that led to the identification of the QTLs. The number of shared genetic variants between *Bglu17* and *Hdlq17* was surprisingly high. Of them, *Sik3*, *Apoa1*, and *Apoc3* are located precisely underneath the linkage peak of *Bglu17* and *Hdlq17*, and they are also functional candidate genes of *Hdlq17*. Indeed, recent GWAS studies have associated these three genes with dyslipidemia or variations in HDL, LDL cholesterol, and triglyceride levels [6, 39, 42]. The finding in this study strongly suggests that one or more of these "lipid genes" might be the causal gene(s) of *Bglu17*, contributing to variation in fasting glucose levels. Although it is unknown how they affect glucose homeostasis, one probable effect path is through the influence on plasma lipid levels, which then predispose variation in glucose-related traits. The current

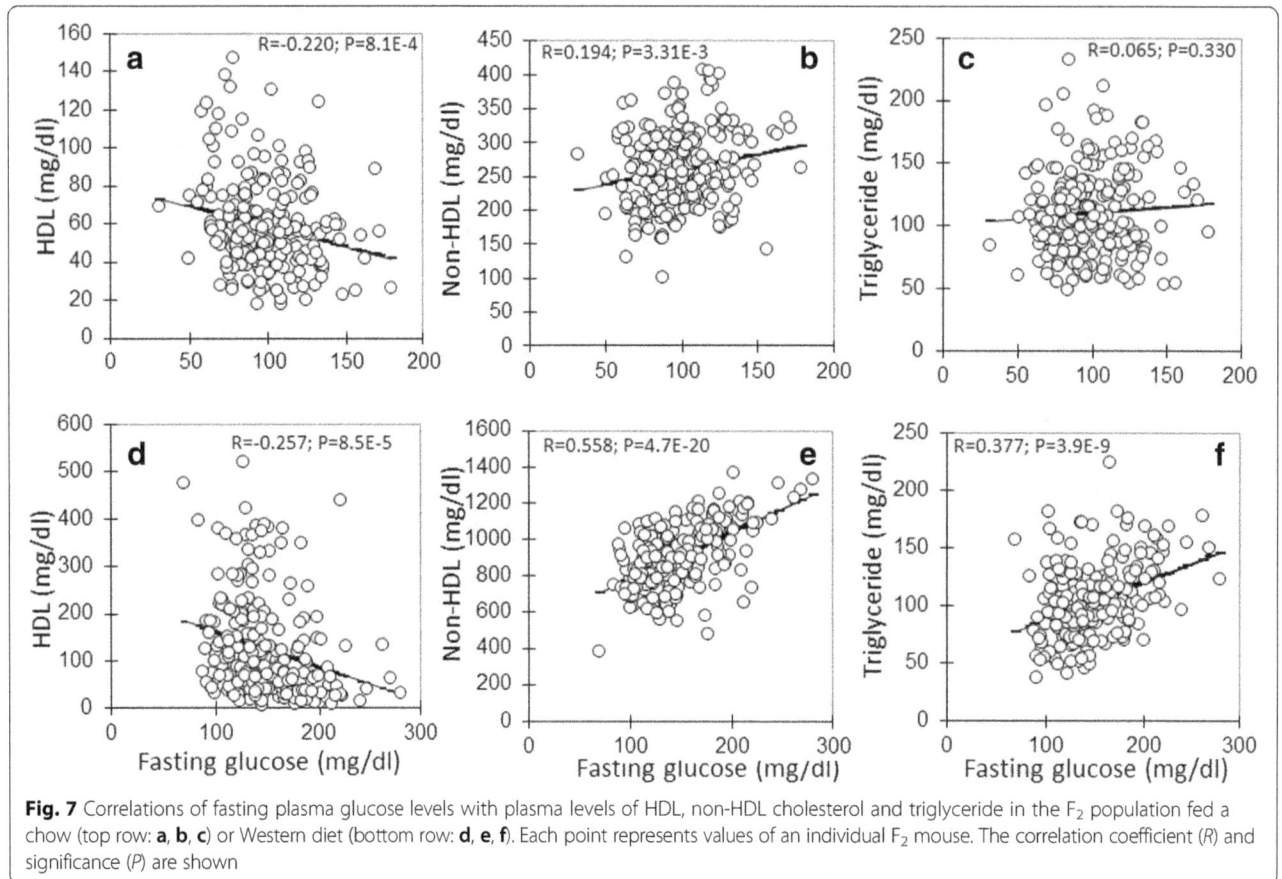

Fig. 7 Correlations of fasting plasma glucose levels with plasma levels of HDL, non-HDL cholesterol and triglyceride in the F₂ population fed a chow (top row: **a**, **b**, **c**) or Western diet (bottom row: **d**, **e**, **f**). Each point represents values of an individual F₂ mouse. The correlation coefficient (R) and significance (P) are shown

observation on the significant correlations of fasting glucose levels with HDL, non-HDL cholesterol, and triglyceride levels in this cross supports this speculation. Plasma lipid levels, especially non-HDL cholesterol, of the F_2 mice were significantly elevated on the Western diet, so were the fasting glucose levels. When fed the Western diet, $Apoe^{-/-}$ mice display a rapid rise in non-HDL cholesterol levels, often reaching their peak within a couple of weeks (unpublished data), whereas their blood glucose levels rise more slowly and gradually within 12 weeks [43, 44]. This difference in onset suggests a causal role for plasma lipids in the rise of blood glucose in the $Apoe^{-/-}$ mouse model.

A significant reverse correlation was observed between plasma HDL cholesterol levels and fasting glucose levels in this cross on either chow or Western diet. This result is consistent with the findings of prospective human studies that low HDL levels can predict the future risk of developing T2D and low HDL levels are more prevalent in diabetic patients than in the normal population [45, 46]. HDL can increase insulin secretion from β-cells, improve insulin sensitivity of the target tissues, and accelerate glucose uptake by muscle via the AMP-activated protein kinase [47]. A significant correlation of non-HDL cholesterol levels with fasting glucose levels was also observed in this cross, and the correlation was

extremely high when mice were fed the Western diet. Emerging human studies have also revealed associations of non-HDL cholesterol and ApoB with fasting glucose levels and incident type 2 diabetes [48–50]. We previously observed that the elevation of non-HDL cholesterol levels in $Apoe^{-/-}$ mice during the consumption of a Western diet induces a chronic, low-grade inflammation state characterized by rises in circulating cytokines and infiltration of monocytes/macrophages in various organs or tissues [13, 17, 20, 43]. Inflammation in the islets impairs β-cell function [20]. LDL can also directly affect function and survival of β-cells [51]. In addition, high levels of LDL can induce insulin resistance due to its lipotoxicity and effect on endoplasmic reticulum stress [1].

Plasma triglyceride levels were strongly correlated with fasting glucose levels in this cross on the Western diet, although no significant correlation was found when mice were fed the chow diet. Despite the strong correlation, no overlapping QTLs were observed for fasting glucose and triglyceride. The reason for the discrepancy between non-HDL cholesterol and triglyceride in terms of the presence or absence of colocalized QTLs is unclear.

A suggestive QTL for fasting glucose near the proximal end of chromosome 9 (2.37 cM) was detected in this cross, initially on the chow diet and then replicated on the Western diet. The LOD score plot for chromosome 9 has shown 2 distinct peaks, one with a suggestive LOD score at the proximal end and one with a significant LOD score at a more distal region, suggesting the existence of two loci for fasting glucose on the chromosome. The bootstrap test, a statistical method for defining the confidence interval of QTLs using simulation [52], also indicated the existence of two QTLs for the trait on chromosome 9. We named the proximal one *Bglu16* to represent a new QTL for fasting glucose in the mouse. Naming a suggestive locus is considered appropriate if it is repeatedly observed [53].

Two suggestive QTLs for fasting glucose on chromosome 5 were identified when mice were fed the Western diet. The proximal one replicated *Bglu13*, recently mapped in the B6-Apoe$^{-/-}$ x BALB-Apoe$^{-/-}$ cross [21]. One probable candidate gene for this QTL is *Hnf1a*, which encodes hepatocyte nuclear factor 1α. In humans, *Hnf1a* mutations are the most common cause of maturity-onset diabetes of the young (MODY) [54]. The suggestive QTL in the distal region was novel.

Most of the QTLs identified for plasma lipids confirm those identified in previous studies, whereas two QTLs for non-HDL are new and named *Nhdlq15* and *Nhdlq16*, respectively. The QTLs on distal chromosome 1 for HDL and triglyceride has been mapped in a number of mouse crosses, and *Apoa2* has been identified as the underlying causal gene [55]. However, the QTL (~90 cM) mapped in this study showed that it was more distal to the *Apoa2* gene (80 cM), thus suggesting a different underlying causal gene.

Conclusion

We have identified multiple QTLs contributing to dyslipidemia and hyperglycemia in a segregating F$_2$ population. The finding on the colocalization of QTLs for fasting glucose, HDL and non-HDL cholesterol levels and the sharing of probable candidate genes has demonstrated genetic connections between dyslipidemia and type 2 diabetes. The close correlations of fasting glucose with HDL, non-HDL cholesterol, and triglyceride support the hypothesis that dyslipidemia plays a causal role in the development of type 2 diabetes [1]. The haplotype analysis has prioritized candidates for either chromosome 9 QTL down to a handful of genes. Nevertheless, functional studies need to be performed to prove causality.

Availability of supporting data

Data are accessible through this link: https://mynotebook.labarchives.com/doi/MTc2Mzk0LjR8MTM1Njg4LzEzNTY4OC9Ob3RlYm9vay80MTAyOTgxMTQ8fDQ0Nzc3c3MC40/10.6070/H4D50K0X.

Abbreviations
QTL: Quantitative trait locus (QTL) analysis; LOD: Logarithm of odds ratio; T2D: Type 2 diabetes; LDL: Low-density lipoprotein; HDL: High-density lipoprotein; GWAS: Genome-wide association studies; Apoe: Apolipoprotein E.

Competing interests
Authors have no financial and nonfinancial competing interests.

Authors' contributions
QW measured plasma lipid and glucose levels. ATG performed haplotype analysis. AM analyzed the genotyping data. EF and SOG genotyped the cross. WS conceived of the study, created the cross, analyzed the data and draft the manuscript. All authors read and approved the final manuscript.

Acknowledgements
This work was supported by NIH grants DK097120 and HL112281.

Author details
[1]Department of Radiology & Medical Imaging, University of Virginia, Snyder Bldg Rm 266, 480 Ray C. Hunt Dr., P.O. Box 801339, Fontaine Research Park, Charlottesville, VA 22908, USA. [2]University of Virginia, Snyder Bldg Rm 266, 480 Ray C. Hunt Dr., P.O. Box 801339, Fontaine Research Park, Charlottesville, VA 22908, USA. [3]Department of Biochemistry & Molecular Genetics, University of Virginia, Charlottesville, VA, USA. [4]University of Virginia, Charlottesville, VA, USA. [5]Center for Public Health and Genomics, University of Virginia, Charlottesville, VA, USA.

References
1. Li N, Fu J, Koonen DP, Kuivenhoven JA, Snieder H, Hofker MH. Are hypertriglyceridemia and low HDL causal factors in the development of insulin resistance? Atherosclerosis. 2014;233(1):130–8.
2. Saleheen D, Nazir A, Khanum S, Haider SR, Frossard PM. R1615P: a novel mutation in ABCA1 associated with low levels of HDL and type II diabetes mellitus. Int J Cardiol. 2006;110(2):259–60.
3. Albert JS, Yerges-Armstrong LM, Horenstein RB, Pollin TI, Sreenivasan UT, Chai S, et al. Null mutation in hormone-sensitive lipase gene and risk of type 2 diabetes. N Engl J Med. 2014;370(24):2307–15.
4. Hu Y, Ren Y, Luo RZ, Mao X, Li X, Cao X, et al. Novel mutations of the lipoprotein lipase gene associated with hypertriglyceridemia in members of type 2 diabetic pedigrees. J Lipid Res. 2007;48(8):1681–8.
5. Mani A, Radhakrishnan J, Wang H, Mani A, Mani MA, Nelson-Williams C, et al. LRP6 mutation in a family with early coronary disease and metabolic risk factors. Science. 2007;315(5816):1278–82.
6. Teslovich TM, Musunuru K, Smith AV, Edmondson AC, Stylianou IM, Koseki M, et al. Biological, clinical and population relevance of 95 loci for blood lipids. Nature. 2010;466(7307):707–13.
7. Global Lipids Genetics Consortium, Willer CJ, Schmidt EM, Sengupta S, Peloso GM, Gustafsson S, et al. Discovery and refinement of loci associated with lipid levels. Nat Genet. 2013;45(11):1274–83.
8. Dupuis J, Langenberg C, Prokopenko I, Saxena R, Soranzo N, Jackson AU, et al. New genetic loci implicated in fasting glucose homeostasis and their impact on type 2 diabetes risk. Nat Genet. 2010;42(2):105–16.

9.   Soranzo N, Sanna S, Wheeler E, Gieger C, Radke D, Dupuis J, et al. Common variants at 10 genomic loci influence hemoglobin A(1)(C) levels via glycemic and nonglycemic pathways. Diabetes. 2010;59(12):3229–39.

10. Manning AK, Hivert MF, Scott RA, Grimsby JL, Bouatia-Naji N, Chen H, et al. A genome-wide approach accounting for body mass index identifies genetic variants influencing fasting glycemic traits and insulin resistance. Nat Genet. 2012;44(6):659–69.

11. Li N, van der Sijde MR, LifeLines Cohort Study Group, Bakker SJ, Dullaart RP, van der Harst P, et al. Pleiotropic effects of lipid genes on plasma glucose, HbA1c, and HOMA-IR levels. Diabetes. 2014;63(9):3149–58.

12. Shi W, Wang NJ, Shih DM, Sun VZ, Wang X, Lusis AJ. Determinants of atherosclerosis susceptibility in the C3H and C57BL/6 mouse model: evidence for involvement of endothelial cells but not blood cells or cholesterol metabolism. Circ Res. 2000;86(10):1078–84.

13. Tian J, Pei H, James JC, Li Y, Matsumoto AH, Helm GA, et al. Circulating adhesion molecules in apoE-deficient mouse strains with different atherosclerosis susceptibility. Biochem Biophys Res Commun. 2005;329(3):1102–7.

14. Nakashima Y, Plump AS, Raines EW, Breslow JL, Ross R. ApoE-deficient mice develop lesions of all phases of atherosclerosis throughout the arterial tree. Arterioscler Thromb. 1994;14(1):133–40.

15. Shi W, Zhang Z, Chen MH, Angle JF, Matsumoto AH. Genes within the MHC region have a dramatic influence on radiation-enhanced atherosclerosis in mice. Circ Cardiovasc Genet. 2010;3(5):409–13.

16. Liu S, Li J, Chen MH, Liu Z, Shi W. Variation in Type 2 Diabetes-Related Phenotypes among Apolipoprotein E-Deficient Mouse Strains. PLoS One. 2015;10(5):e0120935.

17. Zhang Y, Kundu B, Zhong M, Huang T, Li J, Chordia MD, et al. PET imaging detection of macrophages with a formyl peptide receptor antagonist. Nucl Med Biol. 2015;42(4):381–6.

18. Breslow JL. Genetic differences in endothelial cells may determine atherosclerosis susceptibility. Circulation. 2000;102(1):5–6.

19. Su Z, Li Y, James JC, Matsumoto AH, Helm GA, Lusis AJ, et al. Genetic linkage of hyperglycemia, body weight and serum amyloid-P in an intercross between C57BL/6 and C3H apolipoprotein E-deficient mice. Hum Mol Genet. 2006;15(10):1650–8.

20. Li J, Wang Q, Chai W, Chen MH, Liu Z, Shi W. Hyperglycemia in apolipoprotein E-deficient mouse strains with different atherosclerosis susceptibility. Cardiovasc Diabetol. 2011;10(1):117.

21. Zhang Z, Rowlan JS, Wang Q, Shi W. Genetic analysis of atherosclerosis and glucose homeostasis in an intercross between C57BL/6 and BALB/cJ apolipoprotein E-deficient mice. Circ Cardiovasc Genet. 2012;5(2):190–201.

22. Su Z, Li Y, James JC, McDuffie M, Matsumoto AH, Helm GA, et al. Quantitative trait locus analysis of atherosclerosis in an intercross between C57BL/6 and C3H mice carrying the mutant apolipoprotein E gene. Genetics. 2006;172(3):1799–807.

23. Yuan Z, Pei H, Roberts DJ, Zhang Z, Rowlan JS, Matsumoto AH, et al. Quantitative trait locus analysis of neointimal formation in an intercross between C57BL/6 and C3H/HeJ apolipoprotein E-deficient mice. Circ Cardiovasc Genet. 2009;2(3):220–8.

24. Rowlan JS, Li Q, Manichaikul A, Wang Q, Matsumoto AH, Shi W. Atherosclerosis susceptibility Loci identified in an extremely atherosclerosis-resistant mouse strain. J Am Heart Assoc. 2013;2(4):e000260.

25. Wang X, Paigen B. Genetics of variation in HDL cholesterol in humans and mice. Circ Res. 2005;96(1):27–42.

26. Seidelmann SB, De Luca C, Leibel RL, Breslow JL, Tall AR, Welch CL. Quantitative trait locus mapping of genetic modifiers of metabolic syndrome and atherosclerosis in low-density lipoprotein receptor-deficient mice: identification of a locus for metabolic syndrome and increased atherosclerosis on chromosome 4. Arterioscler Thromb Vasc Biol. 2005;25(1):204–10.

27. Ishimori N, Li R, Kelmenson PM, Korstanje R, Walsh KA, Churchill GA, et al. Quantitative trait loci analysis for plasma HDL-cholesterol concentrations and atherosclerosis susceptibility between inbred mouse strains C57BL/6 J and 129S1/SvImJ. Arterioscler Thromb Vasc Biol. 2004;24(1):161–6.

28. Korstanje R, Li R, Howard T, Kelmenson P, Marshall J, Paigen B, et al. Influence of sex and diet on quantitative trait loci for HDL cholesterol levels in an SM/J by NZB/BlNJ intercross population. J Lipid Res. 2004;45(5):881–8.

29. Lyons MA, Wittenburg H, Li R, Walsh KA, Korstanje R, Churchill GA, et al. Quantitative trait loci that determine lipoprotein cholesterol levels in an intercross of 129S1/SvImJ and CAST/Ei inbred mice. Physiol Genomics. 2004;17(1):60–8.

30. Wittenburg H, Lyons MA, Li R, Kurtz U, Wang X, Mossner J, et al. QTL mapping for genetic determinants of lipoprotein cholesterol levels in combined crosses of inbred mouse strains. J Lipid Res. 2006;47(8):1780–90.

31. Sehayek E, Duncan EM, Yu HJ, Petukhova L, Breslow JL. Loci controlling plasma non-HDL and HDL cholesterol levels in a C57BL /6 J x CASA /Rk intercross. J Lipid Res. 2003;44(9):1744–50.

32. Li Q, Li Y, Zhang Z, Gilbert TR, Matsumoto AH, Dobrin SE, et al. Quantitative trait locus analysis of carotid atherosclerosis in an intercross between C57BL/6 and C3H apolipoprotein E-deficient mice. Stroke. 2008;39(1):166–73.

33. Ishimori N, Li R, Kelmenson PM, Korstanje R, Walsh KA, Churchill GA, et al. Quantitative trait loci that determine plasma lipids and obesity in C57BL/6 J and 129S1/SvImJ inbred mice. J Lipid Res. 2004;45(9):1624–32.

34. Stylianou IM, Langley SR, Walsh K, Chen Y, Revenu C, Paigen B. Differences in DBA/1J and DBA/2J reveal lipid QTL genes. J Lipid Res. 2008;49(11):2402–13.

35. Yaguchi H, Togawa K, Moritani M, Itakura M. Identification of candidate genes in the type 2 diabetes modifier locus using expression QTL. Genomics. 2005;85(5):591–9.

36. Lyons MA, Korstanje R, Li R, Walsh KA, Churchill GA, Carey MC, et al. Genetic contributors to lipoprotein cholesterol levels in an intercross of 129S1/SvImJ and RIIIS/J inbred mice. Physiol Genomics. 2004;17(2):114–21.

37. Su Z, Wang X, Tsaih SW, Zhang A, Cox A, Sheehan S, et al. Genetic basis of HDL variation in 129/SvImJ and C57BL/6J mice: importance of testing candidate genes in targeted mutant mice. J Lipid Res. 2009;50(1):116–25.

38. Rowlan JS, Zhang Z, Wang Q, Fang Y, Shi W. New quantitative trait loci for carotid atherosclerosis identified in an intercross derived from apolipoprotein E-deficient mouse strains. Physiol Genomics. 2013;45(8):332–42.

39. Ko A, Cantor RM, Weissglas-Volkov D, Nikkola E, Reddy PM, Sinsheimer JS, et al. Amerindian-specific regions under positive selection harbour new lipid variants in Latinos. Nat Commun. 2014;5:3983.

40. Comuzzie AG, Cole SA, Laston SL, Voruganti VS, Haack K, Gibbs RA, et al. Novel genetic loci identified for the pathophysiology of childhood obesity in the Hispanic population. PLoS One. 2012;7(12):e51954.

41. Purcell-Huynh DA, Weinreb A, Castellani LW, Mehrabian M, Doolittle MH, Lusis AJ. Genetic factors in lipoprotein metabolism. Analysis of a genetic cross between inbred mouse strains NZB/BlNJ and SM/J using a complete linkage map approach. J Clin Invest. 1995;96(4):1845–58.

42. Willer CJ, Mohlke KL. Finding genes and variants for lipid levels after genome-wide association analysis. Curr Opin Lipidol. 2012;23(2):98–103.

43. Li J, Lu Z, Wang Q, Su Z, Bao Y, Shi W. Characterization of Bglu3, a mouse fasting glucose locus, and identification of Apcs as an underlying candidate gene. Physiol Genomics. 2012;44(6):345–51.

44. Zhou W, Chen MH, Shi W. Influence of phthalates on glucose homeostasis and atherosclerosis in hyperlipidemic mice. BMC Endocr Disord. 2015;15(1):13-015-0015-4.

45. Wilson PW, Kannel WB, Anderson KM. Lipids, glucose intolerance and vascular disease: the Framingham Study. Monogr Atheroscler. 1985;13:1–11.

46. Wilson PW, Meigs JB, Sullivan L, Fox CS, Nathan DM, D'Agostino RBS. Prediction of incident diabetes mellitus in middle-aged adults: the Framingham Offspring Study. Arch Intern Med. 2007;167(10):1068–74.

47. Drew BG, Rye KA, Duffy SJ, Barter P, Kingwell BA. The emerging role of HDL in glucose metabolism. Nat Rev Endocrinol. 2012;8(4):237–45.

48. Hwang YC, Ahn HY, Yu SH, Park SW, Park CY. Atherogenic dyslipidaemic profiles associated with the development of Type 2 diabetes: a 3.1-year longitudinal study. Diabet Med. 2014;31(1):24–30.

49. Hwang YC, Ahn HY, Park SW, Park CY. Apolipoprotein B and non-HDL cholesterol are more powerful predictors for incident type 2 diabetes than fasting glucose or glycated hemoglobin in subjects with normal glucose tolerance: a 3.3-year retrospective longitudinal study. Acta Diabetol. 2014;51(6):941–6.

50. Ley SH, Harris SB, Connelly PW, Mamakeesick M, Gittelsohn J, Wolever TM, et al. Association of apolipoprotein B with incident type 2 diabetes in an aboriginal Canadian population. Clin Chem. 2010;56(4):666–70.

51. Rutti S, Ehses JA, Sibler RA, Prazak R, Rohrer L, Georgopoulos S, et al. Low- and high-density lipoproteins modulate function, apoptosis, and proliferation of primary human and murine pancreatic beta-cells. Endocrinology. 2009;150(10):4521–30.

52. Visscher PM, Thompson R, Haley CS. Confidence intervals in QTL mapping by bootstrapping. Genetics. 1996;143(2):1013–20.

53. Abiola O, Angel JM, Avner P, Bachmanov AA, Belknap JK, Bennett B, et al. The nature and identification of quantitative trait loci: a community's view. Nat Rev Genet. 2003;4(11):911–6.

54. Shepherd M, Ellis I, Ahmad AM, Todd PJ, Bowen-Jones D, Mannion G, et al. Predictive genetic testing in maturity-onset diabetes of the young (MODY). Diabet Med. 2001;18(5):417–21.

55. Wang X, Korstanje R, Higgins D, Paigen B. Haplotype analysis in multiple crosses to identify a QTL gene. Genome Res. 2004;14(9):1767–72.

22

Serum bilirubin concentration is modified by *UGT1A1* Haplotypes and influences risk of Type-2 diabetes in the Norfolk Island genetic isolate

M. C. Benton[1], R. A. Lea[1], D. Macartney-Coxson[2], C. Bellis[1,3], M. A. Carless[3], J. E. Curran[3], M. Hanna[1], D. Eccles[1], G. K. Chambers[4], J. Blangero[5] and L. R. Griffiths[1*]

Abstract

Background: Located in the Pacific Ocean between Australia and New Zealand, the unique population isolate of Norfolk Island has been shown to exhibit increased prevalence of metabolic disorders (type-2 diabetes, cardiovascular disease) compared to mainland Australia. We investigated this well-established genetic isolate, utilising its unique genomic structure to increase the ability to detect related genetic markers. A pedigree-based genome-wide association study of 16 routinely collected blood-based clinical traits in 382 Norfolk Island individuals was performed.

Results: A striking association peak was located at chromosome 2q37.1 for both total bilirubin and direct bilirubin, with 29 SNPs reaching statistical significance ($P < 1.84 \times 10^{-7}$). Strong linkage disequilibrium was observed across a 200 kb region spanning the UDP-glucuronosyltransferase family, including *UGT1A1*, an enzyme known to metabolise bilirubin. Given the epidemiological literature suggesting negative association between CVD-risk and serum bilirubin we further explored potential associations using stepwise multivariate regression, revealing significant association between direct bilirubin concentration and type-2 diabetes risk. In the Norfolk Island cohort increased direct bilirubin was associated with a 28 % reduction in type-2 diabetes risk (OR: 0.72, 95 % CI: 0.57-0.91, $P = 0.005$). When adjusted for genotypic effects the overall model was validated, with the adjusted model predicting a 30 % reduction in type-2 diabetes risk with increasing direct bilirubin concentrations (OR: 0.70, 95 % CI: 0.53-0.89, $P = 0.0001$).

Conclusions: In summary, a pedigree-based GWAS of blood-based clinical traits in the Norfolk Island population has identified variants within the UDPGT family directly associated with serum bilirubin levels, which is in turn implicated with reduced risk of developing type-2 diabetes within this population.

Keywords: Norfolk Island, GWAS, Bilirubin, type-2 diabetes, *UGT1A1*

Background

This study examined a large multi-generational pedigree from the isolated population of Norfolk Island to identify genomic variants (SNPs – single nucleotide polymorphisms) associated with routinely collected blood-based clinical traits. The Norfolk Island population is a genetic isolate with strong family groups and a well-documented family genealogy [1]. Norfolk Island is a small volcanic island located in the Pacific Ocean between Australia (about 1600 km north-east of Sydney) and New Zealand (1077 km north-west of Auckland). Alongside geographic isolation, a unique history has shaped the genomic architecture of the current pedigree members resulting in an admixed population with both European and Polynesian ancestry [2]. Recent estimation of the admixture in the Norfolk Island cohort reported 88 % European ancestry and 12 % Polynesian ancestry [2].

---

* Correspondence: lyn.griffiths@qut.edu.au
[1]Genomics Research Centre, Institute of Health and Biomedical Innovation, Queensland University of Technology, Kelvin Grove, QLD 4059, Australia
Full list of author information is available at the end of the article

To date the Norfolk Island Health Study (NIHS) has collected data and samples for 1199 Norfolk Islanders, 52 % ($N = 624$) of whom were found to have direct links to the original founders. Using this in-depth genealogical information a large multi-generational Norfolk pedigree was reconstructed [1]. Several studies have established admixture scores and presence of founder effects within the Norfolk Island pedigree [1–3] and the pedigree has been shown to have sufficient power to detect genetic loci influencing complex traits via linkage and association [4–7].

The Norfolk Island population has high rates of metabolic syndrome [7] and cardiovascular related risk factor traits, especially obesity, compared to mainland Australia. Research on the Norfolk pedigree has shown that traits for obesity, dyslipidaemia, blood glucose and hypertension exhibit a substantial genetic component, with heritability estimates ranging from 30 % for systolic blood pressure (SBP) to 63 % for low density lipoproteins (LDL) cholesterol [1, 4, 5]. In addition, factor analysis identified "composite" phenotypes with high heritability [5], suggesting that common gene(s) underlie cardiovascular disease-related phenotypes. Furthermore, genetic linkage analysis in the Norfolk Island pedigree has successfully identified previously documented regions associated with cardiovascular disease risk traits, the most significant being for SBP on chromosome 1 (1p36) [4].

Reported rates of type-2 diabetes within the Norfolk Island population are similar to mainland Australia (4-8 %). However, a significantly higher proportion of individuals had fasting blood glucose in excess of normal ranges (>5 mmol/L), suggesting a high prevalence of pre-diabetes and possible under-diagnosis of type-2 diabetes [4, 8]. Additionally, clinical diagnosis of type-2 diabetes using AUSDRISK [9] identified that 42 % of the Norfolk Island population were at high-risk of developing the disease [7].

Bilirubin is a component of haemoglobin, formed during metabolic breakdown in the liver. Total serum bilirubin measures both water-soluble (direct-) and fat-soluble (indirect-) bilirubin. Bilirubin is also a potent antioxidant and as such has a vital role in the protection of the body against reactive oxygen species [10–12]. Numerous epidemiological analyses have reported strong negative associations between CVD-risk and serum bilirubin levels. Very few studies investigating the link between type-2 diabetes and serum bilirubin concentration have been conducted [13], although recently an association with mortality in a type-2 diabetic cohort was observed [14]. Serum bilirubin concentration has been shown to be tightly regulated by the UDP-glucuronosyltransferase (UDPGT) enzyme family, with several large GWAS and linkage studies identifying variants within UGT1A in particular [15–18]. This is suggestive of a potentially heritable metabolic disease factor, for which a recent study provides further supportive

evidence; a Mendelian randomization study exploring total bilirubin levels in a prospective study found further evidence for a protective role in type-2 diabetes [19].

The aim of this study was to update the previously calculated heritabilities for a range of blood-based traits relating to CVD risk in the Norfolk Island cohort and to perform genome-wide association studies (GWASs) of the heritable traits using a pedigree-based approach.

Results
Heritability of individual metabolic traits

A description of the blood-based clinical traits investigated in this study, including summary statistics, is shown in Additional file 1. The latest pedigree relationship information and GenABEL were used to calculate heritability (h^2) statistics for all traits profiled in the Norfolk Island cohort. In total, 16 traits (out of 19) yielded statistically significant h^2 values ranging from 0.225 – 0.563 (nominal $P < 0.05$). The average heritability was 0.39 and 8 traits exhibited a higher than average heritability (total protein, globin, total bilirubin, LDL-C, cholesterol, alkaline phosphatase, and urea) the most heritable trait being total protein ($h^2 = 0.563$, $P = 2.26 \times 10^{-4}$). A summary of all significantly heritable major blood-based clinical traits is shown in Table 1.

GWAS of metabolic traits

All 16 heritable blood-based clinical traits were screened for association separately; individual trait GWAS Manhattan plots can be viewed in Additional file 2. There were 2 traits with robustly associated clusters (i.e. SNPs in close

Table 1 Significantly heritable metabolic traits in the Norfolk Island population

Trait	h^2	P value
Total protein	0.563	2.26E-04
Globin	0.525	3.36E-04
Total bilirubin	0.502	4.45E-05
LDL-C	0.454	7.51E-05
Cholesterol	0.426	8.41E-05
Chlorine	0.426	8.41E-05
Alkaline phosphatase	0.425	6.10E-04
Urea	0.424	7.87E-04
GGT	0.369	3.02E-03
Albumin	0.358	9.01E-03
Uric Acid	0.350	2.17E-03
HDL-C	0.348	1.58E-02
Direct Bilirubin	0.327	3.30E-02
Creatinine	0.257	2.43E-02
AST	0.251	3.81E-02
Cholesterol/HDL-C ratio	0.225	2.42E-02

proximity to each other); total bilirubin and direct bilirubin. It should be noted that a number of SNPs passed the adjusted significance threshold for liver function traits (i.e. GGT, AST, ADH). These traits exhibited numerous SNPs passing M_{eff} adjustment, however robust 'peaks'/clusters of SNPs were not observed.

Exploration of the bilirubin association on chromosome 2q37.1

The strongest observed association was seen between a cluster of 29 SNPs on chromosome 2q37.1 passing a M_{eff} adjusted threshold and total serum bilirubin (Fig. 1a, Table 2). The most robustly associated SNP was rs6744284 ($P = 1.87 \times 10^{-16}$). A weaker association was observed for the same cluster of SNPs on chromosome 2q37.1 with direct serum bilirubin levels (Fig. 1b). These 29 SNPs span a region of 189.8 kb, and lie directly on top of a complex locus that codes numerous isoforms of the UDP-glucuronosyltransferase (UGT) family (Fig. 2).

LD block identification

Evidence of strong linkage disequilibrium (LD) across the 29 SNPs was observed in the Norfolk Island population (Fig. 3); summarised LD statistics for the 29 SNPs: r^2 (min = 0.026, 1st Quartile = 0.33, median = 0.49, mean = 0.51, 3rd Quartile = 0.72, max = 1.00), D' (min = 0.24, 1st Quartile = 0.82, median = 0.90, mean = 0.89, 3rd Quartile = 1.00, max = 1.00).. Haploview analysis identified 2 LD blocks across the region; the first block contained 9 SNPs and spanned 88 kb, the second block consisted of 19 SNPs and spanned a region of 74 kb. Further analysis of LD across 3 separate HapMap populations was conducted to compare with that obtained in the Norfolk Island cohort; CEU (European), CHD (Chinese) and JPT (Japanese). Due to the use of different SNP arrays, 25 of the 29 SNPs were available across the 4 populations, thus the LD mapping was restricted to these 25 SNPs. The LD pattern for the Norfolk Island cohort was most similar to the CEU population, and extensively different from both of the Asian HapMap groups used (Additional file 3). LD

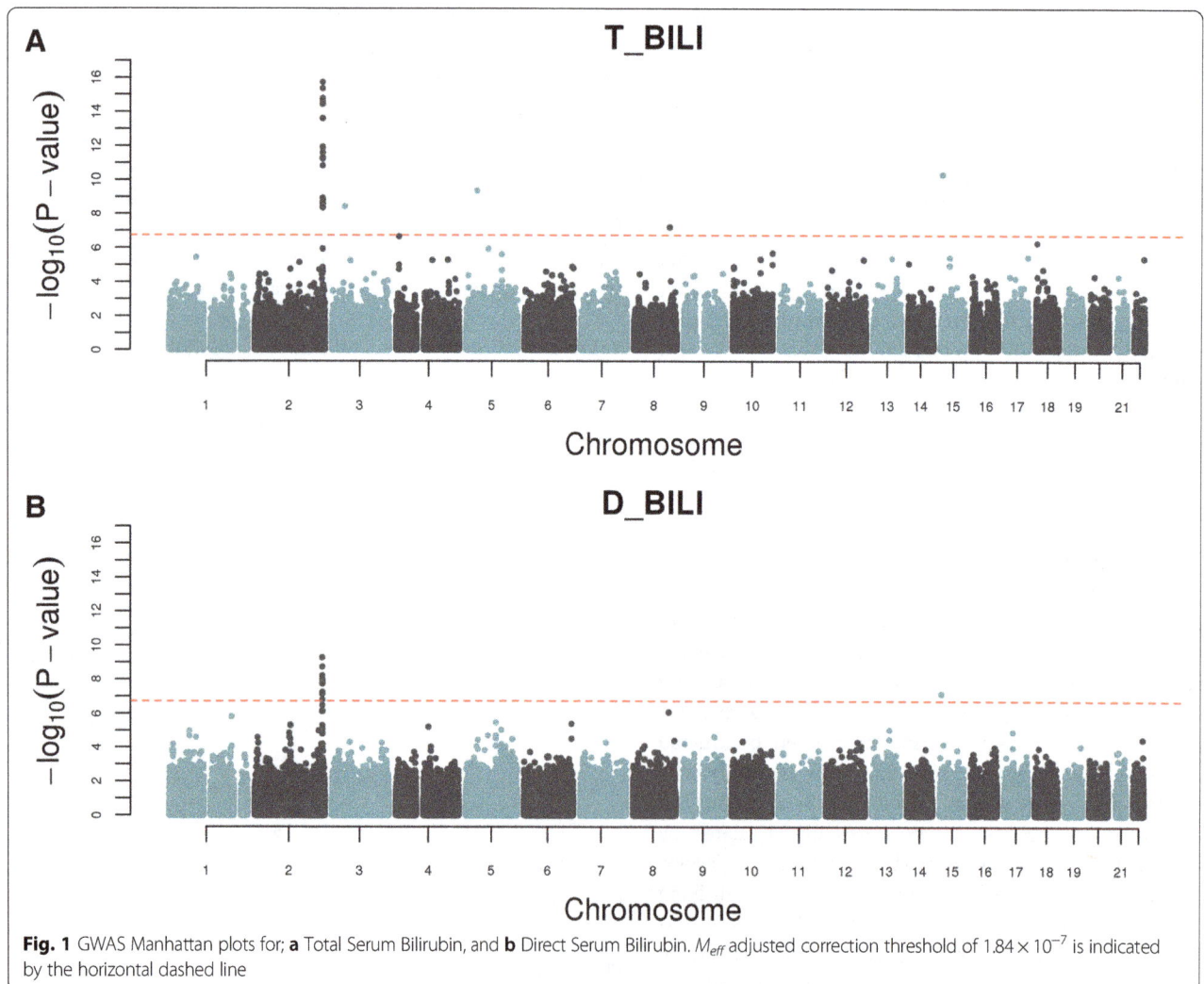

Fig. 1 GWAS Manhattan plots for; **a** Total Serum Bilirubin, and **b** Direct Serum Bilirubin. M_{eff} adjusted correction threshold of 1.84×10^{-7} is indicated by the horizontal dashed line

Table 2 Top M_{eff} corrected SNPs associated with total serum bilirubin

SNP	CHR	Position	A1	A2	effB	se_effB	chi2.1df	P value
rs6744284	2	234625297	C	T	3.278	0.395	68.90	1.87E-16
rs3771341	2	234673239	C	T	3.235	0.395	67.17	4.43E-16
rs17863787	2	234611094	T	G	3.044	0.379	64.48	1.70E-15
rs6742078	2	234672639	G	T	3.025	0.379	63.60	2.64E-15
rs887829	2	234668570	G	A	3.011	0.379	63.01	3.55E-15
rs4148324	2	234672722	T	G	3.011	0.379	63.01	3.55E-15
rs4148325	2	234673309	C	T	3.011	0.379	63.01	3.55E-15
rs1105880	2	234601965	T	C	2.829	0.368	59.13	2.46E-14
rs1105879	2	234602202	T	G	2.829	0.368	59.13	2.46E-14
rs6725478	2	234615400	C	T	2.711	0.378	51.40	1.18E-12
rs2741045	2	234580140	C	T	2.907	0.406	51.17	1.32E-12
rs10168416	2	234597087	C	G	2.827	0.400	49.85	2.55E-12
rs2070959	2	234602191	A	G	2.827	0.400	49.85	2.55E-12
rs2741012	2	234508963	C	T	2.887	0.415	48.43	5.20E-12
rs2741027	2	234518011	G	A	2.887	0.415	48.43	5.20E-12
rs7608175	2	234599089	C	G	2.536	0.364	48.41	5.27E-12
rs10168155	2	234596836	C	T	2.530	0.364	48.21	5.82E-12
rs10171367	2	234597667	C	G	2.530	0.364	48.21	5.82E-12
rs2361502	2	234698790	T	C	2.590	0.380	46.37	1.46E-11
rs2925260	15	25869972	T	C	18.764	2.825	44.11	4.56E-11
rs2930593	15	25877815	C	A	18.764	2.825	44.11	4.56E-11
rs12654591	5	38271582	C	A	13.063	2.075	39.63	4.34E-10
rs2741023	2	234516714	G	A	2.446	0.399	37.67	1.17E-09
rs10179094	2	234597825	T	A	2.347	0.386	37.06	1.58E-09
rs7586110	2	234590527	T	G	2.344	0.385	37.05	1.59E-09
rs2221198	2	234658623	C	T	2.251	0.374	36.33	2.29E-09
rs4124874	2	234665659	A	C	2.194	0.365	36.05	2.63E-09
rs3755319	2	234667582	T	G	2.194	0.365	36.05	2.63E-09
rs4148326	2	234673462	T	C	2.194	0.365	36.05	2.63E-09
rs1876506	3	42547127	A	G	13.734	2.308	35.43	3.61E-09
rs4294999	2	234635467	A	G	2.155	0.364	35.11	4.23E-09
rs2008595	2	234637192	G	A	2.155	0.364	35.11	4.23E-09
rs4663963	2	234650193	T	G	2.155	0.364	35.11	4.23E-09
rs16892701	8	120525983	G	A	11.336	2.077	29.79	6.25E-08
rs16892482	8	120386382	A	G	11.326	2.077	29.74	6.41E-08

appeared slightly stronger in the Norfolk Island SNPs than for CEU. Allele frequencies for the 25 SNPs in these 4 populations are detailed in Additional file 4.

Haplotype mapping and association with bilirubin levels

Haploview association analysis was performed on the individual 29 SNP 'markers', minor allele frequencies (MAF) and association statistics are documented in Table 3 (for additional information see Additional file 5). All 29 SNPs exhibited significantly ($P < 1.0 \times 10^{-4}$) increased MAF in the high serum bilirubin group. The most significantly associated marker was rs17863787; the frequency of the 'G' allele was observed to be 62.3 % in those with high serum bilirubin and 24.9 % in those with normal serum bilirubin ($P = 5.51 \times 10^{-17}$).

To further investigate the association of genomic structure across the chr2q37.1 region with serum bilirubin, a haplotype association analysis was conducted in Haploview. There were a total of 6 haplotypes inferred for LD block 1 and 7 haplotypes for LD block 2 (Additional file 6);

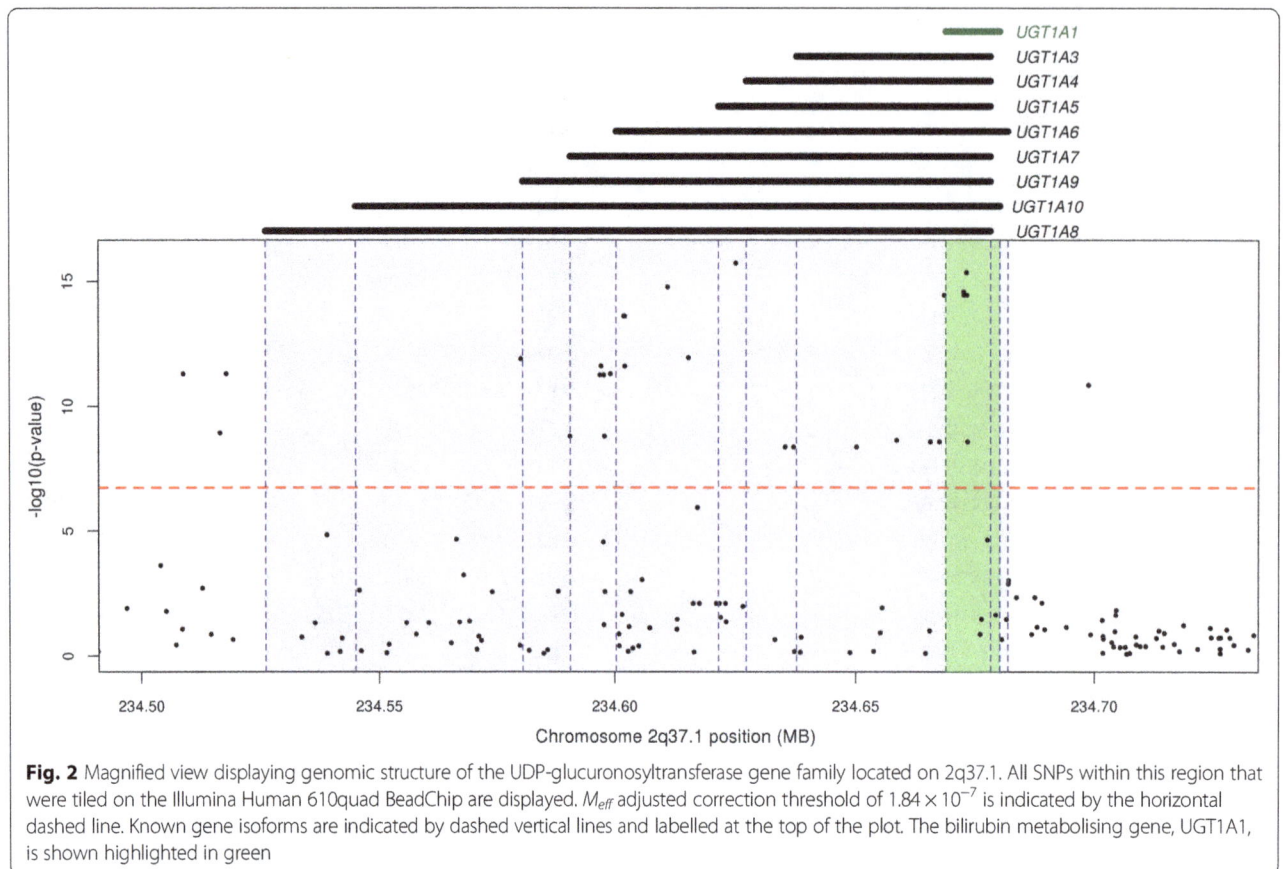

Fig. 2 Magnified view displaying genomic structure of the UDP-glucuronosyltransferase gene family located on 2q37.1. All SNPs within this region that were tiled on the Illumina Human 610quad BeadChip are displayed. M_{eff} adjusted correction threshold of 1.84×10^{-7} is indicated by the horizontal dashed line. Known gene isoforms are indicated by dashed vertical lines and labelled at the top of the plot. The bilirubin metabolising gene, UGT1A1, is shown highlighted in green

haplotypes present in >1 % of the total population are shown. The block 1 haplotype most significantly associated with the high bilirubin group was 'TAAGTGGGA', which is estimated to exist at 20.3 % in the total population. This haplotype was observed in 40.3 % of the high serum bilirubin group, and 17.2 % of the normal group ($P = 4.59 \times 10^{-9}$). The most abundant block 1 haplotype ('CGGTCCACT', 33.6 % of total population) was observed to be significantly associated with the normal serum bilirubin group; 36.9 % normal vs 19 % high ($P = 9.31 \times 10^{-5}$). The LD block 2 haplotype most significantly associated with high serum bilirubin was 'GGGC GTTGTGAGCTTGTTC'; which is estimated to be present in 18.8 % of the total population. This haplotype was observed in 43.5 % of the high serum bilirubin group, and 14.3 % of the normal group ($P = 1.73 \times 10^{-14}$). The most abundant block 2 haplotype ('CAAA TCCACTGTACGTCCT', 49.2 % of total population) was observed to be significantly associated with the normal serum bilirubin group; 54.6 % normal vs 26.1 % high ($P = 3.51 \times 10^{-9}$). Frequency and combination of the block specific haplotypes is illustrated in Fig. 4.

Nine tagging SNPs were identified that capture the allelic variance of the 29 SNPs (Table 4); the tagging analysis captured all 29 alleles at $r^2 >= 0.8$ which contains 100 % of alleles with mean r^2 of 0.963. These SNPs could be used in future replication analyses to tag variation across the region in other populations.

Bilirubin correlations with clinical metabolic syndrome and cardiovascular disease

It is well established that serum bilirubin levels are inversely correlated with risk of developing cardiovascular disease [20–22]. Therefore this was investigated using the cardiovascular disease risk score previously calculated for the Norfolk Island population [7], along with potential relationships between other metabolic risk scores, including metabolic syndrome and type-2 diabetes (scores previously estimated [7]).

A significant inverse relationship was observed between total serum bilirubin and the clinical risk score for metabolic syndrome. Of the 592 individuals with available data 66 % had normal bilirubin levels and no metabolic syndrome, 11.5 % had high bilirubin levels and no metabolic syndrome, 25.3 % had normal bilirubin and metabolic syndrome, 1.2 % had high bilirubin and metabolic syndrome. A chi-squared contingency test followed by Fisher's exact showed that this was a significant observation; $\chi 2 = 4.18$ ($P = 0.04$), Fisher's Exact OR = 2.45 ($P = 0.03$). This correlation suggests that Norfolk

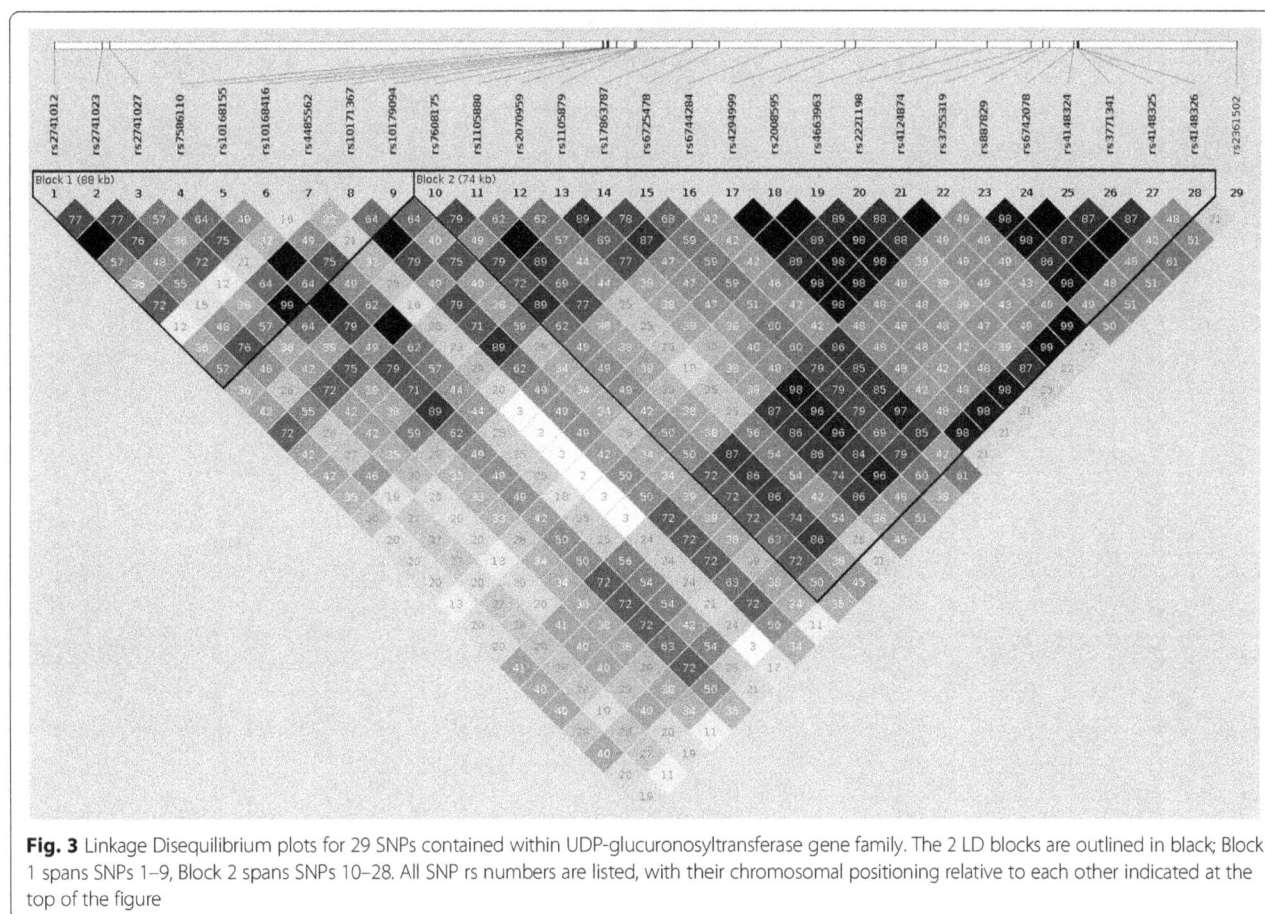

Fig. 3 Linkage Disequilibrium plots for 29 SNPs contained within UDP-glucuronosyltransferase gene family. The 2 LD blocks are outlined in black; Block 1 spans SNPs 1–9, Block 2 spans SNPs 10–28. All SNP rs numbers are listed, with their chromosomal positioning relative to each other indicated at the top of the figure

Island individuals with higher serum bilirubin levels are less likely to develop metabolic syndrome.

Numerous studies have also attributed smoking behaviour to be associated with serum bilirubin levels [23–25]. This was tested in the Norfolk Island population using the students independent t-test, and revealed a significant difference in mean serum bilirubin levels between smokers (6.46 μmol/L) and non-smokers (8.12 μmol/L); t = 3.99 with P = 4.06×10^{-5}.

To further examine potential relationships a series of t-tests between a variety of quantitative metabolic syndrome/cardiovascular disease traits and categorised serum bilirubin group were performed. There were a total of 9 significant ($P < 0.05$) trait correlations with categorised bilirubin level, these were; body mass index (BMI), body fat, cholesterol/HDL-C ratio, total cholesterol, hip circumference, LDL-C, type-2 diabetes risk score, total protein and triglycerides (Table 5). These findings highlight traits that are consistent with previous literature [26, 27].

Body fat was observed to have the strongest correlation with serum bilirubin, with significantly reduced body fat composition in individuals who had high serum bilirubin levels. Unlike previous observations [20, 27, 28],

cardiovascular disease risk score was not significantly reduced in those individuals with higher serum bilirubin, whereas, type-2 diabetes risk did show a significant reduction in the higher bilirubin group, consistent with previous literature [26, 29].

Genotype effects on metabolic syndrome, type-2 diabetes and cardiovascular disease traits

To further explore the above approach, associations between the 29 significantly associated SNPs and metabolic traits other than serum bilirubin were explored. Traits which showed a significant ($P < 0.05$) correlation with total serum bilirubin (Table 5) were selected. Only one trait was observed which showed a significant association with any of the 29 markers, this was type-2 diabetes-risk when categorised: "low"; "intermediate", and "high" [9]. Using a chi-squared test rs2741012 and rs2741027 were significantly associated with type-2 diabetes-risk ($\chi2 = 9.63$, $P = 0.0069$). Again this was followed with a Fisher's Exact test which confirmed significance ($P = 0.0081$). The same observation with the minor allele and suggestive protection was observed.

To further investigate the above associations logistic regression was used to identify a model that predicts outcome

Table 3 Haploview marker associations showing frequencies of the recessive alleles

SNP	Associated allele	High bili allele freq	Normal bili allele freq	OR (95 % CI)	Chi square	P value
rs2741012	T	0.415	0.188	0.33 (0.22 – 0.49)	31.93	1.60E-08
rs2741023	A	0.431	0.239	0.41 (0.28 – 0.61)	20.06	7.52E-06
rs2741027	A	0.415	0.188	0.33 (0.22 – 0.49)	31.93	1.60E-08
rs7586110	G	0.500	0.251	0.33 (0.23 – 0.49)	32.33	1.30E-08
rs10168155	T	0.654	0.326	0.26 (0.17 – 0.38)	48.96	2.62E-12
rs10168416	G	0.492	0.194	0.25 (0.17 – 0.37)	51.93	5.75E-13
rs4485562	G	0.800	0.626	0.42 (0.26 – 0.66)	14.44	1.00E-04
rs10171367	G	0.654	0.326	0.26 (0.17 – 0.38)	48.96	2.62E-12
rs10179094	A	0.500	0.249	0.33 (0.22 – 0.49)	32.83	1.01E-08
rs7608175	G	0.654	0.326	0.26 (0.17 – 0.38)	49.11	2.42E-12
rs1105880	C	0.646	0.271	0.20 (0.14 – 0.30)	68.04	1.60E-16
rs2070959	G	0.492	0.194	0.25 (0.17 – 0.37)	51.93	5.75E-13
rs1105879	G	0.646	0.271	0.20 (0.14 – 0.30)	68.04	1.60E-16
rs17863787	G	0.623	0.249	0.20 (0.13 – 0.30)	70.15	5.51E-17
rs6725478	T	0.631	0.300	0.25 (0.17 – 0.37)	51.58	6.88E-13
rs6744284	T	0.577	0.222	0.21 (0.14 – 0.31)	66.86	2.91E-16
rs4294999	G	0.715	0.429	0.30 (0.21 – 0.45)	35.46	2.61E-09
rs2008595	A	0.715	0.429	0.30 (0.21 – 0.45)	35.46	2.61E-09
rs4663963	G	0.715	0.429	0.30 (0.21 – 0.45)	35.46	2.61E-09
rs2221198	T	0.672	0.402	0.33 (0.22 – 0.49)	31.32	2.18E-08
rs4124874	C	0.715	0.427	0.30 (0.21 – 0.45)	35.86	2.12E-09
rs3755319	G	0.715	0.427	0.30 (0.21 – 0.45)	35.86	2.12E-09
rs887829	A	0.623	0.254	0.20 (0.14 – 0.30)	67.92	1.70E-16
rs6742078	T	0.623	0.253	0.20 (0.14 – 0.30)	68.23	1.46E-16
rs4148324	G	0.623	0.254	0.20 (0.14 – 0.30)	67.92	1.70E-16
rs3771341	T	0.577	0.227	0.22 (0.15 – 0.32)	64.55	9.40E-16
rs4148325	T	0.623	0.254	0.20 (0.14 – 0.30)	67.92	1.70E-16
rs4148326	C	0.715	0.427	0.30 (0.21 – 0.45)	35.86	2.12E-09
rs2361502	C	0.546	0.268	0.30 (0.21 – 0.45)	38.61	5.16E-10

Note: odds ratios are not adjusted for age and sex

(type-2 diabetes) from trait (bilirubin) and factors in potential modifiers (genotype). Logistic regression modelling identified direct bilirubin as being significantly associated with categorised type-2 diabetes risk (r^2: 0.05, p-value: 0.005), suggesting that in the Norfolk Island cohort increased direct bilirubin was associated with a 28 % reduction in type-2 diabetes risk (OR:0.72, 95 % CI: 0.57-0.91). Based on a bi-directional stepwise regression model approach 2 of the 9 tagging SNPs remained significant; rs2741027 and rs6725478. These SNPs effectively tag the two major 'protective'/high bilirubin haplotypes. When included, the adjusted model remained significant (r^2: 0.13, p-value: 0.0001) and confirmed the initial association; direct bilirubin (OR:0.70, 95 % CI: 0.53-0.89, p-value: 0.005): rs2741027 (OR:0.25, 95 % CI: 0.10-0.58, p-value: 0.002), rs6725478 (OR:0.27, 95 % CI: 0.10-0.63, p-value: 0.004).

This indicates that when controlling for bilirubin levels genotype affects risk of type-2 diabetes within the Norfolk Island population. Therefore, inclusion of SNP genotypes when assessing the relationship between direct bilirubin and type-2 diabetes risk increases the accuracy of the 'risk' estimate within the Norfolk Island cohort.

Functional Annotation of UDP-glucuronosyltransferase SNPs

Investigation of the 29 SNPs revealed several of potential functional interest (SNP annotation Table 6). Three SNPs are within the coding region of UGT1A6 (Table 6); rs1105880 (synonymous), rs1105879 and rs2070959 (non-synonymous). Further investigation with SNPnexus (http://www.snp-nexus.org/) revealed rs1105879 had a PolyPhen score of 'possibly damaging', indicating the usually conserved nature of the coded amino acid. Six

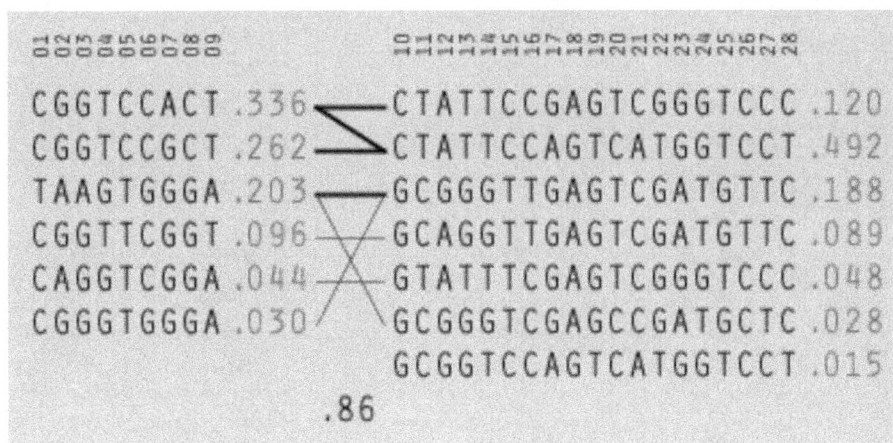

Fig. 4 Haplotype structure across the two identified LD blocks in the Norfolk Island cohort UDP-glucuronosyltransferase gene family. Displayed haplotypes reside at >1 % frequency in the genotyped samples. Connecting lines represent haplotype combinations: thick lines represent haplotype combinations that reside at >10 %, thin lines >1 % of samples

SNPs were observed to reside within 5' prime untranslated regions (5'UTR); UGT1A1 (rs887829, rs3755319), UGT1A3 (rs2008589), UGT1A6 (rs7608175), UGT1A7 (rs7586110), and UGT1A9 (rs2741045).

Discussion

We have identified a significant genomic association at 2q37.1 in the region of the UDP-glucuronosyltransferase (UDPGT) enzyme family members, with direct and total serum bilirubin levels. Correlation analyses between metabolic syndrome related traits and serum bilirubin levels identified significant inverse relationships for numerous traits. Haplotype association testing revealed the presence of potentially protective haplotypes within the Norfolk Island population. Thus this study has identified a complex region which shows interplay between genomic and environmental conditions and has a large effect on overall serum bilirubin levels.

Previous literature has suggested a linkage between bilirubin and metabolic risk with clinical associations observed between cardiovascular disease risk, obesity and bilirubin concentrations [20–22, 27] and more recently metabolic syndrome [30–34]. Therefore, we investigated potential relationships between bilirubin and metabolic traits in the Norfolk Island cohort. An inverse correlation between serum bilirubin and several important metabolic traits was observed, with the most notable being metabolic syndrome and type-2 diabetes risk. Given that metabolic syndrome and type-2 diabetes increase cardiovascular disease risk it is consistent with the current body of literature which documents inverse association between high serum bilirubin and cardiovascular disease risk (review [26]).

Our analysis refined an association with serum bilirubin concentration to a 189.8 kb region on chromosome 2q37.1 with genotypic analyses revealing that the level of serum bilirubin was greatly increased in individuals with the rare allele. This region encodes one of the major drug metabolising families (UDP-glucuronosyltransferase, UDPGT) [35–37]; there are 9 documented

Table 4 Haploview 'Tagger' analysis of the 29 GWAS associated chr2q37.1 SNPs identified 9 SNPs as tagging the allelic variation across the region

SNP tested	Alleles captured
rs4148325	rs3771341,rs887829,rs1105879,rs4148325,rs6742078,rs17863787,rs1105880,rs6744284,rs4148324
rs2008595	rs4294999,rs4148326,rs4663963,rs2221198,rs3755319,rs4124874,rs2008595
rs6725478	rs6725478,rs10168155,rs10171367,rs7608175
rs7586110	rs10179094,rs7586110
rs10168416	rs10168416,rs2070959
rs2741027	rs2741027,rs2741012
rs2741023	rs2741023
rs2361502	rs2361502
rs4485562	rs4485562

Table 5 Metabolic trait correlation with serum bilirubin group

Trait	t	P	High bili mean	Normal bili mean
albumin	0.90	3.70E-01	41.75	41.43
Body Mass Index*	−1.91	3.00E-02	25.06	26.27
body fat*	−3.43	5.30E-04	26.63	30.67
cholesterol/HDL-C ratio*	−1.82	4.00E-02	3.94	4.29
cholesterol*	−2.71	4.00E-03	5.30	5.67
creatinine	0.15	8.80E-01	80.19	79.90
cardiovascular disease risk	0.40	6.90E-01	7.20	6.69
diastolic blood pressure	0.51	6.90E-01	77.69	76.68
globin	1.91	6.10E-02	30.29	29.34
HDL-C	0.62	5.30E-01	1.44	1.41
hip circumference*	−2.04	2.00E-02	99.42	101.92
Inbreeding	−0.42	3.40E-01	0.01	0.01
LDL-C*	−1.66	5.00E-02	2.66	2.86
mean arterial pressure	0.26	8.00E-01	95.01	94.44
Polynesian Admixture	−1.41	8.50E-02	0.21	0.25
pulse pressure	0.02	9.80E-01	52.31	52.26
systolic blood pressure	0.29	6.10E-01	130.00	128.95
type-2 diabetes risk*	−2.31	1.00E-02	9.19	11.22
total protein*	2.28	3.00E-02	72.10	70.79
triglycerides*	−2.15	2.00E-02	1.58	2.00
urea	−0.44	6.60E-01	5.49	5.58
uric acid	0.53	6.00E-01	0.35	0.34
waist circumference	−1.91	1.30E-01	85.04	87.62
waist-hip ratio	0.34	6.30E-01	0.86	0.86
weight	0.20	5.80E-01	75.79	75.31

* significant at $P < 0.05$

UDPGT isoforms; UGT1A1, UGT1A3, UGT1A4, UGT 1A5, UGT1A6, UGT1A7, UGT1A8, UGT1A9 and UG T1A10 (Fig. 2). UGT1A1 is well known to preferentially metabolise bilirubin and has been previously mapped in linkage and GWAS studies [16–18, 38–43]. UGT1A3 and UGT1A4 also have been shown to have potential action with bilirubin [37]. However all gene family members, including UGT1A1, exhibit affinity for numerous substrates and it is therefore possible that the gene effects are not mediated (entirely) by total bilirubin. Such pleotropic effects at this loci are likely to be the case as evidenced by the fact that adjustment for

Table 6 Functional annotation of the chr2q37.1 SNPs significantly associated with total serum bilirubin levels

RefSNP ID	NCBI gene ID	Gene symbol	Function class	Residue	Amino acid position
rs1105879	54578	UGT1A6	coding-nonsynonymous	S	183
rs2070959	54578	UGT1A6	coding-nonsynonymous	A	180
rs1105880	54578	UGT1A6	coding-synonymous	L	104
rs2008595	54659	UGT1A3	5'UTR	-	-
rs2741045	54600	UGT1A9	5'UTR	-	-
rs3755319	54658	UGT1A1	5'UTR	-	-
rs7586110	54577	UGT1A7	5'UTR	-	-
rs7608175	54578	UGT1A6	5'UTR	-	-
rs887829	54658	UGT1A1	5'UTR	-	-

SNPs displayed reside in regions other than introns

serum bilirubin in our modelling did not completely nullify the observed association between genotype and outcome. Future work is required to more fully explore these effects along with associations of other substrates with variants at this genomic region.

Mutations in UGT1A1 have also been associated with Crigler-Najjar syndromes types I and II and in Gilbert syndrome [44–46]. Gilbert Syndrome (GS) is a well-documented benign increase in serum bilirubin, and is caused by the reduced activity of UDPGT [47–51]. In line with the observations that serum bilirubin is inversely correlated with metabolic risk diabetic patients with GS are less likely to develop vascular dysfunctions [52]. Furthermore, the incidence of diabetes and cardiovascular disease risk mortality is lower in GS individuals, with one study exploring the efficacy of increasing serum bilirubin in type-2 diabetic patients [53]. Further evidence confirming the protective role of circulating bilirubin for type-2 diabetes has been reported in a prospective study [19].

Significant difference has been identified between functional polymorphisms within the UGT1A family between Caucasian and other populations [54]. Polymorphisms in the promoter region for UGT1A1 (2 bp TA insertion in the TATA box) increased activity in Caucasian GS patients; this was not observed in Asian and African GS patients or Pacific populations [54]. The authors suggest that due to the complex nature of environmental and genetic factors, unstable polymorphisms within UGT1A1 may act to "fine-tune" plasma bilirubin levels on a population by population basis, meaning that the promoter variation explains the presence of GS in some populations, but in other populations it's more likely a combination of variants in the encoding region along with environmental factors [54], our data supports this hypothesis. Additionally, meta-analysis has demonstrated strong replication for a genetic influence on serum bilirubin levels of the UGT1A1 locus ($P < 5 \times 10^{-324}$), specifically at the proximal promoter region of UGT1A1 tagged by rs6742078 [40]. While we didn't have genotype information for this SNP we were able to impute against the 1000 Genomes panel to extrapolate associations between the two studies. Using imputed information we were able to illustrate that there is tight LD between rs6742078 and the top associated SNP from our study, rs6744284 ($r^2 = 0.85$), suggesting that the Norfolk Island cohort exhibit a similar genetic pattern of association.

We identified strong LD across the region of 2q37.1, potentially suggesting that the Norfolk Island population's unique genomic structure is influencing serum bilirubin concentration. LD across the same region in data available through the HapMap project [55] showed that the Norfolk Island cohort exhibited an LD pattern similar to that observed in the European population (CEU), while both the Asian populations (Chinese and Japanese) exhibited very different genetic structure across this region. This is not unexpected because of the large amount of recent European admixture in the Norfolk population. Additionally, it was noted that haplotypes containing the minor allele(s) in the Norfolk Island population potentially conferred protection to metabolic disorders as measured by clinical metabolic syndrome and type-2 diabetes-risk. It is possible that selection is driving the presence of high serum bilirubin within populations, although this may be achieved by different variation across the region. It appears that in Europeans this variation is often in the promoter region, whereas in Asian and African populations this is not the case, and it is polymorphisms in the gene body that seems to account for the associations with increased bilirubin. This strongly suggests that it is beneficial for a population to have a certain frequency of individuals with naturally high serum bilirubin, and potentially points to a complex interaction between environmental and genomic factors maintaining this.

One significant association between 2 SNPs (rs2741012 and rs2741027) and categorised type-2 diabetes-risk was observed. These two SNPs are just upstream of the promoter and 5'UTR region of the UDPGT family. It is likely that these SNPs are in LD with untyped polymorphisms (SNPs not on the 610quad chip) that reside in these regions and potentially form a LD block/haplotype in the Norfolk Island population which confers protection to type-2 diabetes as well as metabolic syndrome. Interestingly, and in support of our approach, this reduction in risk correlates well with previous work conducted in a large US cohort [13]; these variants (or variants tagged by them) may be functional, i.e. they might directly affect transcription and/or translation of the isoforms encoded by the UDPGT family. It is also possible that there are additional rare variants within the region that further influence serum bilirubin as recently evidenced by an exome sequencing study performed in elderly individuals [56].

Given that bilirubin is a cheap and commonly measured laboratory test, routine screening of serum bilirubin levels could be beneficial in the stratification and treatment of metabolic disorders such as cardiovascular disease and type-2 diabetes. Identification of genes/variants that exhibit pleiotropic effects (effects of the same variant on multiple characteristics or disease risks) is an ultimate goal. The significant interaction observed here provides evidence that bilirubin may be affected by genetic and environmental factors and their interactions.

Conclusions

In summary, this study identified strong associations of variants within the UGT1A family with regulation of serum bilirubin levels in the Norfolk Island population, which replicated previous GWAS and epidemiological findings. This successful implementation of pedigree-based analysis using the unique properties of the Norfolk Island cohort highlights a functional region that offers protective benefit from metabolic disease and further eludes to a potentially heritable component with the Norfolk Island population. Specific haplotype structure was significantly associated with increased serum bilirubin, and as such this study has identified a potential set of 'protective' haplotypes that exist within the Norfolk Island population. Further studies are warranted to validate these findings, with the next step being to explore these associations in larger outbred populations.

Methods

Sample/cohort collection, pedigree information and ethics

The Norfolk Island Health Study (NIHS) is well established with regards to data collection and initial disease prevalence studies [4, 5, 8]. The Norfolk Island pedigree structure has been previously outlined [57], and subsequently updated [1]. The most recent update led to the reconstruction of a core-pedigree consisting of 1388 members coalescing over 11 generations (or 200 years) back to the original founders. [3, 7]. This study focuses on a reduced core-pedigree, meaning that individuals; a) are genetically related to the original founders, and b) have phenotype and genotype information available. The total number of individuals fitting these criteria was 382. All individuals gave written informed consent. Ethical approval was granted prior to the commencement of the study by the Griffith University Human Research Ethics Committee (ethical approval no: 1300000485) and the project was carried out in accordance with the relevant guidelines, which complied with the Helsinki Declaration for human research.

Heritability analysis

All 19 metabolic traits assessed in this analysis are part of the NIHS2000 collection [8]. Traits measured were: fasting plasma glucose, HDL-C, LDL-C, total plasma cholesterol, cholesterol-HDL-C ratio, triglycerides, creatinine, total protein, globin, albumin, urea, uric acid, total serum bilirubin, direct serum bilirubin, and numerous enzymes that are markers for liver/kidney function (ALT, AST, Alkaline Phosphatase, GGT, Lactate Dehydrogenase [LDH]).

Heritability estimates

The R package and genetic analysis program GenABEL [58] was used to calculate heritability estimates for all metabolic syndrome/cardiovascular disease related traits. The genetic kinship matrix derived from the SNP data and reconstructed core-pedigree was used to estimate trait heritability (h^2 [narrow-sense heritability]) by polygenic modelling. All traits were screened for covariant effects of age and sex interactions.

Genome-wide SNP genotyping

EDTA anticoagulated venous blood samples were collected from all participants. Genomic DNA was extracted from blood buffy coats using standard phenol-chloroform procedures. DNA samples were genotyped according to the manufacturer's instructions on Illumina Infinium High Density (HD) Human610-Quad DNA analysis BeadChip version 1. BeadChips were a four-sample format requiring 200 ng of DNA per sample. Samples were scanned on the Illumina BeadArray 500GX Reader. Raw data was obtained using Illumina BeadScan image data acquisition software (version 2.3.0.13). Raw data from Illumina idat files were SNP genotyped in R using the CRLMM package [59]. Genotype data then underwent initial quality control routines using PLINK [60]. SNPs were filtered based on: minor allele frequency >0.01; call rate >0.95, and Hardy-Weinberg equilibrium testing p-value >10^{-5}. After this initial quality control, 590,603 SNPs were exported from PLINK and imported into the CRAN package GenABEL [58]. Further filtering (including Mendelian inheritance violations and sex-checking based on available X and Y markers) in GenABEL lead to the reduction of the SNP set to a total of ~480,000; this included removal of both X and Y chromosome SNPs after gender checking, as well as the removal of mitochondrial and XY SNPs.

Genome-wide association analysis

A pedigree based GWAS analysis of all heritable traits was batched using custom R scripts and the package GenABEL [58]. GenABEL uses an additive approach and the loci are coded as 0, 1, 2 (corresponding to genotypes AA, AB, and BB, respectively). A detailed explanation of the association model and specific GWAS overview as implemented in the Norfolk Island was previously described [7]. Breifly, a correction was made for the relatedness inherent in the Norfolk Island population using the polygenic model with age and sex interactions, as well as genetic structure [the top 2 genomic principal components of the complete SNP set as calculated by KING [61]]. The top two components were chosen as covariates because we found that these explained the majority of the variance in the outcomes being tested and because inclusion of additional, less informative components only served to reduce the parsimony of the models. For association analysis the mmscore function implemented in GenABEL was used. This function represents a mixed model approximation analysis for association between a trait and genetic polymorphism(s), and is specifically designed for association testing in samples of related individuals. This allows for per SNP association testing using a mixed model polygenic approach. After correcting for multiple testing, the study-wide significance was set based on M_{eff} adjustment (P = 1.84×10^{-7}). It should be noted that this M_{eff} threshold is tailored to trait-wise associations, not multi-trait analyses therefore p-values are adjusted on a per trait basis. Association statistics for every SNP for each trait were generated and output to compressed files (.gz.tar) for storage and future reference. GWAS Manhattan plots where generated for each trait association using a custom modified version of the GenABEL plot.scan.gwaa function (for all Manhattan plots see Additional file 2). Annotation of the robustly associated bilirubin SNPs identified as being functional was performed using: http://brainarray.mbni.med.u-mich.edu/Brainarray/Database/SearchSNP/snpfunc.aspx.

LD testing and haplotype association

Genotype data for the chrq37.12 region was phased using SHAPEIT2 [62], which has functionality to deal with complex pedigree structures – implemented through the duoHMM algorithm. From this process we observed no Mendelian errors before moving the phased data over to Haploview analyses. Haplotype/LD testing, SNP tagging and association analyses were all conducted in Haploview 4.2 [63]. LD blocks were determined using the default Haploview settings which infer LD based on a pairwise comparison of correlation (r^2) values between SNPs. Haplotypes were inferred from the genotypes of SNPs which made up the identified LD blocks, and were only recorded if they existed in more than 1 % of the population. Tagging SNPs were determined using the 'tagger' option of Haploview, using a pair-wise tagging method with a minimum observed r^2 between pairs of 0.8. Association analyses were carried out on both markers (SNPs) and haplotypes using the inbuilt Haploview association function. A phenotype column was added to the dataset to allow a 'case'/'control' experimental set-up; where case represented the high bilirubin group and control the normal bilirubin group. There were a total of 65 cases and 317 controls with 124 genotyped individuals missing phenotype information. Permutation testing was run to confirm the above association analyses for both marker and haplotype associations. To ensure the robustness of final P values the number of permutations was set at 1,000,000 (this should lead to a reduction of the FDR). A further exploration of potentially similar structure across the region spanned by the 29 SNPs was tested in 3 HapMap populations; CEU (European), CHD (Chinese), and JPT (Japanese). Due to data being generated on different genotype platforms (SNP chips), a final list of 25 consensus SNPs was retained for Haploview analysis for Norfolk Island and the 3 HapMap sets. Linkage disequilibrium plots across the 25 SNP region was generated for each of the 4 populations (Additional file 3).

Correlations with metabolic traits

Initial exploratory correlations between risk scores for cardiovascular disease and type-2 diabetes, clinically defined Metabolic Syndrome (categorical: 0 (no MetS), 1 (MetS)), and various related traits were conducted in R 2.15.2 [64].

For all analyses total serum bilirubin levels were categorised into 'normal' and 'high' groupings, with 'high' being defined as >14 μmol/L, this approximates a clinical cut-off and allows facilitates interpretation in line with existing clinical guidlines. For all other traits tested a standard student's t-test (as implemented in R) was used to test for a significant difference of means between the given trait and bilirubin level. There were two categorical traits tested for correlation with serum bilirubin levels; smoking and presence of metabolic syndrome. Smoking has been previously well documented to be associated with serum bilirubin levels [23], and was categorised in the Norfolk Island cohort as either 'yes' (smokers $N = 133$) or 'no' (non-smokers $N = 458$). Correlation testing between smoking and bilirubin was carried out using a 2×2 chi-squared contingency test, followed by a Fisher's Exact test. For correlation analysis between total serum bilirubin and metabolic syndrome there were a total of 598 individuals with available matched phenotype data; 'metabolic syndrome' ($N = 156$) and 'no metabolic syndrome' ($N = 442$). The clinical diagnosis of metabolic syndrome previously calculated for the Norfolk Island cohort was used [7]. A 2×2 chi-squared contingency test was used to evaluate the significance, followed by a Fisher's Exact test as one of the tables cells contained a value less than 5 %. Due to the initial exploratory nature of these analyses all tests are unadjusted, so nominal p-values are reported. Additionally, relatedness within the population is accounted for in later formal modelling using GLM regression.

Regression modelling testing association between outcome, trait and genotype

To further explore associations between bilirubin, type-2 diabetes and the genotypic architecture across the *UGT1A1* region regression modelling was conducted in R. To establish an initial association, separate logistic regression was conducted between categorised type-2 diabetes risk and total bilirubin and then direct bilirubin. Additionally a bi-directional stepwise logistic regression model was used to test the significance of each of the 9 tagging SNPs identified in the LD block analysis. The model was not corrected for common covariates (age, sex, smoking, BMI) as these are all accounted for in the calculation of the AUSDRISK type-2 diabetes risk score (as previously calculated in the Norfolk Island cohort [7]). To address the issue of relatedness we included the average pedigree kinship as a covariate in the stepwise regression model. This was excluded from the final model indicating that in this instance relatedness is not a significant issue. Reported r^2 values use the Nagelkerke Index pseudo r^2 as calculated in R. Model p values were generated from an ANOVA using the F distribution, which tests the null hypothesis that the coefficients represented in the overall regression model (represented by R 2) are equal to 0.

Web resources

Online Mendelian Inheritance in Man (http://www.omim.org)

Catalogue of genome-wide association studies (http://www.genome.gov/gwastudies)

BrainArray (http://brainarray.mbni.med.umich.edu/Brainarray/Database/SearchSNP/snpfunc.aspx).

Availability of supporting data

Due to current ethical constraints, restricted data access is in place to anonymise genotypic SNP GWAS and phenotype data. The Norfolk Island Health Study steering committee will assess restricted data access requests via our GRC computational genetics group (interested researchers should contact grccomputationalgenomics@gmail.com).

Additional files

Additional file 1: Summary statistics of phenotypic traits measured in the Norfolk Island population. Table with an overview of the Norfolk Island phenotype data analysed, 16 traits in total. (PDF 96 kb)

Additional file 2: GWAS Manhattan plots for metabolic related traits. GWAS Manhattan plots for all 16 traits. (PDF 7896 kb)

Additional file 3: LD lots for 4 populations across 200 kb of chr2q37.1. Haploview LD plots for 25 SNPs spanning a region of chr2q37.1 for four populations; NI (Norfolk Island), CEU (European); CHD (Chinese), and JPT (Japanese). (PDF 2809 kb)

Additional file 4: Haploview allele frequency data for 25 chr2q37.1 SNPs across 4 populations. Minor allele frequencies for 25 SNPs across five populations; Norfolk Island (NI), European (CEU), Chinese (CHD and CHB), and Japanese (JPT). (PDF 137 kb)

Additional file 5: Detailed Haploview allele frequency data for all 29 SNPs in the NI cohort. Allele frequency data for all 29 SNPs across the chr2q37.1 region for the Norfolk Island samples. (PDF 155 kb)

Additional file 6: Haploview haplotype associations with bilirubin levels. Haplotypes identified by Haploview and their association statistics. (PDF 33 kb)

Abbreviations

BMI: Body mass index; CEU: European Hapmap population; CHD: Chinese Hapmap population; CVD: Cardiovascular disease; DBP: Diastolic blood pressure; GS: Gilbert's syndrome; GWAS: Genome-wide association study; h^2: Heritability (narrow-sense); HDL-C: High-density lipoprotein cholesterol; JPT: Japanese Hapmap population; LD: Linkage disequilibrium; LDL-C: Low-density lipoprotein cholesterol; MAF: Minor allele frequency; NI: Norfolk Island; NIHS: Norfolk Island health study; SBP: Systolic blood pressure; SNP: Single nucleotide polymorphism.

Competing interests

The authors declare that they have no competing interests.

Authors' contributions

CB, MC, and JC carried out the genotype assays. MH curated phenotype and pedigree data. MB, RL, and DE participated in the design of the study and performed the statistical analysis. JB provided detailed statistical expertise and critical evaluation of methodology. MB, RL and LG conceived of the study, and participated in its design. DM and GC helped to draft the manuscript, provided critical revision and intellectual input into the final manuscript design. All authors read and approved the final manuscript.

Acknowledgements

This research was supported by funding from a National Health and Medical Research Council of Australia (NHMRC) Project Grant. It was also supported by infrastructure purchased with Australian Government EIF Super Science Funds as part of the Therapeutic Innovation Australia – Queensland Node project. Also Miles Benton was supported by a Corbett Postgraduate Research Scholarship. We would like to acknowledge Amanda Miotto and also QUT for providing computational support for this project. Lastly, we extend our appreciation to the Norfolk Islanders who volunteered for this study.

Author details

[1]Genomics Research Centre, Institute of Health and Biomedical Innovation, Queensland University of Technology, Kelvin Grove, QLD 4059, Australia. [2]Kenepuru Science Centre, Institute of Environmental Science and Research, Wellington 5240, New Zealand. [3]Texas Biomedical Research Institute, San Antonio, TX 78227-5301, USA. [4]School of Biological Sciences, Victoria University of Wellington, Wellington 6140, New Zealand. [5]South Texas Diabetes and Obesity Institute, University of Texas, Rio Grande Valley School of Medicine, Brownsville, TX 78520, USA.

References

1. Macgregor S, Bellis C, Lea RA, Cox H, Dyer T, Blangero J, et al. Legacy of mutiny on the Bounty: founder effect and admixture on Norfolk Island. Eur J Hum Genet. 2010;18:67–72.

2. McEvoy BP, Zhao ZZ, Macgregor S, Bellis C, Lea RA, Cox H, et al. European and Polynesian admixture in the Norfolk Island population. *Heredity (Edinb)* 2010;105(2):229–234

3. Benton MC, Stuart S, Bellis C, Macartney-Coxson D, Eccles D, Curran JE, et al. "Mutiny on the Bounty": the genetic history of Norfolk Island reveals extreme gender-biased admixture. Investig Genet. 2015;6:11.

4. Bellis C, Cox HC, Dyer TD, Charlesworth JC, Begley KN, Quinlan S, et al. Linkage mapping of CVD risk traits in the isolated Norfolk Island population. Hum Genet. 2008;124:543–52.

5. Cox HC, Bellis C, Lea RA, Quinlan S, Hughes R, Dyer T, et al. Principal Component and linkage analysis of cardiovascular risk traits in the Norfolk Isolate. Hum Hered. 2009;68:55–64.

6. Maher BH, Lea RA, Benton M, Cox HC, Bellis C, Carless M, et al. An X chromosome association scan of the Norfolk Island genetic isolate provides evidence for a novel migraine susceptibility locus at Xq12. PLoS One. 2012;7:e37903.

7. Benton MC, Lea RA, Macartney-Coxson D, Carless MA, Göring HH, Bellis C, et al. Mapping eQTLs in the Norfolk Island Genetic Isolate Identifies Candidate Genes for CVD Risk Traits. Am J Hum Genet. 2013;93:1087–99.

8. Bellis C, Hughes RM, Begley KN, Quinlan S, Lea RA, Heath SC, et al. Phenotypical characterisation of the isolated Norfolk Island population focusing on epidemiological indicators of cardiovascular disease. Hum Hered. 2006;60:211–9.

9. Chen L, Magliano DJ, Balkau B, Colagiuri S, Zimmet PZ, Tonkin AM, et al. Ausdrisk: an australian type 2 diabetes risk assessment tool based on demographic, lifestyle and simple anthropometric measures. Med J Aust. 2010;192:197–202.

10. Stocker R, Yamamoto Y, McDonagh A, Glazer A, Ames B. Bilirubin is an antioxidant of possible physiological importance. Science (80-). 1987;235:1043–6.

11. Yamaguchi T, Terakado M, Horio F, Aoki K, Tanaka M, Nakajima H. Role of bilirubin as an antioxidant in an ischemia-reperfusion of rat liver and induction of heme oxygenase. Biochem Biophys Res Commun. 1996;223:129–35.

12. Baranano DE, Rao M, Ferris CD, Snyder SH. Biliverdin reductase: a major physiologic cytoprotectant. Proc Natl Acad Sci U S A. 2002;99:16093–8.

13. Cheriyath P, Gorrepati VS, Peters I, Nookala V, Murphy ME, Srouji N, et al. High Total Bilirubin as a Protective Factor for Diabetes Mellitus: An Analysis of NHANES Data From 1999–2006. J Clin Med Res. 2010;2:201–6.

14. Cox AJ, Ng MC-Y, Xu J, Langefeld CD, Koch KL, Dawson PA, et al. Association of SNPs in the UGT1A gene cluster with total bilirubin and mortality in the diabetes heart study. Atherosclerosis. 2013;229:155–60.

15. Grant DJ, Bell DA. Bilirubin UDP-glucuronosyltransferase 1A1 gene polymorphisms: susceptibility to oxidative damage and cancer? Mol Carcinog. 2000;29:198–204.

16. Melton PE, Haack K, Göring HH, Laston S, Umans JG, Lee ET, et al. Genetic influences on serum bilirubin in American Indians: The Strong Heart Family Study. Am J Hum Biol. 2011;23:118–25.

17. Chen G, Ramos E, Adeyemo A, Shriner D, Zhou J, Doumatey AP, et al. UGT1A1 is a major locus influencing bilirubin levels in African Americans. Eur J Hum Genet. 2012;20:463–8.

18. Jylhävä J, Lyytikäinen L-P, Kähönen M, Hutri-Kähönen N, Kettunen J, Viikari J, et al. A genome-wide association study identifies UGT1A1 as a regulator of serum cell-free DNA in young adults: the cardiovascular risk in young finns study. PLoS One. 2012;7:e35426.

19. Abbasi A, Deetman PE, Corpeleijn E, Gansevoort RT, Gans ROB, Hillege HL, et al. Bilirubin as a potential causal factor in type 2 diabetes risk: a Mendelian randomization study. Diabetes. 2015;64:1459–69.

20. Schwertner H, Jackson W, Tolan G. Association of low serum concentration of bilirubin with increased risk of coronary artery disease. Clin Chem. 1994;40:18–23.

21. Madhavan M, Wattigney WA, Srinivasan SR, Berenson GS. Serum bilirubin distribution and its relation to cardiovascular risk in children and young adults. Atherosclerosis. 1997;131:107–13.

22. Novotný L, Vítek L. Inverse relationship between serum bilirubin and atherosclerosis in men: a meta-analysis of published studies. Exp Biol Med (Maywood). 2003;228:568–71.

23. Schwertner HA. Association of smoking and low serum bilirubin antioxidant concentrations1 the views expressed in this article are those of the author and do not reflect the official policy of the Department of Defense or other Departments of the US Government 1. Atherosclerosis. 1998;136:383–7.

24. Hoydonck PGV. Serum bilirubin concentration in a Belgian population: the association with smoking status and type of cigarettes. Int J Epidemiol. 2001;30:1465–72.

25. Jo J, Kimm H, Yun JE, Lee KJ, Jee SH. Cigarette smoking and serum bilirubin subtypes in healthy Korean men: the Korea Medical Institute study. J Prev Med Public Health. 2012;45:105–12.

26. Vítek L. The role of bilirubin in diabetes, metabolic syndrome, and cardiovascular diseases. Front Pharmacol. 2012;3:55.

27. McArdle PF, Whitcomb BW, Tanner K, Mitchell BD, Shuldiner AR, Parsa A. Association between bilirubin and cardiovascular disease risk factors: using Mendelian randomization to assess causal inference. BMC Cardiovasc Disord. 2012;12:16.

28. Lin J-P, Schwaiger JP, Cupples LA, O'Donnell CJ, Zheng G, Schoenborn V, et al. Conditional linkage and genome-wide association studies identify UGT1A1 as a major gene for anti-atherogenic serum bilirubin levels–the Framingham Heart Study. Atherosclerosis. 2009;206:228–33.

29. Wu Y, Li M, Xu M, Bi Y, Li X, Chen Y, et al. Low serum total bilirubin concentrations are associated with increased prevalence of metabolic syndrome in Chinese. J Diabetes. 2011;3:217–24.

30. Andersson C, Weeke P, Fosbøl EL, Brendorp B, Køber L, Coutinho W, et al. Acute effect of weight loss on levels of total bilirubin in obese, cardiovascular high-risk patients: an analysis from the lead-in period of the Sibutramine Cardiovascular Outcome trial. Metabolism. 2009;58:1109–15.

31. Guzek M, Jakubowski Z, Bandosz P, Wyrzykowski B, Smoczyński M, Jabloiska A, et al. Inverse association of serum bilirubin with metabolic syndrome and insulin resistance in Polish population. Przegląd Epidemiol. 2012;66:495–501.

32. Choi SH, Yun KE, Choi HJ. Relationships between serum total bilirubin levels and metabolic syndrome in Korean adults. Nutr Metab Cardiovasc Dis. 2013;23:31–7.

33. Kwon K-M, Kam J-H, Kim M-Y, Kim M-Y, Chung CH, Kim J-K, et al. Inverse association between total bilirubin and metabolic syndrome in rural korean women. J Womens Health (Larchmt). 2011;20:963–9.

34. Jo J, Yun JE, Lee H, Kimm H, Jee SH. Total, direct, and indirect serum bilirubin concentrations and metabolic syndrome among the Korean population. Endocrine. 2011;39:182–9.

35. Tephly TR, Burchell B. UDP-glucuronosyltransferases: a family of detoxifying enzymes. Trends Pharmacol Sci. 1990;11:276–9.

36. Burchell B, Brierley CH, Rance D. Specificity of human UDP-Glucuronosyltransferases and xenobiotic glucuronidation. Life Sci. 1995;57:1819–31.

37. Fisher MB, Paine MF, Strelevitz TJ, Wrighton SA: The role of hepatic and extrahepatic UDP-glucuronosyltransferases in human drug metabolism. Drug Metab Rev. 2002;33(3-4):273-97

38. Lin J-P, Cupples LA, Wilson PWF, Heard-Costa N, O'Donnell CJ. Evidence for a gene influencing serum bilirubin on chromosome 2q telomere: a genomewide scan in the Framingham study. Am J Hum Genet. 2003;72:1029–34.

39. Newton-Cheh C, Johnson T, Gateva V, Tobin MD, Bochud M, Coin L, et al. Genome-wide association study identifies eight loci associated with blood pressure. Nat Genet. 2009;41:666–76.

40. Johnson AD, Kavousi M, Smith AV, Chen M-H, Dehghan A, Aspelund T, et al. Genome-wide association meta-analysis for total serum bilirubin levels. Hum Mol Genet. 2009;18:2700–10.

41. Bielinski SJ, Chai HS, Pathak J, Talwalkar JA, Limburg PJ, Gullerud RE, et al. Mayo Genome Consortia: a genotype-phenotype resource for genome-wide association studies with an application to the analysis of circulating bilirubin levels. Mayo Clin Proc. 2011;86:606–14.

42. Dai X, Wu C, He Y, Gui L, Zhou L, Guo H, et al. A genome-wide association study for serum bilirubin levels and gene-environment interaction in a Chinese population. Genet Epidemiol. 2013;37:293–300.

43. Lingenhel A, Kollerits B, Schwaiger JP, Hunt SC, Gress R, Hopkins PN, et al. Serum bilirubin levels, UGT1A1 polymorphisms and risk for coronary artery disease. Exp Gerontol. 2008;43:1102–7.

44. Bosma PJ, Chowdhury NR, Goldhoorn BG, Hofker MH, Oude Elferink RP, Jansen PL, et al. Sequence of exons and the flanking regions of human bilirubin-UDP-glucuronosyltransferase gene complex and identification of a genetic mutation in a patient with Crigler-Najjar syndrome, type I. Hepatol. 1992;15:941–7.

45. Seppen J, Bosma PJ, Goldhoorn BG, Bakker CT, Chowdhury JR, Chowdhury NR, et al. Discrimination between Crigler-Najjar type I and II by expression of mutant bilirubin uridine diphosphate-glucuronosyltransferase. J Clin Invest. 1994;94:2385–91.

46. Kadakol A, Ghosh SS, Sappal BS, Sharma G, Chowdhury JR, Chowdhury NR. Genetic lesions of bilirubin uridine-diphosphoglucuronate glucuronosyltransferase (UGT1A1) causing Crigler-Najjar and Gilbert syndromes: correlation of genotype to phenotype. Hum Mutat. 2000;16:297–306.

47. Black M, Billing BH. Hepatic bilirubin udp-glucuronyl transferase activity in liver disease and gilbert's syndrome. N Engl J Med. 1969;280:1266–71.

48. Koiwai O, Nishizawa M, Hasada K, Aono S, Adachi Y, Mamiya N, et al. Gilbert's syndrome is caused by a heterozygous missense mutation in the gene for bilirubin UDP-glucuronosyltransferase. Hum Mol Genet. 1995;4:1183–6.

49. Borlak J, Thum T, Landt O, Erb K, Hermann R. Molecular diagnosis of a familial nonhemolytic hyperbilirubinemia (Gilbert's syndrome) in healthy subjects. Hepatol. 2000;32(4 Pt 1):792–5.

50. Bosma PJ, Chowdhury JR, Bakker C, Gantla S, de Boer A, Oostra BA, et al. The genetic basis of the reduced expression of bilirubin UDP-glucuronosyltransferase 1 in Gilbert's syndrome. N Engl J Med. 1995;333:1171–5.

51. Bulmer AC, Blanchfield JT, Toth I, Fassett RG, Coombes JS. Improved resistance to serum oxidation in Gilbert's syndrome: A mechanism for cardiovascular protection. Atherosclerosis. 2008;199:390–6.

52. Inoguchi T, Sasaki S, Kobayashi K, Takayanagi R, Yamada T. Relationship between Gilbert syndrome and prevalence of vascular complications in patients with diabetes. JAMA. 2007;298:1398–400.

53. Dekker D, Dorresteijn MJ, Pijnenburg M, Heemskerk S, Rasing-Hoogveld A, Burger DM, et al. The bilirubin-increasing drug atazanavir improves endothelial function in patients with type 2 diabetes mellitus. Arterioscler Thromb Vasc Biol. 2011;31:458–63.

54. Beutler E, Gelbart T, Demina A. Racial variability in the UDP-glucuronosyltransferase 1 (UGT1A1) promoter: a balanced polymorphism for regulation of bilirubin metabolism? Proc Natl Acad Sci. 1998;95:8170–4.

55. Frazer KA, Ballinger DG, Cox DR, Hinds DA, Stuve LL, Gibbs RA, et al. A second generation human haplotype map of over 3.1 million SNPs. Nature. 2007;449:851–61.

56. Oussalah A, Bosco P, Anello G, Spada R, Guéant-Rodriguez R-M, Chery C, et al. Exome-wide association study identifies new low-frequency and rare UGT1A1 coding variants and UGT1A6 coding variants influencing serum bilirubin in elderly subjects: a strobe compliant article. Medicine (Baltimore). 2015;94:e925.

57. Bellis C, Cox HC, Ovcaric M, Begley KN, Lea RA, Quinlan S, et al. Linkage disequilibrium analysis in the genetically isolated Norfolk Island population. Heredity (Edinb). 2008;100:366–73.

58. Aulchenko YS, Ripke S, Isaacs A, Van Duijn CM. GenABEL: an R library for genorne-wide association analysis. Bioinformatics. 2007;23:1294–6.

59. Scharpf RB, Irizarry RA, Ritchie ME, Carvalho B, Ruczinski I. Using the R package crlmm for genotyping and copy number estimation. J Stat Softw. 2011;40:1–32.

60. Purcell S, Neale B, Todd-Brown K, Thomas L, Ferreira MAR, Bender D, et al. PLINK: a tool set for whole-genome association and population-based linkage analyses. Am J Hum Genet. 2007;81:559–75.

61. Manichaikul A, Mychaleckyj JC, Rich SS, Daly K, Sale M, Chen WM. Robust relationship inference in genome-wide association studies. Bioinformatics. 2010;26:2867–73.

62. O'Connell J, Gurdasani D, Delaneau O, Pirastu N, Ulivi S, Cocca M, et al. A general approach for haplotype phasing across the full spectrum of relatedness. PLoS Genet. 2014;10:e1004234.

63. Barrett JC, Fry B, Maller J, Daly MJ. Haploview: analysis and visualization of LD and haplotype maps. Bioinformatics. 2005;21:263–5.

64. R-Development-Core-Team: R: A language and environment for statistical computing. 2015.

Permissions

All chapters in this book were first published in Genetics, by BioMed Central; hereby published with permission under the Creative Commons Attribution License or equivalent. Every chapter published in this book has been scrutinized by our experts. Their significance has been extensively debated. The topics covered herein carry significant findings which will fuel the growth of the discipline. They may even be implemented as practical applications or may be referred to as a beginning point for another development.

The contributors of this book come from diverse backgrounds, making this book a truly international effort. This book will bring forth new frontiers with its revolutionizing research information and detailed analysis of the nascent developments around the world.

We would like to thank all the contributing authors for lending their expertise to make the book truly unique. They have played a crucial role in the development of this book. Without their invaluable contributions this book wouldn't have been possible. They have made vital efforts to compile up to date information on the varied aspects of this subject to make this book a valuable addition to the collection of many professionals and students.

This book was conceptualized with the vision of imparting up-to-date information and advanced data in this field. To ensure the same, a matchless editorial board was set up. Every individual on the board went through rigorous rounds of assessment to prove their worth. After which they invested a large part of their time researching and compiling the most relevant data for our readers.

The editorial board has been involved in producing this book since its inception. They have spent rigorous hours researching and exploring the diverse topics which have resulted in the successful publishing of this book. They have passed on their knowledge of decades through this book. To expedite this challenging task, the publisher supported the team at every step. A small team of assistant editors was also appointed to further simplify the editing procedure and attain best results for the readers.

Apart from the editorial board, the designing team has also invested a significant amount of their time in understanding the subject and creating the most relevant covers. They scrutinized every image to scout for the most suitable representation of the subject and create an appropriate cover for the book.

The publishing team has been an ardent support to the editorial, designing and production team. Their endless efforts to recruit the best for this project, has resulted in the accomplishment of this book. They are a veteran in the field of academics and their pool of knowledge is as vast as their experience in printing. Their expertise and guidance has proved useful at every step. Their uncompromising quality standards have made this book an exceptional effort. Their encouragement from time to time has been an inspiration for everyone.

The publisher and the editorial board hope that this book will prove to be a valuable piece of knowledge for researchers, students, practitioners and scholars across the globe.

List of Contributors

Wen-Yi Yang, Lutgarde Thijs, Zhen-Yu Zhang, Lotte Jacobs, Azusa Hara,Fang-Fei Wei, Yu-Mei Gu, Judita Knez, Nicholas Cauwenberghs, Tatiana Kuznetsova, Thibault Petit and Jan A. Staessen
Research Unit Hypertension and Cardiovascular Epidemiology, KU Leuven Department of Cardiovascular Sciences, University of Leuven, Kapucijnenvoer 35, Box 7001, BE-3000 Leuven, Belgium

Thibault Petit
Cardiology, Department of Cardiovascular Sciences, University of Leuven, Leuven, Belgium

Erika Salvi, Matteo Barcella, Cristina Barlassina and Daniele Cusi
Genomics and Bioinformatics Platform at Filarete Foundation, Department of Health Sciences and Graduate School of Nephrology, Division of Nephrology, San Paolo Hospital, University of Milan, Milan, Italy

Lorena Citterio and Simona Delli Carpini
Division of Nephrology and Dialysis, IRCCS San Raffaele Scientific Institute, University Vita-Salute San Raffaele, Milan, Italy

Paolo Manunta
School of Nephrology, University Vita-Salute San Raffaele, Milan, Italy

Giulia Coppiello, Xabier L. Aranguren, Peter Verhamme and Aernout Luttun
Centre for Molecular and Vascular Biology, Department of Cardiovascular Sciences, University of Leuven, Leuven, Belgium

Jan A. Staessen
R & D VitaK Group, Maastricht University, Maastricht, The Netherlands

Heather L. Norton
Department of Anthropology, University of Cincinnati, 481 Braunstein Hall, PO Box 210380, Cincinnati, OH 45221, USA

Elizabeth Werren
Department of Anthropology, 101 West Hall, University of Michigan, 1085 South University Ave, Ann Arbor, MI 48109, USA

Jonathan Friedlaender
Department of Anthropology, Temple University, Gladfelter Hall, 1115 West Berks Street, Philadelphia, PA 19122, USA

Shinji Sasaki and Mayumi Sugimoto
National Livestock Breeding Center, Odakura, Nishigo, Fukushima 961-8511, Japan

Takayuki Ibi
Graduate School of Environmental and Life Science, Okayama University, Tsushima-naka, Okayama 700-8530, Japan

Tamako Matsuhashi and Kenji Takeda
Gifu Prefectural Livestock Research Institute, Kiyomi, Takayama, Gifu 506-0101, Japan

Shogo Ikeda
Cattle Breeding Development Institute of Kagoshima Prefecture, Osumi, So, Kagoshima 899-8212, Japan

Yoshikazu Sugimoto
Shirakawa Institute of Animal Genetics, Japan Livestock Technology Association, Odakura, Nishigo, Fukushima 961-8061, Japan

Rong Zhou, Yanhong Wu, Min Tao, Chun Zhang and Shaojun Liu
Key Laboratory of Protein Chemistry and Developmental Biology of the State Education Ministry of China, College of Life Sciences, Hunan Normal University, Changsha, Hunan 410081, People's Republic of China

Taro Tsujimura, Ryoko Masuda, Ryuichi Ashino and Shoji Kawamura
Department of Integrated Biosciences, Graduate School of Frontier Sciences, the University of Tokyo, Kashiwanoha 5-1-5, Kashiwa 277-8562 Chiba, Japan

Taro Tsujimura
Department of Advanced Nephrology and Regenerative Medicine, Division of Tissue Engineering, the University of Tokyo Hospital, Hongo 7-3-1, Bunkyo-ku 113-8655Tokyo, Japan

Gary A. Rohrer, Dan J. Nonneman, Ralph T. Wiedmann and James F. Schneider
United States Department of Agriculture, Agricultural Research Service, U.S. Meat Animal Research Center, Clay Center, NE 68933, USA

Melanie B. Prentice
Department of Environmental & Life Sciences, Trent University, 1600 West Bank Drive, Peterborough K9J 7B8, ON, Canada

Jeff Bowman
Wildlife Research and Monitoring Section, Ontario Ministry of Natural Resources and Forestry, 2140 East Bank Drive, Peterborough K9J 7B8, ON, Canada

Paul J. Wilson
Biology Department, Trent University, 1600 West Bank Drive, Peterborough K9J 7B8, ON, Canada

Hugues Aschard and Alexander Gusev
Department of Epidemiology, Harvard School of Public Health, Boston, MA, USA

Robert Brown
Bioinformatics Interdepartmental Program, University of California Los Angeles, Los Angeles, CA, USA

Bogdan Pasaniuc
Bioinformatics Interdepartmental Program, University of California Los Angeles, Los Angeles, CA, USA
Department of Pathology and Laboratory Medicine, University of California Los Angeles, Los Angeles, CA, USA
Department of Human Genetics, University of California Los Angeles, Los Angeles, CA, USA

Kristin L. Young and Misa Graff
Department of Epidemiology, Gillings School of Global Public Health, University of North Carolina, Chapel Hill, NC, USA
Carolina Population Center, Gillings School of Global Public Health, University of North Carolina, Chapel Hill, NC, USA

Kari E. North
Department of Epidemiology, Gillings School of Global Public Health, University of North Carolina, Chapel Hill, NC, USA
Carolina Center for Genome Sciences, Gillings School of Global Public Health, University of North Carolina, Chapel Hill, NC, USA

Kathleen M. Harris
Carolina Population Center, Gillings School of Global Public Health, University of North Carolina, Chapel Hill, NC, USA
Department of Sociology, Gillings School of Global Public Health, University of North Carolina, Chapel Hill, NC, USA

Andrea S. Richardson and Penny Gordon-Larsen
Carolina Population Center, Gillings School of Global Public Health, University of North Carolina, Chapel Hill, NC, USA
Department of Nutrition, Gillings School of Global Public Health, University of North Carolina, Chapel Hill, NC, USA

Karen L. Mohlke, Ethan M. Lange, Leslie A. Lange and Karen L. Mohlke
Carolina Center for Genome Sciences, Gillings School of Global Public Health, University of North Carolina, Chapel Hill, NC, USA
Department of Genetics, Gillings School of Global Public Health, University of North Carolina, Chapel Hill, NC, USA

Kristin L. Young
137 East Franklin Street, Suite 306, Chapel Hill, NC 27514, USA

Yoshinobu Uemoto, Shinji Sasaki, Takatoshi Kojima and Toshio Watanabe
National Livestock Breeding Center, Nishigo, Fukushima 961-8511, Japan

Yoshikazu Sugimoto
Shirakawa Institute of Animal Genetics, Japan Livestock Technology Association, Nishigo, Fukushima 961-8511, Japan

Francesco Vella, Elena Raimondi, Claudia Badiale, Marco Santagostino, Lela Khoriauli, Riccardo Gamba, Ori Klipstein, Francesca M. Piras, Alice Mazzagatti, Solomon G. Nergadze and Elena Giulotto
Dipartimento di Biologia e Biotecnologie "Lazzaro Spallanzani", Università di Pavia, Via Ferrata 1, 27100 Pavia, Italy

Alessandra Russo and Margherita Bonuglia
Laboratorio di Genetica Forense Veterinaria, UNIRELAB srl, Via A. Gramsci 70, 20019 Settimo Milanese (MI), Italy

Paul Stothard and Chinyere Ekine-Dzivenu
Department of Agricultural, Food and Nutritional Science, University of Alberta, Edmonton, AB T6G 2P5, Canada

Liuhong Chen, Carolyn Fitzsimmons and Changxi Li
Department of Agricultural, Food and Nutritional Science, University of Alberta, Edmonton, AB T6G 2P5, Canada
Lacombe Research Centre, Agriculture and Agri-Food Canada, 6000 C&E Trail, Lacombe, AB T4L 1 W1, Canada

Michael Vinsky, Jennifer Aalhus and Mike E. R. Dugan
Lacombe Research Centre, Agriculture and Agri-Food Canada, 6000 C&E Trail, Lacombe, AB T4L 1 W1, Canada

John Basarab
Lacombe Research Centre, Alberta Agriculture and Forestry, 6000 C & E Trail, Lacombe, AB T4L 1 W1, Canada

Gang Wang, Guifang Fu and Christopher Corcoran
Department of Mathematics and Statistics, Utah State University, 3900 Old Main, 84322 Logan, UT, US

Silvia Parolo, Antonella Lisa and Silvia Bione
Computational Biology Unit, Institute of Molecular Genetics-National Research Council, Pavia, Italy

Davide Gentilini and Anna Maria Di Blasio
Molecular Biology Laboratory, Istituto Auxologico Italiano, Milan, Italy

Simona Barlera and Enrico B. Nicolis
Department of Cardiovascular Research IRCCS Mario Negri Institute for Pharmacological Research, Milan, Italy

Giorgio B. Boncoraglio and Eugenio A. Parati
Department of Cerebrovascular Diseases, IRCCS Istituto Neurologico Carlo Besta, Milan, Italy

Jian-Ping Hu, Xiao-Ying Xu, Li-Ying Huang, and Ning-Yuan Fang
The Department of Geriatrics, Ren-Ji Hospital, Shanghai Jiao-Tong University School of Medicine, Shanghai, China

Li-shun Wang
The Division of Translational Medicine, Minhang Hospital, Fudan University, Shanghai, China

Sharon L.R. Kardia and Jennifer A. Smith
Department of Epidemiology, University of Michigan, Ann Arbor, MI, USA

Erin B. Ware
Department of Epidemiology, University of Michigan, Ann Arbor, MI, USA
Institute of Social Research, University of Michigan, 1415 Washington Heights #4614, Ann Arbor, MI 48109, USA

Bhramar Mukherjee
Department of Biostatistics, University of Michigan, Ann Arbor, MI, USA

Yan V. Sun
Department of Epidemiology, Emory University, Atlanta, GA, USA

Ana V. Diez-Roux
Department of Epidemiology and Biostatistics, Drexel University, Philadelphia, PA, USA

Jiaqi Li, Jinlong He, Guang Xiao, Yuanyu Luo, Hao Zhang and Zanmou Chen
National Engineering Research Center for Breeding Swine Industry, Guangdong Provincial Key Lab of Agro-animal Genomics and Molecular Breeding, College of Animal Science, South China Agricultural University, Guangzhou 510642, China

Zhe Zhang and Ning Gao
National Engineering Research Center for Breeding Swine Industry, Guangdong Provincial Key Lab of Agro-animal Genomics and Molecular Breeding, College of Animal Science, South China Agricultural University, Guangzhou 510642, China
Department of Animal Sciences, Animal Breeding and Genetics Group, Georg-August-Universität Göttingen, Göttingen 37075, Germany

Tania Vásquez-Loarte, Milana Trubnykova and Heinner Guio
Laboratorio de Biotecnología y Biología Molecular, Instituto Nacional de Salud, Avenida Defensores del Morro 2268, Lima 9, Peru

Xuejiao Cheng, Lingling Chai, Zhaoyan Chen, Lu Xu, Huijie Zhai, Huiru Peng, Yingyin Yao, Mingshan You, Qixin Sun and Zhongfu Ni
State Key Laboratory for Agrobiotechnology and Key Laboratory of Crop Heterosis and Utilization (MOE) and Key Laboratory of Crop Genomics and Genetic Improvement (MOA), Beijing Key Laboratory of Crop Genetic Improvement, China Agricultural University, Yuanmingyuan Xi Road No. 2, Haidian District, Beijing 100193, China
Department of Plant Genetics and Breeding, China Agricultural University, Beijing 100193, China National Plant Gene Research Centre, Beijing 100193, China

Aiju Zhao
Hebei Crop Genetic Breeding Laboratory Institute of Cereal and Oil Crops, Hebei Academy of Agriculture and Forestry Sciences, Shijiazhuang 050035, China

Jian-Ming Chen, Chuan-Dong Wei, Chun-Fang Wang, Hong-Cheng Luo and Ye-Sheng Wei
Department of Laboratory Medicine, Affiliated hospital of Youjiang Medical University for Nationalities, Baise 533000, Guangxi, China

Yan Lan and Jing Guo
Department of Dermatology, Affiliated Hospital of Youjiang Medical University for Nationalities, Baise 533000, Guangxi, China

Qian Wang and Weibin Shi
Department of Radiology & Medical Imaging, University of Virginia, Snyder Bldg Rm 266, 480 Ray C. Hunt Dr., P.O. Box 801339, Fontaine Research Park, Charlottesville, VA 22908, USA
University of Virginia, Snyder Bldg Rm 266, 480 Ray C. Hunt Dr., P.O. Box 801339, Fontaine Research Park, Charlottesville, VA 22908, USA

Andrew T. Grainger
Department of Biochemistry & Molecular Genetics, University of Virginia, Charlottesville, VA, USA
University of Virginia, Charlottesville, VA, USA

Ani Manichaikul, Emily Farber and Suna Onengut-Gumuscu
Center for Public Health and Genomics, University of Virginia, Charlottesville, VA, USA

M. C. Benton, R. A. Lea, M. Hanna, D. Eccles and L. R. Griffiths
Genomics Research Centre, Institute of Health and Biomedical Innovation, Queensland University of Technology, Kelvin Grove, QLD 4059, Australia

C. Bellis
Genomics Research Centre, Institute of Health and Biomedical Innovation, Queensland University of Technology, Kelvin Grove, QLD 4059, Australia
Texas Biomedical Research Institute, San Antonio, TX 78227-5301, USA

D. Macartney-Coxson
Kenepuru Science Centre, Institute of Environmental Science and Research, Wellington 5240, New Zealand

M. A. Carless and J. E. Curran
Texas Biomedical Research Institute, San Antonio, TX 78227-5301, USA

G. K. Chambers
School of Biological Sciences, Victoria University of Wellington, Wellington 6140, New Zealand

J. Blangero
South Texas Diabetes and Obesity Institute, University of Texas, Rio Grande Valley School of Medicine, Brownsville, TX 78520, USA